Pipeline Maintenance

Level Two

Trainee Guide
Third Edition

PEARSON

Boston Columbus Indianapolis New York San Francisco Amsterdam
Cape Town Dubai London Madrid Milan Munich Paris Montreal Toronto Delhi
Mexico City Sao Paulo Sydney Hong Kong Seoul Singapore Taipei Tokyo

NCCER

President: Don Whyte
Vice President: Steve Greene
Chief Operations Officer: Katrina Kersch
Director of Product Development: Daniele Dixon
Pipeline Mechanical Project Managers: Patty Bird, Elizabeth Schlaupitz, Justin Street, Chris Wilson
Senior Development Manager: Mark Thomas
Senior Production Manager: Tim Davis
Quality Assurance Coordinator: Karyn Payne
Desktop Publishing Coordinator: James McKay
Permissions Specialist: Adrienne Payne
Production Specialist: Adrienne Payne
Editors: Graham Hack, Karyn Payne, Debie Hicks, Jackie Strawbridge, Adrienne Payne

Writing and development services provided by Topaz Publications, Liverpool, NY

Lead Writers/Project Managers: Don Congdon, Veronica Westfall
Desktop Publisher: Joanne Hart
Art Director: Alison Richmond
Permissions Editors: Andrea LaBarge, Briana Morgan
Writers: Don Congdon, Tony Clay, Thomas Burke

Pearson

Director of Alliance/Partnership Management: Andrew Taylor
Editorial Assistant: Collin Lamothe
Program Manager: Alexandrina B. Wolf
Digital Studio Project Managers: Heather Darby, Tanika Henderson, Jose Carchi
Director of Marketing: Leigh Ann Simms
Senior Marketing Manager: Brian Hoehl
Composition: NCCER
Printer/Binder: LSC Communications
Cover Printer: LSC Communications
Text Fonts: Palatino and Univers

Credits and acknowledgments for content borrowed from other sources and reproduced, with permission, in this textbook appear at the end of each module.

PEARSON

ISBN-13: 978-0-13-471653-4
ISBN-10: 0-13-471653-1

ScoutAutomatedPrintCode

Preface

To the Trainee

NCCER is pleased to release the 3rd edition of Pipeline Training and Qualification with updates that will engage you and give you the best training in the industry. The curriculum has been redesigned to better align with the learning objectives, and also includes new end-of-section review questions to compliment the Module Review. Changes to the NCCER's covered task (CT) modules include updates to the latest technology and standards, as well as alignment to the American Petroleum Institute's *API Recommended Practice for Pipeline Operator Qualification (OQ) 1161*, 3rd edition (2014). Each CT module contains task-specific Abnormal Operating Conditions (AOCs) to which the trainee must be prepared to respond. General AOCs for both Field (Liquid and Gas) and Control Center settings are also available as single modules.

CT modules are available as traditional printed levels and as individual module eBooks. Modules are completed by successfully passing the proctored Module Examination, and completing the Performance Profile under the supervision of an NCCER-certified evaluator. These completions are tracked within the NCCER Registry for training delivered through an Accredited Training Sponsor. Covered task module completions in the Registry are also documented in ISNetworld.

The modules are designed to be used for Operational Qualification, and they are each a vital part of NCCER's pipeline workforce development curricula. They provide in-depth training on the fundamental knowledge and skills required for pipeline work, as well as technical training that increases the depth and breadth of a pipeline worker's capability within several specialty areas.

We invite you to visit the NCCER website at **www.nccer.org** for information on the latest product releases and training, as well as online versions of the *Cornerstone* magazine and Pearson's NCCER product catalog.

Your feedback is welcome. You may email your comments to **curriculum@nccer.org** or send general comments and inquiries to **info@nccer.org**.

NCCER Standardized Curricula

NCCER is a not-for-profit 501(c)(3) education foundation established in 1996 by the world's largest and most progressive construction companies and national construction associations. It was founded to address the severe workforce shortage facing the industry and to develop a standardized training process and curricula. Today, NCCER is supported by hundreds of leading construction and maintenance companies, manufacturers, and national associations. The NCCER Standardized Curricula was developed by NCCER in partnership with Pearson, the world's largest educational publisher.

Some features of the NCCER Standardized Curricula are as follows:

- An industry-proven record of success
- Curricula developed by the industry for the industry
- National standardization providing portability of learned job skills and educational credits
- Compliance with the Office of Apprenticeship requirements for related classroom training (*CFR 29:29*)
- Well-illustrated, up-to-date, and practical information

NCCER also maintains the NCCER Registry, which provides transcripts, certificates, and wallet cards to individuals who have successfully completed a level of training within a craft in NCCER's Curricula. *Training programs must be delivered by an NCCER Accredited Training Sponsor in order to receive these credentials.*

Additional Pipeline Resources

The following additional Pipeline Resources are available from NCCER's website at **www.nccer.org/pipeline**:

1. A complete copy of NCCER's Covered Task (CT) list
2. Access to:
   a. An online bookstore to order print and digital versions of NCCER's pipeline curricula.
   b. Instructor resources (Lesson Plans and PowerPoints)
   c. Exams for Operator Qualification
   d. A site to custom build curriculum with pipeline and/or other NCCER-developed modules

Pipeline Career Pathway Training

Entry-level Pipeline Craft Trainee → **Pipeline Fundamentals** → **Level 1**

The newly established pipeline career pathway represented on this page provides the recognition pipeline professionals deserve through standardized training and industry-recognized credentials.

Levels of training have been established by combining training modules into specific focus areas within the industry. This allows trainees to progress through a standardized program and earn a professional credential.

In addition, Levels Two and Three are comprised of covered task training modules that can lead to operator qualifications. *Pipeline Corrosion Control is the only exception which actually provides covered task training starting in Level One.*

The blue boxes denote the professional credential that can be earned with successful completion of each program.

A trained workforce is a safe and qualified workforce!

Pipeline Fundamentals

Core Curriculum

Introduction to the Pipeline Industry

and

a.) Field Abnormal Operating Conditions
Module: AOCFG-17

or

b.) Control Center Abnormal Operating Conditions
Module: AOCCC_17

Level 1

- **Field and Control Center Operations**
- **Pipeline Maintenance and Mechanical**
- **Electrical and Instrumentation**
- **Pipeline Corrosion Control** (CT)

Level 2: Craft-Specific Training with Operator Qualification*	Level 3: Craft-Specific Training with Operator Qualification*	Trained and Qualified Pipeline Professional Credential
Pipeline Operations (Control Center/ Gas/Liquid) CT		Pipeline Operations Technician
Pipeline Mechanical CT	Pipeline Mechanical CT	Pipeline Mechanical Technician
Pipeline Maintenance CT	Pipeline Maintenance CT	Pipeline Maintenance Technician
Pipeline Electrical and Instrumentation CT		Pipeline Electrical and Instrumentation Technician
Pipeline Corrosion Control CT		Pipeline Corrosion Technician

CT *Each of these titles have corresponding operator qualifications. The Covered Task list is available online at www.nccer.org/pipeline-program*

Special Features

In an effort to provide a comprehensive and user-friendly training resource, this curriculum showcases several informative features. Whether you are a visual or hands-on learner, these features are intended to enhance your knowledge of the construction industry as you progress in your training. Some of the features you may find in the curriculum are explained below.

Introduction

This introductory page, found at the beginning of each module, lists the module Objectives, Performance Tasks and Trade Terms. The Objectives list the knowledge you will acquire after successfully completing the module. The Performance Tasks give you an opportunity to apply your knowledge to real-world tasks. The Trade Terms are industry specific vocabulary that you will learn as you study this module.

00104-15
INTRODUCTION TO POWER TOOLS

Objectives

When you have completed this module, you will be able to do the following:

1. Identify and explain how to use various types of power drills and impact wrenches.
   a. Identify and explain how to use common power drills and bits.
   b. Identify and explain how to use a hammer drill.
   c. Identify and explain how to use pneumatic drills and impact wrenches.
2. Identify and explain how to use various types of power saws.
   a. Identify and explain how to use a circular saw.
   b. Identify and explain how to use saber and reciprocating saws.
   c. Identify and explain how to use a portable band saw.
   d. Identify and explain how to use miter and cutoff saws.
3. Identify and explain how to use various grinders and grinder attachments.
   a. Identify and explain how to use various types of grinders.
   b. Identify and explain how to use various grinder accessories and attachments.
4. Identify and explain how to use miscellaneous power tools.
   a. Identify and explain how to use pneumatic and powder-actuated fastening tools.
   b. Identify and explain how to use pavement breakers.
   c. Identify and explain the uses of hydraulic jacks.

Performance Tasks

Under supervision of your instructor, you should be able to do the following:

1. Safely and properly demonstrate the use of three of the following tools:
   - Electric drill
   - Hammer drill or rotary hammer
   - Circular saw
   - Reciprocating saw
   - Portable band saw
   - Miter or cutoff saw
   - Portable or bench grinder
   - Pneumatic nail gun
   - Pavement breaker

Trade Terms

Abrasive	Chuck key	Ground fault circuit interrupter (GFCI)	Revolutions per minute (rpm)
Alternating current (AC)	Countersink	Ground fault protection	Ring test
Arbor	Direct current (DC)	Kerf	Shank
Auger bit	Forstner bit	Masonry bit	Trigger lock
Carbide	Grit	Reciprocating	
Chuck			

Industry Recognized Credentials

If you are training through an NCCER-accredited sponsor, you may be eligible for credentials from NCCER's Registry. The ID number for this module is 00104-15. Note that this module may have been used in other NCCER curricula and may apply to other level completions. Contact NCCER's Registry at 888.622.3720 or go to **www.nccer.org** for more information.

Trade Features

Trade features present technical tips and professional practices based on real-life scenarios similar to those you might encounter on the job site.

Bowline Trivia

Some people use this saying to help them remember how to tie a bowline: "The rabbit comes out of his hole, around a tree, and back into the hole."

Figures and Tables

Photographs, drawings, diagrams, and tables are used throughout each module to illustrate important concepts and provide clarity for complex instructions. Text references to figures and tables are emphasized with *italic* type.

Notes, Cautions, and Warnings

Safety features are set off from the main text in highlighted boxes and categorized according to the potential danger involved. Notes simply provide additional information. Cautions flag a hazardous issue that could cause damage to materials or equipment. Warnings stress a potentially dangerous situation that could result in injury or death to workers.

NOTE
Nameplates must be posted on each material handling device. The nameplate must indicate

CAUTION
It is essential to note the revision designation on a construction drawing and to use only the latest

WARNING!
Saw teeth are very sharp. Use gloves and do not handle the saw teeth with bare hands. When cutting with a saw, ensure that your fingers remain clear of the teeth at all times.

Case History

Case History features emphasize the importance of safety by citing examples of the costly (and often devastating) consequences of ignoring best practices or OSHA regulations.

Case History

Requesting an Outage

An electrical contractor requested an outage when asked to install two bolt-in, 240V breakers in panels in a data processing room. It was denied due to the 24/7 worldwide information processing hosted by the facility. The contractor agreed to proceed only if the client would sign a letter agreeing not to hold them responsible if an event occurred that damaged computers or resulted in loss of data. No member of upper management would accept liability for this possibility, and the outage was scheduled.

The Bottom Line: If you can communicate the liability associated with an electrical event, you can influence management's decision to work energized.

Going Green

Going Green features present steps being taken within the construction industry to protect the environment and save energy, emphasizing choices that can be made on the job to preserve the health of the planet.

GOING GREEN

Reducing Your Carbon Footprint

Many companies are taking part in the paperless movement. They reduce their environmental impact by reducing the amount of paper they use. Using email helps to reduce the amount of paper used,

Did You Know

Did You Know features introduce historical tidbits or interesting and sometimes surprising facts about the trade.

Did You Know?

Safety First

Safety training is required for all activities. Never operate tools, machinery, or equipment without prior training. Always refer to the manufacturer's instructions.

Step-by-Step Instructions

Step-by-step instructions are used throughout to guide you through technical procedures and tasks from start to finish. These steps show you how to perform a task safely and efficiently.

Perform the following steps to erect this system area scaffold:

Step 1 Gather and inspect all scaffold equipment for the scaffold arrangement.

Step 2 Place appropriate mudsills in their approximate locations.

Step 3 Attach the screw jacks to the mudsills.

Trade Terms

Each module presents a list of Trade Terms that are discussed within the text and defined in the Glossary at the end of the module. These terms are presented in the text with **bold, blue** type upon their first occurrence. To make searches for key information easier, a comprehensive Glossary of Trade Terms from all modules is located at the back of this book.

During a rigging operation, the **load** being lifted or moved must be connected to the apparatus, such as a crane, that will provide the power for movement. The connector—the link between the load and the apparatus—is often a sling made of synthetic, chain, or **wire rope** materials. This section focuses on three types of slings:

Section Review

Each section of the module wraps up with a list of Additional Resources for further study and Section Review questions designed to test your knowledge of the Objectives for that section.

Additional Resources

Materials Handling Handbook, The American Society of Mechanical Engineers (ASME) and The International Material Management Society (IMMS), Raymond A. Kulwiec, Editor-in-Chief. 1985. New York, NY: Wiley-Interscience.

Manufacturing Facilities Design & Material Handling, Matthew P. Stevens, Fred E. Meyers. 2013. West Lafayette, IN: Purdue University Press.

1.0.0 Section Review

1. For material handling tasks, it is just as important to be mentally fit as it is to be _____.
   a. physically fit
   b. physically aggressive
   c. closely supervised
   d. over 200 pounds

2. Which of the following is a type of knot that is often used to join the ends of two ropes in non-critical, low-strain applications?
   a. Bowline
   b. Clove hitch
   c. Half hitch
   d. Square knot

Review Questions

The end-of-module Review Questions can be used to measure and reinforce your knowledge of the module's content.

Review Questions

1. Identification tags for slings must include the _____.
   a. type of protective pads to use
   b. type of damage sustained during use
   c. color of the tattle-tail
   d. manufacturer's name or trademark

2. The type of wire rope core that is susceptible to heat damage at relatively low temperatures is the _____.
   a. fiber core
   b. strand core
   c. independent wire rope core
   d. metallic link supporting core

3. Synthetic slings must be inspected _____.
   a. once every month
   b. visually at the start of each work week
   c. before every use
   d. once wear or damage becomes apparent

4. An alloy steel chain sling must be removed from service if there is evidence that _____.
   a. the sling has been used in different hitch configurations
   b. replacement links have been used to repair the chain
   c. the sling has been used for more than one year
   d. strands in the supporting core have weakened

5. A piece of rigging hardware used to couple the end of a wire rope to eye fittings, hooks, or other connections is a(n) _____.
   a. eyebolt
   b. hitch
   c. shackle
   d. U-bolt

6. A lifting clamp is most likely to be used to move loads such as _____.
   a. steel plates
   b. piping bundles
   c. concrete blocks
   d. plastic tubing

7. Chain hoists are able to lift heavy loads by utilizing a _____.
   a. rope and pulley system
   b. rigger's strength
   c. stationary counterweight
   d. gear system

8. Before attempting to lift a load with a chain hoist, make sure that the _____.
   a. hoist is secured to a come-along
   b. load is properly balanced
   c. tag lines are properly anchored
   d. tackle is connected to its power source

9. A hitch configuration that allows slings to be connected to the same load without using a spreader beam is a _____.
   a. double-wrap hitch
   b. choker hitch
   c. bridle hitch
   d. basket hitch

10. To make the emergency stop signal that is used by riggers, extend both arms _____.
    a. horizontally with palms down and quickly move both arms back and forth
    b. directly in front and then move both arms up and down repeatedly
    c. vertically above the head and wave both arms back and forth
    d. horizontally with clenched fists and move both arms up and down

00106-15 Introduction to Basic Rigging Module Six 31

NCCER Standardized Curricula

NCCER's training programs comprise more than 80 construction, maintenance, pipeline, and utility areas and include skills assessments, safety training, and management education.

Boilermaking
Cabinetmaking
Carpentry
Concrete Finishing
Construction Craft Laborer
Construction Technology
Core Curriculum: Introductory Craft Skills
Drywall
Electrical
Electronic Systems Technician
Heating, Ventilating, and Air Conditioning
Heavy Equipment Operations
Highway/Heavy Construction
Hydroblasting
Industrial Coating and Lining Application Specialist
Industrial Maintenance Electrical and Instrumentation Technician
Industrial Maintenance Mechanic
Instrumentation
Insulating
Ironworking
Masonry
Millwright
Mobile Crane Operations
Painting
Painting, Industrial
Pipefitting
Pipelayer
Plumbing
Reinforcing Ironwork
Rigging
Scaffolding
Sheet Metal
Signal Person
Site Layout
Sprinkler Fitting
Tower Crane Operator
Welding

Maritime

Maritime Industry Fundamentals
Maritime Pipefitting
Maritime Structural Fitter

Green/Sustainable Construction

Building Auditor
Fundamentals of Weatherization
Introduction to Weatherization
Sustainable Construction Supervisor
Weatherization Crew Chief
Weatherization Technician
Your Role in the Green Environment

Energy

Alternative Energy
Introduction to the Power Industry
Introduction to Solar Photovoltaics
Introduction to Wind Energy
Power Industry Fundamentals
Power Generation Maintenance Electrician
Power Generation I&C Maintenance Technician
Power Generation Maintenance Mechanic
Power Line Worker
Power Line Worker: Distribution
Power Line Worker: Substation
Power Line Worker: Transmission
Solar Photovoltaic Systems Installer
Wind Turbine Maintenance Technician

Pipeline

Pipeline Fundamentals
Pipeline Corrosion Control
Pipeline Electrical and Instrumentation
Pipeline Maintenance
Pipeline Mechanical
Pipeline Operations (Control Center, Gas, Liquid)

Safety

Field Safety
Safety Orientation
Safety Technology

Supplemental Titles

Applied Construction Math
Tools for Success

Management

Construction Workforce Development Professional
Fundamentals of Crew Leadership
Mentoring for Craft Professionals
Project Management
Project Supervision

Spanish Titles

Acabado de concreto: nivel uno (*Concrete Finishing Level One*)
Aislamiento: nivel uno (*Insulating Level One*)
Albañilería: nivel uno (*Masonry Level One*)
Andamios (*Scaffolding*)
Carpintería: Formas para carpintería, nivel tres (*Carpentry: Carpentry Forms, Level Three*)
Currículo básico: habilidades introductorias del oficio (*Core Curriculum: Introductory Craft Skills*)
Electricidad: nivel uno (*Electrical Level One*)
Herrería: nivel uno (*Ironworking Level One*)
Herrería de refuerzo: nivel uno (*Reinforcing Ironwork Level One*)
Instalación de rociadores: nivel uno (*Sprinkler Fitting Level One*)
Instalación de tuberías: nivel uno (*Pipefitting Level One*)
Instrumentación: nivel uno, nivel dos, nivel tres, nivel cuatro (*Instrumentation Levels One through Four*)
Orientación de seguridad (*Safety Orientation*)
Paneles de yeso: nivel uno (*Drywall Level One*)
Seguridad de campo (*Field Safety*)

Acknowledgments

This curriculum was revised as a result of the farsightedness and leadership of the following sponsors:

American Petroleum Institute
Central NDT
Construction Craft Academy
CP Masters
EMS USA, Inc.
Enbridge Pipelines, Inc.
Enhance Co.
Enterprise
Flint Energy Services, Inc.
Jomax Construction
Kinder Morgan, Inc.
L.E. Bell Construction Co. Inc.
Marathon Pipe Line
Midwestern Contractors
OneOK
Plains All-American Pipeline
Praxair Services
Protech EIS
Safety Advantage LLC
Southland Safety
Sunoco Logistics, L.P.
Sunoco Pipeline
Thorco Holdings, LLC
URS/Flint Energy

This curriculum would not exist were it not for the dedication and unselfish energy of those volunteers who served on the Authoring Team. A sincere thanks is extended to the following:

Gary Affolter
Carl Allebach
Travis Ashby
Steve Barker
Art Birx
James Carroll
Leslie Carroll
Gordon Carson
Bill Emery
Paul Fontenot II
Paula Gates-Rippe
Trobi Hall
Sid Harrell
David Hoffman
Kimberly-Joy Harris
Les Harrison
Lisa Hartnup
Benjamin Johnson
Dan Lake
Tom Magness
Wallace McCracken
Tim Mills
Steve Montoy
Michael Nordstrom
Tom Pickthall
Donnie Pittsinger
William Rado
David Reeder
Tom Reeder
Randall Rogers
Dan Rohret
Dan Ruane
DeWayne Schapansky
Mark Seitz
Blaine Shellhammer
Lynn Simpson
Cameron Strother
Charles Whiten
Rick Wilenzick
Michael Williams
Erin Worthington
Donavon Wright
Will Wright

NCCER Partners

American Council for Construction Education
American Fire Sprinkler Association
Associated Builders and Contractors, Inc.
Associated General Contractors of America
Association for Career and Technical Education
Association for Skilled and Technical Sciences
Construction Industry Institute
Construction Users Roundtable
Design Build Institute of America
GSSC – Gulf States Shipbuilders Consortium
ISN
Manufacturing Institute
Mason Contractors Association of America
Merit Contractors Association of Canada
NACE International
National Association of Women in Construction
National Insulation Association
National Technical Honor Society
National Utility Contractors Association
NAWIC Education Foundation
North American Crane Bureau
North American Technician Excellence
Pearson
Prov
SkillsUSA®
Steel Erectors Association of America
U.S. Army Corps of Engineers
University of Florida, M. E. Rinker Sr., School of Construction Management
Women Construction Owners & Executives, USA

NCCER Business Partners

Contents

CT37_0-17

Install or Repair Support Structures on Existing Above Ground Components

Covers the procedures for repairing or installing pipeline support structures on aboveground pipelines. (Module ID CT37_0-17; 5 Hours)

CT38_1-17

Visually Inspect Pipe and Pipe Components Prior to Installation

Discusses the process and procedure for visually inspecting pipe and pipe components (CT38.1). (Module ID CT38_1-17; 5 Hours)

CT39_0-17

Backfilling a Trench Following Maintenance

Describes the tasks involved in properly backfilling a trench after maintenance has been completed on a pipeline. (Module ID CT39_017; 5 Hours)

CT52_1-17

Conduct Vegetation Survey

Introduces the trainees to leakage survey requirements for gas pipelines, and to the procedures for conducting a vegetation survey. (Module ID CT52_1-17; 5 Hours)

CT52_2-17

Conduct a Leak Survey with a CGD

Presents the purpose and procedures for conducting leakage surveys on gas pipelines using a combustible gas detector. (Module ID CT52_2-17; 5 Hours)

CT52_3-17

Conduct a Leak Survey with a Flame Ionization Unit

Introduces the trainees to leakage survey requirements for gas pipelines, and to the procedures for conducting a leak survey with a flame ionization detector. (Module ID CT52_3-17; 5 Hours)

CT59_0-17

Vault Maintenance (Gas)

Identifies safety requirements and hazards of confined-space entry, and also discusses vault inspections. (Module ID CT59_0-17; 10 Hours)

CTCC-17

Cold Cutting

Describes the safety requirements and tools used for cold cutting, as well as the procedures personnel must use to perform cold cutting. (Module ID CTCC-17; 10 Hours)

CTFB-17

Flange Bolting

Describes how to join pipe segments using flange fittings, as well as how to store and protect components prior to joining. (Module ID CTFB-17; 15 Hours)

CTMP-17

Mud Plugging

Describes the vapor-isolation method known as mud plugging. The procedure for safely and properly mixing the mud, forming the mud balls, and installing the mud plug is presented in this module. (Module ID CTMP-17; 5 Hours)

(continued)

CTTB-17

Tubing

Describes the sizes and types of tubing commonly used in pipeline maintenance and the safe and proper way to store and handle tubing. The text reviews the tools and methods for cutting, bending, joining, and installing tubing. (Module ID CTTB-17; 7.5 Hours)

CTTP-17

Threaded Pipe Fabrication

Describes the materials used in threaded piping systems. The text explains how to determine pipe lengths between threaded pipe fittings, prepare the pipe and fittings for fit-up, and assemble the piping system. (Module ID CTTP-17; 15 Hours)

AOCCC-17

Abnormal Operating Conditions – Control Center

Overview

Pipeline operators and contractors have programs, procedures, training, safety equipment, and warning devices for practically every facet of pipeline operations. Situations that are outside of normal operating conditions may arise at any time, and control center personnel must know what to do to help keep pipeline operations safe for everyone. In this module, personnel will learn how to recognize, properly react to, and properly report AOCs that may occur during pipeline operations.

Trainees with successful module completions may be eligible for credentialing through NCCER's National Registry. To learn more, go to **www.nccer.org** or contact us at **1.888.622.3720**. Our website has information on the latest product releases and training, as well as online versions of our *Cornerstone* magazine and Pearson's product catalog.

Your feedback is welcome. You may email your comments to **curriculum@nccer.org,** send general comments and inquiries to **info@nccer.org**, or fill in the User Update form at the back of this module.

This information is general in nature and intended for training purposes only. Actual performance of activities described in this manual requires compliance with all applicable operating, service, maintenance, and safety procedures under the direction of qualified personnel. References in this manual to patented or proprietary devices do not constitute a recommendation of their use.

AOCCC-17
Abnormal Operating Conditions – Control Center

Objectives

When you have completed this module, you will be able to do the following:

1. Explain how to recognize abnormal operating conditions in a control center for both gas and liquid pipelines.
   a. Explain how to recognize abnormal facility conditions and pipeline system damage.
   b. Explain how to recognize the activation of a safety device.
   c. Explain how to recognize a communications failure and a control system failure.
   d. Explain how to recognize power interruptions.
   e. Explain how to recognize a fire, explosion, and natural disaster occurring with a pipeline.
   f. Explain how to recognize hazardous liquid or gas encounters.
   g. Explain how to recognize an unexplained flow rate/pressure change.
2. Describe the proper reaction to abnormal operating conditions in a control center for both gas and liquid pipelines.
   a. Describe the proper reaction to abnormal facility conditions and pipeline system damage.
   b. Describe the proper reaction to the activation of a safety device.
   c. Describe the proper reaction to a communications failure and a control system failure.
   d. Describe the proper reaction to power interruptions.
   e. Describe the proper reaction to a fire, explosion, and natural disaster occurring with a pipeline.
   f. Describe the proper reaction to hazardous liquid or gas encounters.
   g. Describe the proper reaction to an unexplained flow rate/pressure change.

Performance Tasks

This is a knowledge-based module; there are no performance tasks.

Trade Terms

Abnormal operating condition (AOC)
American National Standards Institute (ANSI)
Breakout tank
Computational pipeline monitoring (CPM)
Disbondment
Encroachment
Highly volatile liquid (HVL)
Lower explosive limit (LEL)
Maximum allowable operating pressure (MAOP)
Maximum operating pressure (MOP)
Naturally occurring radioactive materials (NORMs)
One-Call notification system
Pipeline and Hazardous Materials Safety Administration (PHMSA)
Pipeline facility
Pipeline operator
Redundant system
Resistance temperature device (RTD)
Right-of-way (ROW)
Self-contained breathing apparatus (SCBA)
Supervisory control and data acquisition (SCADA)
Surge pressure
Uninterruptible power system (UPS)

Industry-Recognized Credentials

If you're training through an NCCER-accredited sponsor, you may be eligible for credentials from NCCER's Registry. The ID number for this module is AOCCC-17. Note that this module may have been used in other NCCER curricula and may apply to other level completions. Contact NCCER's Registry at 888.622.3720 or go to **www.nccer.org** for more information.

Contents

Topics to be presented in this module include:

Figures and Tables

SECTION ONE

1.0.0 RECOGNIZING ABNORMAL OPERATING CONDITIONS

Objective

Explain how to recognize abnormal operating conditions in a control center for both gas and liquid pipelines.

a. Explain how to recognize abnormal facility conditions and pipeline system damage.
b. Explain how to recognize the activation of a safety device.
c. Explain how to recognize a communications failure and a control system failure.
d. Explain how to recognize power interruptions.
e. Explain how to recognize a fire, explosion, and natural disaster occurring with a pipeline.
f. Explain how to recognize hazardous liquid or gas encounters.
g. Explain how to recognize an unexplained flow rate / pressure change.

Trade Terms

Abnormal operating condition (AOC): A condition within a pipeline system, identified by the operator, that may indicate (1) a malfunction of a component or (2) a deviation from normal operations that may indicate a condition exceeding design limits or result in a hazard to persons, property, or the environment.

American National Standards Institute (ANSI): An organization that adopts standards for a variety of products, services, and processes in the United States to ensure consistency.

Breakout tank: A tank used to (a) relieve surges in a hazardous liquid pipeline system or (b) receive and store hazardous liquid transported by a pipeline for reinjection and continued transportation by pipeline.

Computational pipeline monitoring (CPM): Technology that provides monitoring tools used by control center personnel to detect temperature, flow, and pressure irregularities in a pipeline.

Disbondment: Separation of the protective coating from the pipeline.

Encroachment: Activity carried out by and/or equipment belonging to outside parties on the pipeline right-of-way. A particular encroachment might be authorized or not, so each situation requires specific handling.

Highly volatile liquid (HVL): a hazardous liquid which will form a vapor cloud when released to the atmosphere and which has a vapor pressure exceeding 276 kPa (40 psia) at 37.8°C (100°F).

Lower explosive limit (LEL): The lowest concentration of a gas or a vapor in the atmosphere capable of producing fire in the presence of an ignition source. Different products have different LELs.

Maximum allowable operating pressure (MAOP): The maximum pressure at which a pipeline or segment of a pipeline may be operated (gas).

Maximum operating pressure (MOP): The maximum pressure at which a pipeline or segment of a pipeline may be normally operated (liquid).

One-Call notification system: A centralized communication system established by utilities, government agencies, and operators of underground facilities. This system provides one telephone number for contractors and the general public to call to notify of their intent to use equipment on or near a pipeline. It is also in place for any individual to call if pipeline damage is discovered. The One-Call system is generally set up on a state-by-state basis.

Pipeline and Hazardous Materials Safety Administration (PHMSA): An agency of the US Department of Transportation responsible for developing and enforcing regulations for pipeline operations in the United States.

Pipeline facility: New and existing pipe; rights-of-way; and any equipment, facility, or building used in the transportation of hazardous liquids or carbon dioxide, or new and existing pipelines; rights-of-way; and any equipment, facility, or building used in the transportation of gas or in the treatment of gas during the course of transportation.

Pipeline operator: The company that operates the pipeline, as opposed to an individual person who works for the company that operates the pipeline.

Redundant system: A system that duplicates the operations of a system so that, in the event of a failure, a backup system is available.

Resistance temperature device (RTD): A device used on pumps that reads the temperature of the pump case and the product in the pipeline. It will sound an alarm when the temperature reaches a critical level and shut down the pump automatically if needed.

Right-of-way (ROW): The area along the pipeline which has been legally permitted to be used by the pipeline operator to construct, operate, and maintain the pipeline. The ROW extends the entire length of the pipeline.

Supervisory control and data acquisition (SCADA): A computer-based system or systems used by a controller in a control room that collects and displays information about a pipeline facility and may have the ability to send commands back to the pipeline facility.

Surge pressure: Pressure produced by a change in velocity of the moving stream that results from shutting down a pump station or pumping unit, closure of a valve, or any other blockage of the moving stream.

Uninterruptible power system (UPS): A system that provides continuity of power when the normal power supply is interrupted. A UPS provides AC power to the system it serves.

An **abnormal operating condition (AOC)** is defined under federal regulations as a condition identified by the **pipeline operator** that may indicate a malfunction of a component or a deviation from normal operations that may indicate a condition exceeding design limits or result in a hazard to persons, property, or the environment. AOCs are defined as such by federal pipeline safety law 49 *United States Code (U.S.C.)* 60102 and pipeline safety regulations 49 *Code of Federal Regulations (CFR)* 192 (for transportation of gas) and 49 *CFR* 195 (for transportation of liquids). The federal agency that regulates pipeline maintenance is the **Pipeline and Hazardous Materials Safety Administration (PHMSA)** in Washington, DC.

In the control room, control center personnel may become aware of AOCs from reports directly from field personnel, by reports from third parties, or by alerts from computer systems in the control room. Some pipeline equipment is fitted with safety devices that automatically alert control room systems to problems. The control center console (see *Figure 1*) consists of computer systems and monitors that allow control center personnel to observe pipeline operations and take actions as needed.

Types of AOCs, along with examples and proper reactions, are listed in *Table 1*. Note that the reactions are general.

1.1.0 Recognizing Abnormal Pipeline Facility Conditions

Any problem with the physical structure of the pipeline is defined as an abnormal **pipeline facility** condition. A pipeline facility includes the pipeline itself; **right-of-way (ROW)**; and any equipment, facility, or building used in the transportation of a liquid or gas. Examples of components of a pipeline facility include compressors for gas pipelines and pump stations for liquid pipelines (this module refers to both as units). The terms *pipeline* or *pipeline system* mean all parts of a pipeline facility through which a hazardous liquid or gas moves in transportation, including, but not limited to, line pipe, valves, and other equipment connected to line pipe; pumping or compressor units and the fabricated assemblies associated with them; metering and delivery stations; and regulator stations. The following are examples of abnormal pipeline facility conditions that can be reported to the control center:

- Exposed pipeline that should be covered
- Damaged pipeline
- Missing or damaged pipeline supports
- Corrosion
- Coating separating from the pipeline (**disbondment**)
- Missing signs or markers
- Pressure reduction for various reasons
- Reports of unauthorized activity and/or equipment on the pipeline right-of-way (unauthorized **encroachment**)
- Sunken floating tank roof

65102-13_F01.EPS

Figure 1 Control center console.

Table 1A Abnormal Operating Conditions: Examples and General Reactions (1 of 2)

Abnormal Operating Condition	Examples	Appropriate Reaction
Abnormal pipeline facility condition	• Exposed pipe • Evidence of corrosion • Incorrect ANSI pressure rating of a component or material type • Disbondment (coatings separating from the pipe) • Unauthorized activity and/or equipment on the right-of-way (unauthorized encroachment) • Incorrect data or records caused by equipment malfunction • Mechanical/hydraulic system failures • Fire and explosion • Pipeline system damage, such as: – Missing or damaged pipeline supports – Damaged or inoperable pipeline components (pipe, valves, pumps, compressors, meters, tank gauges) due to external force – Gouges, dents, pitting, wrinkles, buckles, or cracks – Cracked or corroded flange or bolts – Sunken floating tank roof – Damaged or malfunctioning pontoon – Defective tank roof seal – Damaged tank shell – Damaged pipe due to natural disaster	Follow standard operating procedures and make notifications per company policy. Isolate and/or shut down the pipeline segment or facility if necessary. This may require coordination with field personnel. Complete the required documentation.
Activation of a safety device	• Emergency or abnormal shutdown alarm • High-pressure shutdown alarm • Activation of a pressure relief valve • Activation of a smoke or fire detector • Activation of a hazardous atmosphere alarm • Activation of an overfill protection device • High sump-level shutdown alarm • High tank-level shutdown alarm • High pump-case pressure alarm • Seal failure alarm	Follow standard operating procedures and make notifications per company policy. Isolate and/or shut down the pipeline segment or facility if necessary. This may require coordination with field personnel. Complete the required documentation.
Communications failures	Personnel communication tools: • Lost use of radio communications • Lost use of landline telephone • Lost use of cell phone System communications: • Facility outages • SCADA communication outage • System programming errors/incorrect command logic • Loss of satellite communications • Loss of microwave signals • Stale data (communications going from the control center to the field, but not being received by the control center from the field)	Switch to alternate equipment or backup systems. Follow standard operating procedures and make notifications per company policy. Complete the required documentation.

- Incorrect data or records caused by equipment malfunction
- Incorrect **American National Standards Institute (ANSI)** pressure rating of a component or material type
- Malfunctions due to human error

Control center personnel may be notified of an abnormal facility condition by field personnel. A report may also come in to the control room as

Table 1B Abnormal Operating Conditions: Examples and General Reactions (2 of 2)

Abnormal Operating Condition	Examples	Appropriate Reaction
Power interruptions	• Transformer failure • Downed power lines • Broken leads or wires • Incorrect wire termination • Station or terminal power problems (due to motor overload or a tripped breaker, for example)	Follow standard operating procedures. Switch to an alternate control system or power source if necessary. Notify the power company of the outage. Make additional notifications per company policy. Complete the required documentation.
Fire, explosion, or natural disaster	• Electrical fire • Ignition of a hazardous vapor • Unintentional detonation • Weather conditions, such as an ice storm • Fire, explosion, or pipeline movement caused by natural disaster, such as: – Lightning strike – Flooding – Landslide – Earthquake – Tornado	Follow company emergency action plans. Call local emergency personnel if appropriate. Isolate and/or shut down the pipeline segment or facility if necessary. Limit access to the area as much as possible. Make notifications per company policy. Complete the required documentation.
Unexpected release of a hazardous gas or liquid	Reports of: • Vapor clouds or frost balls • Smell of hydrocarbon release • Dead vegetation/contaminated soil • Dead wildlife • Rainbow sheen on water Hazardous products include: • Hydrogen sulfide • Carbon dioxide • Anhydrous ammonia • Propane • Natural gas • Refined fuels/products Note that the controlled transfer, sampling, venting, draining, and handling of hazardous products are not considered AOCs.	Follow company emergency action plans. Call local emergency personnel if appropriate. Isolate and/or shut down the pipeline segment or facility if necessary. Make notifications per company policy. Complete the required documentation.
Unexplained flow rate / pressure change	Any change from the expected flow or pressure, such as: • Increased pressure • Decreased pressure • No pressure • High flow limit • Low flow limit More often than not, irregularities are detected through a computational pipeline monitoring (CPM) system	Follow standard operating procedures and make notifications per company policy. Isolate and/or shut down the pipeline segment or facility if necessary. Complete the required documentation.

a result of a call made to the **One-Call notification system** or from a person calling the pipeline operator directly. When a report comes in, personnel must obtain complete information so that they can quickly find the problem and address it. Many pipeline operators use a form to capture the following information:

- Caller's name and telephone number
- Date and time of the call
- Was anyone injured? Was 911 notified by the caller?
- Type of product involved, if relevant and if known
- Location of the problem, including latitude and longitude, if possible
- Nearby landmarks such as highways, streams, mile markers, or any other structure
- The nature of the damage and any conditions observed at the site

One of the most important things to remember as a control center operator is to always follow company policy and procedures with any abnormal operating condition.

1.1.1 Pipeline System Damage

The control center may receive the indication of pipeline damage on the control center console as an unexplained change in operating conditions. Pipeline damage or exposed pipe may also be discovered by field personnel or a third party who then reports the damage to the control center. Third-party examples include the aerial patroller of the pipeline ROW who notifies the pipeline operator, or an individual landowner who calls the One-Call notification system.

One way that control center personnel notice damage is by monitoring the pressure and flow changes on the control center console. Control center personnel use monitoring tools, referred to as **computational pipeline monitoring (CPM)** systems, to detect pressure, flow, and temperature irregularities in a pipeline. These irregularities may indicate a release. Most CPM systems provide an alarm and display other data related to pressures, flows, and temperatures that personnel use to detect problems.

Physical signs of damage to the pipeline and its components may include gouges, cracks, pitting, dents, buckles, and inoperable valves, pumps, or tank gauges. Flanges or bolts may be cracked or show significant corrosion. Tank shells or pipeline supports may show damage. A sunken roof or a damaged pontoon also indicates damage. Less obvious signs of damage could include a rainbow sheen on the water near submerged pipelines, which may indicate a release. The release of a **highly volatile liquid (HVL)** may be discovered by observing a frost ball or vapor cloud along the ROW of a liquid pipeline. While the control center personnel are not able to see physical damage to a pipeline, unless viewed through a camera system, they often receive notice of these types of damage indications.

1.1.2 Damaged Pipeline Supports

Control center personnel may receive reports about damage to pipeline supports (see *Figure 2*) from maintenance or inspection personnel who are traversing the pipeline. Damage can include visible cracking, sagging, or buckling of a support, or bolts or other components of the support that have broken or become loose. This type of damage can be a serious problem and can require a shutdown. Sometimes pipeline operations can continue while repairs are made.

65102-13_F02.EPS

Figure 2 Pipeline on pipe supports.

1.1.3 Malfunctions Due to Human Error

Malfunctions are not always due to mechanical or other system failures; they can occur simply due to human error. Control center personnel should be able to recognize the following possible malfunctions as a result of human error:

- Deviation from the design limits
- Valves that are not sequenced correctly
- Cavern upset
- Units or valves that are not aligned properly
- Units that are set incorrectly, which may cause overpressure in the line
- Overflow of a tank

1.2.0 Recognizing the Activation of a Safety Device

Safety devices are part of the pipeline system designed to sense and alert personnel of any potential danger and/or hazardous condition. Hazardous conditions can include changes in pressure that might exceed pipeline design limits as well as the release of harmful gases. Safety devices include, but are not limited to, pressure relief systems, pressure limiting devices, high pump-case-pressure alarms, seal failure alarms, overfill protection devices, tank-level alarms, main-line pressure shutdown alarms, and hazardous atmosphere alarms.

Generally when a safety device activates, a sensor on the device sends a signal to the control center. An alarm may sound in the control room, and the signal appears as an alarm on the control center console. All safety alarms must be verified and the problem corrected before clearing the alarm.

1.2.1 Pressure Limiting Devices

Pressure-limiting devices include control valves. These valves are monitored and controlled by the control center. If the pressure in the pipeline is different than it should be, the control valves are adjusted as needed.

1.2.2 Seal Failure Alarms

Seal failure alarms are activated when a leak is detected in the seal on a pump. There are different types of seals that can activate seal failure alarms when not working properly. One common type of seal is a tandem seal. Tandem seals are a combination of two seals with a barrier liquid between the seals in a chamber that is not pressurized. When the barrier fluid rises to a certain level from process liquid entering the chamber the alarm is activated.

Another type of seal commonly used is a double seal. Double seals consist of two seals that are mounted back-to-back. Usually a seal failure results in product leaking into the atmosphere. This could cause a hazardous atmosphere alarm to activate.

1.2.3 Overfill Protection Devices

Overfill protection devices activate if product reaches a certain level in the tank. As a result of activation, the system may shut off product flow, restrict product flow, or signal an alarm. Two types of overfill protection device are a high and a high-high level alarm that detects the level of product when it reaches a predetermined level. The alarm may also activate a pump shutdown or valve to divert the product flow to prevent a release. Another device related to the overfill protection device is the low-low level alarm. This alarm activates when the product level reaches a predetermined low level. The pump does not shut down with a low-low alarm activation, but monitoring may be required to ensure the levels do not become dangerously low and damage the pump.

1.2.4 Pressure Switches

Pressure switches activate when electrical contacts within the switch close when the pressure reaches a set level. The switch may be designed to make contact on pressure rise or on pressure fall. Pressure switches can have either automatic or manual resets. The automatic resets will reset when the pressure returns to normal operating range. Manual resets need to be reset by personnel.

1.2.5 Activation of a Pressure Relief System

The activation of relief devices along the pipeline system is a time-sensitive occurrence. Relief valves can discharge into tanks or vessels that are designed and constructed in compliance with American Petroleum Institute (API) *Standard 650* or *API 2510* for pressure vessels. Relief valves can also open into the atmosphere. Relief valves (see *Figure 3*) can be installed on a pipeline or **breakout tank**. The purpose of a pressure relief device is to protect the pipeline or tank from overpressure. Breakout tanks can also be used to temporarily absorb **surge pressure** from the main-line and manifold pressure-relief valves to lower pressure on the line. Pressure-relief devices must be provided for each low-pressure and high-pressure breakout tank. There are different specifications for low-pressure and high-pressure tanks.

Some valves can be manually operated by field personnel, and some are remotely operated by control center personnel. When a relief valve is activated, an alarm may sound.

1.2.6 High Pump-Case Temperature/Pressure Alarm

Main-line pumps are fitted with a **resistance temperature device (RTD)** that give temperature readings for the pump case and product. These devices are preset to signal an alarm in the control center

65102-13_F03.EPS

Figure 3 Relief valve.

when the temperature reaches a critical level, and will shut the pump down automatically if some action cannot be taken to lower the temperature. When a pump-case temperature reaches such a critical level and automatically shuts down, it is considered an AOC. *Figure 4* shows RTDs on a pump case.

1.3.0 Recognizing a Communications Failure

Communications refers to the tools that personnel use to communicate with each other and with systems. Communications failures include the failure of communications among personnel, or the failure of communications between the control center and equipment on the pipeline.

1.3.1 Failure of Personnel Communication Tools

Pipeline personnel use landline telephones, cell phones, and radios to communicate with one another on the job. A breakdown in personnel communications equipment may be easy to recognize. Personnel may not hear messages or be able to make phone calls. Personnel must keep their equipment in good working order and test their equipment regularly to ensure that it is working properly.

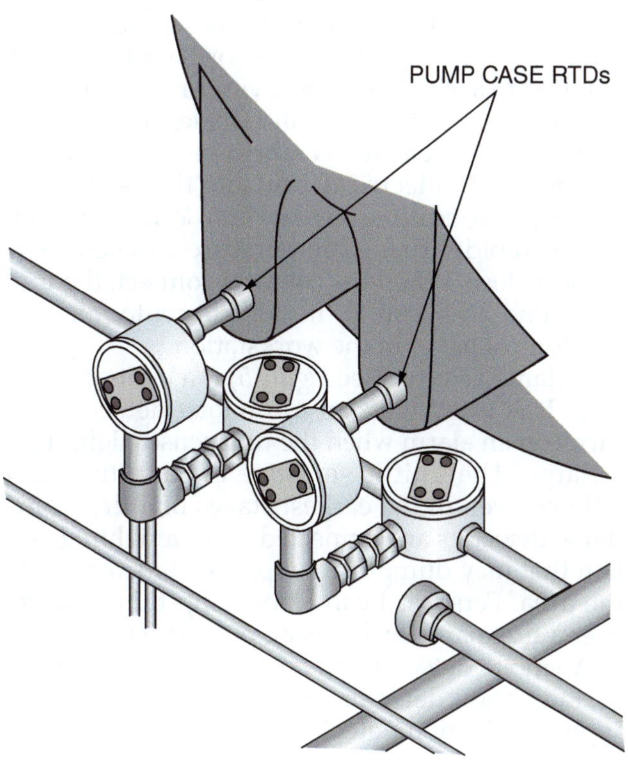

Figure 4 Resistance temperature devices (RTDs) on a pump case.

1.3.2 System Communications Failures

System communications refers to the computer equipment located in the control center and on some pipeline equipment. Personnel in the control center use the **supervisory control and data acquisition (SCADA)** system to manage pipeline operations. The system receives signals from and sends signals to pipeline equipment that may be many miles away. The SCADA system allows control center personnel to monitor pressure, volume, and temperature and keep products flowing safely and on schedule. If the system fails, control center personnel lose the ability to communicate with equipment on the pipeline. Some systems are set up to switch to a **redundant system**—a backup that comes on line if the main system fails.

Some failures of communications systems show up as warnings or messages on the control center console. Another potential sign of a communications failure is stale data showing on the computer monitor. A system failure may include any of the following events:

- Random facility outages
- Partial or complete facility outages
- Loss of the SCADA system
- Loss of satellite communications
- Loss of microwave signals
- Loss of a signal point within the system
- Stale data

If a communication failure occurs, the control center computer signals control center personnel of a problem so that an investigation can begin and the problem can be isolated, if needed.

If an individual remote terminal unit (RTU) loses communication with the control center, control center personnel must follow appropriate company procedures. In an emergency shutdown situation, if communication is lost to a main-line station that must be shut down, control center personnel must use other means to shut down the station. One method is to allow the suction pressure to decrease below the low-suction set point and call the appropriate field personnel.

The discovery that a large group of RTUs is not responding reveals a serious problem. Some SCADA systems are set up to switch to a backup communication system so that an attempt to control the situation can begin as soon as possible. Some companies are equipped with two backup computer systems in case the SCADA fails. Control center personnel should request a switch to one of the backup systems if available.

Control center personnel must consult with appropriate personnel to try to estimate how long the communications system will be in failure. The console monitors will reveal how many systems are affected. If control center personnel are advised that the system will be in failure for a specified amount of time, they will initiate appropriate procedures.

1.3.3 Control System Failures

Mechanical and hydraulic failures, also known as control system failures, are related to system communications failures. For example, the SCADA system sends a signal to a piece of equipment, but the equipment does not respond to the signal. In some cases, the problem is a result of a programming or command logic error. In other cases, equipment may not work because of a mechanical, electrical, or hydraulic problem. Examples include frozen valves, electrical component burnouts, and compressor breakdowns. A valve may have a mechanical failure that prevents it from being operated remotely by the control system. Sometimes valve actuators, electrical components, or connections burn out. This prevents the equipment from responding to a command by the system to start up or shut down.

1.4.0 Recognizing Power Interruptions

Electric power is essential to pipeline operations. Nearly every part of the pipeline depends on electricity in some way. A loss of power impacts the abilities of the control center or facility operations to safely control and monitor the pipeline facility. Power interruptions that result from bad weather or a problem at the local utility company, such as transformer failure or downed power lines, can be common.

Some pipeline operators have an **uninterruptible power system (UPS)** that turns on if electrical power is off for more than a few seconds. UPS systems supply emergency alternating current (AC) power to pipeline components to keep the pipeline operating until power is restored. Instead of electricity, these systems run on engines fueled with gasoline, diesel, or natural gas that are designed to start up if power is off for more than a few seconds. When power fails, an AC power failure alarm may be sent to the control center. To ensure that the backup system has turned on, control center personnel check to see whether they are receiving updated information from the pipeline.

If power is lost, electronic indicators may be lost. The SCADA system may switch off and pumps or compressors may shut down. In addition, lights, heat, and air conditioning may fail. An area-wide power outage may cause a loss of telephone communications. Local power outages may shut down only a portion of the pipeline system. Power outages can also cause pumps or compressors at the receiving terminal to stop operating, causing a liquid or gas shipment to overflow in tanks or be unable to be shipped further.

Some equipment is protected from power failure by battery systems. These systems run on direct current (DC). The following types of equipment can operate on DC power:

- Fire detection equipment and alarms
- Halon dispenser systems
- Fire suppression systems
- Emergency lighting systems
- SCADA
- Alarm panels
- Control modules
- Process control panels
- Specialized compressor units

1.5.0 Recognizing Fire, Explosion, or Natural Disaster

Fire, explosion, or natural disaster can occur at any time. Some events may be anticipated, but many are not. Control center personnel must be trained to recognize these events.

1.5.1 Fire

Detection devices installed along the pipeline and in pipeline facility buildings alert personnel in case of a fire. These devices send an alarm when their sensors detect heat or flame. In most cases, control center personnel will only see or hear the alarm, not the fire itself. Detection devices include thermal detectors, flame detectors, and smoke detectors.

A thermal detector is a heat-sensitive unit. It is fitted with electrical contacts that stay open as long as conditions are normal. When the unit senses rapidly rising temperatures, the contacts start to close. When the contacts connect, the unit may send an alarm to the control center and the fire control panel in the workstation.

A flame detector (see *Figure 5*) is a light-sensitive unit. It is fitted with an ultraviolet detector that turns on an alarm when the unit senses light from a flame. The unit is sensitive enough to detect a flame even in the earliest stages of a fire. Some flame detectors are connected to a firefighting system that may dump fire suppression material into the area. Personnel can also dump this material over a fire manually by using pull switches.

A smoke detector can be a light-sensitive unit, an ionization detector, or a combination of both types. Ionization detectors are more sensitive to fires in the early smoldering stage, while light detectors are more sensitive to the flaming stage of fires than ionization detectors.

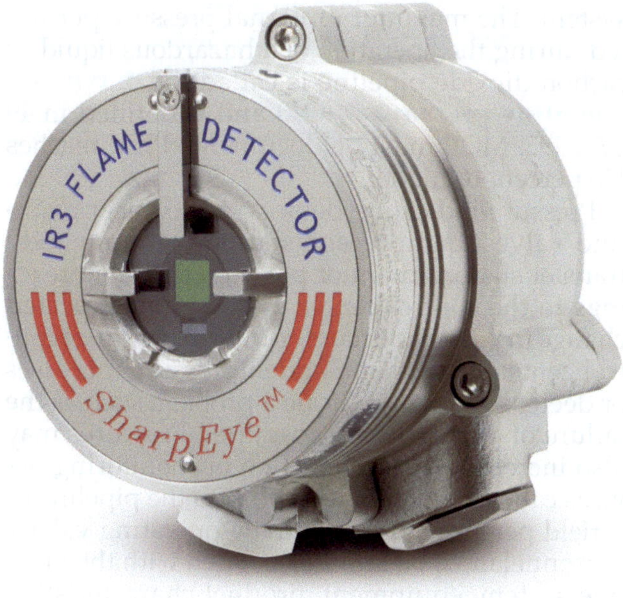

65102-13_F05.EPS

Figure 5 Flame detector.

An ionization smoke detector produces ionization in air; when there is smoke, a difference is detected. If any smoke particles pass into the chamber of the detector, the ions attach to the particles, and become less able to carry the current. An electronic circuit detects the current drop, and sounds the alarm. The light detector operates by detecting light beams. In the absence of smoke, the light passes in front of the detector in a straight line. When smoke enters the chamber across the path of the light beam, light is scattered by smoke particles, directing it at the sensor that triggers the alarm.

Monitoring devices that detect a **lower explosive limit (LEL)** produce alarms to indicate that there is a risk of explosion. The LEL is the lowest concentration of a gas or a vapor in the atmosphere capable of producing fire in the presence of an ignition source. Different products have different LELs. The monitoring device must be set to detect the particular LEL in an enclosed area. Hazardous atmosphere detectors also need to be used in an enclosed area where an LEL has gone off before anyone re-enters the area.

1.5.2 Explosion

Pipeline operators install a number of safety devices to warn personnel about conditions that could cause explosions. However, operators cannot control every circumstance, such as intentional damage, that could lead to an explosion on a pipeline. An explosion on a pipeline registers in the control center as an unexplained change in pressure. An explosion may occur in any enclosed space where hazardous vapors can accumulate to dangerous levels. Examples of such spaces include pipeline facility buildings.

If natural gas leaks into an enclosed space, the odorant added to the gas sometimes alerts people to the problem. However, gas can sometimes accumulate undetected. For example, a building may be vacant or the odors may be too far away from people inside the building. In addition, products may not have an odorant added during transport. Be aware that a spark from electrical equipment can cause the gas to explode.

Whenever a call comes into the control center in which the caller describes a smell related to a pipeline, always make note of it for reporting purposes.

WARNING!

Personnel should never rely on their sense of smell to detect a natural gas leak. Products do not always have a detectable odor.

1.5.3 Natural Disaster

Natural disasters include earthquakes, hurricanes, tornadoes, landslides, wildfires, or flooding. These events may cause soil movement, which may result in forces sufficient enough to cause pipeline failure or product release. Even heavy rains resulting in saturated soil conditions may cause soil movement in some areas.

Pipeline operators monitor for weather that can develop into a natural disaster. Control center personnel may monitor for earthquakes with seismic indicators. In addition, they may obtain real-time information on the internet. Up-to-date earthquake information is available from the National Earthquake Information Center, which is located in Denver, Colorado. Personnel may track weather alerts and warnings through the National Weather Service website. In case of a power failure, personnel can continue to receive information through the National Weather Service's National Weather Radio or other electronic devices. A list of weather and natural disaster information sources is included in *Appendix A*.

1.6.0 Recognizing the Unexpected Release of Hazardous Gas or Liquid

Control center personnel will usually receive reports about an unexpected release from field personnel. The field personnel may notice the physical signs of a release such as a pool of liquid on the ground, a frost ball, vapor cloud or mist in the air, dead or discolored vegetation, or a rainbow sheen on water. Other signs include dead fish in waterways or other dead wildlife. Know company policies and procedures for each type of release and be prepared to act quickly and decisively to protect people, the pipeline, and the environment.

1.7.0 Recognizing Unexplained Flow Rate / Pressure Changes

Flow rate and pressure are important indicators of both normal and abnormal operating conditions. Minor changes in pressure may occur normally on the pipeline. However, significant unexplained pressure changes are considered AOCs. Unexplained pressure changes include absences of, increases in, and decreases in pressure. The maximum pressure a gas pipeline or segment of a gas pipeline may be safely operated is called the **maximum allowable operating pressure (MAOP)** and is defined in 49 *CFR* 192. MAOP is based on the materials transported and the design of the system. The maximum internal pressure permitted during the operation of a hazardous liquid or carbon dioxide pipeline is called the **maximum operating pressure (MOP)** and is defined in 49 *CFR* 195. MOP must be documented if it reaches 110 percent or higher.

Pressure-indicating devices located near mainline valves, compressor stations, and custody transfer stations monitor pressure and send readings to the control center. When an unexpected change in pressure occurs, an alarm may alert control center personnel. Sudden pressure increases or decreases might indicate a problem such as the failure of a valve or a transmitter. Pressures may also increase when inspection or monitoring devices called pigs are launched into the pipeline or if field personnel are manually operating valves. Personnel must be familiar enough with the pipeline system, equipment, product characteristics, and environmental factors to quickly identify any important changes in pressure and flow.

Additional Resources

It is assumed that you will learn to perform these procedures under the direct supervision of your crew leader, foreman, or instructor. For independent study, refer to the recommended Additional Resources at the back of this module.

1.0.0 Section Review

1. Which of the following is an AOC?
   a. Aboveground pipe on pipe supports
   b. Damaged pipeline
   c. Valves operating properly
   d. Authorized encroachment

2. Safety devices include _____.
   a. hazardous atmosphere alarms
   b. tank shells
   c. SCADA
   d. compressors

3. SCADA serves as a _____.
   a. regulatory agency
   b. notification system for all utilities
   c. device worn to provide breathable air where the atmosphere is hazardous
   d. command system that assists control center personnel

4. An uninterruptible power system (UPS) supplies _____.
   a. temperature readings
   b. emergency AC power
   c. emergency DC power
   d. pressure readings

5. A flame detector is designed to activate when it detects _____.
   a. an increase in temperature
   b. light
   c. smoke
   d. pipeline pressure

6. Dead vegetation or dead wildlife may be a sign of an unexpected release of a hazardous gas or liquid.
   a. True
   b. False

7. The maximum operating pressure of a gas pipeline is called _____.
   a. MOP
   b. MAOP
   c. surge pressure
   d. compression

SECTION TWO

2.0.0 REACTING TO ABNORMAL OPERATING CONDITIONS

Objective

Describe the proper reaction to abnormal operating conditions in a control center for both gas and liquid pipelines.

a. Describe the proper reaction to abnormal facility conditions and pipeline system damage.
b. Describe the proper reaction to the activation of a safety device.
c. Describe the proper reaction to communications failure and a control system failure.
d. Describe the proper reaction to power interruptions.
e. Describe the proper reaction to a fire, explosion, and natural disaster occurring with a pipeline.
f. Describe the proper reaction to hazardous liquid or gas encounters.
g. Describe the proper reaction to an unexplained flow rate / pressure change.

Trade Terms

Naturally occurring radioactive materials (NORMs): Materials with radioactive elements that can be extracted during oil and gas extraction from the earth and can get into product streams and become transported in the pipeline.

Self-contained breathing apparatus (SCBA): A device worn to provide breathable air for an individual in an area where the atmosphere is hazardous.

Time is of the essence when AOCs occur, and control center personnel are expected to take decisive action. Personnel must always follow established company-specific policies and procedures. Personnel must make use of their training, skills, and good judgment to maintain safe operation until solutions are found.

2.1.0 Reacting to Abnormal Pipeline Facility Conditions

Some abnormal pipeline facility conditions may require immediate action. Others may require scheduling repairs at a later date. In most cases, if control center personnel are notified of an abnormal pipeline facility condition, they must notify the appropriate personnel so that repairs can be made safely and in a timely manner. If notified of a condition by a third-party call, keep in mind that a caller's description may not be entirely accurate. The problem may be other than what was described. If a report is confirmed and the condition does not require shutdown, make appropriate notifications as required by company policy. Reports of exposed pipe and damaged pipeline supports are examples of conditions that may require immediate action.

2.1.1 Pipeline System Damage

If a report of exposed pipe comes in to the control center, control center personnel should check the data received from the monitors and SCADA and send field personnel to investigate according to company policy. It is possible that communications were destroyed at the time the damage occurred and thus no alarm was signaled. The pipeline may need to be shut down until the investigation is complete, according to company policies and procedures. The decision to do so should be based on the assessment provided by trained and qualified personnel. The sooner qualified personnel can arrive on the scene, the more quickly a practical assessment of conditions can be obtained and control of the situation can begin.

Some incidents may be serious enough to require an emergency shutdown of the pipeline. One example of a serious incident is the loss of control of a vehicle on the ROW, extensively gouging or denting a large section of pipeline or damaging a valve. Another common example is third-party damage to pipeline from excavation. Whenever pipeline damage results in a release and/or explosion, control center personnel should receive a report from field personnel. In some cases, control center personnel may be able to drain a portion of the pipeline into a delivery system to lessen the amount of product that could be released from the pipeline.

Pipeline shutdowns do not have to occur only because of obvious damage to the pipeline. A blocked valve that causes increased pressure in the line can be reason for a shutdown. However, if pressures remain within the normal operating parameters, a shutdown may not be necessary, but the control center personnel must report the status change to the appropriate personnel.

Once the pipeline has resumed normal operations, the pipeline operator will conduct an internal investigation. This investigation focuses on evaluating control systems and mechanical equipment, as well as personnel performance. Personnel actions must be evaluated to determine whether company policies and procedures were followed. The work must also be evaluated to determine the effectiveness of existing company procedures. If the operator suspects that the damage was intentional, the appropriate personnel may contact regulatory and law enforcement agencies that may then conduct a criminal investigation.

2.1.2 Damaged Pipeline Supports

The proper reaction to the discovery of damaged pipeline supports varies depending on the severity of the damage. If the damage to the support is critical or indicative of a possible release, control center personnel may receive a report from field personnel. Notification is especially important if the pipe is in danger of excessive movement. Follow company policy precisely to investigate the condition and dispatch qualified personnel to the scene.

2.2.0 Reacting to the Activation of a Safety Device

Control center personnel must respond to safety device alarms quickly and calmly. They must be able to dispatch the appropriate personnel and inform them of the potential for a hazardous environment. Control center personnel may need to shut down the pipeline, if necessary. Field personnel may need to reset alarms. Company procedures must always be followed.

It is important to ensure that an alarm is not a false alarm. In case of a false alarm, the control center should notify affected personnel and have the alarm equipment checked. Generally, no further action is required when an alarm is false.

For example, if responding to a seal failure alarm, personnel should investigate the validity of the alarm, check the station sump level, and notify the appropriate personnel. This may not necessarily result in an AOC, but it may require control center personnel to document the alarm and notify the appropriate personnel to check and possibly replace the seal. Control center personnel should never attempt to restart a pump before investigating the cause of the alarm and responding to the alarm according to company policy.

During an emergency, personnel must focus on three key areas: ensuring personnel safety, reducing the damage to equipment, and protecting the environment. The general rule is to put people first. The goal is also to prevent small problems from turning into larger incidents. All reactions must be made safely.

WARNING!

For safety reasons, only authorized personnel who have had the proper training may use PPE.

2.2.1 Activation of a Pressure Relief System

Each relief event should be recognized and must be documented by control center personnel. If a relief device is activated, control center personnel must immediately minimize the situation and verify the device is operating correctly. If the alarm is signaled and the pressure remains high, personnel must take immediate action to reduce pressure. A pressure reduction is usually achieved by shutting down the upstream pumping or compressor station or the pipeline system.

2.2.2 High Pump-Case Temperature/Pressure Alarm

Some main-line pumps are set to shut down on low pressure. Pipeline operators fit pumps with differential pressure switches and temperature switches to protect the equipment from damage caused by low pressure, no pressure, or no flow; to avoid exceeding maximum operating pressure (MOP)/maximum allowable operating pressure (MAOP), which means no more than 110 percent MOP/MAOP; and to ensure that the temperatures do not increase in the pipeline facility. Pumps also have high discharge-pressure alarms that will shut down the pump when the discharge pressure exceeds a preset limit. Many times, this indicates a downstream line obstruction that is not allowing the product to flow properly from the pump.

2.3.0 Reacting to a Communications Failure

The proper reaction to a communications failure depends on whether the failure involves a communications tool that personnel use or the failure of system communications.

2.3.1 Failure of Personnel Communication Tools

When communication tools fail, personnel should first check to see if the problem is a simple one that can be quickly and easily fixed. For example, some telephones have a volume control that may accidentally be turned off. The telephone may seem to be dead, but a simple adjustment to the volume corrects the problem. Other simple solutions include replacing a battery or recharging a unit.

Some communications problems are more serious. Weather or problems at the local utility company can knock out communications completely. Regardless of the outage, the best preparation is to ensure personnel are trained to follow these steps:

Step 1 Switch to alternate equipment if necessary.

Step 2 Notify the appropriate personnel per company policy.

Step 3 Complete any required documentation.

2.3.2 System Communications Failures

Control center personnel must recognize an abnormal condition, investigate immediately, and react according to company policy. The condition must be verified, and the cause(s) should be determined. If the failure involves a major pipeline component such as a remote gate valve or block valve, control center personnel must try to solve the problem as quickly as possible to prevent an incident. In some instances, control center personnel might be able to bypass the problem component until field personnel can perform appropriate maintenance. In other cases, they may be required to shut down.

If one of the pumps at a station that is running suddenly indicates a stop status with no apparent changes to the station pressures, control center personnel should notify appropriate field personnel, and investigate as required per company policy. If a control center loses its monitoring capabilities for a remotely unmanned station, but SCADA is showing normal operating pressures and flow rates upstream and downstream of the station, control center personnel should not make multiple unit switches upstream or downstream of the facility before it is confirmed that a problem actually exists. Appropriate reactions include restoring communications with an alternate system, notifying repair personnel, and having the pipeline facility manned.

The control center must notify field personnel and/or systems support about system communications failures so that they can isolate, investigate, and address the problem as needed. If field personnel can operate the component manually, the problem can be alleviated until repairs can be made. If the component cannot be manually operated, the pipeline may have to be shut down. Control center personnel must monitor pressure remotely or shut down. If pressure is remaining within limits or the component is not one that will affect the operation directly, appropriate personnel must decide if the pipeline can safely remain in operation. If the pipeline remains in operation, the condition must be monitored closely.

After normal conditions have been restored, control center personnel must continue to monitor the condition to make certain that the procedures have been successful. If SCADA communication is lost, control center personnel must follow their company's policies and procedures, usually covered in the business continuity plan. These procedures may include switching to a backup system that may be in another location.

2.4.0 Reacting to Power Interruptions

The proper reaction to a power failure depends on which part(s) and how much of the pipeline is affected. Regardless of how the outage affects the pipeline facility, the appropriate personnel should always notify the power company of the outage as soon as possible. If only a single component or station is involved, control center and/or field personnel may be able to adjust line pressure and flow to keep the pipeline in operation. For example, if a power interruption occurs while a valve is in operation, field personnel may have to operate the valve manually. However, if a station is needed to keep pressure within nnormal operating limits, control center personnel may have to isolate a pipeline segment. If a shutdown is not required, the gas or liquid shipment will continue to arrive, and the schedule must be followed and the tasks completed manually.

Power interruptions can happen in the control center itself. If backup power isn't available, the control center personnel should contact field personnel to go to their control stations to maintain pipeline operations.

Each company has specific procedures for responding properly to a power interruption. Always follow company policy and make appropriate notifications per company policy.

2.5.0 Reacting to a Fire, Explosion, or Natural Disaster

The pipeline operator must have procedures in place so that personnel know what to do in case of a fire, explosion, or natural disaster that affects a pipeline. Always react according to company policy, which includes the policies regarding shutting down the pipeline.

When any of these events occur, the safety of personnel is always top priority. Personnel must ensure that any injured individuals receive medical attention as soon as possible. Consider any alarm as an actual alarm and react appropriately and decisively until it has been verified as false. Control center personnel must investigate the cause of any alarms or other notifications and confirm the nature of the problem in order to react appropriately. Personnel must make appropriate notifications and complete the required documentation.

In emergency situations, including evacuation if needed, follow the company emergency procedures manual as appropriate. Personnel must be adequately trained and familiar with emergency procedures in advance.

2.5.1 Emergency Shutdowns

Some events occurring on the pipeline that present immediate hazards require an emergency shutdown. Such an event might be caused by human error, mechanical failure, electrical failure, or an accident on the pipeline. It might also be brought about by threats from weather or civil disobedience. Regardless of the reason, when an emergency shutdown is required the line must be shut down quickly and safely. Each company has a standard operating procedure for emergency shutdowns.

In the case of a fire, explosion, or release, time for discussion may be very limited. Once the need is confirmed, decisive action must be taken. The pipeline must be shut down by control center personnel who will shut down the appropriate pumps. Control center personnel may isolate or authorize the isolation of the system by closing block valves as necessary.

In emergency conditions such as a possible release, the pipeline operator has full responsibility and authority to make shutdown decisions and take appropriate action. These actions are governed by the company release policy above all other considerations.

2.5.2 Evacuations

The safety of pipeline personnel and the public is the first priority when any AOC occurs. If there is any threat of injury or death to any persons, an evacuation of pipeline personnel must occur. Events that could result in the need for an evacuation are fire, explosion, release of hydrocarbon or other hazardous vapors, seismic activity, or other natural disasters.

Companies have standard procedures in place to guide evacuation of personnel. By law, emergency response plans must be in place and evaluated annually. Facility layouts with evacuation maps are required to be posted in every area or department. Safe areas are designated for all personnel. When the command to evacuate is given, trained personnel may act as safety coordinators. Their job is to conduct a roll call to confirm that all personnel are safe. To assist in this process, personnel must report to their designated safe areas and identify themselves to the designated safety coordinator. If any personnel cannot be accounted for, authorized personnel must be informed so that missing individuals can be located and rescued, if necessary.

2.6.0 Reacting to the Unexpected Release of Hazardous Gas or Liquid

When a hazardous gas or liquid is released from a pipeline unexpectedly, personal safety should be the first consideration. Beyond that, the proper reaction to the unexpected release of a product depends on the characteristics of the product and the location of the release. Some gases and liquids can explode under certain conditions, and some are toxic to people and the environment. To identify the hazards associated with the chemical being released, consult the material safety data sheets (MSDSs) / safety data sheets (SDSs).

Once notified by field personnel of a release, the source of the release must be determined quickly. Control center personnel may need to reduce pressure and shut down the pipeline. If power cannot safely be cut off at the switchgear, control center personnel must call the power company and have power cut off from outside the pipeline facility. Releases in outdoor areas include the additional concern for the safety of the general public, as well as the safety of the environment. In all cases, follow company policies and procedures, make the appropriate notifications, and complete the required documentation.

Personnel must take special precautions when dealing with the release of products that can become toxic. Examples of toxic products include carbon dioxide (CO_2), hydrogen sulfide (H_2S), propane, anhydrous ammonia, and natural gas.

Carbon dioxide is a colorless, odorless, gaseous compound used as a propellant in many domestic applications, such as fire extinguishing equipment, and is a by-product of combustion. Because it is heavier than air, it can collect in low-lying areas where it may displace oxygen; exposure can result in suffocation and death. A leak detector can detect a hazardous atmosphere, but without it, personnel can be overcome without realizing the problem. Some pipeline operations pump carbon dioxide commercially, and in such operations the hazards are increased. To enter an area where a carbon dioxide release is suspected, personnel must use the proper gas detection meter, calibrated for the detection of carbon dioxide, and wear the appropriate PPE, such as a **self-contained breathing apparatus (SCBA)** (see *Figure 6*).

65102-13_F06.EPS

Figure 6 Self-contained breathing apparatus.

Hydrogen sulfide (H_2S) is a colorless gas that often smells like rotten eggs. However, in fairly low concentrations, it can deaden the sense of smell. Exposure to a higher concentration can cause death. In addition, it is flammable and may cause a flash fire. When a gas detection unit shows that H_2S is present, personnel must wear the appropriate PPE or immediately leave the area.

Propane is a common natural gas that is colorless and heavier than air, so it can collect in low-lying areas where it may displace oxygen. Exposure can cause death. At high concentrations, it is flammable.

Anhydrous ammonia is a liquefied gas that has a sharp odor. Exposure in high concentration can cause death. Even exposure in low concentrations can cause severe reactions from mild irritation to severe burns. The reaction depends on the length of exposure and concentration amounts. If anhydrous ammonia is released, call emergency personnel.

Naturally occurring radioactive materials (NORMs) may be present when oil and gas is extracted from the earth. These NORMs are then introduced into the product streams and transported in the pipeline. Natural gas liquids (NGLs), along with processing associated with these products, may trap NORMs in the form of sludge, scale, and particulates. Filters, valves, scrapers, and even internal surfaces of pipelines may contain high levels of NORMs. Precautions should be taken to protect personnel by monitoring for NORMs in accordance with all local, state, federal, and company requirements. If high levels are detected, personnel should distance and/or shield themselves from the source and notify the appropriate personnel.

WARNING!

Exposure to hydrogen sulfide (H_2S), carbon dioxide (CO_2), and propane can be fatal. H_2S is flammable and may cause a flash fire. When a gas detection unit shows that H_2S is present, wear the appropriate PPE or immediately leave the area. CO_2 and propane are colorless, odorless gases that are heavier than air, so they can collect in low-lying areas where they may displace oxygen.

2.7.0 Reacting to Unexplained Flow Rate/Pressure Changes

Control center personnel must respond to a flow rate/pressure change alarm immediately to prevent serious damage to the pipeline. The sooner a response is initiated, the more quickly the problem will be solved and the more likely a serious condition will be averted. Control center personnel should check pressure and flow rates upstream and downstream for comparison. If the time frame involved in the change is very short, a release might be indicated and the pipeline shut down per company policy.

2.7.1 Unexplained Flow Rate Changes

Control center personnel must determine if there has been an intentional change in pumping status or compressor status on the pipeline that was not scheduled. A valve might be the cause of the problem. If the main-line flow has dropped, a pump unit or compressor unit may have shut down upstream, and control center personnel can adjust pumping or compressor units upstream in an attempt to compensate. It is the responsibility of control center personnel to initiate corrective action.

If a pipeline block valve has been closed without authorization or if personnel notice a high flow rate at the origin station, personnel should follow established troubleshooting procedures, perform a shutdown as necessary, and make all proper notifications.

If a pipeline is running, and the origin meter flow rate increases without justification, possible causes for this include SCADA malfunction, a pipeline rupture downstream of the origin meter, or maintenance personnel interfering with data transmission prior to notification. Again, control center personnel should follow established troubleshooting procedures, perform a shutdown as necessary, and make all proper notifications.

If there is not a release, it may be possible to safely work around the AOC. Control center personnel should try to adjust the pipeline pressure and flow to achieve a steady-state operation. During line upsets or valve swings, control center personnel must avoid frequent or unnecessary line upsets because data indicating a release is very difficult to identify during these conditions. If the pressure and flow cannot be corrected within a reasonable length of time, control center personnel should shut the pipeline down as quickly as possible without jeopardizing safety. Company policy determines exact procedures that must be followed. Appropriate personnel should be notified of the shutdown as soon as possible.

2.7.2 Unexplained Pressure Changes

Unexplained pressure changes can include an absence of pressure, a sudden increase in pressure, or a sudden decrease in pressure. Deviations in pressure may indicate an unusual condition on the pipeline. Occurrences such as unintended valve closures may cause AOCs that could endanger the pipeline. The condition must be recognized, and the reaction must be quick and decisive.

If a high-pressure alarm is received from a compressor or pumping unit, personnel should investigate the alarm and determine the appropriate action. This could include shutting down the unit or line as needed.

If the condition is a valve closure or a sudden decrease in discharge pressure followed by a sudden increase in discharge pressure, control center personnel must troubleshoot to determine the cause of the closure. If there is not a high-pressure condition downstream and a valve closure is identified between the two stations, then a command to open the valve must be sent by control center personnel.

If the valve cannot be opened automatically or if the condition occurs at a location with personnel available, opening the valve manually may be possible. If the valve does not open, the pipeline may have to be shut down due to increasing pressure. Company policy and procedure must be followed.

If control center personnel conclude that there may be time to make adjustments before the pipeline reaches MOP/MAOP, immediate action must be taken. Personnel must notify all operations upstream to reduce or stop pipeline injections. Personnel may reduce strategic pumping units upstream or initiate other means to relieve the pressure. Appropriate personnel may be consulted if there is time. If the condition can be corrected by such efforts, the pipeline operation may be continued. Control center personnel must monitor the pressure readings closely until satisfied that the pipeline has stabilized.

If there is no time to accomplish the adjustments or if it appears that the efforts are having no effect, control center personnel must perform a shutdown according to company policy and procedure.

At a steady-state, multipump station pipeline, if there is an increased pressure reduction between two stations, a flow increase upstream of the pressure loss, and all analog data indicates the high probability of a release, control center personnel should shut down the line with the lowest possible pressure on the segment with the suspected release. If personnel notice a mainline block valve status has changed from opened to closed with an increase in discharge pressure at the upstream pump station, personnel should shut down the pipeline until the correct valve position is determined.

Additional Resources

It is assumed that you will learn to perform these procedures under the direct supervision of your crew leader, foreman, or instructor. For independent study, refer to the recommended Additional Resources at the back of this module.

2.0.0 Section Review

1. If damaged pipeline is reported to the control center, the proper reaction always includes shutting down the pipeline.
   a. True
   b. False

2. If a safety device activates, personnel should focus *first* on _____.
   a. protecting the pipeline
   b. ensuring the safety of personnel
   c. notifying PHMSA
   d. verifying that the alarm is not false

3. Field personnel may need to operate equipment in the field manually if there is a systems failure.
   a. True
   b. False

4. When a power outage occurs, the appropriate pipeline personnel should contact _____.
   a. the One-Call notification system
   b. the power company
   c. OSHA
   d. local authorities

5. When reacting to a fire, explosion, or natural disaster, the first priority is always to protect _____.
   a. the pipeline facility
   b. the environment
   c. people
   d. the pipeline product

6. One reason that the unexpected release of carbon dioxide can be dangerous is because it is _____.
   a. odorless
   b. a compound
   c. lighter than air
   d. white in color

7. A pipeline rupture downstream of the origin meter could be the cause of _____.
   a. a power outage
   b. an unexplained communication failure
   c. an unexplained flow rate change
   d. damaged pipe

Summary

Pipeline operators have developed effective routines and policies that enable personnel to respond quickly, safely, and effectively to AOCs. Usually an AOC is a routine matter that can be handled safely by pipeline personnel. Sometimes, however, an AOC can result in an emergency situation that may involve working with outside agencies.

In this module, you learned about the different types of AOCs that can occur on the pipeline and the proper reactions to each type of AOC. During your career, you may have to respond to some or all of the AOCs outlined in this module. When responding to these AOCs, follow company policies and procedures. Protecting people, property, and the environment should be of the utmost concern.

Review Questions

1. By definition, an AOC is a condition that may exceed _____.
   a. maximum operating pressure
   b. design limits
   c. safety laws
   d. size limits
2. By definition, a pipeline facility includes the right-of-way.
   a. True
   b. False
3. Which of the following is an example of pipeline system damage?
   a. Gouges or dents in the pipeline
   b. Transformer failure
   c. Decreased pressure
   d. Activation of a pressure-relief valve
4. Pressure-relief systems, tank-level alarms, and main-line pressure shutdown alarms are examples of _____.
   a. prevention devices
   b. warning devices
   c. hazardous atmosphere alarms
   d. fire alarms
5. The loss of SCADA is an example of which type of AOC?
   a. Hydraulic failure
   b. Mechanical failure
   c. Communications failure
   d. Power failure
6. To keep the pipeline running in case of a power failure, some companies use an _____.
   a. operations override power source
   b. uninterruptible power system
   c. interface power system
   d. indirect power system
7. The failure of a valve or a transmitter may be indicated by _____.
   a. ongoing pressure increases or decreases
   b. sudden pressure increases or decreases
   c. high-level pressure increases or decreases
   d. maximum pressure increases or decreases
8. When pipeline support damage is discovered, notification is especially important if the pipe _____.
   a. is in danger of excessive movement
   b. has changed color
   c. is dented
   d. is corroded
9. If a relief device alarm is activated, control center personnel must immediately _____.
   a. shut down or call 911
   b. reduce pressure or shut down
   c. contact field personnel
   d. complete the required reports
10. At a station that is running, if a pump suddenly indicates a stop status with no apparent changes to the station pressures, control center personnel should _____.
    a. notify appropriate field personnel
    b. reduce pressure
    c. shut down the SCADA system
    d. activate safety devices
11. If SCADA fails, it may be necessary for equipment to be operated manually by _____.
    a. PHMSA
    b. field personnel
    c. control center personnel
    d. local authorities
12. When a power interruption occurs, personnel may be able to keep the pipeline in operation by adjusting _____.
    a. MOP/MAOP
    b. line pressure and pipe location
    c. line pressure and flow
    d. product temperature
13. Regardless of the nature of an emergency, if there is a threat of injury or death, pipeline personnel must be _____.
    a. contacted
    b. evacuated
    c. warned
    d. involved

14. The top priority is always the safety of _____.
    a. personnel
    b. the pipeline
    c. the environment
    d. equipment

15. Emergency shutdowns are generally required when _____.
    a. equipment is reported on the ROW
    b. natural disasters are anticipated
    c. there is an immediate hazard to persons
    d. pipeline damage is reported by third parties

16. Product releases in outdoor areas involve an extra concern for _____.
    a. safety of third-party equipment
    b. safety of pipeline equipment
    c. safety of the pipeline
    d. safety of the environment

17. Control center personnel should rely solely on reports of unexpected releases by field personnel or third parties to determine the proper reaction.
    a. True
    b. False

18. AOCs are not necessarily considered safety-related conditions.
    a. True
    b. False

19. All of the following are examples of toxic products, *except* _____.
    a. CO_2
    b. anhydrous ammonia
    c. propane
    d. H_2O

20. What type of device is preset to signal an alarm in the control center when the temperature reaches a critical level?
    a. RTD
    b. seal failure alarm
    c. LEL
    d. high-pressure alarm

Trade Terms Quiz

Fill in the blank with the correct term that you learned from your study of this module.

1. A computer-based system or systems used by a controller in a control room that collects and displays information about a pipeline facility and may have the ability to send commands back to the pipeline facility is called _______.
2. The _______ is the area along the pipeline which has been legally permitted to be used by the pipeline operator to construct, operate, and maintain the pipeline.
3. Activity carried out by and/or equipment belonging to outside parties on the pipeline ROW is referred to as _______.
4. A(n) _______ is a device worn by an individual to provide breathable air in an atmosphere that is hazardous.
5. _______ is the separation of the protective coating from the pipeline.
6. A(n) _______ is a backup system that provides AC power to the systems it serves.
7. The maximum pressure at which a gas pipeline or segment of a gas pipeline may be normally operated is the _______.
8. A(n) _______ is used to relieve surges in a hazardous liquid pipeline system or receive and store hazardous liquid transported by a pipeline for reinjection and continues transportation by pipeline.
9. A device used on pumps that reads the temperature of the pump case and the product in the pipeline that will sound an alarm when the temperature reaches a critical level and shut down the pump automatically if needed is a(n) _______.
10. A(n) _______ duplicates the operations of a system so that a complete backup system is available in the event of systems failure.
11. A(n) _______ is identified by the operator and may indicate (1) a malfunction of a component or (2) a deviation from normal operations that may indicate a condition exceeding design limits or result in a hazard to persons, property, or the environment.
12. The pressure produced by a change in the velocity of the moving stream that can result from the closure of a valve is called _______.
13. The maximum pressure at which a liquid pipeline or segment of a liquid pipeline may be operated is the _______.
14. The _______ is a centralized communication system established by utilities, government agencies, and operators of underground facilities that provides one telephone number for contractors and the general public to call to notify of their intent to use equipment on or near a pipeline.
15. A hazardous liquid that has a vapor pressure exceeding 40 pounds per square inch absolute at 100°F is classified as a(n) _______.
16. The government agency that is responsible for developing and enforcing regulations for pipeline operations in the United States is the _______.
17. The _______ is the company that operates the pipeline.

18. Materials with radioactive elements that can be extracted during oil and gas extraction from the earth and can get into product streams and become transported in the pipeline are called _______.
19. An organization that adopts standards for a variety of products, services, and processes in the United States to ensure consistency is the _______.
20. The lowest concentration of a gas or a vapor in the atmosphere capable of producing fire in the presence of an ignition source is the _______.
21. Technology that provides monitoring tools used by control center personnel to detect temperature, flow, and pressure irregularities in a pipeline is _______.
22. New and existing pipe on the pipeline itself, the right-of-way (ROW), and any equipment, facility, or building used in the transportation of a liquid or gas are defined as part of a(n) _______.

Trade Terms

Abnormal operating condition (AOC)
American National Standards Institute (ANSI)
Breakout tank
Computational pipeline monitoring (CPM)
Disbondment
Encroachment
Highly volatile liquid (HVL)
Lower explosive limit (LEL)
Maximum allowable operating pressure (MAOP)
Maximum operating pressure (MOP)
Naturally occurring radioactive materials (NORMs)
One-Call notification system
Pipeline and Hazardous Materials Safety Administration (PHMSA)
Pipeline facility
Pipeline operator
Redundant system
Resistance temperature device (RTD)
Right-of-way (ROW)
Self-contained breathing apparatus (SCBA)
Supervisory control and data acquisition (SCADA)
Surge pressure
Uninterruptible power system (UPS)

Trade Terms Introduced in This Module

Abnormal operating condition (AOC): A condition identified by the operator that may indicate (1) a malfunction of a component or (2) a deviation from normal operations that may indicate a condition exceeding design limits or result in a hazard to persons, property, or the environment.

American National Standards Institute (ANSI): An organization that adopts standards for a variety of products, services, and processes in the United States to ensure consistency.

Breakout tank: A tank used to (a) relieve surges in a hazardous liquid pipeline system or (b) receive and store hazardous liquid transported by a pipeline for reinjection and continued transportation by pipeline.

Computational pipeline monitoring (CPM): Technology that provides monitoring tools used by control center personnel to detect temperature, flow, and pressure irregularities in a pipeline.

Disbondment: Separation of the protective coating from the pipeline.

Encroachment: Activity carried out by and/or equipment belonging to outside parties on the pipeline right-of-way (ROW). A particular encroachment might be authorized or not, so each situation requires specific handling.

Highly volatile liquid (HVL): a hazardous liquid which will form a vapor cloud when released to the atmosphere and which has a vapor pressure exceeding 276 kPa (40 psia) at 37.8°C (100°F).

Lower explosive limit (LEL): The lowest concentration of a gas or a vapor in the atmosphere capable of producing fire in the presence of an ignition source. Different products have different LELs.

Maximum allowable operating pressure (MAOP): The maximum pressure at which a pipeline or segment of a pipeline may be operated under this part (liquid).

Maximum operating pressure (MOP): The maximum pressure at which a pipeline or segment of a pipeline may be normally operated under this part (gas).

Naturally occurring radioactive materials (NORMs): Materials with radioactive elements that can be extracted during oil and gas extraction from the earth and can get into product streams and become transported in the pipeline.

One-Call notification system: A centralized communication system established by utilities, government agencies, and operators of underground facilities. This system provides one telephone number for contractors and the general public to call to notify of their intent to use equipment on or near a pipeline. It is also in place for any individual to call if pipeline damage is discovered. The One-Call system is generally set up on a state-by-state basis.

Pipeline and Hazardous Materials Safety Administration (PHMSA): An agency of the US Department of Transportation responsible for developing and enforcing regulations for pipeline operations in the United States.

Pipeline facility: New and existing pipe; rights-of-way; and any equipment, facility, or building used in the transportation of hazardous liquids or carbon dioxide, or new and existing pipelines; rights-of-way; and any equipment, facility, or building used in the transportation of gas or in the treatment of gas during the course of transportation.

Pipeline operator: The company that operates the pipeline, as opposed to an individual person who works for the company that operates the pipeline.

Redundant system: A system that duplicates the operations of an entire system so that, in the event of a failure, a backup system is available.

Resistance temperature device (RTD): A device used on pumps that reads the temperature of the pump case and the product in the pipeline. It will sound an alarm when the temperature reaches a critical level and shut down the pump automatically if needed.

Right-of-way (ROW): The area along the pipeline which has been legally permitted to be used by the pipeline operator to construct, operate, and maintain the pipeline. The ROW extends the entire length of the pipeline.

Self-contained breathing apparatus (SCBA): A device worn to provide breathable air for an individual in an area where the atmosphere is hazardous.

Supervisory control and data acquisition (SCADA): A computer-based system or systems used by a controller in a control room that collects and displays information about a pipeline facility and may have the ability to send commands back to the pipeline facility.

Surge pressure: Pressure produced by a change in the velocity of the moving stream that results from shutting down a pump station or pumping unit, closure of a valve, or any other change in the velocity of the moving stream.

Uninterruptible power system (UPS): A system that provides continuity of power when the normal power supply is interrupted. A UPS provides AC power to the system it serves.

Appendix A

Weather and Natural Disaster Information Sources

Emergency Managers Weather Information. US Department of Commerce, National Oceanic and Atmospheric Administration, National Weather Service, 1325 East-West Highway, Silver Spring, MD 20910. **www.nws.noaa.gov**

National Earthquake Information Center, US Geological Survey. Box 25046, DFC, MS 967 Denver, CO 80225. **www.earthquake.usgs.gov**

National Hurricane Center Tropical Prediction Center. 11691 SW 17th Street, Miami, FL 33165-2149. **www.nhc.noaa.gov**

National Weather Service and National Weather Service Weather Radio, National Oceanic and Atmospheric Administration (NOAA). 1325 East-West Highway, Silver Spring, MD 20910. **www.weather.gov**

Appendix B

Organizations of Interest to the Pipeline Industry

American Gas Association (AGA). 400 N. Capitol Street, NW, Washington, DC 20001. **www.aga.org**

American Petroleum Institute (API). 1220 L Street, NW, Washington, DC 20005. **www.api.org**

Interstate Natural Gas Association of America (INGAA). 20 F Street, NW, Suite 450, Washington, DC 20001. **www.ingaa.org**

National Institute for Occupational Safety and Health (NIOSH). 395 E Street, SW, Suite 9200, Patriots Plaza Building, Washington, DC 20201. **www.cdc.gov/niosh**

National Response Center (NRC). c/o United States Coast Guard (CG-5335 - Stop 7581), 2100 2nd Street, SW, Washington, DC 20593-0001. **www.nrc.uscg.mil**

Occupational Safety and Health Administration (OSHA). 200 Constitution Avenue, NW, Washington, DC 20210. **www.osha.gov**

Pipeline Hazardous Material Safety Administration (PHMSA). East Building, 2nd Floor, 1200 New Jersey Ave., SE, Washington, DC 20590. **www.phmsa.dot.gov**

Additional Resources

The following recommended resources can provide additional helpful information related to the requirements for this covered task.

Code of Federal Regulations 49, Parts 192 and 195.

OSHA *Occupational Safety and Health Standards* 1910, Subpart L, Standard 1910.157.

Figure Credits

Courtesy of Sunoco Logistics, Module Opener, Figure 1

U.S. Geological Survey/photo by Don Becker, Figure 2

PORV image courtesy of Farris Engineering, Figure 4

Spectrex Inc., Figure 5

Courtesy of MSA, Figure 6

Section Review Answer Key

Answer	Section Reference	Objective
Section One		
1. b	1.1.0	1a
2. a	1.2.0	1b
3. d	1.3.2	1d
4. b	1.4.0	1d
5. b	1.5.1	1e
6. a	1.6.0	1f
7. b	1.7.0	1g
Section Two		
1. b	2.1.1	2a
2. b	2.2.0	2b
3. a	2.3.2	2d
4. b	2.4.0	2e
5. c	2.5.0	2f
6. a	2.6.0	2h
7. c	2.7.1	2i

NCCER CURRICULA — USER UPDATE

NCCER makes every effort to keep its textbooks up-to-date and free of technical errors. We appreciate your help in this process. If you find an error, a typographical mistake, or an inaccuracy in NCCER's curricula, please fill out this form (or a photocopy), or complete the online form at **www.nccer.org/olf**. Be sure to include the exact module ID number, page number, a detailed description, and your recommended correction. Your input will be brought to the attention of the Authoring Team. Thank you for your assistance.

Instructors – If you have an idea for improving this textbook, or have found that additional materials were necessary to teach this module effectively, please let us know so that we may present your suggestions to the Authoring Team.

NCCER Product Development and Revision

13614 Progress Blvd., Alachua, FL 32615

Email: curriculum@nccer.org

Online: www.nccer.org/olf

❏ Trainee Guide ❏ Lesson Plans ❏ Exam ❏ PowerPoints Other ______________

Craft / Level: Copyright Date:

Module ID Number / Title:

Section Number(s):

Description:

Recommended Correction:

Your Name:

Address:

Email: Phone:

AOCFG-17

Abnormal Operating Conditions – Field and Gas

Overview

Pipeline operators and contractors have programs, procedures, training, safety equipment, and warning devices for practically every facet of pipeline operations. Situations that are outside of normal operating conditions may arise at any time, and field personnel must know what to do to help keep pipeline operations safe for everyone. In some cases, an abnormal operating condition (AOC) is a routine matter that can be quickly and safely addressed by personnel. In other cases, an AOC can result in an emergency situation. In this module, personnel will learn how to recognize, properly react to, and properly report AOCs that may occur during pipeline operations.

Trainees with successful module completions may be eligible for credentialing through the NCCER Registry. To learn more, go to **www.nccer.org** or contact us at **1.888.622.3720**. Our website has information on the latest product releases and training, as well as online versions of our *Cornerstone* magazine and Pearson's product catalog.

Your feedback is welcome. You may email your comments to **curriculum@nccer.org**, send general comments and inquiries to **info@nccer.org**, or fill in the User Update form at the back of this module.

This information is general in nature and intended for training purposes only. Actual performance of activities described in this manual requires compliance with all applicable operating, service, maintenance, and safety procedures under the direction of qualified personnel. References in this manual to patented or proprietary devices do not constitute a recommendation of their use.

AOCFG-17

Abnormal Operating Conditions – Field and Gas

Objectives

When you have completed this module, you will be able to do the following:

1. Explain how to recognize abnormal operating conditions for both gas and liquid pipelines.
   a. Explain how to recognize abnormal pipeline facility conditions.
   b. Explain how to recognize the activation of a safety device.
   c. Explain how to recognize a communications failure and a control system failure.
   d. Explain how to recognize power interruptions.
   e. Explain how to recognize a fire, explosion, and natural disaster occurring with the pipeline.
   f. Explain how to recognize the unexpected release of hazardous liquid or gas.
   g. Explain how to recognize unexplained pressure/flow rate changes.
2. Describe the proper reaction to abnormal operating conditions for both gas and liquid pipelines.
   a. Describe the proper reaction to abnormal pipeline facility conditions.
   b. Describe the proper reaction to the activation of a safety device.
   c. Describe the proper reaction to a communications failure and a control system failure.
   d. Describe the proper reaction to power interruptions.
   e. Describe the proper reaction to a fire, explosion, and natural disaster occurring with the pipeline.
   f. Describe the proper reaction to the unexpected release of hazardous liquid or gas.
   g. Describe the proper reaction to unexplained pressure/flow rate changes.

Performance Tasks

This is a knowledge-based module; there are no performance tasks.

Trade Terms

Aboveground span
Abnormal operating condition (AOC)
American National Standards Institute (ANSI)
Breakout tank
Disbondment
Encroachment
Flame ionization detection unit
Highly volatile liquid (HVL)
Incipient fire
Intrinsically safe/explosionproof
Lower explosive limit (LEL)
Maximum allowable operating pressure (MAOP)
Maximum operating pressure (MOP)
Naturally occurring radioactive materials (NORMs)
One-Call notification system
Pipeline and Hazardous Materials Safety Administration (PHMSA)
Pipeline facility
Pipeline operator
Redundant system
Right-of-way (ROW)
Self-contained breathing apparatus (SCBA)
Supervisory control and data acquisition (SCADA)
Surge pressure
Uninterruptible power system (UPS)

Industry-Recognized Credentials

If you're training through an NCCER-accredited sponsor, you may be eligible for credentials from NCCER's Registry. The ID number for this module is AOCFG-17. Note that this module may have been used in other NCCER curricula and may apply to other level completions. Contact NCCER's Registry at 888.622.3720 or go to **www.nccer.org** for more information.

Contents

Topics to be presented in this module include:

Figures and Tables

Section One

1.0.0 Recognizing Abnormal Operating Conditions

Objective

Explain how to recognize abnormal operating conditions for both gas and liquid pipelines.

a. Explain how to recognize abnormal pipeline facility conditions.
b. Explain how to recognize the activation of a safety device.
c. Explain how to recognize a communications failure and a control system failure.
d. Explain how to recognize power interruptions.
e. Explain how to recognize a fire, explosion, and natural disaster occuring with the pipeline.
f. Explain how to recognize the unexpected release of hazardous liquid or gas.
g. Explain how to recognize unexplained pressure/flow rate changes.

Trade Terms

Aboveground span: A length of pipeline that is entirely aboveground and supported by pipe supports.

Abnormal operating condition (AOC): A condition identified by the operator that may indicate (a) a malfunction of a component or (b) a deviation from normal operations that may indicate a condition exceeding design limits or result in a hazard to persons, property, or the environment.

American National Standards Institute (ANSI): An organization that adopts standards for a variety of products, services, and processes in the United States to ensure consistency.

Breakout tank: A tank used to (a) relieve surges in a hazardous liquid pipeline system or (b) receive and store hazardous liquid transported by a pipeline for reinjection and continued transportation by pipeline.

Disbondment: Separation of the protective coating from the pipeline.

Encroachment: Activity carried out by, and/or equipment belonging to outside parties on the pipeline right-of-way (ROW). A particular encroachment might be authorized or not, so each situation requires specific handling.

Flame ionization detection unit: A device that detects some hazardous gases in the atmosphere.

Highly volatile liquid (HVL): A hazardous liquid which will form a vapor cloud when released to the atmosphere and which has a vapor pressure exceeding 276 kPa (40 psia) at 37.8°C (100°F).

Intrinsically safe/explosion-proof: Describes electrical equipment that is designed specifically for use in explosive or flammable locations so that a spark cannot ignite the atmosphere.

Lower explosive limit (LEL): The lowest concentration of a gas or a vapor in the atmosphere capable of producing fire in the presence of an ignition source. Different products have different LELs.

Maximum allowable operating pressure (MAOP): The maximum pressure at which a pipeline or segment of a pipeline may be operated (gas).

Maximum operating pressure (MOP): The maximum pressure at which a pipeline or segment of a pipeline may be normally operated (liquids).

One-Call notification system: A centralized communication system established by utilities, government agencies, and operators of underground facilities. This system provides one telephone number for contractors and the general public to call to notify of their intent to use equipment on or near a pipeline. It is also in place for any individual to call if pipeline damage is discovered. The One-Call system is generally set up on a state-by-state basis.

Pipeline and Hazardous Materials Safety Administration (PHMSA): An agency of the US Department of Transportation responsible for developing and enforcing regulations for pipeline operations in the United States.

Pipeline facility: New and existing pipe, rights-of-way, and any equipment, facility, or building used in the transportation of hazardous liquids or carbon dioxide, or new and existing pipelines, rights-of-way, and any equipment, facility, or building used in the transportation of gas or in the treatment of gas during the course of transportation.

Pipeline operator: The company that operates the pipeline, as opposed to an individual person who works for the company that operates the pipeline.

Redundant system: A system that duplicates the operations of a system so that, in the event of a failure, a backup system is available.

Right-of-way (ROW): The area along the pipeline which has been legally permitted to be used by the pipeline operator to construct, operate, and maintain the pipeline. The ROW extends the entire length of the pipeline.

Supervisory control and data acquisition (SCADA): A computer-based system or systems used by a controller in a control room that collects and displays information about a pipeline facility and may have the ability to send commands back to the pipeline facility.

Surge pressure: Pressure produced by a change in velocity of the moving stream that results from shutting down a pump station or pumping unit, closure of a valve, or any other blockage of the moving stream.

Uninterruptible power system (UPS): A system that provides continuity of power when the normal power supply is interrupted. A UPS provides AC power to the system it serves.

An **abnormal operating condition (AOC)** is defined under federal regulations as a condition identified by the **pipeline operator** that may indicate (1) a malfunction of a component or (2) a deviation from normal operations that may indicate a condition exceeding design limits or result in a hazard to persons, property, or the environment. AOCs are defined as such by federal pipeline safety law 49 *United States Code (U.S.C.)* 60102 and pipeline safety regulations 49 *Code of Federal Regulations (CFR)* 192 (for transportation of gas) and 49 *CFR* 195 (for transportation of liquids). The federal agency that regulates pipeline maintenance is the **Pipeline and Hazardous Materials Safety Administration (PHMSA)** in Washington, DC.

Personnel in the field are responsible for being aware of pipeline surroundings at all times so they can recognize AOCs when they occur. Types of AOCs, along with examples and proper reactions, are listed in *Table 1*. Note that the reactions are general.

Some pipeline equipment is fitted with warning devices to alert personnel to problems. Sometimes field workers will be notified of AOCs by control center personnel. In other cases, field workers must use their skills and training to observe the physical signs of AOCs. Field personnel play a critical role as the "eyes, ears, and nose" of the pipeline operator. They must be able to recognize the signs of AOCs and report them appropriately. They must stay in close communication with control center personnel who may need to take the proper action remotely. In the most serious cases, the control center may need to shut down the pipeline.

One of the most important responsibilities of field personnel is to be able to recognize the unexpected release of hazardous gases or liquids. In some cases, early recognition can prevent releases from turning into fires and explosions. Dead vegetation, dead wildlife (*Figure 1*), and an oily sheen on water are just a few of the signs that a release has occurred. *Table 2* illustrates the ways that field personnel must use their senses to be aware of the signs of releases.

1.1.0 Recognizing Abnormal Pipeline Facility Conditions

Any problem with the physical structure of the pipeline is defined as an abnormal **pipeline facility** condition. A pipeline facility is defined as new and existing pipe, **right-of-way (ROW)**, and any equipment, facility, or building used in the transportation of hazardous liquids or carbon dioxide, or new and existing pipelines, rights-of-way, and any equipment, facility, or building used in the transportation of gas or in the treatment of gas during the course of transportation. Examples of components of a pipeline facility include compressors (for gas pipelines) and pump stations (for liquid pipelines).

The following are examples of abnormal pipeline facility conditions:

- Exposed pipeline that should be covered
- Pipeline system damage
- Missing or damaged pipeline supports
- Corrosion
- Coating separating from the pipeline (**disbondment**)
- Unauthorized activity and/or equipment on the pipeline right-of-way (unauthorized **encroachment**)
- Missing signs or markers
- Sunken floating tank roof
- Incorrect data or records caused by equipment malfunction
- Incorrect **American National Standards Institute (ANSI)** pressure rating of a component or material type
- Damage caused by fire or explosion
- Malfunctions due to human error

Work crews, line patrols, or other field personnel are often the first to discover abnormal pipeline facility conditions. However, a report may come in as a result of a call made to the **One-Call notification system** or from a person calling the pipeline operator directly. When a report comes in, personnel must obtain complete information so that they can quickly find the problem and address it. Many pipeline operators use a form to capture the following information:

- Caller's name and telephone number
- Date and time of the call

Table 1A Abnormal Operating Conditions: Examples and General Reactions (1 of 2)

Abnormal Operating Condition	Examples	Appropriate Reaction
Abnormal pipeline facility condition	• Exposed pipe • Evidence of corrosion • Incorrect ANSI pressure rating of a component or material type • Disbondment (coatings separating from the pipe) • Missing signs or markers • Unauthorized activity and/or equipment on the right-of-way (unauthorized encroachment) • Incorrect data or records caused by equipment malfunction • Mechanical/hydraulic system failures • Fire and explosion • Pipeline system damage, such as: – Missing or damaged pipeline supports – Damaged or inoperable pipeline components (pipe, valves, pumps, compressors, meters, tank gauges) due to external force – Gouges, dents, pitting, wrinkles, buckles, or cracks – Cracked or corroded flange or bolts – Sunken floating tank roof – Damaged or malfunctioning pontoon – Defective tank roof seal – Damaged tank shell – Damaged pipe due to natural disaster	Follow standard operating procedures and make notifications per company policy. Isolate and/or shut down the pipeline segment or facility if necessary. Complete repairs as necessary. Complete the required reporting.
Activation of a safety device	• Emergency or abnormal shutdown alarm • High-pressure shutdown alarm • Activation of a pressure relief valve • Activation of a smoke or fire detector • Activation of a hazardous atmosphere alarm • Activation of an overfill protection device • High sump-level shutdown alarm • High tank-level shutdown alarm • High pump-case pressure alarm • Seal failure alarm	Follow standard operating procedures and make notifications per company policy. Isolate and/or shut down the pipeline segment or facility if necessary. Complete the required reporting.
Communications failures	Personnel communication tools: • Lost use of radio communications • Lost use of landline telephone • Lost use of cell phone System communications: • Facility outages • SCADA communication outage • System programming errors/incorrect command logic • Loss of satellite communications • Loss of microwave signals • Stale data (communications going from the control center to the field, but not being received by the control center from the field)	Switch to alternate equipment or backup systems. Follow standard operating procedures and make notifications per company policy. Complete the required reporting.

Table 1B Abnormal Operating Conditions: Examples and General Reactions (2 of 2)

Power interruptions	• Transformer failure • Downed power lines • Broken leads or wires • Incorrect wire termination • Station or terminal power problems (due to motor overload or a tripped breaker, for example) • Weather conditions, such as an ice storm	Follow standard operating procedures. Switch to an alternate control system or power source if necessary. Notify the power company of the outage. Make additional notifications per company policy. Complete the required reporting.
Fire, explosion, or natural disaster	• Electrical fire • Ignition of a hazardous vapor • Unintentional detonation • Fire, explosion, or pipeline movement caused by natural disaster, such as: – Lightning strike – Flooding – Landslide – Earthquake –Tornado	Follow company emergency action plans. Call 911 if appropriate. Isolate and/or shut down the pipeline segment or facility if necessary. Limit access to the area as much as possible. Make notifications per company policy. Complete the required reporting.
Unexpected release of a hazardous gas or liquid	Signs include: • Vapor clouds or frost balls • Smell of hydrocarbon release • Dead vegetation/contaminated soil • Dead wildlife • Rainbow sheen on water Hazardous products include: • Hydrogen sulfide • Carbon dioxide • Propane • Natural gas • Refined fuels/products Note that the controlled transfer, sampling, venting, draining, and handling of hazardous products are not considered AOCs.	Follow company emergency action plans. Call 911 if appropriate. Isolate and/or shut down the pipeline segment or facility if necessary. Make notifications per company policy. Complete the required reporting.
Unexplained flow rate/ pressure change	Any change from the expected flow or pressure, such as: • Increased pressure • Decreased pressure • No pressure • High-pressure limit • Low-pressure limit • High flow • Low flow • No flow	Follow standard operating procedures and make notifications per company policy. Isolate and/or shut down the pipeline segment or facility if necessary. Complete the required reporting.

- Location of the problem (obtain the latitude/ longitude of the location from a global positioning system [GPS] device or a phone, if possible)
- Nearby landmarks such as highways, streams, mile markers, or any other structure that can help repair crews pinpoint the location
- The nature of the damage and any conditions observed at the site

1.1.1 Pipeline System Damage

Accidents that can damage the pipeline and its components occasionally happen. Personnel working on the right-of-way may strike the pipeline without realizing it. An excavator operator may accidentally hit the line. Although the locations of pipelines are clearly marked with signs or markers, construction work near the pipeline may hide markings. Ships' anchors may hit offshore

67107-13_F01.EPS

Figure 1 Dead wildlife may be a sign of an unexpected release.

pipelines. In some parts of the country, hunters have accidentally shot pipelines. Intentional damage due to vandalism, terrorist activity, or other criminal activity may also occur.

Pipeline damage may be discovered by field personnel, control center personnel, or a third party. Field personnel may notice physical signs of damage while out in the field. The control center may receive the indication of pipeline damage on the control center console as an unexplained change in operating conditions.

Physical signs of damage to the pipeline and its components may include gouges, cracks, pitting, dents, buckles, and inoperable valves, pumps, or tank gauges. Flanges or bolts may be cracked or show significant corrosion. Tank shells or pipeline supports may show damage. A sunken roof or a damaged pontoon can also indicate damage. Less obvious signs of damage could include a rainbow sheen on the water near submerged pipelines, which may indicate a release. The release of a

Table 2 Recognizing Warning Signs of Hazardous Gases or Liquids

Sense	Examples of Warning Signs
Sight	• Dead, discolored vegetation, often near healthy vegetation • Dead wildlife • Dry or frozen spots in soil during warm weather • Oily sheen on water/"rainbow" sheen • Vapor fog • Blowing dirt around the pipeline • Bubbles in standing water
Sound	Hissing, blowing, or roaring sound
Smell	Gaseous, skunk-like, or "rotten egg" odor is sometimes recognizable. Never rely on sense of smell only; a release can be completely odorless.

highly volatile liquid (HVL) may be discovered by observing a frost ball or vapor cloud along the ROW of a liquid pipeline.

1.1.2 Damaged Pipeline Supports

Damage to pipeline supports can be a serious problem and can lead to a shutdown. In other cases, pipeline operations may continue while repairs are made. Personnel may recognize pipeline support damage while performing maintenance or inspection tasks in the field. Examples of damage include evidence of corrosion, visible cracking, sagging, or buckling of a support, or bolts or other components of the support that have broken or become loose. Damage to supports may be caused by corrosion, settling of soil, erosion due to wind or water, landslides, washouts, or earthquakes.

Any aboveground span must adhere to specific regulations. These spans refer to lengths of pipeline that are entirely aboveground and supported by pipe supports (see *Figure 2*). 49 *CFR* Part 195.110 requires that pipeline systems be designed to handle anticipated external loads from such conditions as earthquakes, vibration, thermal expansion, and contraction. In addition, the pipe and other components must be supported in such a way that the support does not cause excess stress to the pipe wall. 49 *CFR* Part 195.254 states each component of the pipeline that is aboveground must be protected from the forces exerted by the anticipated loads.

1.2.0 Recognizing the Activation of a Safety Device

Safety devices are a part of the pipeline system and are designed to sense and alert personnel to potential danger and/or hazardous conditions.

67107-13_F02.EPS

Figure 2 Aboveground span of pipeline on pipe supports.

Hazardous conditions can include changes in pressure that may exceed pipeline design limits as well as the release of harmful gases. The purpose of a safety device is to prevent injury or accidents. Safety devices include but are not limited to pressure relief systems, pressure limiting devices, high pump-case-pressure alarms, seal failure alarms, overfill protection devices, tank-level alarms, main-line pressure shutdown alarms, and hazardous atmosphere alarms.

When a safety device activates, it may sound an alarm. The device may ring loudly, beep continuously, or flash a light. A sensor on the device may also send a signal to the control center, where it appears as an alert on the control center console. Never ignore a safety alarm and never turn off a safety alarm until the problem has been checked and cleared by authorized personnel. Notify the control center when an alarm sounds, to maintain communication about appropriate steps to take.

1.2.1 Activation of a Pressure Relief System

The activation of relief devices along the pipeline system is a time-sensitive occurrence. Relief valves reduce pressure by discharging product. Depending on the product, it may be discharged into the atmosphere, tanks, flares, or sumps. Relief valves (see *Figure 3*) can be installed on pipeline or **breakout tanks**. The purpose of a pressure relief device is to protect the pipeline or pipeline facility from overpressure. Pressure relief valves are usually self-actuating at preset limits determined by the operator. However, according to DOT regulations, the pressure in a pipeline cannot exceed 110 percent of the operating pressure, so all preset limits must be within acceptable parameters.

1.2.2 Unintended Valve Closure or Shutdown

The unintended closure of a main-line block valve, remote block valve, manual block valve, pump-station suction or discharge valve, and certain terminal valves (manifold/tank) may cause a pressure change which could activate a pressure-relief system. Breakout tanks can be used to temporarily absorb **surge pressure** from the main-line and manifold pressure-relief valves to lower pressure on the line. According to 49 *CFR* 195, relief venting must be provided for each breakout tank. Pressure relief devices must be provided for each low-pressure and high-pressure breakout tank. There are different specifications for low-pressure and high-pressure tanks.

The unintended closure of the valves could also result in a shutdown of the pipeline. In these cases, personnel should inspect the equipment to determine why closure occurred, and reset, open, or close the valve when it is safe to do so. Such occurrences may not cause an AOC, but if the event causes the pipeline to exceed design limits in any way, its classification changes to abnormal.

1.3.0 Recognizing a Communications Failure

Communications refers to both the tools that personnel use to communicate with each other and with communication systems. Communications failures include the failure of communications among personnel, or the failure of communications between the control center and equipment on the pipeline.

1.3.1 Failure of Personnel Communication Tools

Pipeline personnel use landline telephones, cell phones, and radios to communicate with one another on the job. If using a cell phone near the pipeline facility, personnel must make sure it is **intrinsically safe/explosion-proof**. Personnel must keep their equipment in good working order and test their equipment regularly to ensure that it is working properly.

A breakdown in personnel communications equipment is easy to recognize. Personnel cannot hear messages or make phone calls. Some equipment may beep or flash a liquid crystal display to alert users to a problem. These tools are usually reliable, but breakdowns can occur. If radio communication goes down, personnel must address the problem immediately. A communications breakdown can also occur due to weather-related conditions and location.

67107-13_F03.EPS

Figure 3 Relief valve.

1.3.2 System Communications Failures

System communications refers to the computer equipment located in the control center and on some pipeline equipment. Personnel in the control center use the **supervisory control and data acquisition (SCADA)** system to manage pipeline operations. They receive signals from and send signals to pipeline equipment that may be many miles away. The SCADA system allows them to monitor pressure, volume, and temperature and keep products flowing safely and on schedule. If the system fails, control center personnel lose the ability to communicate with equipment on the pipeline. Some systems are set up to switch to a **redundant system**—a backup that comes on line if the main system fails.

Some failures of communications systems show up as warnings or messages on the control center console. A system failure may include any of the following events:

- Random facility outages
- Partial or complete facility outages
- Loss of the SCADA system
- Loss of satellite communications
- Loss of microwave signals
- Stale data
- Loss of signal point within the system

Mechanical and hydraulic failures are related to system failures. For example, the SCADA system sends a signal to a piece of equipment, but the equipment does not respond to the signal. In some cases, the problem is a result of a programming or command logic error. In other cases, equipment may not work because of a mechanical, electrical, or hydraulic problem. Examples of monitoring device failures include, but are not limited to, frozen valves, electrical component burnouts, and compressor breakdowns. A valve may have a mechanical failure that prevents it from being operated remotely by the control system. Sometimes electrical components or connections burn out, preventing the equipment from responding to a command by the system to start up or shut down.

1.4.0 Recognizing Power Interruptions

Electric power is essential to pipeline operations. Nearly every part of the pipeline depends on electricity in some way. A loss of power impacts the abilities of the control center or pipeline facility operations to safely control and monitor the pipeline facility. Power interruptions that result from bad weather or a problem at the local utility company, such as transformer failure or downed power lines, are not uncommon.

Some pipeline operators have an **uninterruptible power system (UPS)** that turns on if electrical power is off for more than a few seconds. UPS systems supply emergency alternating current (AC) power to pipeline components to keep the pipeline operating until power is restored. Instead of electricity, these systems run on engines fueled with gasoline, diesel, or natural gas. When power fails, an AC power failure alarm may be sent to the control center. To ensure that the backup system has turned on, control center personnel check to see whether they are receiving updated information from the pipeline.

If power is lost, electronic indicators will turn off. Some compressors and remote control valves may not operate. The SCADA system may switch off and pumps may shut down. In addition, lights, heat, and air conditioning may fail. An area-wide power outage may cause a loss of telephone communications. Local power outages may shut down only a portion of the pipeline system. Power outages can also cause pumps at the receiving terminal to stop operating, causing a gas or liquid shipment to overflow in tanks or be unable to be shipped further without pumps.

Some equipment is protected from power failure by batteries that supply temporary low voltage power. These batteries run on direct current (DC).

The following types of equipment can operate on DC power:

- Fire detection equipment and alarms
- Fire suppression systems
- Emergency lighting systems
- SCADA
- Alarm panels
- Control modules
- Process control panels
- Specialized compressor units

> **NOTE**
> Equipment plugged into electric sockets uses AC power. AC power reverses its direction at regular intervals. Batteries supply DC power. DC power flows in one direction.

1.5.0 Recognizing Fire, Explosion, or Natural Disaster

Fire, explosion, or natural disaster can occur at any time. Some events may be anticipated, but many cannot. Field personnel must be trained to recognize these events.

1.5.1 Fire

Personnel must take steps to minimize the likelihood of fire and explosions in a pipeline facility. Two examples include using intrinsically safe/explosion-proof electronic equipment and checking atmospheric conditions before starting any hot work. Always follow company procedures related to the prevention of fire and explosion. Detection devices installed along the pipeline and in pipeline facility buildings alert personnel in case of a fire. These devices send an alarm when their sensors detect heat or flame. Such equipment includes thermal detectors, flame detectors, and smoke detectors.

> **NOTE**
> Before performing any work that may produce an ignition source, verify that a hazard analysis has been performed. Examples of such work include welding or cutting, opening electrical boxes, or operating non-explosionproof tools in hazardous classified areas.

A thermal detector is a heat-sensitive unit. It is fitted with electrical contacts that stay open as long as conditions are normal. When the unit senses rapidly rising temperatures, the contacts start to close. When the contacts connect, the unit may send an alarm, which may be audible, to the control center and the fire control panel in the workstation. If a thermal detector activates, personnel should evacuate the area and, investigate the cause of the alarm (if appropriately trained according to company procedures).

A flame detector (*Figure 4*) is a light-sensitive unit. It is fitted with an ultraviolet detector that turns on an alarm when the unit senses light from a flame. The unit is sensitive enough to detect a flame even in the earliest stages of a fire. Some flame detectors are connected to a firefighting system that may dump fire suppression material into the area. Personnel can also dump this material over a fire manually by using pull switches.

A smoke detector can be a light-sensitive unit or an ionization detector, or it can be a combination of both types. Ionization detectors are more sensitive to fires in the early smoldering stage. Light detectors are more sensitive to the flaming stage of fires.

An ionization smoke detector detects ionization in air caused by smoke. If any smoke particles pass into the chamber of the detector, the ions will attach to the particles, and so become less able to carry the current. An electronic circuit detects the current drop and sounds the alarm. The light detector operates by detecting light beams. In the absence of smoke, the light passes in front of the detector in a straight line. When smoke enters the chamber across the path of the light beam, light is scattered by smoke particles, directing it at the sensor that triggers the alarm.

67107-13_F04.EPS

Figure 4 Flame detector.

CAUTION

Before using any fire suppression material, personnel must know how to handle it. Be sure to read the material safety data sheet (MSDS)/safety data sheet (SDS) for information about how to use the material safely.

Monitoring devices that detect a lower explosive limit (LEL) produce alarms to indicate that there is a risk of explosion. The LEL is the lowest concentration of a gas or a vapor in the atmosphere capable of producing fire in the presence of an ignition source. Different products have different LELs. The monitoring device must be set to detect the particular LEL in an enclosed area. If an LEL alarm activates while welding is taking place, the first action is to stop the welding immediately. Always check atmospheric conditions before reentering an area after an LEL alarm has activated.

1.5.2 Explosion

Pipeline operators install a number of warning devices to warn personnel about conditions that could cause explosions. However, operators cannot control every circumstance, such as intentional damage, that could lead to an explosion on the pipeline. An explosion on a pipeline registers in the control center as an unexplained change in pressure. An explosion may occur in any enclosed space where flammable vapors can accumulate to dangerous levels. Examples of such spaces include deep pipeline trenches and pipeline facilities.

A natural gas explosion may occur when natural gas leaks into an enclosed space. The odorant added to the gas sometimes alerts people to the problem. However, gas can sometimes accumulate undetected. For example, the building may be vacant or the odors may be too far away from people inside the building. In addition, products may not have an odorant added during transport. Be aware that a spark from electrical equipment can cause the gas to explode.

WARNING!

Personnel should never rely on their sense of smell to detect a natural gas leak. Products do not always have a detectable odor.

For an explosion to take place in an enclosed space, there must be a source of ignition and fuel. The vapors creating the fuel for an explosion can be of various types. Each type of flammable vapor or gas has a different lower and upper explosive limit (LEL/UEL). For example, when natural gas escapes into the air, its molecules generally dissipate or scatter because natural gas is lighter than air. However, in enclosed areas natural gas can accumulate, and if there is an ignition source an explosion may occur.

1.5.3 Natural Disasters

Natural disasters include earthquakes, hurricanes, tornadoes, landslides, wildfires, and flooding. These events may cause soil movement, which may result in forces sufficient to cause pipeline failure or product release. Even heavy rains resulting in saturated soil conditions may cause soil movement in some areas.

Pipeline operators generally have designated personnel to monitor weather conditions that can develop into a natural disaster. Control center personnel monitor for earthquakes with seismic indicators. In addition, they may obtain real-time information on the internet. Up-to-date earthquake information is available from the National Earthquake Information Center, which is located in Denver, Colorado. Personnel may track weather alerts and warnings through the National Weather Service website. In case of a power failure, personnel can continue to receive information through the National Weather Service's National Weather Radio or other electronic devices. A list of weather and natural disaster information sources is included in *Appendix A*.

Operators should also have field personnel designated to patrol the ROW to monitor for soil movement and to ensure the accessibility of pipeline components, such as valves, that are needed in case of emergency response actions such as shutdown, isolation, and containment in case of a pipeline failure.

1.6.0 Recognizing the Unexpected Release of Hazardous Gas or Liquid

Pipeline releases are often noticed when someone sees the physical signs. Depending on the product, the release may result in a pool of liquid on the ground, a frost ball, vapor cloud, mist in the air, dead or discolored vegetation, or a rainbow sheen on water. Other signs include dead fish in waterways or other dead wildlife. Gas leaks may also cause a whistling or hissing sound. Natural gas leaks may be identified by a loud, roaring noise. Note that the controlled transfer, sampling, venting, draining, and handling of hazardous products are not AOCs.

Personnel may use a combustible gas indicator or a flame ionization detection unit to detect hazardous gases such as hydrogen sulfide (H_2S), propane, natural gas, and in some cases, carbon dioxide (CO_2). An unexpected release may lead to a hazardous atmosphere, an explosion, or other serious condition. Additional PPE may be required with an unexpected release. Always follow company procedures when dealing with any release.

Personnel must not rely on their sense of smell to detect a hazardous gas for the following reasons:

- Some hazardous gases are odorless. Products may not have an odorant added to alert workers to the presence of gas.
- A low concentration of some hazardous gases can deaden the sense of smell.
- A person's sense of smell may be deadened by a medical problem.

1.7.0 Recognizing Unexplained Flow Rate / Pressure Changes

Flow rate and pressure are important indicators of both normal and abnormal operating conditions. Minor changes in pressure may occur normally in the pipeline. However, significant deviations from the expected flow or pressure are considered AOCs. Unexplained pressure changes include absences of, increases in, and decreases in pressure. The maximum internal pressure permitted during the operation of a gas pipeline is called the **maximum allowable operating pressure (MAOP)** and is defined in 49 *CFR* 192. The maximum internal pressure permitted during the operation of a liquid pipeline is called the **maximum operating pressure (MOP)** and is defined in 49 *CFR* 195.

Pressure-indicating devices located near mainline valves, compressor stations, and custody transfer stations monitor pressure and send readings to the control center. When an unexpected change in pressure occurs, an alarm may alert control center personnel, who then notify field personnel. Sudden pressure increases or decreases might indicate a problem such as the failure of a valve or a transmitter. Personnel must be familiar enough with the pipeline system, equipment, product characteristics, and environmental factors to quickly identify any important changes in pressure and flow.

Internal Combustion Engines

Pipeline facilities/systems that process or store flammable liquids or gases are particularly vulnerable to the risk of fire and explosion. But did you know that the internal combustion engines that are typically used in pipeline facilities can ignite flammable liquids and gases? Internal combustion engines include vehicles of any kind (trucks, cranes, excavators, fire engines, and ambulances) as well as stationary or portable engines such as compressors, generators, pumps, and welders.

Be sure that you know and understand your company's specific safety policies about internal combustion engines to ensure your own safety, the safety of other personnel, and the safety of the entire pipeline facility. To prevent internal combustion engines from becoming an ignition source, pipeline operators are responsible for implementing safety measures such as:

- Ensuring that materials and equipment are stored and used in accord with applicable Occupational Safety and Health Administration (OSHA) standards
- Ensuring that permanently mounted internal combustion engines are not stored where flammable vapors or gases could exist
- Establishing and marking safe boundaries and traffic routes for vehicles and portable equipment through hazardous areas
- Installing flammable gas and vapor detectors in areas where hazards could exist
- Providing training to personnel about the signs and hazards of flammable vapors and gas, the safe handling of internal combustion engines, and all relevant emergency procedures.

Additional Resources

It is assumed that you will learn to perform these procedures under the direct supervision of your crew leader, foreman, or instructor. For independent study, refer to the recommended Additional Resources at the back of this module.

1.0.0 Section Review

1. Which of the following is an abnormal pipeline facility condition?
   a. Aboveground pipe on pipe supports
   b. Missing signs or markers
   c. Valves operating properly
   d. Authorized encroachment

2. Safety devices include _____.
   a. hazardous atmosphere alarms
   b. tank shells
   c. SCADA
   d. compressors

3. SCADA serves as a _____.
   a. regulatory agency
   b. notification system for all utilities
   c. device worn to provide breathable air where the atmosphere is hazardous
   d. command system that assists control center personnel

4. An uninterruptible power system (UPS) supplies _____.
   a. temperature readings
   b. emergency backup power
   c. nuclear power
   d. pressure readings

5. A thermal detector, unlike a flame detector, senses _____.
   a. rapidly rising temperatures
   b. ultraviolet light
   c. flames
   d. pipeline pressure

6. Which of the following is a reason that personnel should *not* rely on their sense of smell to detect a hazardous gas release?
   a. Hazardous gases always have an odor.
   b. Low concentrations of hazardous gases heighten the sense of smell.
   c. A person's sense of smell may be deadened by a medical problem.
   d. Personnel can hear a gas release.

7. The maximum operating pressure of a gas pipeline is called _____.
   a. MOP
   b. MAOP
   c. surge pressure
   d. compression

Section Two

2.0.0 Reacting to Abnormal Operating Conditions

Objective

Describe the proper reaction to abnormal operating conditions for both gas and liquid pipelines.

a. Describe the proper reaction to abnormal pipeline facility conditions.
b. Describe the proper reaction to the activation of a safety device.
c. Describe the proper reaction to a communications failure and a control system failure.
d. Describe the proper reaction to power interruptions.
e. Describe the proper reaction to a fire, explosion, and natural disaster occuring with the pipeline.
f. Describe the proper reaction to the unexpected release of hazardous liquid or gas.
g. Describe the proper reaction to unexplained pressure / flow rate changes.

Trade Terms

Incipient fire: A fire in its beginning stage.

Naturally occurring radioactive materials (NORMs): Materials with radioactive elements that can be extracted during oil and gas extraction from the earth and can get into product streams and become transported in the pipeline.

Self-contained breathing apparatus (SCBA): A device worn to provide breathable air for an individual in an area where the atmosphere is hazardous.

Time is of the essence when AOCs occur and field personnel are expected to take decisive action. Personnel must always follow established company-specific policies and procedures. As regulation 49 *CFR* 195.402 states, each pipeline operator must ensure its personnel maintain a thorough knowledge of its operations, maintenance, and emergency manuals. Personnel must make use of their training, skills, and good judgment to maintain safe operation until solutions are found.

2.1.0 Reacting to Abnormal Facility Conditions

Some abnormal pipeline facility conditions may require immediate action. Others may require scheduling repairs for a later date. If field personnel recognize an abnormal pipeline facility condition, they must notify the appropriate personnel so that repairs can be made safely and in a timely manner. If notified of a condition by a third-party call, it is important that personnel record information received from the third party. They must also keep in mind that a caller's description may not be entirely accurate. The problem may be different than described. In rare cases, a call might not be valid. After a report is confirmed, field personnel should generally follow these steps:

Step 1 Investigate and evaluate the report to determine the extent of the damage, if any.

Step 2 Carry out company safety procedures to protect people, property, and the environment if necessary.

Step 3 Notify the appropriate personnel so that proper notifications or repairs can be made.

2.1.1 Pipeline System Damage

When field personnel notice or accidently cause pipeline damage, they investigate the extent of the damage, note the location on the pipeline, and isolate the damage, as applicable. They must also contact the control center, according to company procedures. If an immediate hazard exists, contact 911 emergency personnel.

> **NOTE**
> Before excavation or other digging near a pipeline begins, responsible personnel must contact the One-Call notification system.

Some incidents may be serious enough to require an emergency shutdown of the pipeline. One example of a serious incident is someone losing control of a vehicle on the ROW, extensively gouging or denting a large section of pipeline or damaging a valve. Whenever pipeline damage results in a release and/or explosion, field personnel must notify the control center as soon as possible. In some cases, control center personnel may be able to drain a portion of the pipeline into a delivery system to reduce the amount of product released from the pipeline.

Once the pipeline has resumed normal operations, the pipeline operator conducts an internal investigation. This investigation focuses on evaluating control systems and mechanical equipment as well as personnel performance.

Each company has specific procedures for reacting to pipeline damage. These procedures explain how and when to contact the control center, emergency personnel, and local, state, and federal law enforcement agencies. Always follow company policy, make appropriate notifications per company policy, and complete the required reporting.

2.1.2 Damaged Pipeline Supports

The proper reaction to the discovery of damaged pipeline supports varies depending on the severity of the damage. If the damage to the support is critical or indicative of a possible release, the individual making the discovery should contact the control center or other appropriate personnel immediately. Notification is especially important if the pipe is in danger of excessive movement. The appropriate personnel must then follow company policy precisely to investigate the condition and dispatch qualified personnel to the scene. Until the arrival of qualified personnel, all other personnel should stay a safe distance away from the site, evacuate the area if necessary, and make the appropriate notifications as required by company policy.

In some cases, the pipeline operator may need to isolate the pipeline segment. Whether or not the segment needs to be isolated, a damage report must be submitted so that the repair work can be scheduled. The decision to continue or halt operations is a judgment call made by management and qualified inspection personnel. If the pipeline segment must be isolated, affected personnel must be notified.

Field personnel must know and follow their company's procedures. Those procedures are especially important when a condition requires a decision about whether to isolate a segment of the pipeline facility. Make notifications and complete the reporting as required by company policy.

2.2.0 Reacting to the Activation of a Safety Device

Personnel must respond to safety device alarms quickly and calmly. To react to an alarm, personnel should follow these steps:

Step 1 Notify affected personnel.

Step 2 Carry out company warning procedures to protect people, property, and the environment.

Step 3 Troubleshoot the problem and make necessary repairs if qualified to do so.

Step 4 Reset and test the warning device alarm.

Step 5 Make the appropriate notifications and complete the required reporting.

It is important to ensure that the alarm is not a false alarm. In case of a false alarm, the control center should notify affected personnel and have the alarm equipment checked. Generally, no further action is required when an alarm is false.

When a hazardous atmosphere alarm sounds, the proper reaction is somewhat different. In this case, personnel wearing the appropriate personal protective equipment (PPE) must follow these steps:

Step 1 Evacuate the area and don the appropriate PPE if available.

Step 2 Make the appropriate notifications per company policy.

During an emergency, personnel must focus on three key areas: ensuring personnel safety, reducing the damage to equipment, and protecting the environment. The general rule is to put people first. The goal is also to prevent small problems from turning into larger incidents. All reactions must be made safely. Personnel investigating the alarm must wear the appropriate PPE. If told by authorized personnel to evacuate an area, they must leave immediately.

If reacting to a seal failure alarm, personnel should investigate the alarm, check the station sump level if it is equipped with one, and notify the appropriate personnel.

If a pressure-relief system activates, an audible alarm may activate. Immediate action must be taken to reduce pressure. A pressure reduction is usually achieved by shutting down the upstream pumping station, closing a block valve, or shutting down the pipeline system. Field personnel may need to inspect the relief system, test it, and recalibrate it as necessary.

2.3.0 Reacting to a Communications Failure

The proper reaction to a communications failure depends on whether the failure involves a communications tool that personnel use or the failure of system communications.

2.3.1 Failure of Personnel Communication Tools

When communication tools fail, personnel should first check to see if the problem is a simple one that can be quickly and easily fixed. Immediately correct any communications failure, if possible. For example, some telephones have a volume control that may accidentally be turned off. The telephone may seem to be dead, but a simple adjustment to the volume corrects the problem. Other simple solutions include replacing a battery or recharging a unit.

Some communications problems are more serious. Weather or problems at the local utility company can knock out communications completely. Regardless of the outage, the best preparation is to ensure personnel are trained to follow these steps:

Step 1 Switch to alternate equipment if necessary.

Step 2 Notify the appropriate personnel per company policy.

Step 3 Complete any required reporting or documentation.

If communications are interrupted for long periods, it may not be possible to ask for instructions from a remote location. Training before an emergency situation occurs is the best way to ensure that field personnel know how to react properly. However, field personnel must do everything that they can to re-establish communications as soon as possible.

2.3.2 System Communications Failures

The control center will generally notify field personnel about system communications failures so that they can isolate, investigate, and react according to company policy. The proper reactions for field personnel may include some or all of the following activities:

- Switch to alternate or backup systems
- Isolate a pipeline segment if necessary
- Determine the scope of the problem
- Take corrective actions to restore the system
- Estimate how long the main system will be down
- Notify affected personnel about how long repairs will take
- Complete the required reporting or documentation

If the problem is with the SCADA system, or equipment cannot otherwise be operated from the control center, field personnel may have to operate equipment manually. The pipeline may be able to continue to operate while repairs are made. Operations are dependent on company procedures.

2.4.0 Reacting to Power Interruptions

The proper reaction to a power failure depends on which part(s) and how much of the pipeline is affected. Regardless of how the outage affects the pipeline facility, the appropriate personnel should always notify the power company of the outage as soon as possible. If only a single component or station is involved, control center and/or field personnel may be able to adjust line pressure and flow to keep the pipeline in operation. For example, if a power interruption occurs while a valve is in operation, field personnel may have to operate the valve manually. However, if a station is needed to keep pressure within normal operating limits, for example, control center personnel may have to isolate a pipeline segment. If a shutdown is not required, the gas or liquid shipment will continue to arrive, and the schedule must be followed and the tasks completed manually.

Each company has specific procedures for responding to a power interruption. Always follow company policy, make appropriate notifications per company policy, and complete the required documentation.

2.5.0 Reacting to a Fire, Explosion, or Natural Disaster

The pipeline operator must have a training program in place so that personnel know what to do in case of a fire, explosion, or natural disaster that affects a pipeline. Many companies conduct regular drills so that personnel can practice their responses to such events. Drills help ensure that personnel are prepared to handle emergencies safely and calmly. Each operator has its own procedures for emergencies. In general, trained personnel should know the following information:

- The location of fire suppression equipment and how to operate it; certain fire suppression equipment and PPE may require additional training
- The process for shutting down equipment manually if communication with the control center is lost
- The company's evacuation plan
- The procedures for contacting local emergency response agencies, if necessary

Depending on the situation, pipeline personnel may be able to safely handle an emergency without calling for outside help. Follow company procedures for contacting local emergency response agencies.

2.5.1 Fire

When a fire occurs, the safety of personnel is always top priority, so personnel should evacuate the area. Protection of property is always second.

Personnel must ensure that any injured personnel receive medical attention as soon as possible. Consider any alarm as an actual alarm and react appropriately and decisively.

Personnel must also eliminate the source of the fire, but only as authorized. For example, only qualified personnel are authorized to turn off valves. Company policy dictates procedures for emergency actions and must be followed. Procedures are site-specific and may allow for some flexibility in responding. However, personnel should not need to consult a manual in case of an alarm. They must be adequately trained and familiar with emergency procedures in advance. It may even be necessary to shut down the pipeline if required to control the emergency. Always notify the appropriate personnel and emergency agencies and complete any required documentation or reporting.

According to 49 *CFR* 192 and 195, operators shall establish and conduct a continuing training program to instruct emergency personnel in the proper use of firefighting procedures and equipment, in the identification of hazardous conditions likely to cause emergencies, in the steps necessary to control accidental releases, and in the characteristics of hazardous liquids or carbon dioxide being transported.

The proper reaction differs according to the type of fire in progress. If the fire is small or just beginning (**incipient fire**), personnel should use the available fire extinguishing equipment to control the situation. According to *OSHA 1910.157*, companies must provide designated personnel with training on fire extinguishing equipment as well as the hazards involved with incipient firefighting.

Use the proper type of fire extinguisher for each class of fire. It is important to know the extinguisher's fire class rating before using it. The fire class rating is printed at the bottom of the label on a fire extinguisher. There are four fire class ratings (see *Table 3*). Most extinguishers can be used for multiple types of fires.

If the fire is larger than the immediate workforce can contain, personnel should try to mitigate the situation, notify appropriate personnel and emergency agencies, and complete the required documentation and/or reporting. For example, during an electrical storm, if lightning strikes a full tank at a facility and the roof seal catches on fire, personnel should follow company policy and contact the fire department to contain the fire.

Once the fire is out, the air is cleared, and any injured personnel have received medical attention, the following actions are taken:

Table 3 Fire Class Ratings

Fire Class	Materials and Proper Fire Extinguisher
A	Involves ordinary combustibles, such as wood or paper. This type of fire is suppressed by cooling the fire. Class A extinguishers contain water. A Class A fire extinguisher can never be used on a Class B, C, or D fire.
B	Involves grease, flammable liquids, and gases. Class B extinguishers contain carbon dioxide or another material that smothers the fire by removing oxygen.
C	Involves electrical equipment. Class C extinguishers are designed to protect the firefighter from electric shock. This type of extinguisher smothers the fire.
D	Involves metals. Class D extinguishers contain a powder that either forms a crust around the burning metal or gives off gases that prevent oxygen from reaching the fire. Some metals may keep burning even after they have been coated with powder from a Class D extinguisher. The best way to fight these fires is to keep using the extinguisher so that the fire does not spread.

- Alarms are reset and tested.
- Equipment damage is assessed and reported to the appropriate personnel.
- Work orders are issued and required permits are obtained for repairs.
- Personnel performance is evaluated and corrective action is determined if necessary.
- Injuries or deaths are reported to the appropriate regulatory agencies if required.

The evaluation of personnel performance during a fire may reveal problems in personnel response. Management generally recommends retraining so that personnel can improve their responses to a fire.

A false alarm or an expected alarm due to maintenance activities requires no further action unless the control center determines that the system needs to be investigated. If an alarm has been verified as a false alarm, the appropriate personnel follow these steps:

Step 1 Notify affected personnel.

Step 2 Test alarms to ensure that they are operating properly.

Step 3 Reset alarms or ensure that they have automatically reset.

2.5.2 Explosion

Each operator has specific procedures for reacting to an explosion. In addition, the correct order of the steps to take may vary from operator to operator. A response to an explosion must be appropriate, decisive, and calm. If the main-line valves do not close automatically, designated personnel must initiate actions to ensure that the valves are closed. In general, the proper reactions to an explosion are:

- Evacuate and notify appropriate personnel
- Evaluate and assess the situation.
- Carry out company safety procedures to protect people, property, and the environment.
- Call local emergency response agencies if necessary.
- Confirm the location of all pipeline personnel near the explosion location.
- Confirm the location of any members of the public near the explosion location.
- Call local, state, or federal law enforcement if the cause indicates criminal or terrorist activity.
- Make the appropriate notifications and complete the required documentation and/or reporting.
- Conduct an incident investigation.

2.5.3 Natural Disaster

Much of the pipeline's ability to withstand extreme weather that can result in natural disaster is already present in its design and installation.

Each company has specific procedures for dealing with natural disasters. These procedures reflect local conditions. For example, hurricanes occur more frequently in Florida or the Gulf states. Therefore, companies located in these areas include hurricane response in their procedures.

The following are examples of proper reactions to weather-related problems:

- Monitor weather conditions such as tropical storms that may become hurricanes
- Carry out company safety procedures for protecting people, property, and the environment
- Adjust work schedules as needed to respond to emergency conditions
- Follow company procedures for responding to fire, explosion, or damage resulting from natural disasters
- Set up a command post if conditions warrant such a step
- Work with local emergency response agencies if required
- Conduct investigations when appropriate

2.5.4 Emergency Shutdowns

Events occurring on the pipeline that present immediate hazard to persons or property require an emergency shutdown. An event might be caused by human error, mechanical failure, electrical failure, or an accident on the pipeline. It might also be brought about by threats from weather or civil disobedience. Regardless of the reason, the line must be shut down quickly and safely. Each company has a standard operating procedure for emergency shutdowns.

In the case of a fire, explosion, or release, time for discussion may be very limited. Once the need is confirmed, decisive action must be taken. The pipeline must be isolated by control center personnel by safely shutting down the originating station and continuing downstream in a safe manner until all pumps are offline. Control center personnel will then close block valves to complete the shutdown and notify other personnel as needed. In case of a pump shutdown from low pressure, there may be some time to discuss the decision with appropriate management if the line has enough flexibility to withstand the stress.

Keep in mind that threats from natural disaster or civil disobedience may come without warning. If field personnel witness these kinds of threats to the pipeline, they should contact their supervisor or control center immediately. Affected facilities may need to be isolated. Once facilities are isolated, pipeline shutdown may follow.

In emergency conditions such as a possible release, the pipeline operator has full responsibility and authority to make shutdown decisions and take appropriate action. These actions are governed by the company release policy above all other considerations.

2.5.5 Evacuations

The safety of pipeline personnel and the public is the first priority when any AOC occurs. If there is any threat of injury or death to any persons, an evacuation of pipeline personnel must occur. Events that could result in the need for an evacuation are fire, explosion, release of hydrocarbon or other hazardous vapors, seismic activity, or other natural disasters.

Companies have standard procedures in place to guide the evacuation of personnel. By law, emergency response plans must be in place and evaluated annually. Facility layouts with evacuation maps are required to be posted in every area or department. Safe areas are designated for all personnel. When the command to evacuate is given, trained personnel may act as safety coordinators. Their job is to conduct a roll call to confirm that all personnel are safe. To assist in this process, personnel must report to their designated safe areas and identify themselves to the designated safety coordinator. If any personnel cannot be accounted for, authorized personnel must locate and rescue them if necessary.

Some emergencies may require an evacuation of the public located near the pipeline. The pipeline operator may work with the local authorities to provide details about the incident, determine whether a general evacuation is required, provide contact information for people residing or working near the pipeline, and provide personnel to help staff a command center if requested.

NOTE

Local emergency planning committees (LEPCs) exist in every jurisdiction in the United States. Law enforcement agencies, firefighting companies, local government agencies, hospitals, companies, and citizen's groups make up the memberships of LEPCs. Committee members meet on a regular basis to plan for emergencies. Planning for a general evacuation is just one example of the projects handled by an LEPC.

2.6.0 Reacting to the Unexpected Release of Hazardous Gas or Liquid

When a hazardous gas or liquid is released from a pipeline unexpectedly, personal safety should be the first consideration. Beyond that, the proper reaction to the unexpected release of a product depends on the characteristics of the product and the location of the release. Some gases and liquids can explode under certain conditions, and some are toxic to people and the environment.

NOTE

To identify the hazards associated with the chemical being released, consult the material safety data sheets (MSDS)/safety data sheets (SDS).

Regardless of location or the type of product, when a product is released personnel *must* take the following steps:

Step 1 Evacuate the area.

Step 2 Identify the source of the release and eliminate potential ignition sources.

Step 3 Vent the area if possible.

Step 4 Prevent anyone who is not wearing the appropriate PPE from entering the area (personnel must be trained to use PPE).

Step 5 Call for emergency rescue and medical help, if required.

Step 6 Contact the control center; appropriate personnel may need to isolate the pipeline.

Step 7 Have the power company turn off the power, if appropriate.

WARNING

When witnessing a release and there is a downed person in the area of the release, call 911 and evacuate the area. Do not attempt to rescue personnel.

NOTE

OSHA stipulates that employers must ensure that contract personnel are trained in the work practices necessary to perform their job safely.

Of equal importance are actions personnel must avoid. When reacting to a release, personnel *must not*:

- Open switchgear doors.
- Operate electrical switches or equipment in the facility, including lights, doors, louvers, or vents.
- Operate personal electrical equipment (including cell or landline telephones) unless the equipment is classified as intrinsically safe/explosion-proof.

Intrinsically safe/explosion-proof electrical equipment is designed so that a spark cannot ignite the atmosphere. It is made to meet the following specifications:

- Outside air cannot penetrate the casing.
- The casing must be strong enough to contain an explosion inside the equipment without igniting the surrounding atmosphere.
- The interior must be filled with nonflammable powder or inert gas to prevent ignition.

This equipment is determined to be safe by the National Electrical Manufacturers Association (NEMA), and is certified according to the standards of the *National Electric Code®*. It must also meet the requirements of Underwriters Laboratories (UL). Equipment that is intrinsically safe/explosion-proof should always be labeled as such.

> **WARNING**
> Electronic equipment must not be used in the presence of hazardous gases/vapors unless it is classified as intrinsically safe/explosion-proof. Failure to use intrinsically safe/explosion-proof equipment in a hazardous atmosphere may result in an explosion.

2.6.1 Releases in Enclosed Areas

If a product release occurs in an enclosed area, the source of the release must be determined quickly. It is also important that personnel move to a safe location as quickly as possible and then notify the appropriate personnel, who can operate equipment remotely. If trained, personnel may close the cutoff valve to mitigate the situation.

2.6.2 Releases in Outdoor Areas

Any release must be approached by personnel from the upwind side. Releases in outdoor areas include the additional concern for the safety of the general public, as well as the safety of the environment. Personnel should evacuate the area, go upwind, and take additional appropriate action per company policy. If an odor of hydrocarbon is detected even a few hundred feet from the source, contact control center personnel to shut down the pipeline. Report the incident and any action taken, and complete the required documentation.

2.6.3 Toxic Products

Personnel must take special precautions when dealing with the release of products that can become toxic. Examples include carbon dioxide (CO_2), hydrogen sulfide (H_2S), propane, anhydrous ammonia, and natural gas.

Carbon dioxide is a colorless, odorless, gaseous compound used as a propellant in many domestic applications, such as fire extinguishing equipment, and is a byproduct of combustion. Because it is heavier than air, it can collect in low-lying areas where it may displace oxygen; exposure can result in suffocation and death. A leak detector can detect a hazardous atmosphere, but without it, personnel can be overcome without realizing the problem. Some pipeline operations pump carbon dioxide commercially, and in such operations, the hazards are increased. To enter an area where a carbon dioxide release is suspected, personnel must use the proper gas detection meter, calibrated for the detection of carbon dioxide, and wear the appropriate PPE, such as **self-contained breathing apparatus (SCBA)** (see *Figure 5*).

67107-13_F05.EPS

Figure 5 Self-contained breathing apparatus.

> **WARNING**
> Carbon dioxide (CO_2) is a colorless, odorless gas that is heavier than air, so it can collect in low-lying areas where it may displace oxygen. Exposure can be fatal.

Hydrogen sulfide (H_2S) is a colorless gas that smells like rotten eggs. However, in fairly low concentrations, it can deaden the sense of smell. Exposure to high concentrations can be fatal.

> **WARNING**
> Exposure to high concentrations of hydrogen sulfide (H_2S) can be fatal. Even in low concentrations, it can cause headaches or eye irritation. In addition, it is flammable and may cause a flash fire. When a gas detection unit shows that H_2S is present, wear the appropriate PPE or immediately leave the area.

Anhydrous ammonia is a liquefied gas that has a sharp odor. Exposure in high concentration can cause death. Even exposure in low concentrations can cause severe reactions from mild irritation to severe burns. The reaction depends on the length of exposure and concentration amounts. If anhydrous ammonia is released, call emergency personnel.

WARNING

Exposure to high concentrations of anhydrous ammonia can cause death. Even in low concentrations, it can cause severe reactions. If detected, wear the appropriate PPE or immediately leave the area. If a release occurs, call emergency personnel.

Propane is a common natural gas that is colorless and heavier than air, so it can collect in low-lying areas where it may displace oxygen. Exposure can cause death. At high concentrations, it is flammable.

WARNING

Exposure to propane gas can cause death. At high concentrations, it is flammable. If detected, wear the appropriate PPE or immediately leave the area.

Naturally occurring radioactive materials (NORMs) may be present when oil and gas is extracted from the earth. These NORMs are then introduced into the product streams and transported in the pipeline. Natural gas liquids (NGLs), along with the processing associated with these products, may trap NORMs in the form of sludge, scale, and particulates. Filters, valves, scrapers, and even internal surfaces of pipelines may contain high levels of NORMs. Precautions should be taken to protect personnel by monitoring for NORMs in accordance with all local, state, federal, and company requirements. If high levels are detected, personnel should distance and/or shield themselves from the source and notify the appropriate personnel.

Only appropriately trained personnel may enter an area where a hazardous gas has been detected. These personnel must be trained and qualified to use SCBA. Other personnel must stay a safe distance away until the deadly atmosphere is cleared. Warning signs must be posted at access points so that personnel do not enter until the area has been inspected and tested.

2.7.0 Reacting to Unexplained Flow Rate/ Pressure Changes

If a change in pressure or flow rate occurs that is different from what is expected, alarms may be initiated to alert personnel to the problem. Personnel must act quickly to address the problem. Qualified personnel will investigate and evaluate the problem. Some problems may require field personnel to manually adjust components, such as valves and transmitters, or make repairs. More serious problems may result in the need to isolate a segment of the pipeline.

Additional Resources

It is assumed that you will learn to perform these procedures under the direct supervision of your crew leader, foreman, or instructor. For independent study, refer to the recommended Additional Resources at the back of this module.

2.0.0 Section Review

1. If a damaged pipeline support is discovered and the damage appears to be critical, the individual who made the discovery should _____.
   a. begin repairs immediately
   b. stay close to the site
   c. report it to the appropriate personnel
   d. assume it is safe
2. If a hazardous atmosphere alarm sounds, field personnel should first _____.
   a. call 911
   b. evacuate the area
   c. notify control center personnel
   d. verify that the alarm is not false
3. Field personnel may need to operate equipment in the field manually if there is a system failure.
   a. True
   b. False
4. When a power outage occurs, the appropriate pipeline personnel should contact _____.
   a. the One-Call notification system
   b. the power company
   c. OSHA
   d. local authorities
5. Fire extinguishing equipment can be used on an incipient fire.
   a. True
   b. False
6. Electronic equipment must not be used in the presence of hazardous gases unless it is classified as _____.
   a. OSHA approved
   b. intrinsically safe/explosion-proof
   c. NRC approved
   d. mechanically safe/disruption proof
7. When an unexplained flow rate or pressure change occurs, field personnel may need to _____.
   a. notify the One-Call system
   b. contact the power company
   c. make repairs
   d. remove pipe supports

SUMMARY

Pipeline operators have developed effective procedures and policies that enable personnel to respond quickly, safely, and effectively to AOCs. Usually an AOC is a routine matter that can be handled safely by pipeline personnel. Sometimes, however, an AOC can result in an emergency situation that may involve working with outside agencies.

In this module, you learned about the different types of AOCs that can occur on the pipeline and the proper reactions to each type of AOC. During your career, you may have to respond to some or all of the AOCs outlined in this module. When responding to these AOCs, you must follow your company's policies and procedures. Protecting people, property, and the environment should be your utmost concern.

Review Questions

1. By definition, an AOC is a condition that may exceed _____.
   a. company operating budget
   b. design limits
   c. safety laws
   d. size limits

2. The federal agency that regulates pipeline maintenance is _____.
   a. OSHA
   b. GAO
   c. EPA
   d. PHMSA

3. By definition, a pipeline facility includes the right-of-way.
   a. True
   b. False

4. Which of the following is *not* always an AOC?
   a. Exposed pipe
   b. Reduced pressure
   c. Lost radio communication
   d. Unexpected release of a hazardous gas

5. Pressure relief systems, tank-level alarms, and main-line pressure shutdown alarms are examples of _____.
   a. prevention devices
   b. safety devices
   c. hazardous atmosphere alarms
   d. fire alarms

6. The loss of SCADA is an example of which type of AOC?
   a. Hydraulic failure
   b. Mechanical failure
   c. Communications failure
   d. Power failure

7. In the event of a power failure, pipeline companies may be able to keep the pipeline running by using an _____.
   a. operations override power source
   b. uninterruptible power system
   c. interface power system
   d. indirect power system

8. In the control center, one indication of a pipeline explosion could be a(n) _____.
   a. change in product temperature
   b. valve malfunction
   c. power outage
   d. unexplained change in pressure

9. Which of the following is *not* true of hazardous gas?
   a. Odorants are often added.
   b. High concentrations can deaden a person's sense of smell.
   c. Some gases are odorless.
   d. Low concentrations can deaden a person's sense of smell.

10. Which of the following best defines an unexplained change in flow rate or pressure?
    a. Reduced pressure only
    b. No flow only
    c. High flow rate
    d. Any change from the intended flow or pressure

11. When pipeline support damage is discovered, notification is especially important if the pipe _____.
    a. is in danger of excessive movement
    b. has changed color
    c. is metal
    d. is corroded

12. When a safety device sounds an alarm, field personnel must _____.
    a. turn it off
    b. never ignore it
    c. reset it
    d. contact the local emergency agency

13. If personnel communication tools fail, personnel should first _____.
    a. check for problems that have simple solutions
    b. contact the control center
    c. call 911
    d. complete any required reporting

14. If SCADA fails, equipment may have to operated manually by _____.

 a. PHMSA
 b. field personnel
 c. control center personnel
 d. local authorities

15. When a power interruption occurs, personnel may be able to keep the pipeline in operation by adjusting the _____.

 a. MAOP
 b. MOP
 c. line pressure and flow
 d. product temperature

16. A false alarm may trigger an investigation of the system, which is determined by _____.

 a. the control center
 b. field personnel
 c. the job site supervisor
 d. SCADA

17. Before using a fire extinguisher, it is important to know its _____.

 a. fire class rating
 b. OSHA rating
 c. combustibility
 d. capacity

18. The pipeline operator may work with local authorities to facilitate an evacuation of the area surrounding a pipeline.

 a. True
 b. False

19. An area where a hazardous gas has been detected may only be entered by personnel trained in rescue operations and qualified to use a _____.

 a. combustible gas indicator
 b. flame ionization indicator
 c. self-contained breathing apparatus
 d. fire suppression kit

20. All of the following are examples of a pipeline facility *except* _____.

 a. compressors
 b. pump stations
 c. control center
 d. rights-of-way

Trade Terms Quiz

Fill in the blank with the correct term that you learned from your study of this module.

1. The computer-based system(s) used by a controller in a control room that collects information about a pipeline facility and may have the ability to send commands back to the pipeline facility is called_______.
2. The _______ is the area along the pipeline which has been legally permitted to be used by the pipeline to construct, operate, and maintain the pipeline.
3. A fire in its beginning stage is called a(n) _______.
4. _______ is the separation of the protective coating from the pipeline.
5. A(n) _______ is a system that provides continuity of power when the normal power supply is interrupted.
6. The maximum pressure at which a pipeline or segment of a pipeline may be operated under this part (gas) is called _______.
7. A(n) _______ is used to relieve surges in a hazardous liquid pipeline system or to receive and store hazardous liquid transported by a pipeline for injection and continued transportation by pipeline.
8. The lowest concentration of a gas or a vapor in the atmosphere capable of producing fire in the presence of an ignition source is the _______.
9. A _______ duplicates the operations of a system so that, in the event of a failure, a backup system is available.
10. A(n) _______ indicates a malfunction of a component or a deviation from normal operations that may indicate a condition exceeding design limits or result in a hazard to persons, property, or the environment.
11. A device that detects some hazardous gases in the atmosphere is a(n) _______.
12. The pressure produced by a change in the velocity of the moving stream that results from shutting down a pump station or pumping unit, closure of a valve, or any other blockage of the moving stream is called _______.
13. _______ is the maximum pressure at which a pipeline or segment of a pipeline may be normally operated under this part (liquids).
14. _______ is a centralized communication system established by utilities, government agencies, and operators of underground facilities.
15. A hazardous liquid which will form a vapor cloud when released to the atmosphere and which has a vapor pressure exceeding 276 kPa (40 psia) at 37.8°C (100°F) is classified as a(n) _______.
16. The government agency that is responsible for developing and enforcing regulations for pipeline operations in the United States is the _______.
17. Electrical equipment that is _______ is designed specifically for use in explosive or flammable locations so that a spark cannot ignite the atmosphere.
18. An organization that adopts standards for a variety of products, services, and processes in the United States to ensure consistency is the _______.
19. A _______ is a device worn by an individual to provide breathable air in an atmosphere that is hazardous.
20. The _______ is the company that operates the pipeline, as opposed to an individual person who works for the company that operates the pipeline.

21. New and existing pipe on the pipeline itself, the right-of-way (ROW), and any equipment, facility, or building used in the transportation of a liquid or gas are defined as part of a(n) _______.
22. Activity carried out by, and/or equipment belonging to outside parties on the pipeline ROW is referred to as _______.
23. A length of pipeline that is entirely aboveground and supported by pipe supports is called a(n) _______.
24. Materials with radioactive elements that can be extracted during oil and gas extraction from the earth and can get into product streams and become transported in the pipeline are called _______.

Trade Terms

Aboveground span
Abnormal operating condition (AOC)
American National Standards Institute (ANSI)
Breakout tank
Disbondment
Encroachment
Flame ionization detection unit
Highly volatile liquid (HVL)
Incipient fire
Intrinsically safe/explosion-proof
Lower explosive limit (LEL)
Maximum allowable operating pressure (MAOP)
Maximum operating pressure (MOP)
Naturally occurring radioactive materials (NORMs)
One-Call notification system
Pipeline and Hazardous Materials Safety Administration (PHMSA)
Pipeline facility
Pipeline operator
Redundant system
Right-of-way (ROW)
Self-contained breathing apparatus (SCBA)
Supervisory control and data acquisition (SCADA)
Surge pressure
Uninterruptible power system (UPS)

Trade Terms Introduced in This Module

Aboveground span: Length of pipeline that is entirely aboveground and supported by pipe supports.

Abnormal operating condition (AOC): A condition within a pipeline system, identified by the operator, that may indicate (1) a malfunction of a component or (2) a deviation from normal operations that may indicate a condition exceeding design limits or result in a hazard to persons, property, or the environment.

American National Standards Institute (ANSI): An organization that adopts standards for a variety of products, services, and processes in the United States to ensure consistency.

Breakout tank: a tank used to (a) relieve surges in a hazardous liquid pipeline system or (b) receive and store hazardous liquid transported by a pipeline for reinjection and continued transportation by pipeline.

Disbondment: Separation of the protective coating from the pipeline.

Encroachment: Activity carried out by, and/or equipment belonging to outside parties on the pipeline right-of-way (ROW). A particular encroachment might be authorized or not, so each situation requires specific handling.

Flame ionization detection unit: A device that detects some hazardous gases in the atmosphere.

Highly volatile liquid (HVL): A hazardous liquid which will form a vapor cloud when released to the atmosphere and which has a vapor pressure exceeding 276 kPa (40 psia) at 37.8°C (100°F).

Incipient fire: A fire in its beginning stage.

Intrinsically safe/explosion-proof: Describes electrical equipment that is designed specifically for use in explosive or flammable locations so that a spark cannot ignite the atmosphere.

Lower explosive limit (LEL): The lowest concentration of a gas or a vapor in the atmosphere capable of producing fire in the presence of an ignition source. Different products have different LELs.

Maximum allowable operating pressure (MAOP): The maximum pressure at which a pipeline or segment of a pipeline may be operated under this part (gas).

Maximum operating pressure (MOP): The maximum pressure at which a pipeline or segment of a pipeline may be normally operated under this part (liquids).

Naturally occurring radioactive materials (NORMs): Materials with radioactive elements that can be extracted during oil and gas extraction from the earth and can get into product streams and become transported in the pipeline.

One-Call notification system: A centralized communication system established by utilities, government agencies, and operators of underground facilities. This system provides one telephone number for contractors and the general public to call to notify of their intent to use equipment on or near a pipeline. It is also in place for any individual to call if pipeline damage is discovered. The One-Call system is generally set up on a state-by-state basis.

Pipeline and Hazardous Materials Safety Administration (PHMSA): An agency of the US Department of Transportation responsible for developing and enforcing regulations for pipeline operations in the United States.

Pipeline facility: New and existing pipe, rights-of-way and any equipment, facility, or building used in the transportation of hazardous liquids or carbon dioxide, or new and existing pipelines, rights-of-way, and any equipment, facility, or building used in the transportation of gas or in the treatment of gas during the course of transportation.

Pipeline operator: The company that operates the pipeline, as opposed to an individual person who works for the company that operates the pipeline.

Redundant system: A system that duplicates the operations of a system so that, in the event of a failure, a backup system is available.

Right-of-way (ROW): The area along the pipeline which has been legally permitted to be used by the pipeline operator to construct, operate, and maintain the pipeline. The ROW extends the entire length of the pipeline.

Self-contained breathing apparatus (SCBA): A device worn to provide breathable air for an individual in an area where the atmosphere is hazardous.

Supervisory control and data acquisition (SCADA): a computer-based system or systems used by a controller in a control room that collects and displays information about a pipeline facility and may have the ability to send commands back to the pipeline facility.

Surge pressure: Pressure produced by a change in velocity of the moving stream that results from shutting down a pump station or pumping unit, closure of a valve, or any other blockage of the moving stream.

Uninterruptible power system (UPS): A system that provides continuity of power when the normal power supply is interrupted. A UPS provides AC power to the system it serves.

Additional Resources

The following recommended resources can provide additional helpful information related to the requirements for this covered task.

American National Standard Institute (ANSI), 1899 L Street, NW, 11th Floor, Washington, DC 20036. **www.ansi.org**

American Petroleum Institute (API), 1220 L Street, NW, Washington, DC 20005. **www.api.org**

American Society for Mechanical Engineers (ASME), Two Park Avenue, New York, NY 10016-5990. **www.asme.org**

American Society for Testing and Materials (ASTM) Specification D396.

Association of Oil Pipe Lines (AOPL), 1808 Eye Street, NW, Suite 300, Washington, DC 20006. **www.aopl.org**

Code of Federal Regulations 49, Parts 192 and 195. **www.ecfr.gov**

International Society of Automation (ISA), 67 TW Alexander Drive, Research Triangle Park, NC 27709. **www.isa.org**

National Association of Corrosion Engineers (NACE), 1440 South Creek Drive, Houston, TX 77084. **www.nace.org**

National Electrical Code®. **www.nfpa.org**

Occupational Safety and Health Administration (OSHA), 200 Constitution Avenue, NW, Washington, DC 20210. **www.osha.gov**

Pipeline and Hazardous Material Safety Administration (PHMSA), East Building, 2nd Floor, 1200 New Jersey Avenue, SE, Washington, DC 20590. **www.phmsa.dot.gov**

US Bureau of Labor Statistics, Postal Square Building, 2 Massachusetts Avenue, NE, Washington, DC 20212. **www.bls.gov**

US Energy Information Administration, 100 Independence Avenue, SW, Washington DC 20585. **www.eia.gov**

Figure Credits

U.S. Geological Survey/photo by Don Becker, Module Opener

Energy Information Administration, Office of Oil & Gas, Natural Gas Division, Gas Transportation Information System, 2009, Figure 1

Photograph No. 5998083 (Photographer Eric Vance), September 2006; Records of the Environmental Protection Agency, 1944–2006, Record Group 412; National Archives and Records Administration–College Park, Figure 9

Cameron Strother, Southland Safety, Figure 15

Photo courtesy of STV, Figure 34

Courtesy of Liquid Controls, LLC, Figure 35

Photographs and illustrations provided courtesy of Daniel Measurement and Control, Inc. Daniel Senior and Simplex are registered trademarks of Daniel Measurement and Control, Figure 36

Courtesy of Sunoco Logistics, Figure 39

Appendix A

Weather and Natural Disaster Information Sources

Emergency Managers Weather Information. US Department of Commerce, National Oceanic and Atmospheric Administration, National Weather Service, 1325 East-West Highway, Silver Spring, MD 20910.**www.nws.noaa.gov/emwin/index.htm**

National Earthquake Information Center, US Geological Survey. Box 25046, DFC, MS 967 Denver, CO 80225. **www.earthquake.usgs.gov/regional/neic**

National Hurricane Center Tropical Prediction Center. 11691 SW 17th Street, Miami, FL 33165-2149. **www.nhc.noaa.gov**

National Weather Service and National Weather Service Weather Radio, National Oceanic and Atmospheric Administration (NOAA). 1325 East-West Highway, Silver Spring, MD 20910. **www.weather.gov**

Appendix B

Organizations of Interest to the Pipeline Industry

American Gas Association (AGA). 400 N. Capitol Street, NW, Washington, DC 20001. **www.aga.org**

American Petroleum Institute (API). 1220 L Street, NW, Washington, DC 20005. **www.aga.org**

Interstate Natural Gas Association of America (INGAA). 20 F Street, NW, Suite 450, Washington, DC 20001. **www.ingaa.org**

National Institute for Occupational Safety and Health (NIOSH). 395 E Street, SW, Suite 9200, Patriots Plaza Building, Washington, DC 20201. **www.cdc.gov/niosh**

National Response Center (NRC). c/o United States Coast Guard (CG-5335 - Stop 7581), 2100 2nd Street, SW, Washington, DC 20593-0001. **www.nrc.uscg.mil**

Occupational Safety and Health Administration (OSHA). 200 Constitution Avenue, NW, Washington, DC 20210. **www.osha.gov**

Pipeline Hazardous Material Safety Administration (PHMSA). East Building, 2nd Floor, 1200 New Jersey Ave., SE, Washington, DC 20590. **www.phmsa.dot.gov**

Section Review Answer Key

Answer	Section Reference	Objective
Section One		
1. b	1.1.0	1a
2. a	1.2.0	1b
3. d	1.3.2	1c
4. b	1.4.0	1d
5. a	1.5.1	1e
6. c	1.6.0	1f
7. b	1.7.0	1g
Section Two		
1. c	2.1.2	2a
2. b	2.2.0	2b
3. a	2.3.2	2c
4. b	2.4.0	2d
5. a	2.5.1	2e
6. b	2.6.0	2f
7. c	2.7.0	2g

NCCER CURRICULA — USER UPDATE

NCCER makes every effort to keep its textbooks up-to-date and free of technical errors. We appreciate your help in this process. If you find an error, a typographical mistake, or an inaccuracy in NCCER's curricula, please fill out this form (or a photocopy), or complete the online form at **www.nccer.org/olf**. Be sure to include the exact module ID number, page number, a detailed description, and your recommended correction. Your input will be brought to the attention of the Authoring Team. Thank you for your assistance.

Instructors – If you have an idea for improving this textbook, or have found that additional materials were necessary to teach this module effectively, please let us know so that we may present your suggestions to the Authoring Team.

NCCER Product Development and Revision

13614 Progress Blvd., Alachua, FL 32615

Email: curriculum@nccer.org
Online: www.nccer.org/olf

❑ Trainee Guide ❑ Lesson Plans ❑ Exam ❑ PowerPoints Other ____________

Craft / Level: Copyright Date:

Module ID Number / Title:

Section Number(s):

Description:

Recommended Correction:

Your Name:

Address:

Email: Phone:

CT15_1-17

Visually Inspect Surface Conditions of Right-of-Way

Overview

The right-of-way (ROW) is the area along the pipeline that can be legally used by the pipeline company to construct, operate, and maintain the pipeline. To maintain the safety of the pipeline, right-of-ways are inspected by aerial, vehicle, and foot patrols. When patrolling the pipeline ROW, personnel check for certain conditions that require attention, and follow reporting procedures according to federal regulations and company-specific policies.

Trainees with successful module completions may be eligible for credentialing through the NCCER Registry. To learn more, go to **www.nccer.org** or contact us at 1.888.622.3720. Our website has information on the latest product releases and training, as well as online versions of our *Cornerstone* magazine and Pearson's product catalog.

Your feedback is welcome. You may email your comments to **curriculum@nccer.org**, send general comments and inquiries to **info@nccer.org**, or fill in the User Update form at the back of this module.

This information is general in nature and intended for training purposes only. Actual performance of activities described in this manual requires compliance with all applicable operating, service, maintenance, and safety procedures under the direction of qualified personnel. References in this manual to patented or proprietary devices do not constitute a recommendation of their use.

CT15_1-17

Visually Inspect Surface Conditions of Right-of-Way

Objectives

When you have completed this module, you will be able to do the following:

1. Describe the purpose of right-of-way inspections, how to locate the line to be inspected, and methods used to perform inspections.
   a. Describe the regulatory requirements for surface right-of-way inspections.
   b. Describe how to locate the pipeline to be inspected.
   c. Describe aerial patrol and vehicle/foot patrol methods of inspection.
2. Identify reportable conditions that may be observed during a visual inspection of ROW surface conditions, and describe reporting procedures.
   a. Identify the reportable conditions that may be observed during a visual inspection.
   b. Describe types of reporting procedures.
3. Describe how to perform a leak survey and monitor for pressure loss.
   a. Describe the procedures for conducting a leak survey.
   b. Describe how to monitor the pipeline for pressure loss.
4. Describe how to visually inspect surface conditions of the right-of-way (CT15_1-17).
   a. Identify the prerequisites, competencies, and abnormal operating conditions associated with visually inspecting surface conditions of the right-of-way.
   b. Describe procedures for visually inspecting surface conditions of the right-of-way.

Performance Task

Under the supervision of your instructor, you should be able to do the following:

1. Visually inspect surface conditions of right-of-way (CT15_1-17).
   - Identify potential abnormal operating conditions that may occur during performance of this CT, and know the appropriate actions to take in response to them.
   - Utilize the appropriate personal protective equipment according to relevant company procedures.
   - Accurately identify the right-of-way to be inspected from alignment sheets and/or pipeline maps.
   - Perform the visual inspection/patrol of the right-of-way.
   - Make proper notifications.
   - Complete appropriate documentation as required by operator's procedures.

Trade Terms

Abnormal Operating Conditions (AOCs)
Aerial patrol
Emergency situation
Linewalker
Patroller

Industry-Recognized Credentials

If you're training through an NCCER-accredited sponsor, you may be eligible for credentials from NCCER's Registry. The ID number for this module is CT15_1-17. Note that this module may have been used in other NCCER curricula and may apply to other level completions. Contact NCCER's Registry at 888.622.3720 or go to **www.nccer.org** for more information.

Contents

Figures and Tables

[illegible]

SECTION ONE

1.0.0 RIGHT-OF-WAY SURFACE CONDITION INSPECTIONS

Objective

Describe the purpose of right-of-way inspections, how to locate the line to be inspected, and methods used to perform inspections.

a. Describe the regulatory requirements for surface right-of-way inspections.
b. Describe how to locate the pipeline to be inspected.
c. Describe aerial patrol and vehicle/foot patrol methods of inspection.

Trade Terms

Aerial patrol :The most common form of information gathering performed from an overhead position on a weekly or biweekly schedule; used to report developments or changes on, and adjacent to, the right-of-way.

Linewalker: The ground patroller who walks or uses a vehicle to inspect the line.

Patroller: The person who performs a surface right-of-way inspection, regardless of whether the method used is aerial, vehicle, or foot patrol.

The pipeline right-of-way (ROW) runs the entire length of the line. It is wide enough to allow for construction equipment, and may also be wide enough for access roads. ROWs can range in width from 10 to 100 feet wide. Personnel must know the actual width of the ROW before inspections begin. The ROW must be patrolled regularly by plane, by vehicle, or on foot so personnel can identify any change in conditions such as construction activity, encroachments, markers, soil condition, uncovered pipe, facility conditions and security, leaks, or other indications of problems.

After the pipeline is initially built, the ROW is returned to its original condition as much as possible, including restoring fences across it, so that the use of the adjoining land is not obstructed. The ROW itself is not usually fenced in, except for aboveground spans and facilities. Before an inspection begins, the appropriate personnel should notify affected landowners according to company policy.

This section describes the regulatory requirements related to right-of-way inspections, how to locate the line, and the methods used to visually inspect surface conditions of the pipeline ROW.

1.1.0 Regulatory Requirements for Inspections

Federal regulations for liquid pipeline ROW inspections are not the same as regulations for gas pipelines. According to *49CFR(Code of Federal Regulations)* Part 195.412(a), the number of all aerial, vehicle, and foot patrols for hazardous liquid pipelines must be at least 26 times per year and at least every 3 weeks for each liquid pipeline ROW.

Gas pipelines involve additional considerations that determine the required frequency of inspections. According to *49CFR* Part 192.705, natural gas or other gas transmission line right-of-ways must be inspected periodically to ensure that they are being maintained in a safe manner. The frequency of gas pipeline ROW inspections is determined by several factors, including the size of the line, the operating pressures, the class location of the line, the terrain where the line is installed, and the weather conditions since the last inspection. Some gas pipeline ROW inspections occur with greater frequency depending on the product and the population density. Also, there are some conditions such as floods, significant increases in third-party construction or agricultural activities, or evidence or threats of vandalism that may warrant more frequent inspections.

While all of these factors are taken into account when determining how often a right-of-way should be inspected, the timeframe between inspections should be no longer than the time intervals presented in *Table 1*.

Class location designation is based on population density. More specifically, the class location of a line is defined by the number of dwellings intended for human occupancy that exist within 220 yards on either side of the centerline of any continuous one-mile length of pipeline. Note that each separate dwelling unit in a multiple dwelling unit building is counted as a separate building intended for human occupancy. The four class locations are defined as follows:

- *Class 1 location* – An off-shore area or an area containing 10 or fewer buildings intended for human occupancy.
- *Class 2 location* – An area containing between 10 and 46 buildings intended for human occupancy.

Table 1. Maximum Intervals Between Gas Pipeline ROW Inspections

Class Locations of Lines	At Highway and Railway Crossings	All Other Places
1, 2	$7^1/_2$ months, but at least twice each calendar year	15 months, but at least once each calendar year
3	$4^1/_2$ months, but at least four times each calendar year	$7^1/_2$ months, but at least twice each calendar year
4	$4^1/_2$ months, but at least four times each calendar year	$4^1/_2$ months, but at least four times each calendar year

- *Class 3 location* – An area containing 46 or more buildings intended for human occupancy, an area where the pipeline lies within 100 yards of a building, or a small, well-defined outside area that is occupied by 20 or more persons at least 5 days a week for 10 weeks in any 12-month period.
- *Class 4 location* – An area where buildings with four or more stories above ground are prevalent.

NOTE

Class locations may be adjusted by other factors. Refer to 49 *CFR* Part 192.5 for further details about class locations.

1.2.0 Locating the Pipeline

The pipeline is located by referencing existing drawings. Alignment sheets, also known as *strip maps*, show the pipeline route, and are used to help locate the pipeline (or other line) with relation to surface features and by depth. A pipeline alignment sheet is shown in *Figure 1*.

Alignment sheets show the location and size of each pipe installed; all crossings of public roads, railroads, rivers, and streams; as well as buried utilities and pipelines operated by other companies. The alignment sheet may show views of the ground from an aerial or plan view, and side or elevation views that show the ground profile along a specific line. These sheets will have the pipeline superimposed on the plan view with indications of lengths and angles. Special features such as rivers, highways, and railroads require additional information and planning.

The strips of each type of map or drawing are laid out side by side on the same sheet so that they line up along the route. These alignment sheets will be maintained and updated throughout the life of the pipeline. Alignment sheets can vary greatly from company to company.

NOTE

Always ensure that all maps, survey records, and drawings are up to date before using them for inspections.

1.3.0 Visual Inspection Methods

There are two primary methods used to visually inspect the right-of-way. Inspections may be done from the air (**aerial patrol**), or from the ground. Personnel on the ground may walk or use a vehicle to perform a visual inspection of the ROW.

1.3.1 Aerial Patrol Inspections

The most common form of information gathering is done by aerial patrol on a weekly or biweekly schedule, according to federal regulations as well as company policy. During these patrols, the pilots watch for and report developments or changes on and adjacent to the ROW that may result in the need for maintenance or investigation. Aerial markers may be used to help the pilot locate the pipeline and to identify specific locations along the pipeline.

Finding leaks, reporting their locations, and estimating their size and nature are the prime responsibilities of pilots. In addition to regularly scheduled patrols, additional flights may be made to investigate conditions when potential leaks are suspected by the operator. Efficient communication between pilots and the repair crew facilitates rapid response in case of an emergency.

The first sign of an underground leak is often an area of stained or discolored vegetation. Soil erosion, streambed changes, weathering of markers and signs, and growth of bushes and trees are gradual changes that will eventually require maintenance. Construction work near the ROW should be noted and reported so that local pipeline personnel can investigate these activities and make arrangements to prevent pipeline damage.

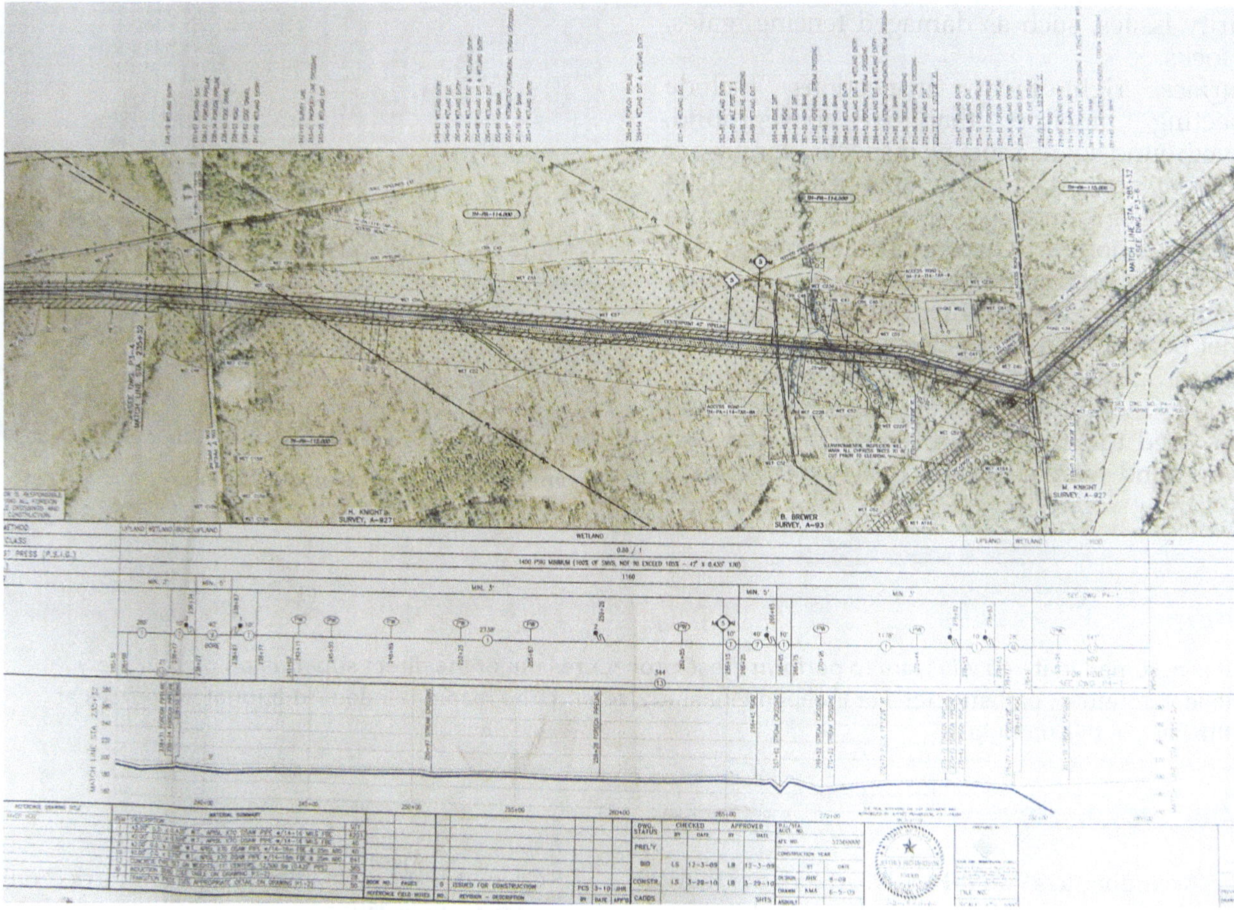

62107-13_F01.EPS

Figure 1 Sample alignment sheet.

If suspicious activity is observed, the pilot will communicate a warning and/or instructions to the appropriate personnel. Any suspicious activity is reported immediately and according to company policy.

Pilots must be trained to recognize conditions that must be reported, and follow company reporting procedures. They must develop good communications with pipeline personnel to ensure an effective working relationship. Discussions with personnel before each aerial patrol about possible problems or areas of interest can increase the effectiveness of aerial patrols. It is sometimes necessary to have personnel accompany the pilot to make and validate observations.

Because of the variety of geographic and physical circumstances, details of speed, altitude, weather, time of day, direction of flights, flight schedules, and other factors must be determined by the pilot and local requirements. Photographs of observations are encouraged where practical, and the photos should be used to document any leak or discharge, damage to the pipeline or ROW, or any unusual conditions. The types of equipment normally required for aerial patrols include relevant pipeline maps, a cell phone, a camera, and Global Positioning System (GPS) equipment.

1.3.2 Vehicle and Foot Patrol Inspections

Although aerial patrol has taken over much of line inspection, linewalkers are still needed in certain areas. A **linewalker** is a ground **patroller** who walks or uses a vehicle to inspect the pipeline. Linewalkers are able to more closely observe conditions where heavy vegetation exists and in densely populated areas, where flying at a low altitude may not be allowed. Covering the ROW at ground level may also be required when emergencies occur during darkness or when bad weather prevents aerial patrol. Vehicle and foot patrols are an effective way to check the conditions closely.

Vehicle/foot patrol duties include observing the ROW conditions, construction activity at or near the line, encroachments, line-marker conditions, soil erosion, uncovered pipe, pipeline facility condition, leaks, and any other condition that would place the pipeline or public in danger. A patroller should also check for signs of possible security issues, such as damaged fencing, gates, and locks.

Surface right-of-way inspections include inspecting pipeline that is aboveground. Aboveground spans must be inspected for proper alignment and attachment points from bank to bank. Aboveground pipe spans are designed to be flexible enough to accommodate stresses due to vibration, product load, and relevant weather impacts such as temperature, wind, and ice; personnel performing the inspection must therefore check for any related impacts to the integrity of the pipe, including checking all supports and cables used to secure and position the span. An aboveground span of pipeline is shown in *Figure 2*.

Figure 2 Aboveground span of pipeline.

Additional Resources

It is assumed that you will learn to perform these procedures under the direct supervision of your crew leader, foreman, or instructor. For independent study, refer to the recommended Additional Resources at the back of this module.

1.0.0 Section Review

1. According to 49 *CFR* Part 195.412(a), the ROW of a hazardous liquid pipeline must be inspected _____.
   a. when there is a threat of suspicious activity
   b. at least every three weeks
   c. monthly
   d. ten times per year

2. Alignment sheets are maintained and updated _____.
   a. after each inspection
   b. as required by federal regulations
   c. twice a year
   d. for the life of the pipeline

3. Ground-level ROW inspections are carried out by personnel called _____.
   a. rangers
   b. linewalkers
   c. road runners
   d. pipewalkers

SECTION TWO

2.0.0 REPORTABLE CONDITIONS AND REPORTING PROCEDURES

Objective

Identify reportable conditions that may be observed during a visual inspection of ROW surface conditions, and describe reporting procedures.

a. Identify the reportable conditions that may be observed during a visual inspection.
b. Describe types of reporting procedures.

Trade Term

Emergency situation: A leak, evidence of a possible leak, or activities that pose an immediate threat to the pipeline or to public safety.

This section describes how to identify reportable conditions along the ROW. It also describes types of reporting procedures for conditions such as evidence of a release, pipeline damage, vandalism, and other unsafe conditions.

2.1.0 Reportable Conditions

Personnel performing a visual inspection of surface conditions on or adjacent to the pipeline ROW may observe reportable conditions that can immediately or eventually impact the safety of the pipeline. Reportable conditions fall into one of three general categories based on the conditions observed and the potential safety hazards. These conditions are described as follows:

- Conditions that could pose an immediate threat to persons, property, or the environment, resulting in a typical reaction that would include stopping right-of-way surveillance activities, moving to a safe distance, and notifying appropriate pipeline personnel.
- Conditions that could impact the safety or integrity of the pipeline, resulting in a typical reaction that would mandate prompt notification.
- Conditions that could impact the pipeline, resulting in a typical response.

This section describes common reportable conditions and the appropriate actions that should be taken by a patroller observing these conditions.

2.1.1 Reporting Emergency Situations/ Evidence of Release

The definition of an **emergency situation** is a leak, evidence of a possible leak, or activities that pose an immediate threat to persons, property, or the environment. Any evidence of a leak requires an immediate verbal report, as well as written reports. Company policy will dictate response and repair.

Observable indications of a leak could include any one of the following:

- Vapor cloud or frost ball on or near a right-of-way
- Sheen on or bubbles in the water on a right-of-way
- Dead vegetation or wet spot
- Dead or sick animals near the pipeline
- Odor of hazardous gas or liquids
- Audible cues (hissing, roaring, etc.)

Personnel should be aware that if buried pipe is leaking, it usually takes some time for the evidence to appear on the ground.

WARNING!

If unexpected hazardous liquid or carbon dioxide is encountered, local emergency personnel should be notified immediately. Make proper company notifications and try to identify the source of the leak, if possible.

The following observations should also be considered emergency situations:

- Flooding, washouts, erosion, or exposure that could immediately damage or affect the stability of the pipeline
- Fire or explosion near the right-of-way

WARNING!

If a fire or explosion is observed near the right-of-way, local emergency personnel should be notified immediately. Limit access to the area as much as possible and make proper company notifications.

2.1.2 Reporting Pipeline Integrity Issues

Patrollers must be alert for conditions that threaten the safety or integrity of the pipeline, but which do not present an emergency situation. Often involving structural issues, common reportable conditions for this category include the following:

- Construction or excavation equipment or other signs of construction activity on or near a right-of-way
- Soil movement, such as a landslide, mudslide, or sinkhole
- Sagging aboveground pipe at a span
- Damaged, leaning, or failing pipe support system
- Unusual materials, equipment, and/or foreign objects on or near the right-of-way
- Damage to pipeline facilities or suspicious activity that might indicate vandalism or terrorist actions

Pipeline exposure may be caused by environmental conditions, such as weather or earth movements. Exposed pipe should be reported for inspection and reburial. Whenever exposed pipe is discovered, it must be inspected for evidence of external damage, corrosion, and the extent of the damage or corrosion. Corrective action must be taken according to company policy. An example of exposed pipe is shown in *Figure 3*.

On aboveground pipe, damage or exposure may be obvious. Some evidence to look for on the pipe or structure includes wrinkling, buckling, dents, gouges, or any signs of unusual stress on the pipe.

Damage may also occur on pipe supports. These supports are required on various aboveground pipeline structures to provide protection from the forces exerted by anticipated loads. They are often found at overhead crossings, spans over ditches, pig traps, block valves, manifold valves and piping, and other structures. To be effective, these supports must be undamaged and in the correct position. Damaged or missing supports must be reported and repaired according to company policy. *Figure 4* shows pipe on supports.

Damage to supports may be an emergency situation if there is a threat of imminent collapse. Temporary supports may need to be installed, and the area may need to be evacuated. Attempts should be made to determine if the damage occurred naturally, resulted from vandalism, or resulted from third-party activity.

Encroachment is activity on the company's ROW carried out by outside parties. A particular encroachment might be authorized or not, so each situation requires specific handling. Unauthorized activities on the pipeline ROW that may affect the pipeline are often spotted by patrollers doing line inspection and reported to management. The unauthorized encroaching party is advised to cease activity until consent can be given by the operator. The operator will then investigate the activity and communicate with the encroaching party about a final resolution.

If an immediate danger is posed by such activities, they must be stopped, and an emergency call must be made to the control center. Foreign crossings are reviewed and approved. Specific company procedures will apply to the crossing to ensure that the company pipeline is not damaged or compromised.

Besides encroachment, other nearby activities that may impact the safety or integrity of the pipeline include the following:

- Blasting or surface mining within one-quarter mile of the pipeline
- Excavations within the ROW or excavations threatening the pipeline

62107-13_F03.EPS

Figure 3 Exposed pipe.

62107-13_F04.EPS

Figure 4 Pipe on supports.

- Clearing of timberland over, across, or along pipelines
- Building of terraces, water ponds, or pits within an area defined by company policy
- Construction of buildings, well derricks, residences, fences, or other structures within an area defined by company policy
- Brush or tree clearing, excavation, or ditching for new roads, utilities, or structures
- Grading, construction, or resurfacing of roads
- Fires occurring within an area defined by company policy
- Agricultural deep-ripping or subsoil plowing

Evidence of activity around a block valve or any other place along the pipeline ROW might indicate vandalism, theft of product, or sabotage. All surface facilities such as pump stations and breakout tanks should be provided with protection against vandalism or other unauthorized entry. Fencing may be present and should be checked by patrollers for damage. Signs are required around these exposed facilities and should also be checked.

2.1.3 Reporting Typical Conditions

Reportable conditions that could impact the pipeline but which do not present an immediate threat to the safety or integrity of the pipeline include the following types of observations:

- Vegetation overgrowth/excessive canopy that may obstruct view of right-of-way
- Damaged or missing line marker(s)
- Soil movement such as subsidence or settling
- Tripped anchors on aboveground pipe
- Damage to coatings or insulation on aboveground pipe or components

ROW policies limit or ban trees growing on the ROW. Any vegetation that grows near or on the ROW may hamper inspection or threaten to block access for maintenance and repair. Excessive vegetation growth must be reported per company policy.

Signs and/or markers are required at all aboveground pipeline facilities; all road, rail, and water crossings; and at sufficient intervals along the ROW to clearly identify the pipeline location. They should be clearly legible and securely installed directly over the pipeline, if possible. The information on the signs or markers must be accurate and up to date. Aerial markers may also be present for use by aerial patrols. Marker locations may be indicated on company maps.

Signs may become damaged by weather or other natural conditions, or they may be vandalized or stolen. During inspection patrols, any damaged, missing, improperly located, or inaccurate sign must be reported and replaced according to company policy.

2.1.4 Other Reportable Conditions

Other unsafe conditions that may be encountered along the ROW must also be reported. These include, but are not limited to the following:

- Localized or general corrosion
- Pipe movement due to environmental causes
- Material defects or physical damage
- Equipment malfunction

Other events impacting the pipeline cover, such as airplane crashes, train wrecks, and heavy equipment traffic, must also be reported.

2.2.0 Reporting Procedures

Federal regulations and company policies require that records of ROW inspections be made and maintained. Written inspection reports must be completed after each aerial, vehicle, or foot patrol, and additional reports must be made after any further investigation is made, if applicable. If an emergency situation is discovered, verbal reports must be made immediately to the appropriate personnel.

The inspection records provided by the patroller must be retained for at least five years, or longer if required by company policy. Inspection results are compared with existing records to determine if any changes have occurred and if such changes require repairs. If any actions are required, they must be completed as quickly as possible.

The reporting procedures include completing the following documents:

- Written inspection reports
- Emergency and verbal inspection reports, if applicable
- ROW investigation reports, if applicable

2.2.1 Written Inspection Reports

A written report of the inspection should be made as soon as practical following the inspection. The report is then distributed to the appropriate personnel, per company policy. The report usually includes the following information:

- Date
- Patroller's name
- Reason for inspection
- Location of observation

- Time of observation
- Identification of pipeline(s), section, or segment
- Pipe and material conditions
- Defect, damage, or corrosion observed, if applicable
- Report number
- Any actions taken
- Notification of any verbal reports made

2.2.2 Emergency and Verbal Inspection Reports

An emergency situation, such as a possible leak, requires the patroller to make an immediate telephone call to the appropriate contact. If the situation affects excavation or other activities performed by construction workers on or near the pipeline, the patroller should first warn the appropriate personnel. The verbal report must be followed up with a written report.

Possible leak observations include petroleum product, vapor, or other discharge; dead vegetation over the pipeline; dead livestock or other dead animals near the pipeline; or petroleum sheen on nearby water. Emergency observations also include activities or circumstances, such as construction activity, that may pose a threat to the integrity of the pipeline or to public safety.

Emergency verbal reports should include the following information:

- Patroller's name
- Identification of pipeline(s), section, or segment
- Flight identification (for aerial patrols)
- Location of the observation referenced to landmarks, geography, or city
- GPS location
- Description of the observation
- Indication that the observation is deemed an emergency

WARNING!

In case of a fire or explosion caused by the ignition of hydrocarbons, local emergency personnel should be notified immediately. Access to the area should be limited as much as possible, and the proper company notifications should be made.

2.2.3 ROW Investigation Reports

If the initial inspection requires further investigation, the appropriate personnel are dispatched to investigate the reported conditions and to complete ROW investigation reports according to company policy. Investigation reports provide further detail about the condition and the result of the investigation and any action taken. The investigator will take photos or create drawings to show unusual conditions, such as exposed pipe or soil erosion. These ROW investigation reports are then filed according to federal regulations.

Additional Resources

It is assumed that you will learn to perform these procedures under the direct supervision of your crew leader, foreman, or instructor. For independent study, refer to the recommended Additional Resources at the back of this module.

2.0.0 Section Review

1. Patrollers who observe evidence of a possible leak should consider it _____.
   a. an indication of possible vandalism
   b. safe until proven otherwise
   c. an emergency situation
   d. easily rectified

2. After an inspection, a written report should be submitted _____.
   a. within 24 hours
   b. in one week or less
   c. as soon as its practical to do so
   d. before the next scheduled inspection

3. In the case of an emergency situation, the initial report must be _____.
   a. verbal
   b. documented in detail
   c. written in accordance with company policy
   d. submitted only by qualified personnel

4. A right-of-way investigation report may include _____.
   a. a step-by-step description of the investigation
   b. photos or drawings of unusual conditions
   c. a summary of verbal reports
   d. estimated costs for repairs

SECTION THREE

3.0.0 POSTBLAST LEAK SURVEILLANCE

Objective

Describe how to perform a leak survey and monitor for pressure loss.

a. Describe the procedures for performing a leak survey.
b. Describe how to monitor the pipeline for pressure loss.

Trade Term

Abnormal operating conditions (AOCs): Conditions within a pipeline system, identified by the operator, that may indicate (1) a malfunction of a component or (2) a deviation from normal operations that may indicate a condition exceeding design limits or result in a hazard to persons, property, or the environment.

After blasting operations have taken place near any segment of pipeline, a leak survey must be performed. A leak survey is a visual inspection of the pipeline right-of-way directly affected by a blasting operation. An additional segment of pipe both upstream and downstream of the blast impact area must also be surveyed. The survey checks for evidence of leaks indicative of damage to the pipeline. Pipeline maintenance personnel must be able to perform a leak survey in accordance with their company policy after any blasting has taken place in proximity of the pipeline.

This section provides a general description of the procedures for conducting a leak survey. Detailed information on covered tasks related to leak surveys can be found in NCCER modules CT52_1-17, *Conduct Vegetation Survey*; CT52_2-17, *Conduct a Leak Survey with a Combustible Gas Detector*; and CT52_3-17, *Conduct a Leak Survey with a Flame Ionization Unit.*

WARNING!

While performing a leak survey, personnel may encounter an explosion or fire, such as burning hydrocarbon vapors. Always protect all personnel and the public. Contact the local fire and police departments if necessary. Immediately initiate and execute line shutdown, and block, depressurize, and/or drain the affected line segment. Only qualified and trained personnel may control and contain leaking hydrocarbon liquids.

3.1.0 Perform a Leak Survey

A leak survey may include the use of gas detection equipment. It may also include exposure of the pipeline if seismographic vibration readings recorded during the blasting operation are in excess of the allowable limit as stated in the operator's specifications/policy.

WARNING!

Pipeline system damage, such as broken, bent, or burnt pipeline, excessive disturbance of earthen support, and dislodged pipeline, may be encountered during a leak survey. Immediately initiate and execute line shutdown, and block, depressurize, and/or drain the affected line segment. Qualified and trained personnel should perform site assessment.

The following steps outline the general procedure for performing a leak survey:

Step 1 Determine where the visual inspection should begin and end.

Step 2 Determine to what the extent the inspection should be performed.

Step 3 Determine any equipment necessary to expedite the inspection. The terrain and environment along right-of-ways can vary greatly. Difficult or blocked access to and egress from a site can interfere with proper inspection.

Step 4 Visually inspect the impact area for indication of a leak:

- Product or crude or frost on the ground
- Discolored or melted snow
- Hydrocarbon smell
- Vapor clouds or shimmery heat-wave-like product evaporation
- Blowing, whistling, or trickling sound

> **WARNING!**
>
> While performing a leak survey, personnel may discover an unexpected hazardous liquid or carbon dioxide. Crude, product, or a chemical may be blowing, leeching, boiling, or vaporizing into the ground, water, or air. Immediately take steps to initiate and execute line shutdown, and block, depressurize, and/or drain the affected line segment. Take any action to protect and minimize the danger to the public and environment, such as notifying residents and local police and fire departments to accomplish evacuation and to control traffic. Only qualified and trained personnel should perform site assessment.

Step 5 Collect data about the existing ground surface condition.

Step 6 Determine if the postblast surface condition indicates that exposure of the pipeline is necessary.

Step 7 If exposure is not necessary, report and/or document surface conditions as required by company policy. If exposure is necessary, use a probing bar to create holes in the ground surface along the pipeline or within the inspection area.

Step 8 Insert a gas-detector drawtube in the probed holes. Do this especially in gullies and low-lying areas. Examine utility system manholes for indications of hydrocarbons.

Step 9 Excavate the pipeline.

3.2.0 Monitor for Pressure Loss

Control center or pipeline maintenance personnel must be able to monitor instrumentation for indications of a pressure loss during or after extensive earthmoving operations, such as excavation, blasting, boring, or tunneling, that are performed in the vicinity of the pipeline. Use this general procedure to monitor for pressure loss:

Step 1 Identify the pipeline and segment to be monitored upon notification that an extensive earthmoving operation is taking place in the field.

Step 2 Determine expected or normal trend pressure variance.

> **NOTE**
>
> Requirements for trending/pressure alarms vary from line to line and vary if a line is active or inactive. There are, however, general ranges of pressure that appear normal or reasonable based on the hydraulics involved.

Step 3 Investigate any low-pressure trend outside of normal conditions or pressure deviation, and review historical trend data per system parameters.

Step 4 Determine why the pressure deviated from the normal trend pressure variance.

Step 5 Communicate the deviation and its cause to the appropriate personnel.

Step 6 Shut down the line according to procedures for **abnormal operating conditions (AOCs).**

> **WARNING!**
>
> While monitoring for pressure loss, personnel may encounter a pressure that is different from the expected pressure. The pressure may be too high, too low, or even absent. The change must be immediately investigated. If the cause is not determined, shut down the operating lines according to the AOC shutdown procedure, and proceed with emergency response notifications.

Step 7 Communicate the deviation to the appropriate personnel for follow-up investigation/response. It is necessary to initiate investigation or activate emergency response from operation, engineering, and field personnel per company policy.

3.0.0 Section Review

1. If a broken, bent, or burnt pipeline is discovered during a leak survey, personnel should _____.
   a. file a written report within 24 hours
   b. reduce pressure in the affected section
   c. immediately initiate and execute a line shutdown
   d. move back at least 500 feet and call in emergency services

2. If the pressure deviates from the expected or normal trend pressure variance when monitoring for pressure loss, _____.
   a. discuss the change with qualified personnel
   b. immediately investigate the change
   c. quickly shut down the affected line
   d. submit a maintenance request

SECTION FOUR

4.0.0 VISUALLY INSPECT SURFACE CONDITIONS OF RIGHT-OF-WAY

Objective

Describe how to visually inspect surface conditions of the right-of-way (CT15_1-17).

a. Identify the prerequisites, competencies, and abnormal operating conditions associated with visually inspecting surface conditions of the right-of-way.
b. Describe procedures for visually inspecting surface conditions of the right-of-way.

Performance Task

Visually inspect surface conditions of right-of-way (CT15_1-17).

- Identify potential abnormal operating conditions that may occur during performance of this CT, and know the appropriate actions to take in response to them.
- Utilize the appropriate personal protective equipment according to relevant company procedures.
- Accurately identify the right-of-way to be inspected from alignment sheets and/ or pipeline maps.
- Perform the visual inspection/patrol of the right-of-way.
- Make proper notifications.
- Complete appropriate documentation as required by operator's procedures.

This section describes the procedures for visually inspecting the surface conditions of the right-of-way. It also provides the prerequisites and competencies required for this covered task, and describes the abnormal operating conditions (AOCs) that may be encountered. This task begins with accurately identifying the right-of-way to be inspected, and ends when proper documentation and notifications are completed.

4.1.0 Prerequisites, Competencies, and Abnormal Operating Conditions

Personnel must possess a variety of skills in order to visually inspect the surface conditions of the right-of-way. The various operations require a series of steps, which personnel must be competent to perform. Personnel must also be aware of the particular abnormal operating conditions that may occur during the installation procedure and know the proper responses to them.

4.1.1 Prerequisites and Competencies

Personnel assigned to visually inspect the surface conditions of the right-of-way must possess the following basic knowledge and skills prior to performing this task:

- Familiarity with the pipeline route
- Ability to recognize and respond to leak indications
- Ability to recognize conditions that could pose an immediate threat to persons, property, or the environment
- Ability to recognize conditions that could impact the safety or integrity of the pipeline
- Ability to recognize conditions that could impact the pipeline
- Ability to accurately report the results of the inspection according to operator's procedures

4.1.2 Abnormal Operating Conditions

Before performing a visual inspection of the surface conditions of the right-of-way, personnel must be able to identify and properly respond to potential abnormal operating conditions, should they occur. The following are potential AOCs, and the appropriate action personnel should take in response to them:

- *There is an abnormal facility condition, or unauthorized construction or equipment on the right-of-way* – Make proper company notifications.
- *Unexpected hazardous liquid or carbon dioxide is encountered* – Notify local 911 emergency personnel if appropriate, make proper company notifications, and attempt to identify the source of the leak, if possible.
- *A fire or explosion occurs* – Limit access to the area as much as possible, and make proper company notifications.

WARNING!

Always follow company procedures regarding potential AOCs, and how to respond to them. Failure to do so could be fatal.

4.2.0 Procedure

Personnel assigned to visually inspect surface conditions of the right-of-way must always follow company policy and procedures. The basic steps for this task are outlined in this section.

> **WARNING!** Always use the proper personal protective equipment when patrolling a right-of-way. Stay alert to the possibility of unexpected leaks. In the event of a leak, safely eliminate all ignition sources and immediately notify the appropriate personnel.

The following steps provide a general procedure for visually inspecting surface conditions of the right-of-way:

Step 1 Identify potential abnormal operating conditions that may occur during performance of this CT, and know the appropriate actions to take in response to them.

Step 2 Utilize the appropriate personal protective equipment according to relevant company procedures.

Step 3 Accurately identify the right-of-way to be inspected by referencing alignment sheets and/or pipeline maps. Once the ROW is identified, the method of inspection is determined (aerial, vehicle, or foot patrol).

Step 4 Perform the visual inspection of the right-of-way, documenting all reportable observations. All regulatory-required inspections must be documented.

Step 5 Follow company procedures for submitting reports and making proper notifications.

Step 6 Complete appropriate documentation as required by operator's procedures.

Additional Resources

It is assumed that you will learn to perform these procedures under the direct supervision of your crew leader, foreman, or instructor. For independent study, refer to the recommended Additional Resources at the back of this module.

4.0.0 Section Review

1. If there is a fire or explosion while patrolling the ROW, the proper response is to immediately _____.
   a. shut down the pipeline
   b. attempt to control the fire if it is safe to do so
   c. limit access to the area
   d. call 911

2. The method used to inspect the right-of-way is determined by _____.
   a. the control center
   b. its location
   c. qualified personnel
   d. the anticipated hazards

SUMMARY

The area along the pipeline that can be legally used by the pipeline company to construct, operate, and maintain the pipeline is called the right-of-way. Pipeline ROWs must be inspected on a regular basis, using aerial, vehicle, and/or foot patrols. When performing a visual inspection, the patroller must report conditions that threaten the safety and integrity of the pipeline, including conditions such as evidence of release, damaged or exposed pipeline, missing or damaged signs/markers, encroachment, and vandalism.

Federal regulations and company policies require that records of ROW inspections be made and maintained. Written reports of each inspection should be submitted per company policy. These reports are compared with existing records to determine if any changes have occurred and if such changes warrant repairs.

Review Questions

1. The right-of-way of a pipeline _____.
   a. refers to a road that crosses the pipeline's path
   b. applies only to the underground sections of the pipeline
   c. applies only to aboveground facilities
   d. runs the entire length of the pipeline

2. The class location that applies to offshore areas is _____.
   a. Class 1
   b. Class 2
   c. Class 3
   d. d.Class 4

3. A drawing used to help locate a pipeline with relation to surface features and by depth is known as a(n) ____.
   a. topographic map
   b. route survey
   c. alignment sheet
   d. easement plan

4. The ground profile along a specific line may be shown on a strip map as a(n) _____.
   a. isometric drawing
   b. elevation view
   c. orthographic projection
   d. plan view

5. Aerial patrols are commonly made _____.
   a. daily
   b. weekly or biweekly
   c. monthly
   d. annually

6. The first sign of a leak in an underground pipeline is often _____.
   a. sick or dead wildlife
   b. a pool of oil
   c. a plume of vapor
   d. stained or discolored vegetation

7. Equipment normally needed for aerial patrols includes relevant maps, cell phone, camera, and _____.
   a. binoculars
   b. GPS device
   c. radar sender/receiver
   d. oxygen mask

8. One advantage of foot patrols over aerial patrols is that problems can be found _____.
   a. where there is little vegetation
   b. where there is heavy vegetation
   c. in sparsely populated areas
   d. when it is daylight

9. Aboveground pipeline spans must be flexible enough to withstand stress from vibration, weather conditions, and _____.
   a. suspension cables
   b. product load
   c. metal fatigue
   d. airborne pollution and contamination

10. To be effective, pipeline supports must be undamaged and _____.
    a. in the correct position
    b. wrapped to prevent corrosion
    c. made from treated timbers
    d. able to withstand high winds

11. The word *encroachment* describes _____.
    a. excavation or blasting done within 250 feet of the pipeline right-of-way
    b. unauthorized activities being carried out on the pipeline right-of-way
    c. any activity, authorized or not, that takes place on the pipeline right-of-way by outside parties
    d. tunneling beneath the pipeline right-of-way

12. The locations of aerial markers along a pipeline route may be indicated on _____.
    a. navigation charts
    b. geodetic surveys
    c. topographic maps
    d. company maps

13. A patroller's written inspection report must include _____.
    a. notification of any verbal reports made
    b. GPS coordinates of the defect or damage
    c. mileage driven during the inspection
    d. date and time the report was written

14. ROW investigation reports are filed according to _____.

    a. date prepared
    b. federal regulations
    c. the degree of urgency
    d. alignment sheet number

15. Control and containment of any hydrocarbon product leak must be done only by _____.

    a. local fire department personnel
    b. the worker(s) who discovered the leak
    c. qualified and trained personnel
    d. a federal quick-response team

16. If the cause of a change in pressure cannot be immediately determined, the line must be shut down as directed in the _____.

    a. Federal Emergency Management Agency regulations
    b. company's emergency procedures handbook
    c. US Pipeline Safety Administration guidelines
    d. abnormal operating condition procedure

17. Personnel assigned to patrol the right-of-way must have _____.

    a. familiarity with the pipeline route
    b. a pilot's license
    c. experience with excavation equipment
    d. d.training as an emergency responder

Trade Terms Quiz

Fill in the blank with the correct term that you learned from your study of this module.

1. The ground patrol who walks or uses a vehicle to inspect the line is a(n) ________.
2. Conditions within a pipeline system, identified by the operator, that may indicate (1) a malfunction of a component or (2) a deviation from normal operations that may indicate a condition exceeding design limits or result in a hazard to persons, property, or the environment are called ________.
3. A leak, evidence of a possible leak, or activities that pose an immediate threat to the pipeline or to public safety is a(n) ________.
4. The most common form of information gathering performed from an overhead position on a weekly or biweekly schedule is a(n) ________.
5. The person who performs a surface right-of-way inspection, regardless of whether the method used is aerial, vehicle, or foot patrol, is called a(n) ________.

Trade Terms

Abnormal operating conditions (AOCs)
Aerial patrol
Emergency situation
Linewalker
Patroller

Trade Terms Introduced in This Module

Abnormal operating conditions (AOCs): Conditions within a pipeline system, identified by the operator, that may indicate (1) a malfunction of a component or (2) a deviation from normal operations that may indicate a condition exceeding design limits or result in a hazard to persons, property, or the environment.

Aerial patrol : The most common form of information gathering performed from an overhead position on a weekly or biweekly schedule; used to report developments or changes on, and adjacent to, the right-of-way.

Emergency situation: A leak, evidence of a possible leak, or activities that pose an immediate threat to the pipeline or to public safety.

Linewalker: The ground patroller who walks or uses a vehicle to inspect the line.

Patroller: The person who performs a surface right-of-way inspection, regardless of whether the method used is aerial, vehicle, or foot patrol.

Additional Resources

The following recommended resources can provide additional helpful information related to the requirements for this covered task:

API Recommended Practice 1161, Third Edition, © 2014 American Petroleum Institute. API Publishing Services, 1220 L Street NW, Washington, DC 20005. Distributed by Thompson Reuters (Scientific) LLC, **www.techstreet.com**.

NCCER Module AOCCC-17, *Abnormal Operating Conditions Control Center*.

NCCER Module AOCFG-17, *Abnormal Operating Conditions Field and Gas*.

NCCER Module 62105-14, *Introduction to Pipeline Documents*.

NCCER Module CT52_1-17, *Conduct Vegetation Survey*.

NCCER Module CT52_2-17, *Conduct a Leak Survey with a Combustible Gas Detector*.

NCCER Module CT52_3-17, *Conduct a Leak Survey with a Flame Ionization Unit*.

Code of Federal Regulations 49, Parts 192.705 and 195.412(a). **www.ecfr.gov**.

Figure Credits

Cameron Strother, Southland Safety, Figure 1
Courtesy Bureau of Land Management, Figure 2
U.S. Air Force PHOTO by AIRMAN First Class Kyle R. Fidel, Figure 3
U.S. Geological Survey/photo by Don Becker, Figure 4

Section Review Answer Key

Answer	Section Reference	Objective
Section One		
1. b	1.1.0	1a
2. d	1.2.0	1b
3. b	1.3.2	1c
Section Two		
1. c	2.1.1	2a
2. c	2.2.1	2b
3. a	2.2.2	2b
4. b	2.2.3	2b
Section Three		
1. c	3.1.0	3a
2. b	3.2.0	3b
Section Four		
1. c	4.1.2	4a
2. b	4.2.0	4b

NCCER CURRICULA — USER UPDATE

NCCER makes every effort to keep its textbooks up-to-date and free of technical errors. We appreciate your help in this process. If you find an error, a typographical mistake, or an inaccuracy in NCCER's curricula, please fill out this form (or a photocopy), or complete the online form at **www.nccer.org/olf**. Be sure to include the exact module ID number, page number, a detailed description, and your recommended correction. Your input will be brought to the attention of the Authoring Team. Thank you for your assistance.

Instructors – If you have an idea for improving this textbook, or have found that additional materials were necessary to teach this module effectively, please let us know so that we may present your suggestions to the Authoring Team.

NCCER Product Development and Revision

13614 Progress Blvd., Alachua, FL 32615

Email: curriculum@nccer.org

Online: www.nccer.org/olf

❑ Trainee Guide ❑ Lesson Plans ❑ Exam ❑ PowerPoints Other ______________

Craft / Level: Copyright Date:

Module ID Number / Title:

Section Number(s):

Description:

Recommended Correction:

Your Name:

Address:

Email: Phone:

CT16_1-17

Inspect Navigable Waterway Crossing

Overview

Navigable waterways fall under the jurisdiction of the federal government. Waters are designated as navigable by 33 *CFR* Part 329, Section 4. Navigable waterway crossings include the right-of-way (ROW) immediately adjacent to either side of the pipeline. This module introduces the elements of a navigable waterway crossing inspection and the associated reporting that is required.

Trainees with successful module completions may be eligible for credentialing through the NCCER Registry. To learn more, go to **www.nccer.org** or contact us at 1.888.622.3720. Our website has information on the latest product releases and training, as well as online versions of our *Cornerstone* magazine and Pearson's product catalog.

Your feedback is welcome. You may email your comments to **curriculum@nccer.org**, send general comments and inquiries to **info@nccer.org**, or fill in the User Update form at the back of this module.

This information is general in nature and intended for training purposes only. Actual performance of activities described in this manual requires compliance with all applicable operating, service, maintenance, and safety procedures under the direction of qualified personnel. References in this manual to patented or proprietary devices do not constitute a recommendation of their use.

CT16_1-17
Inspect Navigable Waterway Crossing

Objectives

When you have completed this module, you will be able to do the following:

1. Explain how to inspect and report on navigable waterway crossings.
   a. Explain how to locate the pipeline.
   b. Explain how to perform a surface and submarine inspection.
   c. Identify reportable observations for navigable waterway crossing inspections.
   d. Identify reporting procedures for navigable waterway crossing inspections.
2. Describe the procedure used to inspect or simulate inspecting a navigable waterway crossing (CT16_1-17).
   a. Identify the prerequisites, competencies, and abnormal operating conditions associated with inspecting a navigable waterway crossing.
   b. Describe the general procedure used to inspect a navigable waterway crossing.

Performance Task

Under the supervision of your instructor, you should be able to do the following:

1. Inspect or simulate inspecting a navigable waterway crossing (CT16_1-17).
   - Identify potential abnormal operating conditions that may occur during performance of this CT, and know the appropriate actions to take in response to them.
   - Utilize the appropriate personal protective equipment according to relevant company procedures.
   - Locate the landfall of the pipeline crossing to establish a reference point.
   - Determine the amount of cover by using a probe or specialized equipment to verify the depth of the submerged pipeline.
   - Document the findings as required by operator's procedures and 49 *CFR* 195.404.

Trade Terms

Abnormal operating conditions (AOCs)
Aerial patrol
As-built river profile
Benchmarks
Depth of cover
Navigable waterways
Probe rod
Submersible

Industry-Recognized Credentials

If you're training through an NCCER-accredited sponsor, you may be eligible for credentials from NCCER's Registry. The ID number for this module is CT16_1-17. Note that this module may have been used in other NCCER curricula and may apply to other level completions. Contact NCCER's Registry at 888.622.3720 or go to **www.nccer.org** for more information.

Contents

Figures

SECTION ONE

1.0.0 REQUIREMENTS FOR NAVIGABLE WATERWAY CROSSING INSPECTIONS

Objective

Explain how to inspect and report on navigable waterway crossings.

a. Explain how to locate the pipeline.
b. Explain how to perform a surface and submarine inspection.
c. Identify reportable observations and reporting procedures for navigable waterway crossing inspections.
d. Identify reporting procedures for navigable waterway crossing inspections.

Trade Terms

Aerial patrol: The most common form of information gathering performed from an overhead position on a weekly or biweekly schedule to report developments or changes on, and adjacent to, the right-of-way.

As-built river profile: Information that includes the pipeline location, the river-bottom profile of previous surveys, the water levels of previous surveys, and the description and location of two benchmarks on each side of the river, as applicable.

Benchmarks: Points established by the surveyor on or close to the inspection site. They are used as a reference for determining elevations during the inspection.

Depth of cover: The vertical distance from the top of the pipe to the surface of the soil/water.

Navigable waterways: As defined by 33 CFR Part 329, Section 4, those waters that are subject to the ebb and flow of the tide and/or are presently used, or have been used in the past, or may be susceptible for use to transport interstate or foreign commerce.

Probe rod: A T-shaped hand tool used to locate and determine the depth of underground or submerged pipelines.

Submersible: A two-man minisubmarine.

As defined by 33 *CFR* Part 329, Section 4, navigable waterways are those waters that are currently used, or could be used, for commerce. Inspections are performed by persons who are qualified to carry out such inspections and to prepare permanent records of their findings. The inspections determine the pipeline location and any exposed pipe within the limits of the navigable waterway and the ROW immediately adjacent on each side. The permanent records must be of sufficient detail to compare with the previous crossing records to determine any change in the location of the pipeline. They must also detail the length and location of any exposed pipeline. The person performing the inspection prepares both plan and profile drawings detailing the findings of the inspection.

Under 49 *CFR* Part 195.412(b), all navigable waterway crossings for liquid pipelines must be inspected at least every five years. To ensure the integrity of the pipeline, waterway crossings may require inspections more frequently to determine the impact of flooding, upstream construction, or other activities. Inspections determine any changes in the horizontal alignment of the pipe and the river-bottom profile, verifying that the pipe has adequate cover.

Under 49 *CFR* Part 192.705, natural gas or other gas transmission line ROWs must be inspected periodically to ensure that they are being maintained in a safe manner. The frequency of inspections for gas pipelines is determined by several factors, including the size of the line, the operating pressures, the class location of the line, the terrain where the line is installed, and the weather conditions since the last inspection.

Some conditions, such as floods, significant increases in third-party construction or agricultural activities, or evidence or threats of vandalism, may warrant scheduling more frequent inspections. Some inspections occur more often, depending on the product and population density. The results of all patrols are reported and documented.

Navigable waterway crossing inspection includes verification of the pipeline location, surface inspection, and submarine inspection.

1.1.0 Locating the Pipeline

A pipeline is located by using existing drawings as well as probing/sonar equipment. Methods of locating the pipeline may depend on the circumstances, and may be accomplished with a combination of these methods.

1.1.1 Existing Drawings

Waterway inspections are performed with specialized equipment and techniques. The results are compared with existing drawings, such as alignment sheets and the **as-built river profile**.

Alignment sheets (*Figure 1*), also known as *strip maps*, show the pipeline route. They are used to help locate the pipeline by depth and in relation to surface features. Alignment sheets show the location and size of each pipe installed; all crossings of public roads, railroads, rivers and streams; and buried utilities and pipelines operated by other companies. These alignment sheets are maintained and updated throughout the life of the pipeline. Alignment sheets can vary greatly from company to company.

The as-built river profile includes the pipeline location, the river-bottom profile of previous surveys, the water levels of previous surveys, and the description and location of two **benchmarks** on each side of the waterway, as applicable. The person performing the inspection re-establishes the baseline indicated on the plan drawings and accurately determines the route of the pipeline through the river with reference to this line. The length of this traverse extends between convenient points on the banks that are above the water level.

The river bottom and banks between reference points are profiled over the pipe, with elevations referenced to the existing benchmarks shown on the drawings. If an existing benchmark has been destroyed, a new benchmark is set as permanently as possible. Profile data also includes the elevation of the water level and the top of pipe at all exposed and suspended sections. Vertical measurements for the profile may be made with a portable Fathometer or weighted sounding chain. Location of the pipeline may be assisted by visual examination of adjoining banks to search for exposed pipe, crossing signs, and benchmarks.

NOTE

Ensure that all maps, survey records, and drawings are up-to-date.

1.1.2 Probing / Sonar Equipment

Probing equipment used to locate the pipeline may include mechanical (line locator), induced-tone, or transmitter/receiver types. Specialized

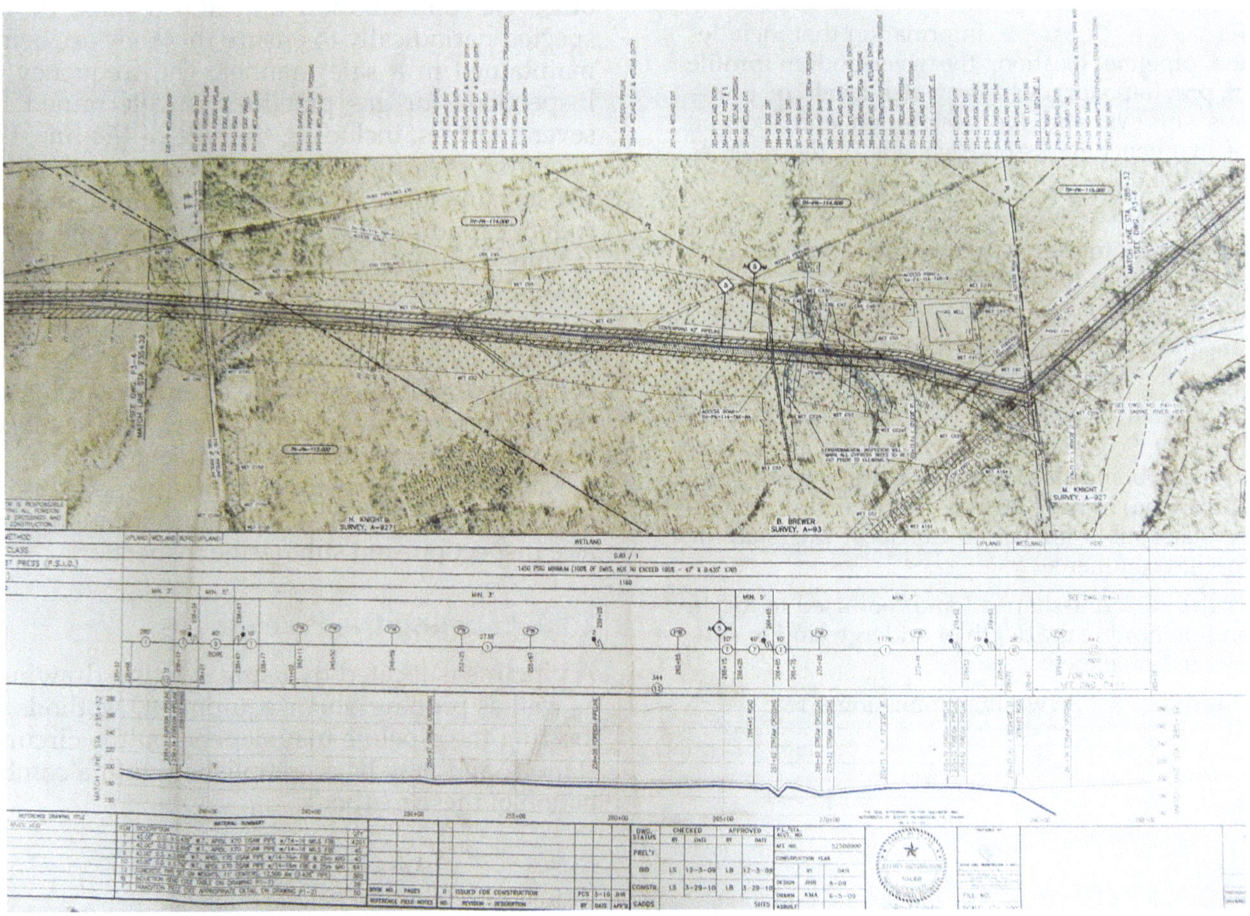

Figure 1 Sample alignment sheet.

sonar equipment transmits a sound signal and monitors the received signal. A **probe rod** (*Figure 2*) is used to physically probe the submerged pipeline to determine the **depth of cover** and to verify the location of the pipeline.

1.2.0 Inspection Types

The person performing the inspection checks the surface conditions on or adjacent to the pipeline ROW and the navigable waterway crossings and reports their observations. In cases other than emergencies, the report is written from the inspection log.

1.2.1 Adjacent Bank / ROW Inspection

Inspection of waterways includes inspection of adjacent banks and the ROW. Always be alert to indications of possible security issues, such as damaged fencing, gates, and locks. Report any unusual conditions or activity.

Any time pipeline problems, excavations, or repairs occur near a water source, the water must be inspected for any contamination by the pipeline product. Crude oil and other less-refined products produce a blackish sheen on water and form a consistent sheen on the surface, even after being disturbed.

1.2.2 Submarine Inspection

During submarine inspection, a diver passes over each pipe at intervals of 20 feet or less, and if exposed or suspended pipe is located, the diver follows these sections with continuous passes. The size and location of any buildup of debris against exposed or suspended pipe sections, the condition of pipe and/or pipe coating, the nature of the river bottom at the pipe, and any other underwater condition that may appear to adversely affect the safety of the pipeline crossing are also determined and reported. Two such underwater conditions are scouring of the river bottom and undercutting of the banks. Sonar equipment, instead of a diver, may be used to detect exposed pipe.

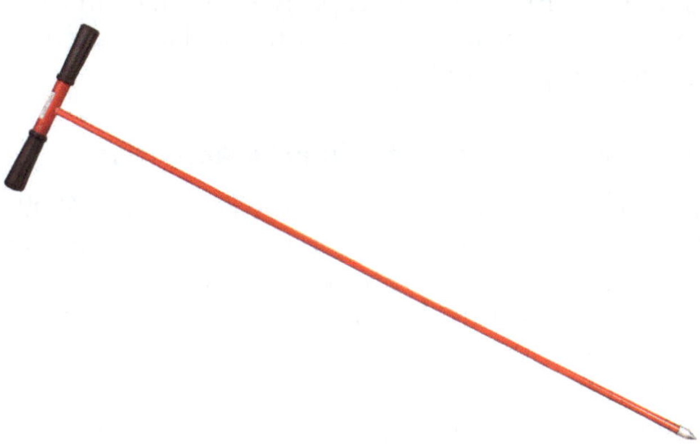

Figure 2 Probe rod.

A **submersible** is a device used for undersea inspection. This is essentially a two-person minisubmarine. The chief advantage of using a submersible instead of divers is that it allows personnel to dive deeper and for longer periods. The submersibles can also carry photographic and sonar equipment.

1.3.0 Reportable Observations

Conditions encountered during a waterway inspection that must be reported include the following:

- Emergency situations / evidence of release
- Environmental changes
- Damaged or exposed pipeline
- Missing or damaged signs/markers
- Encroachment / nearby activities
- Other unsafe conditions

1.3.1 Emergency Situations / Evidence of Release

The definition of an emergency situation is a leak, evidence of a possible leak, or activities that pose an immediate threat to the pipeline or to public safety. A possible leak observation might be petroleum on the ground, vapors in the air, wilted or dead vegetation over the pipeline, dead or sick animals near the pipeline, the smell of hydrocarbon, or petroleum sheen on water. If the leak stems from submerged pipe, it usually takes some time for the evidence to show on the surface. Any evidence of a leak requires an immediate verbal report, as well as written reports. Company policy will dictate response and repair.

> **WARNING!**
> If an unexpected hazardous liquid or carbon dioxide is encountered, local emergency personnel should be notified as quickly as possible. Additionally, the proper company notifications should be made, and the location of the leak should be identified, if possible.

1.3.2 Environmental Changes

Environmental changes may be expected due to local conditions. These types of changes include:

- Washouts over or along the pipeline
- Riverbank erosion
- Excessive vegetation along the right-of-way

- Creek or riverbed location changes
- Earth movement or subsidence
- Trees and debris clogging waterways

Soil erosion caused by water and wind and washouts caused by flood waters may expose or threaten to expose the pipeline. Preventive measures may need to be taken to control the erosion. These measures include terraces for surface erosion, fences for drifting sand, and bulkheads for stream and river bank erosion. Evidence of erosion must be reported so that these measures can be put in place before the pipe is exposed.

ROW policies limit or ban trees growing on the ROW. Any vegetation that grows near or on the ROW may hamper inspection or threaten to block access for maintenance and repair. Excessive vegetation growth must be reported per company policy.

1.3.3 Pipeline Damage or Exposure

Evidence of pipeline damage or leaks may include the following:

- Oil or product on a water surface adjacent to the pipeline
- Waste oil, salt ponds, or drainage
- Downed livestock
- Wilted or discolored vegetation
- Vapor cloud

Any oil, petroleum product, vapor, or other evidence of spills or discharges from the pipeline, main-line block valves, interface detectors, and other facilities must be reported at once.

Pipeline exposure may be caused by environmental conditions, such as weather or earth movements. Exposed pipe should be reported for inspection and reburial. Whenever exposed pipe is discovered, it must be inspected for evidence of external damage (CT5_1-17, CT5_3-17), corrosion (CT5_2-17), and the extent of the damage or corrosion. Corrective action is taken according to company policy. *Figure 3* shows exposed pipe.

Supports are required on various aboveground pipeline structures to provide protection from the forces exerted by anticipated loads. *Figure 4* shows pipe on supports. Supports may be found at overhead crossings, spans over ditches, pig traps, block valves, manifold valves and piping, and other structures. To be effective, these supports must be undamaged and in the correct position. Damaged or missing supports must be reported and repaired according to company policy. Some support problem situations may be emergencies if there is a threat of imminent collapse. These situations may require temporary supports to be installed and/or evacuation of the area. The cause of the damage should be investigated; it may have occurred naturally, resulted from vandalism, or resulted from third-party activity.

Figure 3 Exposed pipe.

The pipeline structure that should be supported must also be examined for damage. There may be evidence of unusual stress on the pipe or structure, or it may have been damaged by whatever caused damage to the support. Some evidence to look for on the pipe or structure includes wrinkling, buckling, dents, and gouges. These types of damage must also be reported and repaired.

1.3.4 Missing or Damaged Signs/Markers

Signs and/or markers are required at all waterway crossings and at sufficient intervals along the ROW to identify the pipeline location. When possible, they should be directly over the pipeline, securely installed, and clearly legible. The information on the signs or markers must be up-to-date and correct. Aerial markers may also be present for use by an **aerial patrol**. Marker locations may be indicated on company maps.

Signs may become damaged by weather or other natural conditions, or they may be vandalized or stolen. During inspection patrols, any damaged, missing, improperly located, or inaccurate sign must be reported and replaced according to company policy.

1.3.5 Encroachment / Nearby Activities

Encroachment is activity on the company's ROW by outside parties. A particular encroachment may or may not be authorized, so assess each situation carefully. Unauthorized activities on the pipeline ROW that may affect the pipeline are often spotted by patrols doing line inspection and are reported to management. The unauthorized

Figure 4 Pipe on supports.

encroaching party is advised to cease activity until consent can be given by the operator. The operator then investigates the activity and communicates with the encroaching party about a final resolution.

Evidence of activity around a block valve or any other place along the pipeline ROW might indicate vandalism, theft of product, or sabotage. If an immediate danger is posed by such activities, they must be stopped, and an emergency call must be made to the control center.

Foreign crossings are reviewed and approved. Specific company procedures will apply to the crossing to ensure that the company pipeline is not damaged or compromised.

Reportable activities include any that might cause hazards or restrictions to the safety, integrity, and operation of the pipelines and easements. Besides encroachment, other activities that may threaten the pipeline security include:

- Blasting or surface mining within one-quarter mile of the pipeline
- Excavations within the ROW or excavations threatening the pipeline
- Clearing of timberland over, across, or along pipelines
- Building of terraces, water ponds, or pits within an area defined by company policy
- Construction of buildings, well derricks, residences, fences, or other structures within an area defined by company policy
- Brush or tree clearing, excavation, or ditching for new roads, utilities, or structures
- Grading, construction, or resurfacing of roads
- River dredging or watercraft anchored over river crossings
- Fires occurring within an area defined by company policy
- Agricultural deep-ripping or subsoil plowing

1.3.6 Other Unsafe Conditions

Other unsafe conditions that may be encountered along the ROW must also be reported. These include, but are not limited to, localized or general corrosion, pipe movement caused by environmental causes, material defects or physical damage, and equipment malfunction.

1.4.0 Reporting Procedures

Federal regulations and company policies require that records of navigable waterway crossings be made and maintained. Written inspection reports must be completed after each inspection, and additional reports must be made after any further investigation is made, as applicable. If an emergency situation is discovered, verbal reports must be made immediately to the appropriate personnel. The inspection records provided by the inspector must be retained for at least five years, or longer if required by company policy. Inspection results are compared with existing records to determine if any changes have occurred and if such changes require repairs. If any actions are required, they must be completed as quickly as possible.

Proper reporting procedures include completing the following documents:

- Written inspection reports
- Emergency and verbal inspection reports, if applicable
- ROW investigation reports, if applicable

1.4.1 Written Inspection Reports

The person performing the inspection must make a written report of the inspection as soon as practical following the inspection, per company policy. The report is then distributed to all relevant offices. The report usually includes the following information:

- Date
- Patroller's name
- Reason for inspection
- Location of observation
- Time of observation
- Identification of pipeline(s), section, or segment
- Pipe and material conditions
- Defect, damage, or corrosion observed, if applicable
- Report number
- Any actions taken
- Notification of any verbal reports made

The inspection and any actions are recorded on a pipeline inspection report form. A sketch of the damaged pipe is also required. The form is dated and signed by a person performing the inspection.

For navigable waterway crossings, reports of the survey and inspection are forwarded in individual written reports for each crossing site. Reports are submitted in triplicate to management and contain the following:

- A list of all exposed and suspended pipe sections, located by station numbers and showing clearance between the pipe and river bottom
- Identification of changes from the last inspection
- Plan and profile sheets of each pipeline surveyed at the crossing site (one used for each pipeline)

1.4.2 Emergency and Verbal Reports

An emergency situation, such as a possible leak, requires the person performing the inspection to make an immediate cell telephone or radio telephone call to the appropriate contact. If the situation involves excavation or other activities performed by construction workers on the ground, the person performing the inspection should first warn the workers. After the first call, a second call is made to the company personnel designated to receive such calls. The verbal report is followed up with a written report.

Possible leak observations include petroleum product, vapor, or other discharge; dead vegetation over the pipeline; dead livestock or other dead animals near the pipeline; or petroleum sheen on nearby water. Emergency observations include activities or circumstances, such as construction activity, that may pose a threat to the integrity of the pipeline or to public safety.

Emergency verbal reports should include the following information:

- Patroller's name
- Identification of pipeline(s), section, or segment
- Location of the observation referenced to landmarks, geography, or city
- Global Positioning System (GPS) location
- Description of the observation
- Indication that the observation is deemed an emergency

WARNING!

In case of a fire or explosion caused by the ignition of hydrocarbons, local emergency personnel should be notified immediately. Access to the area should be limited as much as possible, and the proper company notifications should be made.

1.4.3 ROW Investigation Reports

If the initial inspection requires further investigation, the appropriate personnel are dispatched to investigate the reported conditions and to complete ROW investigation reports according to company policy. Investigation reports provide further detail about the condition and the result of the investigation and any action taken. In addition, the investigator takes photos or creates drawings to show unusual conditions. These ROW investigation reports are then filed according to federal regulations.

Additional Resources

It is assumed that you will learn to perform these procedures under the direct supervision of your crew leader, foreman, or instructor. For independent study, refer to the recommended Additional Resources at the back of this module.

1.0.0 Section Review

1. Federal regulations require inspection of navigable waterway crossings at least every _____.
   a. six months
   b. year
   c. three years
   d. five years

2. Alignment sheets are updated and maintained _____.
   a. for the life of the pipeline
   b. until the next inspection
   c. for one year
   d. for five years

3. During submarine inspection, a diver passes over each pipe at intervals of _____.
   a. 20 feet or less
   b. 40 feet or less
   c. 100 feet or less
   d. 200 feet or less

4. Observed activities that pose an immediate threat to the pipeline or public safety are classified as _____.
   a. intrusions
   b. critical situations
   c. emergencies
   d. reportable incidents

5. If patrollers encounter an unauthorized encroaching party, that party must be _____.
   a. reported to the police as trespassers
   b. advised to cease their activities
   c. photographed and fingerprinted
   d. escorted off the premises

6. Emergency situations must be reported _____.
   a. to local emergency services
   b. and documented with photographs
   c. and immediately acted upon by the inspector
   d. verbally to appropriate personnel

Section Two

2.0.0 Inspecting Navigable Waterway Crossings

Objective

Describe the procedure used to inspect or simulate inspecting a navigable waterway crossing (CT16_1-17).

a. Identify the prerequisites, competencies, and abnormal operating conditions associated with inspecting a navigable waterway crossing.
b. Describe the general procedure used to inspect a navigable waterway crossing.

Performance Task

1. Inspect or simulate inspecting a navigable waterway crossing (CT16_1-17).
   - Identify potential abnormal operating conditions that may occur during performance of this CT, and know the appropriate actions to take in response to them.
   - Utilize the appropriate personal protective equipment according to relevant company procedures.
   - Locate the landfall of the pipeline crossing to establish a reference point.
   - Determine the amount of cover by using a probe or specialized equipment to verify the depth of the submerged pipeline.
   - Document the findings as required by operator's procedures and 49 *CFR* 195.404.

Trade Term

Abnormal operating conditions (AOCs): Conditions within a pipeline system, identified by the operator, that may indicate (1) a malfunction of a component or (2) a deviation from normal operations that may indicate a condition exceeding design limits or result in a hazard to persons, property, or the environment.

Pipelines must be examined at regular intervals to identify any issues that may lead to eventual system failure. This section describes some of the basic skills needed by pipeline maintenance technicians and identifies **abnormal operating conditions (AOCs)** that may occur during navigable waterway inspections. It also describes general procedures for examining a submerged pipeline.

2.1.0 Prerequisites, Competencies, and Abnormal Operating Conditions

Personnel must possess a variety of skills in order to successfully inspect for damage around and on submerged pipelines. The various operations require a series of steps, which personnel must be competent to perform. Personnel must also be aware of the particular abnormal operating conditions that may occur during the procedure and know the proper responses to them.

2.1.1 Prerequisites and Competencies

Personnel must possess the following basic knowledge and skills prior to inspecting a navigable waterway crossing:

- Be able to understand and interpret pipeline routes , as-built drawings, and alignment sheets
- Be competent with specialized locating equipment to determine depth of cover
- Be familiar with inland waterway locations

Personnel must also be familiar with the company safety, evacuation, and response procedures, as well as the proper use of personal protective equipment (PPE).

2.1.2 Abnormal Operating Conditions

Personnel inspecting a navigable waterway crossing must be able to identify potential abnormal operating conditions (AOCs) and make proper operator notifications in response. The following are examples of AOCs that may be encountered:

- *Vessel anchored over the pipeline*
- *Visible sheen or other indication of product release*
- *Waterway bank erosion*
- *Pipe movement or suspension of pipeline*

2.2.0 Procedure

This section details a basic procedure of how to inspect a navigable waterway crossing. Remember that these steps are very general. Company-specified procedures for pipeline inspection must be followed at all times. If you have any questions, contact your supervisor.

The following steps outline the general procedure for inspecting a navigable waterway crossing:

WARNING!

Always obtain and use the appropriate personal protective equipment (PPE) for each task. Stay alert to the possibility of unexpected leaks from the exposed pipeline, such as hazardous liquids or gases. In the event of a leak, safely eliminate all ignition sources and immediately notify the appropriate personnel.

CAUTION

Throughout this procedure, you must identify any potential abnormal operating conditions (AOCs) and initiate the appropriate response.

Step 1 Identify potential abnormal operating conditions that may occur, and know the appropriate actions to take in response to them.

Step 2 Utilize the appropriate personal protective equipment according to relevant company procedures.

Step 3 Locate the landfall of the pipeline crossing to establish a reference point. Ensure that permanent horizontal and vertical control monuments are established on both banks. Accurately measure and record the pipeline horizontal and vertical alignment, depths of cover, and waterway profile and alignment.

Step 4 Determine the amount of cover and verify the location of the pipeline using the appropriate probing equipment, such as a probe rod or sonar technology. Probe for the location of the river bottom and the top of the pipe.

Step 5 Document the findings for comparison to previous inspections, including location of exposed or unsupported pipe, depth of cover, pipeline damage, and debris hung up on the pipeline. Detail the condition of the crossing, any remedial action taken, and problems that may need additional consideration.

WARNING!

Make the proper company notifications if unsupported submerged pipe is discovered.

Step 6 Determine if additional underwater diver inspections are required.

Step 7 Complete all required documentation according to company procedure.

Additional Resources

It is assumed that you will learn to perform these procedures under the direct supervision of your crew leader, foreman, or instructor. For independent study, refer to the recommended Additional Resources at the back of this module.

2.0.0 Section Review

1. If debris is found lodged against the pipeline, the appropriate response is to _____.
   a. remove it
   b. write up a report
   c. make proper operator notifications immediately
   d. submit a sketch of the location

2. When inspecting a navigable waterway crossing, the first step is to _____.
   a. measure the horizontal and vertical alignment
   b. verify the depth of cover
   c. establish a reference point
   d. record the waterway profile

SUMMARY

As defined by 33 *CFR* Part 329, Section 4, navigable waterways are those waters that are used, or could be used, for commerce. Navigable waterway crossings for liquid pipelines must be inspected at least every five years. Inspections for gas pipelines must be performed periodically, but can vary according to different factors.

The person performing the inspection must check for conditions that include emergency situations/evidence of release, environmental changes, damaged or exposed pipeline, missing or damaged signs/markers, encroachment/nearby activities, and other unsafe conditions.

Federal regulations require that records of navigable waterway crossing inspections be made and maintained. Written reports of each inspection must be submitted per company policy. These reports are compared with existing records to determine if any changes have occurred and if such changes warrant repairs.

Review Questions

1. The locations of two benchmarks on each side of the river may be shown on the _____.
   a. alignment sheet
   b. right-of-way diagram
   c. waterway navigation chart
   d. as-built river profile

2. Vertical measurements of the underwater pipeline location can be made with _____.
   a. probe/sonar equipment
   b. a weighted sounding chain
   c. surveying equipment
   d. an aluminum folding rule

3. A probe rod is used to identify the pipeline location and determine _____.
   a. the depth of cover
   b. composition of the riverbed
   c. the location of nearby utility lines
   d. the slope of the pipeline

4. The operation of a sonar system is based on _____.
   a. electronic pulses
   b. flashes of light
   c. sound signals
   d. voltage fluctuations

5. The task of inspecting waterways also includes the inspection of _____.
   a. marine life
   b. tidal influences
   c. water vessels nearby
   d. the adjacent banks and ROW

6. Evidence of an oil pipeline leak is a sheen on water that is _____.
   a. iridescent
   b. blackish
   c. reddish brown
   d. yellow or green

7. Riverbank erosion that could endanger a pipeline is often controlled by installing _____.
   a. flumes
   b. diversion structures
   c. bulkheads
   d. wing dams

8. The information on pipeline signs or markers must be _____.
   a. up-to-date and correct
   b. readable from a distance of 10 feet
   c. in large, red type
   d. in both English and Spanish

9. Inspection results are compared with existing records _____.
   a. to verify their accuracy
   b. to determine if any changes have occurred
   c. to compare riverbank erosion rates
   d. to prevent unnecessary duplication

10. A written inspection report usually includes _____.
    a. the date and time the report was written
    b. serial number of the vehicle used
    c. mileage driven during the inspection
    d. a record of any verbal reports made

Trade Terms Quiz

Fill in the blank with the correct term that you learned from your study of this module.

1. The vertical distance from the top of the pipe to the surface of the soil or water is called the _________.
2. Conditions within a pipeline system, identified by the operator, that may indicate (1) a malfunction of a component or (2) a deviation from normal operations that may indicate a condition exceeding design limits or result in a hazard to persons, property, or the environment are called _________.
3. Waters that are subject to the ebb and flow of the tide and/or are presently used, or have been used in the past, or may be susceptible for use to transport interstate or foreign commerce are defined as _________.
4. Points established by the surveyor on or close to the inspection site used as a reference for determining elevations during the inspection are called _________.
5. The most common form of information gathering performed from an overhead position to report developments or changes on, and adjacent to, the right-of-way is called a(n) _________.
6. A two-man minisubmarine is called a(n) _________.
7. A T-shaped hand tool used to locate and determine the depth of underground or submerged pipelines is called a(n) _________.
8. Information that includes the pipeline location, the river-bottom profile of previous surveys, the water levels of previous surveys, and the description and location of two benchmarks on each side of the river as applicable is called the _________.

Trade Terms

Abnormal operating conditions (AOCs)
Aerial patrol
As-built river profile
Benchmarks
Depth of cover
Navigable waterways
Probe rod
Submersible

Trade Terms Introduced in This Module

Abnormal operating conditions (AOCs): Conditions within a pipeline system, identified by the operator, that may indicate (1) a malfunction of a component or (2) a deviation from normal operations that may indicate a condition exceeding design limits or result in a hazard to persons, property, or the environment.

Aerial patrol: The most common form of information gathering performed from an overhead position on a weekly or biweekly schedule to report developments or changes on, and adjacent to, the right-of-way.

As-built river profile: Information that includes the pipeline location, the river-bottom profile of previous surveys, the water levels of previous surveys, and the description and location of two benchmarks on each side of the river, as applicable.

Benchmarks: Points established by the surveyor on or close to the inspection site. They are used as a reference for determining elevations during the inspection.

Depth of cover: The vertical distance from the top of the pipe to the surface of the soil/water.

Navigable waterways: As defined by 33 CFR Part 329, Section 4, those waters that are subject to the ebb and flow of the tide and/or are presently used, or have been used in the past, or may be susceptible for use to transport interstate or foreign commerce.

Probe rod: A T-shaped hand tool used to locate and determine the depth of underground or submerged pipelines.

Submersible: A two-man minisubmarine.

Additional Resources

The following recommended resources can provide additional helpful information related to the requirements for this covered task:

API Recommended Practice 1161, Third Edition, © 2014 American Petroleum Institute. API Publishing Services, 1220 L Street, NW, Washington, DC 20005. Distributed by Thompson Reuters (Scientific) LLC, **www.techstreet.com**.

33*CFR* 329.4, **www.ecfr.gov.**

49 *CFR* 195.412, **www.ecfr.gov.**

49 *CFR* 192.705, **www.ecfr.gov.**

NCCER module AOCCC-17, *Abnormal Operating Conditions Control Center*

NCCER module AOCFG-17, *Abnormal Operating Conditions Field and Gas*

NCCER module CT5_1-17, *Examine for Mechanical Damage on Buried or Submerged Pipe*

NCCER module CT5_2-17, *Examine for External Corrosion on Buried or Submerged Pipe*

NCCER module CT5_3-17, *Inspect the Condition of External Coating on Buried or Submerged Pipe*

NCCER module CT14_1-17, *Locate Line*

The following websites provide resources for products and training:

US Army Corps of Engineers, **www.usace.army.mil**

US Coast Guard, **www.uscg.mil**

Figure Credits

Robert Schwemmer, Module Opener

Cameron Strother, Southland Safety, Figure 1

Courtesy of Humboldt Mfg. Co., Figure 2

U.S. Air Force PHOTO by AIRMAN First Class Kyle R. Fidel, Figure 3

U.S. Geological Survey/photo by Don Becker, Figure 4

Section Review Answer Key

Answer	Section Reference	Objective
Section One		
1. d	1.0.0	1
2. a	1.1.1	1a
3. a	1.2.2	1b
4. c	1.3.1	1c
5. b	1.3.5	1c
6. d	1.4.0	1d
Section Two		
1. c	2.1.2	2a
2. c	2.2.0	2b

NCCER CURRICULA — USER UPDATE

NCCER makes every effort to keep its textbooks up-to-date and free of technical errors. We appreciate your help in this process. If you find an error, a typographical mistake, or an inaccuracy in NCCER's curricula, please fill out this form (or a photocopy), or complete the online form at **www.nccer.org/olf**. Be sure to include the exact module ID number, page number, a detailed description, and your recommended correction. Your input will be brought to the attention of the Authoring Team. Thank you for your assistance.

Instructors – If you have an idea for improving this textbook, or have found that additional materials were necessary to teach this module effectively, please let us know so that we may present your suggestions to the Authoring Team.

NCCER Product Development and Revision

13614 Progress Blvd., Alachua, FL 32615

Email: curriculum@nccer.org
Online: www.nccer.org/olf

❑ Trainee Guide ❑ Lesson Plans ❑ Exam ❑ PowerPoints Other ____________________

Craft / Level: Copyright Date:

Module ID Number / Title:

Section Number(s):

Description:

Recommended Correction:

Your Name:

Address:

Email: Phone:

CT27_1-17

Routine Inspection of Breakout Tanks (API 653 Monthly or DOT Annual)

Overview

Periodic inspection of all pipeline systems and facilities is necessary for maintenance and is required by Pipeline Hazardous Materials and Safety Administration (PHMSA) regulations. Routine inspections of breakout tanks are necessary to ensure that the tanks are kept up to safety standards and to discourage vandalism and crime. This module describes the routine inspection of breakout tanks that must be performed on a monthly or annual basis.

Trainees with successful module completions may be eligible for credentialing through the NCCER Registry. To learn more, go to **www.nccer.org** or contact us at 1.888.622.3720. Our website has information on the latest product releases and training, as well as online versions of our Cornerstone magazine and Pearson's product catalog.

Your feedback is welcome. You may email your comments to **curriculum@nccer.org**, send general comments and inquiries to **info@nccer.org**, or fill in the User Update form at the back of this module.

This information is general in nature and intended for training purposes only. Actual performance of activities described in this manual requires compliance with all applicable operating, service, maintenance, and safety procedures under the direction of qualified personnel. References in this manual to patented or proprietary devices do not constitute a recommendation of their use.

CT27_1-17

Routine Inspection of Breakout Tanks (API 653 Monthly or DOT Annual)

Objectives

When you have completed this module, you will be able to do the following:

1. Identify the purpose of breakout tanks, and describe breakout tank inspections.
   a. Identify the purpose of a breakout tank.
   b. Describe the conditions that are examined during a tank inspection.
2. Describe the requirements and procedures for performing breakout tank inspections (CT27_1-17).
   a. Identify the prerequisites, competencies, and abnormal operating conditions associated with inspecting breakout tanks.
   b. Describe the procedures used to perform a breakout tank inspection.

Performance Task

Under the supervision of your instructor, you should be able to do the following:

1. Perform routine inspection of a breakout tank (API 653 monthly or DOT annual) (CT27_1-17).
   - Identify potential abnormal operating conditions (AOCs) that may occur during performance of this CT, and know the appropriate actions to take in response to them.
   - Utilize the appropriate personal protective equipment according to relevant company procedures.
   - Perform a walk-around inspection to verify security and to determine that no AOCs exist.
   - Visually inspect the foundation for defects.
   - Visually inspect the shell for defects.
   - Visually inspect shell appurtenances for defects, including manways, nozzles, tank piping manifolds, and the autogauge system.
   - Visually inspect roofs for defects, including deck plate external corrosion, roof deck drainage, floating roof seal systems, and pontoons.
   - Visually inspect roof and tank shell appurtenances for defects, including the sample hatch, autogauge inspection hatch, autogauge float well cover, external floating roof ladder, and the stairway.
   - Inspect the area for buildup of trash, vegetation, or other flammable material that could pose a fire or other safety hazard.
   - Complete appropriate documentation as required by operator's procedures.

Trade Terms

Abnormal operating conditions (AOCs)
Breakout tank
Deck legs
Geodesic dome
Guide pole
Overflow drain
Pontoons
Primary and secondary seals
Rim vent
Roof drains
Sketch plate
Vacuum breakers
Wind girders

Industry-Recognized Credentials

If you're training through an NCCER-accredited sponsor, you may be eligible for credentials from NCCER's Registry. The ID number for this module is CT27_1-17. Note that this module may have been used in other NCCER curricula and may apply to other level completions. Contact NCCER's Registry at 888.622.3720 or go to **www.nccer.org** for more information.

Contents

Figures

Section One

1.0.0 Breakout Tank Conditions

Objective

Identify the purpose of breakout tanks, and describe breakout tank inspections.

a. Identify the purpose of a breakout tank.
b. Describe the conditions that are examined during a tank inspection.

Trade Terms

Abnormal operating conditions (AOCs): Conditions within a pipeline system, identified by the operator, that may indicate (1) a malfunction of a component or (2) a deviation from normal operations that may indicate a condition exceeding design limits or result in a hazard to persons, property, or the environment.

Breakout tank: A tank used to relieve surges in a hazardous liquid pipeline system and/or to receive and store hazardous liquid transported by a pipeline for reinjection and continued transportation by the pipeline.

Deck legs: Steel pipes mounted on the floating roof that provide support for the roof when the liquid is drained below a set level. They are typically adjustable to allow for a change in supported roof height from the bottom of the tank and are often referred to as high or low legs.

Geodesic dome: A self-supporting dome roof, typically made of aluminum, which is intended to prevent rainwater from entering the tank.

Guide pole: A vertical pole device that extends through the floating roof of a tank to maintain the roof in a centered position and to prevent the roof from rotating.

Overflow drain: A nozzle on the shell of a tank located approximately one foot below the top of the tank. They are always larger than tank inlet nozzles.

Pontoons: External or internal compartments on a floating roof that provide buoyancy, allowing the roof to float.

Primary and secondary seals: Flexible sealing devices mounted on the floating roof of a tank to provide a reasonably close fit between the roof and the internal wall of the tank.

Rim vent: A mechanical vent mounted on the floating roof of a tank that vents the vapor space between the primary and secondary rim seals.

Roof drains: Steel pipes or hoses mounted on a floating roof to drain rainwater.

Sketch plate: The bottom plate on which a tank shell rests.

Vacuum breakers: Valves mounted on top of the tank roof to prevent a vacuum under the roof.

Wind girders: Stiffening rings mounted on open-top tanks to maintain shell roundness against wind loads.

Tanks hold enormous amounts of petroleum products. A **breakout tank** is a tank used to relieve surges in a hazardous liquid pipeline system and/or to receive and store hazardous liquid transported by a pipeline for reinjection and continued transportation by the pipeline. If a breakout tank fails, catastrophic damage can occur. Therefore, it is critical that tanks be inspected carefully and regularly.

Tanks are containers used for liquid storage, relief, and breakout. Tanks may exist along the length of the pipeline, at the production site, at pump stations, at terminals, and at delivery points. Although the types and sizes of tanks at each of these sites may vary, they are usually round and made of steel. All atmospheric storage tanks are constructed per *API Standard 650* and repaired per *API Standard 653.*

Tanks must be calibrated for capacity. The dimensions are strapped, or measured, to be used in strapping tables. From these tables, the volume of liquid contained in the tank can be determined. Tank gauges show the tank number and gauge reading from which the corresponding volume in barrels can be determined.

1.1.0 Excessive Pressure

Pressure is the force that a fluid (gas or liquid) exerts over a particular area. For example, within a tank, the fluid exerts a force that acts on the tank's inner surface. If this force becomes too great, it can exceed the material's strength, rupturing the tank.

Tanks can be damaged or even fail due to internal overpressure. A sudden temperature rise caused by a heat wave could cause the stored product to expand, making the tank bulge. A routine task like filling the tank could then destroy it. The rising product level within the tank acts like a piston in a cylinder, compressing the air and product vapor at the top. Eventually, this pressure could reach the point that the tank might rupture, resulting in product release, environmental damage, and possibly an explosion.

Tanks can also fail due to reduced pressure—a vacuum—within the tank. For example, pumping out a tank could create sufficient vacuum to cause the tank to collapse inwardly (implode). The same effect could happen naturally when a tank has heated up in the hot sun and then been rapidly cooled by a cold rain. The sudden cooling causes an abrupt drop in the internal pressure that could lead to collapse.

1.1.1 Managing Pressure

The best way to prevent pressure damage is to remove or reduce the pressure difference. This is how many protective devices deal with overpressure conditions. If a tank has excessive internal pressure, the device releases product or trapped air and vapor from the tank, reducing the pressure difference. If a tank has a vacuum inside, the protective device lets air into the tank from the outside.

A closely related strategy relies on control rather than relief. Sometimes pressure buildup happens when product is flowing too fast for the system to handle. A control system that monitors pressure and adjusts flow can keep the pressure from growing to dangerous levels.

1.2.0 Tank Conditions to Inspect

Breakout tanks (*Figure 1*) are inspected externally on monthly or annual schedules, and any problems are recorded and corrected as necessary. The tank is inspected internally when practical.

Inspection results are recorded on a tank inspection report form. If an internal inspection is done, this should be noted on the inspection form, as well. Required information for the form includes the location of the tank and a description of the condition of each of the physical parts of the tank and all associated equipment. Repairs that are required and/or made must also be described. The form is dated and signed by the inspector. *Figure 2* shows a sample tank inspection report form.

Figure 1 Breakout tanks.

1.2.1 Surrounding Area

A routine walk-around inspection ensures that the proper security is in place and that no signs of leaks or other **abnormal operating conditions (AOCs)** exist. Verifying security is an important part of any inspection. Tanks must be kept secure at all times to protect against unauthorized access or vandalism. All aboveground equipment may be fenced or protected by guardrails. Locks are installed on all gates and facility fences, limiting access of unauthorized persons. Company signs with emergency numbers must be displayed at the tank location, and the site must be free of excessive vegetation that could impede access and/or trash and flammable materials that could pose a fire hazard.

Before performing a routine walk-around inspection, notify the control center, and then verify the security and accessibility of the tank, the tank number, capacity, and product contained within. In addition, be sure to check the secondary containment device, such as a dike, that is designed to be liquid-tight and contain any product release.

1.2.2 Foundation

Due to the massive weight exerted on the ground from a breakout tank it's important to identify any potential problems with the land and structures upon which the tank rests. The force of the tank can possibly cause some settling of the structure, similar to the settling of a house over time. This action can result in an uneven foundation which could prevent proper operation and drainage. Corrosion at the bottom and on the underside of a tank can also lead to hazardous leaks, which should be immediately corrected.

Any settlement noted in the foundation of a tank may indicate active ground movement or insufficient sub-grade strength. Continued settlement could result in damage to the tank shell or could result in tank failure. Document the settlement location, and estimate the magnitude of the settling or displacement and the impact, if any, on the tank shell or equipment. Photographs are recommended to document the progression of settlement over time. Report the condition to the appropriate personnel for further inspection or repair.

MONTHLY TANK INSPECTION REPORT

Date:	Company:	Station:
Tank #:	Capacity:	Inspector:
Location:		
Tank Dimensions		

TANK INFORMATION		Yes	No	N/A	Remarks
Tank Construction:	Welded				
	Bolted				
Tank Type:	Fixed / Coned				
	IFR				
	EFR				
Foundation Type:	Asphalt				
	Sand Base				
	Concrete / Ringwall				
	Tank Band				
	Crushed Rock				
Tank Foundation / Bottom Observations:	Signs of Bottom Leakage				
	Is the shell to bottom joint visible				
	Signs of Settlement around the Perimeter Tank				
	Is the foundation damaged				
Tank Shell Observations:	Shell Coating failure / pitting / corrosion				
	Shell to bottom weld area have corrosion defects or thinning				
	Signs of shell leakage / corrosion				
	Tank shell insulated				
	Condition of Insulation				
	Shell Distortions				
Tank Piping / Appurtenances:	Stress,strain, or shifted piping				
	Signs of corrosion, defects around manways / nozzles / attachments				
	Signs of flange leaks and leaks around bolts				
	Signs of leaks, thermal relief , damage on piping and valves				
	Are gauging devices operational				
	Is there adequate reinforcement around nozzles and appurtenances				
	Does the mixer have adequate supports				
	Does the mixer shows signs of leakage or defects				
Fixed Roof Inspection	Is the fixed roof insulated				
	Is the roof insulation in good condition				
	Are there low spots on the fixed roof				
	Does water accumulate on the roof / insulation				
	Are roof nozzles / manways in good condition				
	Roof coating failure				

Comments:

Figure 2 Tank inspection report form.

When inspecting the foundation, look for the following conditions:

- Broken concrete, spalling, and cracks
- Indications of bottom leakage from the drain openings in the foundation ring, the back of waterdraw basins, the perimeter of the sketch plate, and the top surface of the chime ring
- Cavities under the foundation
- Vegetation against the bottom of the tank
- Problems with rainwater not draining from the shell away from the tank
- Settlement around the perimeter of the tank
- Bonding cables

1.2.3 Shell

If the foundation of a tank helps support the weight of the structure, the shell is the means by which the product is contained to that one particular location. Damage to any portion of the shell can result in a failure point caused by leaks, paint failure, pitting, and corrosion. The anomalies should be documented by sketch or photograph to ensure repair occurs in the correct location. Visually inspect the shell for the following conditions:

- Rivet and seam leaks
- Corrosion damage on the wind girder and handrail, especially at tack-welded junctions
- Broken welds
- Leaks around the telltale/weep hole

1.2.4 Shell Appurtenances

An appurtenance is an object that functions as an accessory item in conjunction with a particular object. A shell appurtenance refers to the components attached to the outside of a tank, such as connection points, valves, and other gauges that are used to access and monitor the product inside. Leaks, component failure, and corrosion can weaken the support around these objects and cause a tank failure. When inspecting the shell appurtenances, look for the following conditions:

- Cracks or signs of leakage on weld joints at nozzles, manways, and reinforcing plates
- Flange leaks and leaks around bolting
- Leaks from manifold piping, flanges, and valves
- Absence of, or problems with, fire-fighting system components
- Leaks from sample connections
- Valve operation problems
- Leaks from the autogauge tape guide and lower sheave housing
- A damaged autogauge head
- Problems with the autogauge head checker tape movement
- Lack of freedom of movement of the marker and float
- Leaks from the **overflow drain**

WARNING!

A component or equipment failure could pose a safety or environmental risk to personnel, the public, or the property adjacent to a breakout tank. If it is safe to do so, close, isolate, or remove the failed component from operation. Immediately notify control center personnel and other personnel per company policy. Document the condition for further inspection or repair. If the failed component presents an imminent hazard condition, evacuate the area immediately, and notify the emergency response team.

1.2.5 Roof

Cylindrical tanks may have flat external/opentop floating roofs, fixed cone-shaped roofs, or both. The first type of roof can be used for gasolines and aviation-grade kerosenes, while the second type can be used for fuel oils. Some tanks have a fixed **geodesic dome** covering a floating pan. Spheroid tanks are used for volatile products because of their high vapor-pressure capacity. A third type of tank structure is one in which the outside cone roof tank covers a floating roof that raises and lowers according to the amount of product inside the tank.

The floating roof tank (*Figure 3*) is designed to reduce losses through evaporation of volatile components of crude oil and to reduce the corrosion from sour crude by eliminating the vapor space between the oil and the roof. Seals around the perimeter of the roof prevent vapor loss and rain contamination. The wind girder on the outside rim of the tank adds strength to the shell.

When inspecting a tank roof, look for the following conditions:

- Paint failure, holes, pitting, and corrosion on the roof deck, **guide pole,** and **roof drains**.
- Indications of standing water or product caused by inadequate drainage, an unlevel roof, or leaking **pontoons**
- Sagging of the fixed-roof deck indicating potential rafter failure
- Deterioration, holes, tears, and cracks in the fabric of the roof seal system, including around the **deck legs**, and **rim vent**
- Openings in the **primary and secondary seals** that would permit vapor emissions
- Protruding bolt or rivet heads against the shell
- Incorrect positioning of the pontoon covers on the external roof

1.2.6 Roof Appurtenances

Inspection of the components mounted to or on the roof is also necessary to ensure proper operation of the tank and the safety of personnel performing maintenance. Corrosion and obstructions can cause failures leading to a release of product. When conducting an inspection of roof appurtenances, look for the following conditions:

- Corrosion or other problems on the sample hatch cover
- Corrosion or missing bolts on the autogauge inspection hatch
- Corrosion on the autogauge float well cover
- Obstructions of the wheels and runners that could cause the ladder to get off track and damage the roof
- Significant damage to and cracked or broken welds in stair treads, handrails, and supports
- Corrosion or other problems around the **vacuum breakers**

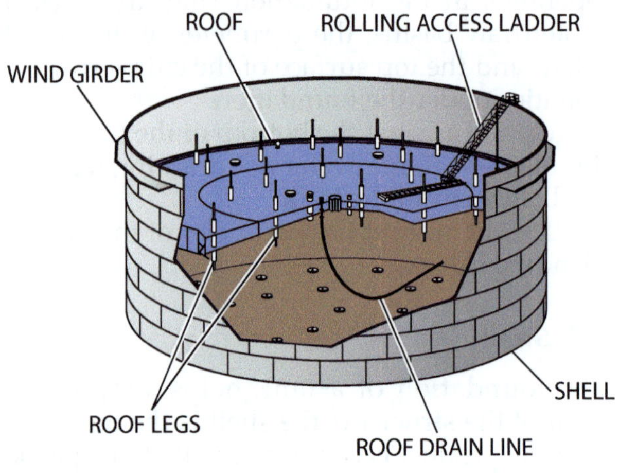

Figure 3 Open floating roof tank.

Additional Resources

It is assumed that you will learn to perform these procedures under the direct supervision of your crew leader, foreman, or instructor. For independent study, refer to the recommended Additional Resources at the back of this module.

1.0.0 Section Review

1. A hot storage tank that's rapidly cooled could _____.
   a. explode
   b. bulge
   c. implode
   d. surge

2. Looking for rivet and seam leaks is part of inspecting the tank _____.
   a. foundation
   b. shell
   c. shell appurtenances
   d. roof

Section Two

2.0.0 Inspecting Breakout Tanks

Objective

Describe the requirements and procedures for performing breakout tank inspections (CT27_1-17).

a. Identify the prerequisites, competencies, and abnormal operating conditions (AOCs) associated with inspecting breakout tanks.
b. Describe the procedures used to perform a breakout tank inspection.

Performance Task

1. Perform routine inspection of a breakout tank (API 653 monthly or DOT annual) (CT27_1-17).
   - Identify potential abnormal operating conditions (AOCs) that may occur during performance of this CT, and know the appropriate actions to take in response to them.
   - Utilize the appropriate personal protective equipment according to relevant company procedures.
   - Perform a walk-around inspection to verify security and to confirm that no AOCs exist.
   - Visually inspect the foundation for defects.
   - Visually inspect the shell for defects.
   - Visually inspect shell appurtenances for defects, including manways, nozzles, tank piping manifolds, and the autogauge system.
   - Visually inspect roofs for defects, including deck plate external corrosion, roof deck drainage, floating roof seal systems, and pontoons.
   - Visually inspect roof and tank shell appurtenances for defects, including the sample hatch, autogauge inspection hatch, autogauge float well cover, external floating roof ladder, and the stairway.
   - Inspect the area for buildup of trash, vegetation, or other flammable material that could pose a fire or other safety hazard.
   - Complete appropriate documentation as required by operator's procedures.

Routine inspections of a breakout tank are performed on a schedule that is consistent with the conditions at the particular site. The purpose of the task is to visually inspect the tank's exterior surface and tank components to identify any potential problems, faulty equipment, or conditions that could result in a safety or environmental hazard.

American Petroleum Institute (API) standard 653 requires inspections of breakout tanks on a monthly basis. PHMSA requires inspections on an annual basis. This module describes the routine inspections that meet these requirements.

As personnel perform an inspection, they must always use their senses to detect:

- Loud noises indicating improperly operating or failing equipment
- Excessive heat produced by a bearing, motor, or other equipment-indicating failure
- Strong odors indicating product vapors leaking or equipment failure

Personnel must also make certain to wear all appropriate personal protective equipment (PPE) and follow company policy and procedures.

2.1.0 Prerequisites, Competencies, and Abnormal Operating Conditions

Personnel must possess a variety of skills in order to successfully inspect breakout tanks. The various operations require a series of steps, which a pipeline maintenance technician must be competent to perform. Personnel must also be aware of the particular abnormal operating conditions (AOCs) that may occur during the procedure and know the proper responses to them.

2.1.1 Prerequisites and Competencies

Personnel must possess the following basic knowledge and skills prior to inspecting tanks:

- Be competent with standard hand tools and any specialized tools recommended by the manufacturer.
- Be able to follow instructions and interpret drawings provided in the manufacturer's service manual.
- Be familiar with the basic construction of tanks and associated equipment, such as mechanical and electrical gauges, seals, and vapor recovery systems.
- Be familiar with different tank types and roof styles.

- Be familiar with entry permits and procedures for a confined space.
- Know how to properly use personal protective equipment (PPE)

2.1.2 Abnormal Operating Conditions

Personnel inspecting tanks must be able to identify the following potential AOCs and know the proper response to each, should they occur:

- *The foundation has settled* – Document the settlement location, estimate magnitude of settling or displacement, and impacts, if any, to the tank shell or equipment. Take photographs to document the progression of settlement over time. Report the condition to appropriate personnel for further inspection or repair.
- *The tank's level is inappropriate or abnormal* – Notify the control center or other appropriate personnel to take the required action.
- *The tank roof, seals, or roof-mounted equipment is damaged* – Notify the process operator and initiate the company procedure for repair or replacement.
- *Free product or other inappropriate material is on the tank roof, shell, or bottom plate* – Notify the process operator to take the required action.
- *Equipment, appurtenances, and/or components are faulty* – If equipment presents an immediate safety hazard, the tank or tank location should be isolated, tagged, or cordoned off to prevent access to the area. Report the condition to appropriate company personnel for further inspection or repair. If a failed component presents an imminent hazardous condition, evacuate the area immediately and notify the emergency response team.
- *A fire and/or explosion or ignition of hazardous, flammable liquids or gases occurs* – Evacuate the area and take only defensive action using the proper PPE. Contact the control center to isolate or shut down the line as quickly as possible. Engage the proper emergency response services.

2.2.0 Breakout Tank Inspections

Inspection procedures for breakout tanks will vary depending on the type and location. When you prepare to perform an inspection be sure that you have a current copy of the manufacturer's service manual. You will need this information not only for guidelines and procedures but also for information on replacing worn or damaged components.

Inspecting Tank Farms

Tank farms must be inspected and the inspections documented according to company policy. The following guidelines detail some areas of concern:

- Observe all system lines, valves, and controls for leaks and proper operation.
- Inspect electrical equipment for damaged or loose conduits, junction-box covers that are out of place, exposed wiring, and any lights that are not operating.
- Verify that dike drain valves are closed and locked.
- Drain water from dikes as needed, and document such action on the appropriate form.
- Note any fencing repair and trash cleanup needed.
- Note any sign of vandalism, forced entry, or theft.
- Check for excessive growth of vegetation.
- Check the secondary liners or containers in the tank.
- Report any abnormal condition to the shift supervisor, and prepare any documentation required.

WARNING!

Perform any preliminary tasks required to ensure safe working conditions. Always follow company policy and procedures both before and during the inspection.

2.2.1 PHMSA Requirements

When an annual inspection is performed according to PHMSA regulations, personnel must document the results on the PHMSA Breakout Tank Inspection Form (Form 10). On the form, the inspector must make notes and respond to a detailed list of questions that addresses compliance with regulations in the following areas:

- Design and new construction
- Tank repair
- Protection of the tank
- Pressure test procedures
- Inspection procedures
- Corrosion
- Inspection field review
- Inspection records review

2.2.2 Procedure

This section presents general information for breakout tank inspection. Company policies and procedures should always be followed. The following steps outline the general procedure for inspecting breakout tanks:

Step 1 Before performing a walk-around inspection, notify the control center. Then, verify the security and accessibility of the tank, and confirm that no AOCs exist.

Step 2 Utilize the appropriate personal protective equipment according to relevant company procedures.

Step 3 Visually inspect the foundation to identify abnormal conditions, such as settlement, underside corrosion, or leaks, and to determine foundation levelness and bottom elevations.

Step 4 Visually inspect the shell to identify leaks, paint failure, pitting, and corrosion. Document any leak locations by sketch or photograph.

Step 5 Visually inspect shell appurtenances to identify any leaks, component failures, or corrosion.

WARNING!

A component or equipment failure could pose a safety or environmental risk to personnel, the public, or the property adjacent to a breakout tank. If it is safe to do so, close, isolate, or remove the failed component from operation. Immediately notify control center personnel and other personnel per company policy. Document the condition for further inspection or repair. If the failed component presents an imminent hazard condition, evacuate the area immediately, and notify the emergency response team.

Step 6 Check the integrity of roof access stairs.

Step 7 Visually inspect the tank roof to identify problems with paint failure, corrosion, standing water, and a non-level roof. Identify any excessive gaps between the shell and the seals on the tanks that may be subject to air quality control regulations.

Step 8 Inspect tank roof appurtenances for proper operation and maintenance.

Step 9 Inspect the area for a buildup of trash, vegetation, and other flammable materials. Be sure to check for a sheen on standing water in the diked area.

Step 10 Remove any debris, vegetation, or flammable materials that could contribute to a fire or create a safety hazard.

Step 11 Document the results of the tank inspection on the appropriate company report form and/or PHMSA Form 10. The report should identify items that were visually inspected and items that need immediate attention.

Additional Resources

It is assumed that you will learn to perform these procedures under the direct supervision of your crew leader, foreman, or instructor. For independent study, refer to the recommended Additional Resources at the back of this module.

2.0.0 Section Review

1. A pipeline maintenance technician who discovers product on a tank roof during inspection should _____.
   a. notify the process operator
   b. try to return the product to the tank
   c. activate the proper emergency response service
   d. note it in the work documentation

2. According to API standard 653, inspections of breakout tanks should occur _____.
   a. daily
   b. monthly
   c. semi-annually
   d. annually

SUMMARY

Routine inspections of breakout tanks are necessary to ensure that the tanks are kept up to safety standards and to discourage vandalism and crime. Monthly inspections are required by API standard 653, and PHMSA requires inspections on an annual basis. Based on findings from the inspections, questionable items are noted, repairs are scheduled, and appropriate reports and documentation are submitted.

Review Questions

1. One purpose of a breakout tank is to _____.
   a. create product surges
   b. receive and store liquid during continued transportation
   c. regulate product temperature
   d. eliminate product contamination

2. Routine inspections of breakout tanks consist of _____.
   a. a visual inspection only
   b. a visual inspection and an internal inspection
   c. a visual inspection, and an internal inspection when practical
   d. an internal inspection only

3. One example of a condition personnel check for when inspecting shell appurtenances is _____.
   a. a damaged autogauge head
   b. cavities under the foundation
   c. corrosion damage on the wind girder
   d. incorrect positioning of the pontoon

4. Standing water or product on the tank roof may be an indication of _____.
   a. seam leaks
   b. settlement around the perimeter of the tank
   c. protruding bolts
   d. leaking pontoons

5. During an inspection, if the foundation of a tank is discovered to have settled, the appropriate response is to _____.
   a. document the settlement location
   b. b.evacuate the area immediately
   c. c.contact PHMSA personnel
   d. d.ignore the problem; it's not important

6. When an annual inspection is performed to meet PHMSA regulations, personnel must document the results on _____.
   a. a company-specific inspection report form
   b. PHMSA Form 10
   c. the Federal Pipeline Inspection Form
   d. API Form 653

Trade Terms Quiz

Fill in the blank with the correct term that you learned from your study of this module.

1. Located on the shell of a tank, approximately one foot below the top of the tank, is a device called a(n) ________.
2. Stiffening rings mounted on open-top tanks to maintain shell roundness against loads are called ________.
3. Adjustable steel pipes mounted on the floating roof that provide support for the roof when the liquid is drained below a set level are called ________.
4. The component on which a tank shell rests is called a(n) ________.
5. A vertical pole device that extends through the floating roof of a tank to maintain the roof in a centered position and to prevent the roof from rotating is called a(n) ________.
6. Conditions within a pipeline system, identified by the operator, that may indicate (1) a malfunction of a component or (2) a deviation from normal operations that may indicate a condition exceeding design limits or result in a hazard to persons, property, or the environment are called ________.
7. External or internal compartments on a floating roof that provide buoyancy are called ________.
8. Valves mounted on top of the tank roof to prevent negative pressure conditions under the roof are called ________.
9. A container used to relieve surges in a hazardous liquid pipeline system and/or to receive and store hazardous liquid transported by a pipeline for reinjection and continued transportation by the pipeline is called a ________.
10. Flexible sealing devices mounted on the floating roof of a tank to provide a reasonably close fit between the roof and the internal wall of the tank are called ________.
11. Steel pipes or hoses mounted on a floating roof to shed rainwater are called ________.
12. A self-supporting roof that is intended to prevent rainwater from entering the tank is called a(n) ________.
13. A mechanical opening mounted on the floating roof of a tank that controls the vapor space between the primary and secondary rim seals is called the ________.

Trade Terms

Abnormal operating conditions (AOCs)
Breakout tank
Deck legs
Geodesic dome
Guide pole
Overflow drain
Pontoons
Primary and secondary seals
Rim vent
Roof drains
Sketch plate
Vacuum breakers
Wind girders

Trade Terms Introduced in This Module

Abnormal operating conditions (AOCs): Conditions within a pipeline system, identified by the operator, that may indicate (1) a malfunction of a component or (2) a deviation from normal operations that may indicate a condition exceeding design limits or result in a hazard to persons, property, or the environment.

Breakout tank: A tank used to relieve surges in a hazardous liquid pipeline system and/or to receive and store hazardous liquid transported by a pipeline for reinjection and continued transportation by the pipeline.

Deck legs: Steel pipes mounted on the floating roof that provide support for the roof when the liquid is drained below a set level. They are typically adjustable to allow for a change in supported roof height from the bottom of the tank and are often referred to as high or low legs.

Geodesic dome: A self-supporting dome roof, typically made of aluminum, which is intended to prevent rainwater from entering the tank.

Guide pole: A vertical pole device that extends through the floating roof of a tank to maintain the roof in a centered position and to prevent the roof from rotating.

Overflow drain: A nozzle on the shell of a tank located approximately one foot below the top of the tank. They are always larger than tank inlet nozzles.

Pontoons: External or internal compartments on a floating roof that provide buoyancy, allowing the roof to float.

Primary and secondary seals: Flexible sealing devices mounted on the floating roof of a tank to provide a reasonably close fit between the roof and the internal wall of the tank.

Rim vent: A mechanical vent mounted on the floating roof of a tank that vents the vapor space between the primary and secondary rim seals.

Roof drains: Steel pipes or hoses mounted on a floating roof to drain rainwater.

Sketch plate: The bottom plate on which a tank shell rests.

Vacuum breakers: Valves mounted on top of the tank roof to prevent a vacuum under the roof.

Wind girders: Stiffening rings mounted on open-top tanks to maintain shell roundness against wind loads.

Additional Resources

The following recommended resources can provide additional helpful information related to the requirements for this covered task:

API Recommended Practice 1161, Third Edition, © 2014 American Petroleum Institute. API Publishing Services, 1220 L Street, NW, Washington, DC 20005. Distributed by Thompson Reuters (Scientific) LLC, **www.techstreet.com**.

49 *CFR* 192, **www.ecfr.gov**

49 *CFR* 195, **www.ecfr.gov**

NCCER module AOCCC-17, *Abnormal Operating Conditions Control Center*

NCCER module AOCFG-17, *Abnormal Operating Conditions Field and Gas*

NCCER module 66101-02, *Introduction to the Pipeline Industry*

NCCER module CT15_1-17, *Surface Right-of-Way Inspection*

NCCER module CT22_1-17, *Inspect Tank Pressure and Vacuum Breakers*

The following websites provide resources for products and training:

American Petroleum Institute (API), **www.api.org**

Pipeline and Hazardous Material Safety Administration (PHMSA), **www.phmsa.dot.gov**

Figure Credits

Courtesy of S4Carlisle Publishing Services, Module Opener, Figure 1

Section Review Answer Key

Answer	Section Reference	Objective
Section One		
1. c	1.1.0	1a
2. b	1.2.3	1b
Section Two		
1. a	2.1.2	2a
2. b	2.0.0	2b

NCCER CURRICULA — USER UPDATE

NCCER makes every effort to keep its textbooks up-to-date and free of technical errors. We appreciate your help in this process. If you find an error, a typographical mistake, or an inaccuracy in NCCER's curricula, please fill out this form (or a photocopy), or complete the online form at **www.nccer.org/olf**. Be sure to include the exact module ID number, page number, a detailed description, and your recommended correction. Your input will be brought to the attention of the Authoring Team. Thank you for your assistance.

Instructors – If you have an idea for improving this textbook, or have found that additional materials were necessary to teach this module effectively, please let us know so that we may present your suggestions to the Authoring Team.

NCCER Product Development and Revision

13614 Progress Blvd., Alachua, FL 32615

Email: curriculum@nccer.org

Online: www.nccer.org/olf

❑ Trainee Guide ❑ Lesson Plans ❑ Exam ❑ PowerPoints Other ____________

Craft / Level: Copyright Date:

Module ID Number / Title:

Section Number(s):

Description:

Recommended Correction:

Your Name:

Address:

Email: Phone:

CT28_0-17

Provide Security for Pipeline Facilities

Overview

The practices required to maintain security of a pipeline facility include monitoring, inspecting, and repair. Routine inspection of remote facilities discourages vandalism and ensures that safety and security devices are in good working order. This module describes security inspections, including the documentation of observations and the testing of alarms, cameras, and intrusion devices at a pipeline facility.

Trainees with successful module completions may be eligible for credentialing through the NCCER Registry. To learn more, go to **www.nccer.org** or contact us at 1.888.622.3720. Our website has information on the latest product releases and training, as well as online versions of our Cornerstone magazine and Pearson's product catalog.

Your feedback is welcome. You may email your comments to **curriculum@nccer.org**, send general comments and inquiries to **info@nccer.org**, or fill in the User Update form at the back of this module.

This information is general in nature and intended for training purposes only. Actual performance of activities described in this manual requires compliance with all applicable operating, service, maintenance, and safety procedures under the direction of qualified personnel. References in this manual to patented or proprietary devices do not constitute a recommendation of their use.

CT28_0-17
Provide Security for Pipeline Facilities

Objectives

When you have completed this module, you will be able to do the following:

1. Describe how to provide security to pipeline facilities.
   a. Identify pipeline facility security requirements.
   b. Explain how to safely perform a security inspection.
2. Describe the procedures used to provide security to pipeline facilities (CT28_0-17).
   a. Identify the prerequisites, competencies, and abnormal operating conditions (AOCs) associated with providing security to a pipeline facility.
   b. Describe the general procedure used to provide security to a pipeline facility.

Performance Task

Under the supervision of your instructor, you should be able to do the following:

1. Provide security for pipeline facilities (CT28_0-17).
   - Identify potential abnormal operating conditions (AOCs) that may occur during performance of this CT, and know the appropriate actions to take in response to them.
   - Utilize the appropriate personal protective equipment according to relevant company procedures.
   - Deactivate the security alarm prior to entering the pipeline facility, if applicable, and notify appropriate personnel prior to entering a remotely monitored pipeline facility.
   - Inspect the pipeline facility perimeter.
   - Inspect all locks for possible damage and ensure that locks are in working order.
   - Inspect signage for proper placement, visibility to the public, and legible and correct information.
   - Inspect local monitoring devices (motion sensors, alarms, and cameras) for operability, damage, or required maintenance.
   - Test intrusion devices to ensure operability.
   - Perform repairs to facility security equipment as appropriate or prepare a facility repair order.
   - Reactivate facility security system or notify appropriate personnel upon leaving remotely monitored facilities.
   - Verify closure to prevent call-out.
   - Complete appropriate documentation as required by operator's procedures.

Trade Terms

Abnormal operating conditions (AOCs)
Control center
Intrusion devices
Remote monitoring
Transportation Security Administration (TSA)

Industry-Recognized Credentials

If you're training through an NCCER-accredited sponsor, you may be eligible for credentials from NCCER's Registry. The ID number for this module is CT28_0-17. Note that this module may have been used in other NCCER curricula and may apply to other level completions. Contact NCCER's Registry at 888.622.3720 or go to **www.nccer.org** for more information.

Contents

Figures

Section One

1.0.0 Pipeline Facility Security Measures

Objective

Describe how to provide security for pipeline facilities.

a. Identify pipeline facility security requirements.
b. Explain how to safely perform a security inspection.

Trade Terms

Abnormal operating conditions (AOCs): Conditions within a pipeline system, identified by the operator, that may indicate (1) a malfunction of a component or (2) a deviation from normal operations that may indicate a condition exceeding design limits or result in a hazard to persons, property, or the environment.

Control center: The operations center of a pipeline, usually located in a terminal or main pipeline facility, where all the control consoles and computer-operated systems are located.

Intrusion devices: Equipment capable of detecting entry into a facility, such as a camera, motion sensor, gate alarm, or other equipment.

Remote monitoring: Monitoring at a location other than the facility being inspected.

Transportation Security Administration (TSA): Federal agency under the US Department of Homeland Security tasked with protecting US transportation systems, including pipelines, to ensure freedom of movement for people and commerce. As a counterterrorism organization, the agency oversees the security of US pipelines and has established pipeline security programs, initiatives, and guidelines.

Pipeline systems may be subject to willful damage by outside parties, as well as accidental damage. As required by 49 *CFR* 195.436, pipeline operators must provide protection from vandalism and unauthorized entry for each pumping station, breakout tank area (*Figure 1*), and other exposed pipeline facility.

1.1.0 Security Requirements

The Transportation Security Administration (TSA) is the federal agency that has responsibility for overseeing the security of US pipelines. The TSA has established a set of measures for pipeline operators to follow in order to ensure the security of pipeline facilities. The baseline TSA pipeline facility security measures are as follows:

Figure 1 Secured tank farm.

- *Physical security and access controls* – If used, fences or other barriers around pipeline facilities should be properly maintained and a clear zone around those barriers maintained. Measures to deter unauthorized access to pipeline facilities should be implemented. Gates and entrances in a pipeline facility should be closed when not in use. "No Trespassing" warning signs should be posted at entry points where they are easily visible.
- *Personnel security* – Identification and badging policies and procedures for on-site personnel should be established. Policies and procedures for pre-employment screening of applicants as well as employee behavior criteria should also be established.
- *Equipment maintenance and testing* – Maintenance programs should be implemented to ensure security systems are in good working order. Security equipment malfunctions or failures should be identified and responded to in a timely manner.
- *Design and construction* – Security measures should be integrated into the design, construction, or renovation of a pipeline facility.
- *Communication* – Internal and external notification procedures regarding security events should be established.
- *Personnel training* – Security awareness briefings should be held for personnel with unescorted access upon hire and every two years thereafter. Security training records should be maintained.
- *Exercises and drills* – Periodic security exercises or drills, including unannounced tests and incident plans, should be conducted. Reports regarding these exercises should be maintained.
- *Security incident procedures* – Procedures for responding to security incidents or emergencies

should be implemented, including appropriate reporting requirements.
- *Recordkeeping* – Policies and procedures for the recordkeeping of security information should be developed and documented. Documents such as corporate security plans, training records, exercise reports, security testing and audits, and security equipment maintenance and testing records should be retained. Security information records should be made available to TSA upon request.

1.1.1 Types of Security Measures

A pipeline facility's security measures are evaluated annually at a minimum, and reliability of the system depends on monitoring, regular inspections, and repairs of the system. Security measures for pipeline facilities may involve any or all of the following measures:

- Fences
- Gates
- Permits
- Locks
- Crew/employee awareness
- Signs
- Cameras
- Guards

Stations are generally completely fenced, and the fences well-maintained. Main-line valves and other exposed equipment may also be fenced or protected by guardrails (*Figure 2*). Gates are located at entry points to facilities and should be locked at all times or controlled by security guards. All main-line valves must be chain-locked in the proper position. All block valves in the relief piping must be chain-locked as well. Locks are installed on all facility doors, limiting access to authorized persons.

Figure 2 Properly-secured facility.

Many larger pipeline facilities have security systems and other **intrusion devices** that have **remote monitoring** capabilities, enabling the **control center** to watch multiple pipeline facilities simultaneously. Security systems may include motion detectors, cameras, lighting, and alarms, as well as security guards. The ability to continuously view all parts of a facility is vital for security. Video cameras and monitors can reveal equipment failure as well as break-ins and other security problems. Surveillance cameras can also focus on critical equipment and locations so that security and field personnel can monitor these areas.

According to 49 *CFR* 195.434, each pipeline operator must maintain signs visible to the public around each exposed pipeline facility. The signs must be positioned so they can be read by people coming towards the pipeline facility from any direction. Each sign must contain the name of the operator and a telephone number where the operator can be reached at all times. Terminals and storage facilities should also display signs visible to the public. Signs must be appropriately sized and laid out to prominently display the information. They must be periodically inspected to ensure that they are in good condition, secure, and legible.

1.2.0 Pipeline Facility Security Inspection

Personnel should be familiar with the layout of the pipeline facility before beginning an inspection. Different pipeline facilities and work areas are often similar and can become confusing in emergency situations. Escape routes should be rehearsed and locations of safe areas identified before entering the pipeline facility or area of work. New personnel must be informed of evacuation procedures to ensure that they are able to reach safe areas without assistance. All personnel must observe site-specific safety procedures at all times.

When performing an inspection, personnel must have the appropriate personal protective equipment (PPE). The minimum PPE required for an inspection includes the following items:

- Hard hat
- Safety glasses
- Fire retardant clothing
- Safety shoes

Additional PPE, including hearing protection, may be required based on site-specific needs. Always follow company policy and procedures for personal protective equipment.

1.2.1 Outside Inspection

Upon arrival at the facility, personnel should observe the general condition and appearance from the outside, including the following:

- Appearance of buildings and installations
- Evidence of vandalism or trespassing
- Condition of fences and gates
- Condition of locks on gates, doors, and mechanical items
- Placement and condition of signs (*Figure 3*)

Figure 3 Signs must be visible and legible.

Personnel should also be alert to indications of possible **abnormal operating conditions (AOCs)**, and be prepared to respond appropriately. Some examples are as follows:

- Loud noises, which might indicate improperly operating or failing equipment
- Excessive heat produced by a bearing, motor, or other equipment, which might indicate a failure
- Strong odors, which can indicate product vapors or equipment failure

> **CAUTION**
>
> Not all pipeline products have odors. Do not rely exclusively on your sense of smell to detect a possible product release.

1.2.2 Inside Inspection

Once inside an installation, personnel should be prepared to observe and document the conditions inside the facility. An inside inspection includes examining mechanical components such as valves, manifold areas (*Figure 4*), and pumps. An inspection inside a pipeline facility typically includes observation of the following:

Figure 4 Tank valve manifold.

- Condition of buildings, including windows, roofs, railings, stairs, floors, and other construction
- Condition of local monitoring devices for operability
- Condition of intrusion devices to ensure operability

A variety of buildings may be included in a security inspection, including control center pump stations, booster stations, offices, and terminals. When inspecting security conditions, check buildings for the following:

- Warning and hazard signs are posted where required.
- Exterior building or trailer identification numbers are posted for emergency response.
- Emergency exits are correctly marked, visible, and accessible.
- There is sufficient lighting to allow for visibility of any intrusion via security cameras.

Additional Resources

It is assumed that you will learn to perform these procedures under the direct supervision of your crew leader, foreman, or instructor. For independent study, refer to the recommended Additional Resources at the back of this module.

1.0.0 Section Review

1. At a minimum, a facility's security measures must be evaluated _____.
   a. annually
   b. monthly
   c. weekly
   d. daily

2. When performing an outside security inspection, personnel should document _____.
   a. indications of corrosion
   b. cleanliness of the area
   c. evidence of vandalism
   d. loud noises

Section Two

2.0.0 Providing Security for Pipeline Facilities

Objective

Describe the procedures used to provide security to pipeline facilities (CT28_0-17).

a. Identify the prerequisites, competencies, and abnormal operating conditions (AOCs) associated with providing security to a pipeline facility.

b. Describe the general procedure used to provide security to a pipeline facility.

Performance Task

Provide security for pipeline facilities (CT28_0-17).

- Identify potential abnormal operating conditions (AOCs) that may occur during performance of this CT, and know the appropriate actions to take in response to them.
- Utilize the appropriate personal protective equipment according to relevant company procedures.
- Deactivate the security alarm prior to entering the pipeline facility, if applicable, and notify appropriate personnel prior to entering a remotely monitored pipeline facility.
- Inspect the pipeline facility perimeter.
- Inspect all locks for possible damage and ensure that locks are in working order.
- Inspect signage for proper placement, visibility to the public and legible and correct information.
- Inspect local monitoring devices (motion sensors, alarms, and cameras) for operability, damage, or required maintenance.
- Test intrusion devices to ensure operability.
- Perform repairs to facility security equipment as appropriate or prepare a facility repair order.
- Reactivate facility security system or notify appropriate personnel upon leaving remotely monitored facilities.
- Verify closure to prevent call-out.
- Complete appropriate documentation as required by operator's procedures.

Personnel conducting pipeline facility security inspections are required to have some basic competencies. This section describes some of the basic skills needed by personnel in order to provide security for pipeline facilities. It identifies abnormal operating conditions (AOCs) that may occur during the performance of this task, and the appropriate responses to them. The basic procedure to follow in order to provide security for a pipeline facility is also described.

2.1.0 Prerequisites, Competencies, and Abnormal Operating Conditions

Personnel must possess a variety of skills in order to successfully provide security for pipeline facilities. The various operations require a series of steps, which a pipeline maintenance technician must be competent to perform. Personnel must also be aware of the particular abnormal operating conditions (AOCs) that may occur during the procedure and know the proper responses to them.

2.1.1 Prerequisites and Competencies

Personnel must possess the following basic knowledge and skills prior to performing facility security inspections:

- Be competent with standard hand tools and any specialized tools
- Have knowledge of pipeline facility construction and layout
- Be familiar with company operations and maintenance manuals
- Be familiar with the procedures for testing facility security systems

Personnel must also be familiar with the company safety, evacuation, and response procedures, as well as the proper use of personal protective equipment (PPE).

2.1.2 Abnormal Operating Conditions

Personnel assigned to provide security for pipeline facilities must be able to identify abnormal operating conditions (AOCs) and know the proper response to each, should they occur. The following are some potential AOCs, and the appropriate action personnel should take in response to them:

- *Perimeter breach, access control breach, or failure of security equipment occurs* – Secure the site if safe to do so and make appropriate notifications and/or repairs according to company policy.

- *Vandalism to facility equipment is discovered* – Notify appropriate company personnel, arrange for or make required repairs, and reassess facility security to determine appropriate improvements.
- *Improper or missing facility signage is discovered* – Assess appropriate requirements for facility and add signage as necessary to ensure visibility.
- *Unexpected hazardous liquid or carbon dioxide leak is encountered* – Identify and isolate the source of the leak if possible and if trained to do so. Proper notifications should be made as soon as possible, including calling 911 if necessary.
- *Fire or explosion occurs* – Limit access to the area and make the proper notifications, including calling 911 if necessary.

2.2.0 Providing Pipeline Facility Security

Regularly scheduled inspections are necessary and required by the Pipeline and Hazardous Materials Safety Administration (PHMSA) to ensure that pipeline facilities are maintained in a secure condition. This section describes procedures for inspecting the outside of pipeline facilities, and for inspecting installations and equipment inside the pipeline facility, including buildings.

2.2.1 Procedure

Personnel performing a security inspection should be aware of any signs of suspicious activity. For example, outside parties are not permitted to take photos of pipeline facilities; anyone attempting to do so must be reported according to company policy.

Crews in the field are responsible for maintaining line security, but every employee must be informed of the importance of security as well as specific company security procedures. While procedures vary from company to company, personnel must generally perform the following steps to provide security for a pipeline facility:

> **CAUTION**
> Throughout this procedure, personnel must identify any potential abnormal operating conditions (AOCs) and initiate the appropriate response.

Step 1 Identify potential AOCs that may occur, and know the appropriate actions to take in response to them.

Step 2 Utilize the appropriate personal protective equipment according to relevant company procedures.

Step 3 Deactivate the security alarm prior to entering the pipeline facility, if applicable, and notify appropriate personnel prior to entering a remotely monitored pipeline facility.

Step 4 Inspect the pipeline facility perimeter fencing, gates, or other entry barriers for integrity and/or possible damage.

Step 5 Inspect all locks for possible damage, and ensure that the locks are in working order.

Step 6 Inspect signs for proper placement and visibility to the public and to ensure that the information on the signs is accurate and legible.

Step 7 Inspect local monitoring devices, such as motion detectors, alarms, and cameras, for operability, damage, or required maintenance.

Step 8 Test security devices to ensure operability.

Step 9 Perform repairs to security equipment as appropriate, or prepare a facility repair order.

Step 10 Notify appropriate personnel upon leaving the facility and reactivate the security system if needed.

Step 11 Verify closure to prevent call-out.

Step 12 Complete appropriate documentation as required by operator's procedures.

Additional Resources

It is assumed that you will learn to perform these procedures under the direct supervision of your crew leader, foreman, or instructor. For independent study, refer to the recommended Additional Resources at the back of this module.

2.0.0 Section Review

1. If a perimeter breach is discovered during an inspection, the proper response is to first _____.
   a. make appropriate company notifications
   b. secure the site if it is safe to do so
   c. evacuate the area
   d. shut down the pipeline

2. The responsibility for maintaining line security lies with _____.
   a. emergency personnel
   b. control center personnel
   c. crews in the field
   d. the utility company

SUMMARY

Performing periodic security inspections of facilities is an extremely important part of pipeline maintenance. Operators employ specific security measures to protect their facilities, and they must periodically verify that these measures are maintained. Facility inspections include observations noted outside of the facility as well as testing and observation performed inside. Inspections inside a facility involve examination of specific buildings and mechanical components. While inspection procedures for each company vary somewhat, the purpose of all inspections is to ensure the security and integrity of the entire pipeline facility.

Review Questions

1. The federal agency that oversees the security of US pipelines is the _____.
   a. Transportation Security Administration (TSA)
   b. Environmental Protection Agency (EPA)
   c. Bureau of Land Management (BLM)
   d. Federal Emergency Management Agency (FEMA)

2. The implementation of identification and badging policies is a measure intended for _____.
   a. personnel security
   b. physical security and access controls
   c. equipment maintenance and testing
   d. design and construction

3. According to pipeline facility security measures set forth by the TSA, operators should provide security awareness briefings for personnel with unescorted access _____.
   a. upon hire
   b. upon hire and every year thereafter
   c. upon hire and every two years thereafter
   d. every two years

4. Gates at pipeline facility entry points should be _____.
   a. left open for safety during emergencies
   b. locked or guarded at all times
   c. closed but not locked
   d. locked at night only

5. Main-line valves may be fenced or protected by _____.
   a. security guards
   b. netting
   c. chains
   d. guardrails

6. Before personnel enter a pipeline facility or area of work, escape routes and locations of safe areas should be _____.
   a. locked
   b. guarded
   c. closed
   d. rehearsed

7. The presence of a strong odor during a security inspection may indicate _____.
   a. an abnormal operating condition
   b. a dead animal
   c. unsanitary conditions
   d. sulfur in the area

8. Examination of alarms, monitors, and intrusion devices are part of a(n) _____.
   a. outside inspection
   b. mechanical inspection
   c. inside inspection
   d. control center security check

9. The mechanical and equipment inspection requirements of each pipeline facility are _____.
   a. always the same
   b. specific to the facility
   c. determined by PHMSA
   d. determined by the inspector

10. PHMSA requires security inspections to be performed _____.
    a. by a PHMSA employee
    b. annually
    c. on a regularly scheduled basis
    d. if the facility perimeter is compromised

11. Outside parties found taking pictures of a pipeline facility should be considered to be _____.
    a. harmless and can be ignored
    b. suspicious and should be reported
    c. PHMSA personnel performing a security check
    d. an immediate threat

12. Before leaving a facility after a security inspection, personnel should _____.
    a. deactivate the security system
    b. reactivate the security system
    c. test each security device
    d. perform as-needed repairs to security equipment

Trade Terms Quiz

Fill in the blank with the correct term that you learned from your study of this module.

1. Monitoring at a location other than the facility being inspected is considered ________.
2. The federal agency that oversees the security of US pipelines and has established pipeline security programs, initiatives, and guidelines is the ________.
3. Cameras, motion sensors, gate alarms, or other equipment capable of detecting entry into a facility are called ________.
4. Conditions within a pipeline system that may indicate (1) a malfunction of a component or (2) a condition exceeding design limits or resulting in a hazard to persons, property, or the environment are called ________.
5. The operations center of a pipeline, usually located in a terminal or main pipeline facility, where all the consoles and computer-operated systems are located is called the ________.

Trade Terms

Abnormal operating conditions (AOCs)
Control center
Intrusion devices
Remote monitoring
Transportation Security Administration (TSA)

Trade Terms Introduced in This Module

Abnormal operating conditions (AOCs): Conditions within a pipeline system, identified by the operator, that may indicate (1) a malfunction of a component or (2) a deviation from normal operations that may indicate a condition exceeding design limits or result in a hazard to persons, property, or the environment.

Control center: The operations center of a pipeline, usually located in a terminal or main pipeline facility, where all the control consoles and computer-operated systems are located.

Intrusion devices: Equipment capable of detecting entry into a facility, such as a camera, motion sensor, gate alarm, or other equipment.

Remote monitoring: Monitoring at a location other than the facility being inspected.

Transportation Security Administration (TSA): Federal agency under the US Department of Homeland Security tasked with protecting US transportation systems, including pipelines, to ensure freedom of movement for people and commerce. As a counterterrorism organization, the agency oversees the security of US pipelines and has established pipeline security programs, initiatives, and guidelines.

Additional Resources

The following recommended resources can provide additional helpful information related to the requirements for this covered task:

API Recommended Practice 1161, Third Edition, © 2014 American Petroleum Institute. API Publishing Services, 1220 L Street NW, Washington, DC 20005. Distributed by Thompson Reuters (Scientific) LLC, **www.techstreet.com**

49 *CFR* 192, **www.ecfr.gov**.

49 *CFR* 195, **www.ecfr.gov**.

NCCER module AOCCC-17, *Abnormal Operating Conditions Control Center*

NCCER module AOCFG-17, *Abnormal Operating Conditions Field and Gas*

NCCER Module 62105-14, *Introduction to Pipeline Documents*

NCCER Module 63104-02, *Piping and Mechanical Blueprint Reading*

NCCER Module 66101-14, *Introduction to the Pipeline Industry*

NCCER Module CT15_1-17, *Visually Inspect Surface Conditions of Right-of-Way*

The following websites provide resources for products and training:

Transportation Security Administration (TSA) Pipeline Security Guidelines, **www.tsa.gov**

Figure Credits

Courtesy of S4Carlisle Publishing Services, Module Opener, Figures 1 and 2

Photo Courtesy of STV, Figure 4

Section Review Answer Key

Answer	Section Reference	Objective
Section One		
1. a	1.1.1	1a
2. c	1.2.1	1b
Section Two		
1. b	2.1.2	2a
2. c	2.2.1	2b

NCCER CURRICULA — USER UPDATE

NCCER makes every effort to keep its textbooks up-to-date and free of technical errors. We appreciate your help in this process. If you find an error, a typographical mistake, or an inaccuracy in NCCER's curricula, please fill out this form (or a photocopy), or complete the online form at **www.nccer.org/olf**. Be sure to include the exact module ID number, page number, a detailed description, and your recommended correction. Your input will be brought to the attention of the Authoring Team. Thank you for your assistance.

Instructors – If you have an idea for improving this textbook, or have found that additional materials were necessary to teach this module effectively, please let us know so that we may present your suggestions to the Authoring Team.

NCCER Product Development and Revision

13614 Progress Blvd., Alachua, FL 32615

Email: curriculum@nccer.org

Online: www.nccer.org/olf

❑ Trainee Guide ❑ Lesson Plans ❑ Exam ❑ PowerPoints Other ______________

Craft / Level: Copyright Date:

Module ID Number / Title:

Section Number(s):

Description:

Recommended Correction:

Your Name:

Address:

Email: Phone:

CT32_0-17

Observation of Excavation Activities

Overview

Excavating is considered one of the most hazardous tasks in the pipeline industry. This is why companies designate a worker who is specifically responsible for observing the excavation in order to protect the pipeline. This module describes the responsibilities of the designated observer/spotter prior to the start of an excavation project, as well as during the excavation itself.

Trainees with successful module completions may be eligible for credentialing through the NCCER Registry. To learn more, go to **www.nccer.org** or contact us at 1.888.622.3720. Our website has information on the latest product releases and training, as well as online versions of our *Cornerstone* magazine and Pearson's product catalog.

Your feedback is welcome. You may email your comments to **curriculum@nccer.org**, send general comments and inquiries to **info@nccer.org**, or fill in the User Update form at the back of this module.

This information is general in nature and intended for training purposes only. Actual performance of activities described in this manual requires compliance with all applicable operating, service, maintenance, and safety procedures under the direction of qualified personnel. References in this manual to patented or proprietary devices do not constitute a recommendation of their use.

CT32_0-17

Observation of Excavation Activities

Objectives

When you have completed this module, you will be able to do the following:

1. Describe the tasks required of the designated excavation observer/spotter that must occur before excavation begins.
   a. Describe the appropriate permits and notifications required before excavation begins.
   b. Describe the pipeline marking process that must be completed before excavation begins.
   c. Describe the clearance requirements for underground pipelines.
   d. Explain the Occupational Safety and Health Administration (OSHA) regulations that pertain to excavation.
2. Describe the tasks required of the designated excavation observer/spotter during an excavation.
   a. Explain the dangers associated with excavation.
   b. Identify the tools used for excavation and the appropriate use for each.
   c. Describe the technique of sloping during excavation work.
   d. Identify methods used to protect the pipeline during excavation.
3. Describe the procedures for measuring pipe clearance and observing an excavation (CT32_0-17).
   a. Identify the prerequisites, competencies, and abnormal operating conditions associated with observation of excavation activities.
   b. Describe the general procedures for measuring pipe clearance.
   c. Describe the general procedures for observing excavation activities.

Performance Tasks

Under the supervision of your instructor, you should be able to do the following:

1. Observe excavation activities (CT32_0-17).
   - Identify potential abnormal operating conditions that may occur during performance of this CT, and know the appropriate actions to take in response to them.
   - Utilize the appropriate personal protective equipment according to relevant company procedures.
   - Notify the control center of the beginning and end of work as required by company procedures.
   - Identify and mark hazards related to the pipeline that surround the excavation site (including underground hazards such as appurtenances to the pipeline).
   - Determine the method of installation for the pipe or foreign structure. If the directional drilling or boring method was used, determine the depths of the pipelines or other structures and compare the depths to determine clearance.
   - Ensure that the spoil pile is placed in a location that does not affect the integrity of the pipeline.
   - Require proper preparation of mechanical equipment for safe digging around the pipeline.
   - Ensure that soil excavated above, below, or on either side of the pipeline meets the company-specified tolerance zone.
   - Require hand or vacuum excavation of the soil nearest the pipeline.
   - Ensure that the pipe is supported as necessary while soil is being removed.
   - Keep area safe from encroachment.
   - Verify the excavation is safe to enter.
   - If structures are in an open trench, measure the distance between the pipeline and nearest foreign structure.
   - Approve the excavation for backfilling by notifying a supervisor, local crew leader, or other appropriate company personnel.
   - Complete appropriate documentation and notifications as required by operator's procedures.

Trade Terms

Abnormal operating conditions (AOCs)
Angle of repose
Backfilling
Bench
Fill
Hydro-vac
Line marker
One-Call System
Pot-holing
Sloping
Spoil
Support systems
Tolerance zone
Well points

Industry-Recognized Credentials

If you're training through an NCCER-accredited sponsor, you may be eligible for credentials from NCCER's Registry. The ID number for this module is CT32_0-17. Note that this module may have been used in other NCCER curricula and may apply to other level completions. Contact NCCER's Registry at 1.888.622.3720 or go to **www.nccer.org** for more information.

Contents

Figures and Tables

SECTION ONE

1.0.0 PREPARATION FOR EXCAVATION

Objective

Describe the tasks required of the designated excavation observer/spotter that must occur before excavation begins.

a. Describe the appropriate permits and notifications required before excavation begins.
b. Describe the pipeline marking process that must be completed before excavation begins.
c. Describe the clearance requirements for underground pipelines.
d. Explain the Occupational Safety and Health Administration (OSHA) regulations that pertain to excavation.

Trade Terms

Abnormal operating conditions (AOCs): Conditions within a pipeline system, identified by the operator, that may indicate (1) a malfunction of a component or (2) a deviation from normal operations that may indicate a condition exceeding design limits or result in a hazard to persons, property, or the environment.

Line marker: A highly visible device used to communicate the location of the pipeline.

One-Call System: An organization that excavators may contact (required in some states) to determine the presence of underground substructures within the area of their proposed excavation activities, and to advise operators of such substructures of the pending activity that may impact their facility.

Spoil: Soil removed during an excavation.

Support systems: Methods used to ensure the stability of adjoining buildings, walls, or other structures endangered by an excavation. (Examples of support systems include shoring, bracing, or underpinning.)

Tolerance zone: Area surrounding underground pipeline or cables within which no power digging for excavation is allowed; personnel must hand dig only.

In every pipeline excavation project, a worker must be specifically designated as the person responsible for the observation of the excavation. The overall responsibility of this designated observer/spotter (sometimes called the *swamper*) is to prevent damage to the pipeline by observing progress and watching for potential dangers. In this role, the spotter must also complete a number of specific tasks prior to excavation as well as during the excavation.

Prior to excavation, the spotter must confirm that all required permits have been obtained and proper notifications have been made. The spotter must also be well informed about the Occupational Safety and Health Administration (OSHA) regulations that pertain to excavations. Excavating is one of the most dangerous tasks performed in pipeline maintenance.

In order to avoid damage to the pipeline, the spotter performs several additional duties. The spotter must identify the appropriate location where **spoil** (excavated material) will be placed in order to protect the pipeline. If not placed an adequate distance away from the excavation, the excavated material could cause stress on the pipeline or result in a cave-in.

To ensure that no excavation tools and equipment used do any damage to the pipeline, the spotter is also responsible for determining the required **tolerance zone** and communicating it to the excavator(s). The tolerance zone is the area surrounding underground pipeline or cables; in this zone, no power digging for excavation is allowed, so personnel must hand dig only. Each state has a defined tolerance zone.

1.1.0 Permits and Notifications

Before excavation near a pipeline can begin, the control center must be notified so that personnel prepare to closely monitor pipeline pressure and flow during excavation. The spotter must confirm that this notification has been made. The spotter must also ensure that required permits have been obtained. It is not uncommon for more than one permit to be required. The number and types of permits may vary with the location; in general, permissions may be required from the following entities:

- Property owner
- PHMSA
- State
- City/local authorities

- Company permits
- Railroads
- Corps of engineers

Prior to excavation, companies performing the excavation must also notify the local utility companies to have the underground line locations identified. The most common method used is an operator-sponsored centralized program known as the **One-Call System**. The One-Call System is a communication system established by two or more utilities, governmental agencies, or other operators of underground facilities to provide one telephone number for excavating contractors and the general public to notify of their intent to use equipment for excavation, demolition, or other work. It requires that the participating members identify and give the location of their facilities. In most states, calls are made to the telephone number 811. In some areas, the One-Call system is also referred to as *811 Call Before You Dig*.

The One-Call System is generally set up state by state. Some states require that the excavation work be completed within a 14-day time period, but this timeframe varies. In some states, if the required time period ends before the work is completed, operators must re-notify the One-Call System. The spotter must confirm that a current One-Call Ticket (excavation notice) is in place before excavation can begin. An example of a One-Call Ticket is shown in *Figure 1*.

> **NOTE**
> Always follow policies and procedures for the company whose line is being exposed.

1.1.1 Operator Response to Standard Notifications

Pipeline operator offices have communication systems connected to the One-Call system. A dedicated phone line and printer may be set up to receive One-Call tickets. Notification may also be received via fax, email, or other online methods. When notices are received, the operator researches the area to confirm the location of the pipe. If the pipe is not threatened, the excavation will be approved. If the pipe may be threatened, the operator will arrange to provide temporary markings and oversee the excavation. Operators must respond to standard notifications within 48 hours (not including weekends and holidays).

1.1.2 Operator Response to Emergency Notifications

In the case of emergencies or problem situations, operators are informed of the problem through the One-Call system and asked to provide immediate assistance in locating their underground facilities. The One-Call system records all communications on the One-Call log. Operators must respond to emergency notifications from the One-Call system within a specific period of time, which varies from state to state. In some states, the required response time is as short as two hours. If necessary, authorized persons may have to begin emergency excavation or demolition of pipeline facilities to protect life, health, or property before the line can be marked. Operators are expected to exercise reasonable precautions to protect underground facilities.

1.2.0 Pipeline Locating and Marking

Before any work at the excavation site begins, it is critical for personnel to precisely locate the pipeline and all other buried structures nearby. Improper excavation may cause pipeline damage or even lead to death. It is the spotter's responsibility to ensure that the pipe and other structures have been located and properly marked before excavation begins.

As part of the line-locating process, a qualified individual must verify the pipe depth of cover. This task is often performed by the spotter, but that may vary from company to company. Depth of cover for pipe and cable is usually specified on pipeline maps and typically varies according to the location and use of the area over the installed system. However, the actual cover of a pipeline or cable may have changed after installation due to soil erosion. Depth of cover is determined using probe rods.

Unless specifically exempted, all new construction of oil and gas pipelines must have the pipe buried below the level of cultivation. However, the operator may have deeper depth-of-cover standards than those indicated in 49 *CFR* 192 or 49 *CFR* 195. *Table 1* and *Table 2* show examples of depth requirements.

> **NOTE**
> Many operators require more cover than these standards. Always check the operator's specific depth-of-cover requirements.

Status: Dispatched to ONG2001 on 07/17/2013 **Type: Standard**

Customers: KMEP03,

```
Send To: KMEP03        Seq No:    1   Map Ref:
Ticket No:   13288068                 STANDARD                NEW

Transmit      Date:   7/17/13    Time: 10:56 AM      Op: webusr
Original Call Date:   7/17/13    Time:  7:08 AM      Op: webusr
Work to Begin Date:   7/20/13    Time: 12:15 AM

Company       : WEBBERS                               Best Time:  AM
Caller Name : JOHN DOE                                Phone: (563)555-5555
Alt. Contact: JOHN DOE                                Phone: (563)555-5555
Caller Address: 5630 HWY 130                          Fax Phone: (563)555-5555
                 TIPTON, IA 52772

Email Address:   doe.john@webbers.com
Type of Work: PIPE LINE MAINTENANCE

Explosives: N          Duration: 2 WEEKS               Trenchless Ex.: N
Work Being Done For: ONEOK

State: KS              County: MCPHERSON                  Place: LINDSBORG
Address:               Street: W CORONADO BLUFF RD
Nearest Intersecting Street: S FOSSE RD

Location of Work: FROM THE INTERSECTION GO EAST APPROX. 300 FEET TO THE PIPE
: LINE CROSSING.  FROM THAT POINT GO SOUTH 100 FEET TO THE SURVEY STAKE IN
THE
: FIELD.  FROM THE STAKE MARK IN A 500 FOOT RADIUS MARKING ALL MAIN, SERVICE
: AND ABANDON LINES.  DIG 110BC-12-F14. ALL DIGGING IS S OF THE COUNTY LINE
: ROAD.

Remarks:

Twp: 16S      Rng: 4W        Sect-Qtr: 34-SW,33-SE                      MP:
N
Twp: 17S      Rng: 4W        Sect-Qtr: 4-NE,3-NW
Caller Lat: 38.6093        Lon: -97.75673             Zone:         Nad: 83
Map NW Lat: 38.609370      Lon: -97.763433        SE Lat: 38.606188 Lon: -
97.752050

Kansas One-Call will notify the following underground facility operator(s):

Link To Map for KMEP03: http://ks.itic.occinc.com/B6ZD-WNH-S2F-2HZ
```

Ticket History (All times Eastern)

Date	Time	Screener	Remark
07/17/13	11:58:23	DigTrack	Ticket received from the One Call Center.
07/17/13	12:13:47	ONG4540	Ticket dispatched to ONG2001
07/17/13	12:14:05	DigTrack	Ticket sent to EM2001 using oke12001@oneok.com.

Figure 1 Sample One-Call Ticket.

Table 1 New Construction Liquid Pipeline Depth Requirements

Location	Cover (inches)	Notes
Industrial, commercial, and residential areas	36	
Inland water of at least 100 ft width high water to high water	48	Water crossings
Drainage ditches at public roads and railroads	36	Bar-Ditch
Deepwater port safety zone	48	
Gulf of Mexico and offshore water less than 12 ft at low tide	36	
All other areas	30	Railroads, highways, others
Underwater	up to 96	

Table 2 New Construction Gas Pipeline Depth Requirements

Location	Normal Soil (inches)	Consolidated Rock (inches)
Class 1 locations	30	18
Class 2, 3, and 4 locations	36	24
Drainage ditches at public roads and railroads	36	24

After the pipeline is precisely located, a highly visible device called a temporary line marker must be used. These markers are installed at the excavation site to allow excavators to more precisely locate the pipeline between permanent markers. Acceptable types of temporary line markers include wires, flags, stakes, or paint. The choice may depend on circumstances and the length of line to be marked. Temporary markers should be durable enough to last throughout the proposed work. Temporary markers must also follow the Uniform Color Code for Temporary Marking of Underground Facilities established by the American Public Works Association (APWA). *Figure 2* shows the colors of the markers as they correspond to the utilities.

NOTE

Procedures for locating and marking pipelines are covered in more detail in NCCER Module CT14_1-17, *Locate Line.* Additional information about color codes for pipeline markers is covered in NCCER Module CT14_2-17 (*Install, Inspect, and Maintain Permanent Marker*), as well as NCCER Module CT14_5-17 (*Install, Inspect, and Maintain Temporary Marker*).

1.3.0 Underground Pipeline Clearance Requirements

The regulation 49 *CFR* 195.250 states that any pipeline installed underground must have a minimum of 12" clearance between the outside of the pipe and the extremity of any other underground structure. In situations where 12" of clearance is not practical, the distance may be reduced if the provisions for corrosion control are adequate. Clearance for drainage tile may be under 12" but not less than 2". According to 49 *CFR* Part 192.325, each main must be installed with enough clearance to allow proper maintenance and to protect it from damage that could result from being too close to another structure. You must determine clearance to ensure that separation between company assets and foreign assets is adequate to prevent electrical interference and physical contact between the two.

WHITE:	PROPOSED EXCAVATION
PINK:	TEMPORARY SURVEY MARKINGS
RED:	ELECTRIC POWER LINES, CABLES, CONDUIT AND LIGHTING CABLES
YELLOW:	GAS, OIL, STEAM, PETROLEUM OR GASEOUS MATERIALS
ORANGE:	COMMUNICATION, ALARM OR SIGNAL LINES, CABLES OR CONDUIT
BLUE:	POTABLE WATER
PURPLE:	RECLAIMED WATER, IRRIGATION AND SLURRY LINES
GREEN:	SEWER AND DRAIN LINES

Figure 2 Uniform Color Code for Temporary Marking of Underground Facilities.

Figure 3 shows a typical report form used to document the general condition of pipe clearance between a company pipeline and a foreign pipe or structure.

Pipeline condition and any **abnormal operating conditions (AOCs)** discovered should be documented and reported to appropriate company personnel. Be sure to follow your company's written and approved procedure for reporting and AOCs, including:

- Corrosion
- Fire and explosion
- Coating damage
- Pipe damage
- Hazardous material release

In order to measure clearance from existing pipe to underground structure, you must first determine if the pipe or foreign structure was installed by excavation, boring, or direction drilling. This information can be found by consulting as-built records or maps for the foreign pipeline owner, such as direction drilling records. Determining the type of installation is important because it establishes the method of clearance measurement that must be utilized.

1.4.0 OSHA Regulations for Excavation

To safeguard personnel at the excavation site, the spotter must have a strong understanding of the OSHA regulations that pertain to excavation. Digging in the vicinity of any underground structure requires extreme care and caution. In fact, all decisions affecting safety on excavation job sites must be made by a person specifically defined by OSHA as a *competent person*. According to OSHA regulations, a competent person is an individual who is capable of identifying existing and predictable hazards in the surroundings, or working conditions that are unsanitary, hazardous, or dangerous to personnel and who has authorization to take prompt corrective measures to eliminate them. An excavation site must be inspected by a competent person before any personnel may enter the site. The competent person must also inspect the excavation site daily or when hazards or site conditions change.

OSHA has strict and specific requirements that apply to any open excavation work, including trenches. The purpose of these requirements is to ensure the safety of excavation workers, because accidents during excavation are among the leading causes of injuries and deaths in the construction industry. These requirements are explained in detail in OSHA 29 *CFR* 1926. The following is a summary of the general requirements:

- Any obstacles near the excavation site that pose a potential hazard to personnel must be removed.
- Before personnel may enter the excavation, atmospheric testing must be performed to ensure safe atmospheric conditions inside the excavation.
- Personnel should not enter excavations in which there is excess water unless the relevant precautions have been taken.
- Appropriate **support systems**, such as shoring or bracing, must be used when needed to ensure the stability of the excavation site.
- Ramps, both those that serve as exits for personnel and those that allow equipment to be brought into the excavation, must be designed by a competent person.
- Personnel should wear reflective vests or other visible clothing, particularly at sites where vehicle traffic is close by.
- Personnel should stand clear of vehicles and any equipment carrying heavy loads of soil.
- Appropriate barricades or other warning devices must be used when mobile equipment is operated near an excavation.
- Walkways or bridges designed to allow personnel to cross over excavations must have guardrails.
- Protection must be put in place to prevent loose soil or rock from injuring personnel.
- If the site is deemed unsafe by a competent person, personnel must not be allowed to work at the site until conditions are corrected.

XYZ PIPELINE COMPANY
REPORT OF FOREIGN LINE CROSSING

DATE ______

LOCATION OF CROSSING:
LOCATION NO. ______
MAP NO. ______
PROPERTY NO. ______
BPL STATION NO. ______
(AT CROSSING)

DESCRIPTION OF FOREIGN LINE OR STRUCTURE:
SIZE, O.D. ______
TYPE ______
COMPOSITION ______
COATING ______
BOND YES ______ NO ______
TEST STATION INSTALLED YES ____ NO ____

OWNER:
NAME ______
ADDRESS ______
______
TELEPHONE ______

CONTRACTOR:
NAME ______
ADDRESS ______
______
TELEPHONE ______

INSPECTION BY:
______
XYZ PIPE LINE COMPANY

DATE RECEIVED DIVISION ______
DATE SUBMITTED TO ______
ENGINEERING ______

BY: ______

CATHODIC PROTECTION COMMENTS:

PIPE TO SOIL READINGS, ETC. ______
______

REMARKS:
______
______
______
______

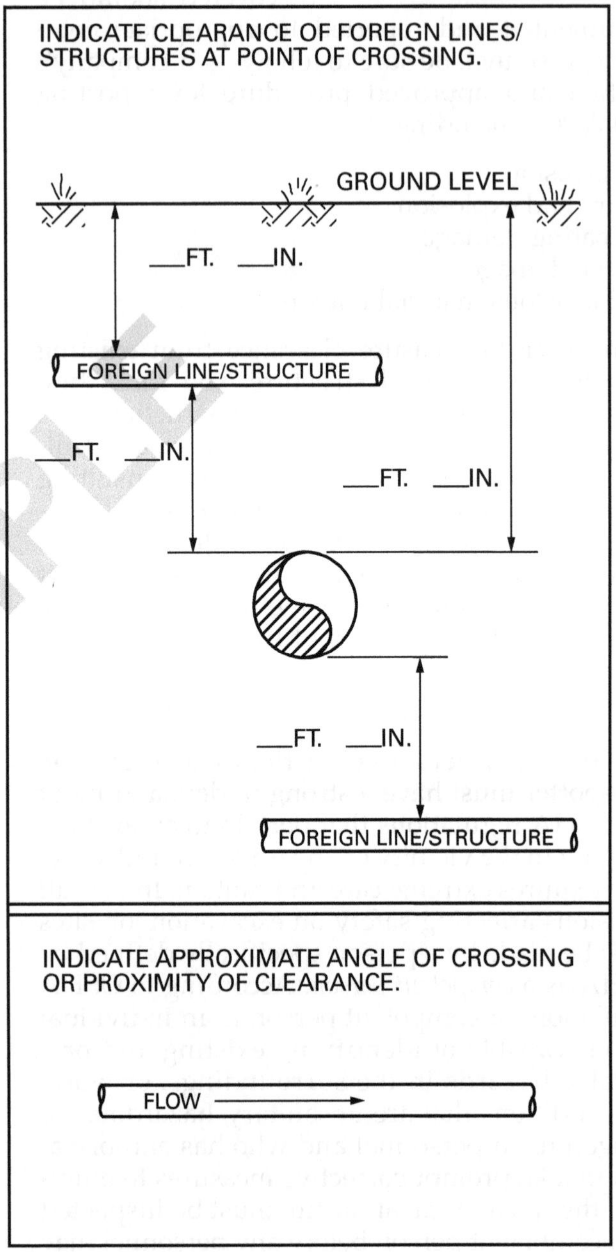

Figure 3 Sample pipeline and structure clearance report.

Additional Resources

It is assumed that you will learn to perform the purging procedure under the direct supervision of your crew leader, foreman, or instructor. For independent study, refer to the recommended Additional Resources at the back of this module.

1.0.0 Section Review

1. Before excavating any pipeline sections, personnel must obtain any required _____.
   a. maps
   b. tools
   c. markers
   d. permits

2. Since depth of cover may have changed since installation, it must be verified by _____.
   a. a qualified individual
   b. hand digging
   c. an independent testing company
   d. an electronic depth finder

3. Personnel should generally not enter an excavation site where there is _____.
   a. a support system
   b. a ramp
   c. excess water
   d. shoring

Section Two

2.0.0 Excavation Responsibilities

Objective

Describe the tasks required of the designated excavation observer/spotter during an excavation.

a. Explain the dangers associated with excavation.
b. Identify the tools used for excavation and the appropriate use for each.
c. Describe the technique of sloping during excavation work.
d. Identify methods used to protect the pipeline during excavation.

Trade Terms

Angle of repose: The greatest angle above the horizontal plane at which the soil will rest without sliding. This angle varies depending on factors such as soil type and weather conditions, and is usually between 0 and 90 degrees.

Backfilling: The technique for covering a completed pipeline so that adequate fill material is underneath and over the pipe to protect it from loose rocks, abrasion, shifting, and washouts.

Bench: Excavation formed with two or more sections that look like stairs in cross section; used to prevent cave-ins.

Fill: Amount of material needed to bring existing ground to a desired grade.

Hydro-vac: A tool used to hydraulically remove soil and expose a pipeline, minimizing the risk associated with other sharp metal tools. (Also called a *water-vac*.)

Pot-holing: A procedure used to locate underground installations by digging a small hole in the ground.

Sloping: A technique used to cut the side walls of a trench at an angle that will be more stable; the angle should start at the top of the trench wall and continue to the floor of the trench.

Well points: Devices installed in waterlogged soil to dry out areas along the pipe ditch line. Hundreds may be needed in some areas.

After the spotter has determined that all precautions have been taken and all required preparation tasks have been completed, excavation can begin. The spotter must be extremely knowledgeable about the dangers specific to excavation and how to avoid them. During excavation, the spotter must continuously watch for any circumstances that could jeopardize the integrity of the excavation site or the pipeline itself. Abnormal operating conditions can also occur at any time during an excavation. The spotter must also be able to recognize and properly react to any AOCs that are encountered.

It is the spotter's duty to ensure that the tolerance zone is maintained throughout the excavation, and that the appropriate tools are used accordingly. The spotter must therefore understand the types of tools used and the appropriate use for each. It is the spotter's responsibility to use the permanent and temporary markers to ensure that the equipment used at the excavation site does not produce unnecessary strain on any buried structures.

Some examples of permanent markers are shown in *Figure 4*. Proper support for the pipe must be provided for at all times as well. A strong understanding of **sloping**, excavation techniques, and soil characteristics, is also required.

For the safety of personnel as well as the pipeline, the spotter must make certain that if the excavation site is left unattended at any time, it is properly secured. Temporary fencing is typically used, but the appropriate methods for securing the site vary and must follow company-specific policies and procedures.

When an excavation is complete, the spotter must notify the control center. Control center personnel should continue closely monitoring the pipeline after excavation activities are completed to ensure proper pressure and flow. The spotter must document the excavation after its completion, per company procedures. Information to be documented typically includes:

- Date and location
- Name of excavator
- Purpose and scope of excavation
- One-Call information
- Depth of cover

Figure 4 Examples of permanent markers.

2.1.0 Excavation Dangers

Excavation is one of the most hazardous parts of pipeline maintenance. Any resulting damage to the pipeline could cause a deadly explosion. Even something as simple as slipping, tripping, or falling can be deadly if it happens in or near a trench. In excavations that exceed 4 feet in depth, access/egress must be provided within 25 feet of each worker. Trenches must be inspected daily before work begins, after rainstorms and other weather events, and when there is any sign of change in stability. Any time the excavation is left unattended, it must be re-inspected. Appropriate personal protective equipment (PPE) must be used at all times.

Some company procedures require pressure reduction in the pipeline before excavation begins. Lowering the pressure may reduce the hazards during excavation or movement of the line.

It is critical for personnel to be aware of the hazards involved in excavation work and how to avoid them. This section provides more specific information about the following hazards:

- Cave-ins
- Gases
- Electricity
- Falling objects
- Fire or explosion

2.1.1 Cave-Ins

Cave-ins are the most obvious dangers involved in excavations and are the single most hazardous part of excavation work. If a worker is buried by collapsing trench walls, the weight alone can be fatal. Being buried up to the waist can provide enough pressure to result in death. If a worker's face is buried, the lack of air may result in suffocation. Just one cubic yard of earth can weigh up to 3,000 pounds. (This is illustrated in *Figure 5*.)

To prevent cave-ins, the walls of earth that are left after digging must be adequately supported. Excavations of five feet or deeper must use shielding devices (*Figure 6*) and shoring devices or sloping to prevent cave-ins. A shoring or shielding system provides a structure such as a metal hydraulic, mechanical, or timber system that supports the sides of an excavation. This structure may be sloping, vertical, or horizontal. OSHA 29 *CFR* Part 1926 defines acceptable trench protective devices and how they are to be used. In general, the trench is no wider than it is deep. Any excavation 20 feet or deeper requires a professional engineer to design an appropriate system.

Personnel must observe the following precautions/considerations to protect themselves from potential cave-ins:

- An excavation is not safe to enter unless the proper support structure or sloping system is provided.
- Excavations in or near backfilled or previously excavated ground are especially dangerous because the soil is loose and will not support itself.
- Saturated soil increases the possibility of a cave-in. The increased pressure of the water exerted on the soil can be the factor that causes the walls to collapse. Personnel should not be allowed in a trench that contains water because it could cause erosion of the trench.
- Clay can be extremely dangerous if dried by the sun. Large chunks of material can break off a trench wall after having been stable and solid for a long period of time (*Figure 7*).
- It is not safe to assume that frozen earth is stable; frozen ground is not an alternative to shoring.
- Shoring must be adequate to overcome additional pressures from piles of excavated material, adjoining structures, vehicular traffic, and nearby equipment.

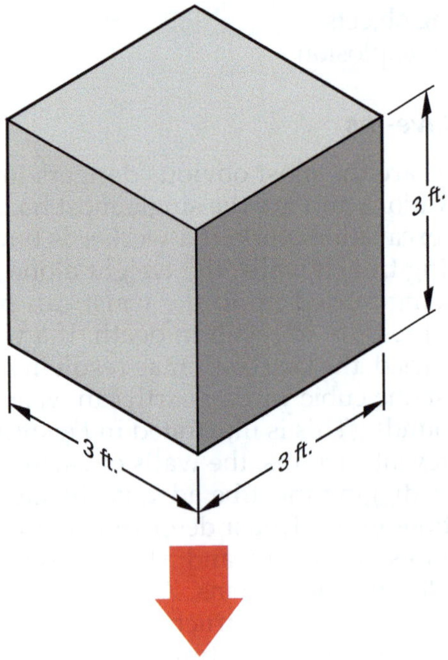

Figure 5 Dimensions of one cubic yard of earth.

An open cut or trench can become more unstable as work is being done. There are some signs to watch for that indicate an unstable trench. Signs such as these must be reported immediately. Work must be stopped until corrective measures are taken. This is necessary even if a support system is being used. Refer to OSHA regulations regarding the safety requirements for trench support systems.

Follow these rules to ensure safety against cave-ins:

- Never enter an excavation to install shoring or trench boxes. These must be in place before personnel are allowed to enter the trench.
- Check that shoring uprights and cross braces are properly installed.
- Never enter an excavation behind the trench box after it has been moved; this should be backfilled as soon as the box is moved.
- Ensure that an exit is available (within 25 feet) where any work is to be done. Ladders should have rails that extend three feet above the landing.

2.1.2 Gases

In some situations, an excavation should be considered a confined space. This means that all appropriate evaluation and controls must be undertaken to ensure against a contaminated atmosphere or toxic substances. Excavations can fill with dangerous gases, or oxygen can be displaced from the bottom of the excavation if located near facilities that store large volumes of chemicals. The liquids or gases, even exhaust fumes from vehicles, could flow or leak into the excavation. When excavating a pipeline for repairs, the liquids or gases could have already leaked into the ground around the pipe. An excavation should never be entered until the atmosphere is determined to be safe by authorized persons and methods.

(A)

(B)

Figure 6 Examples of shielding devices.

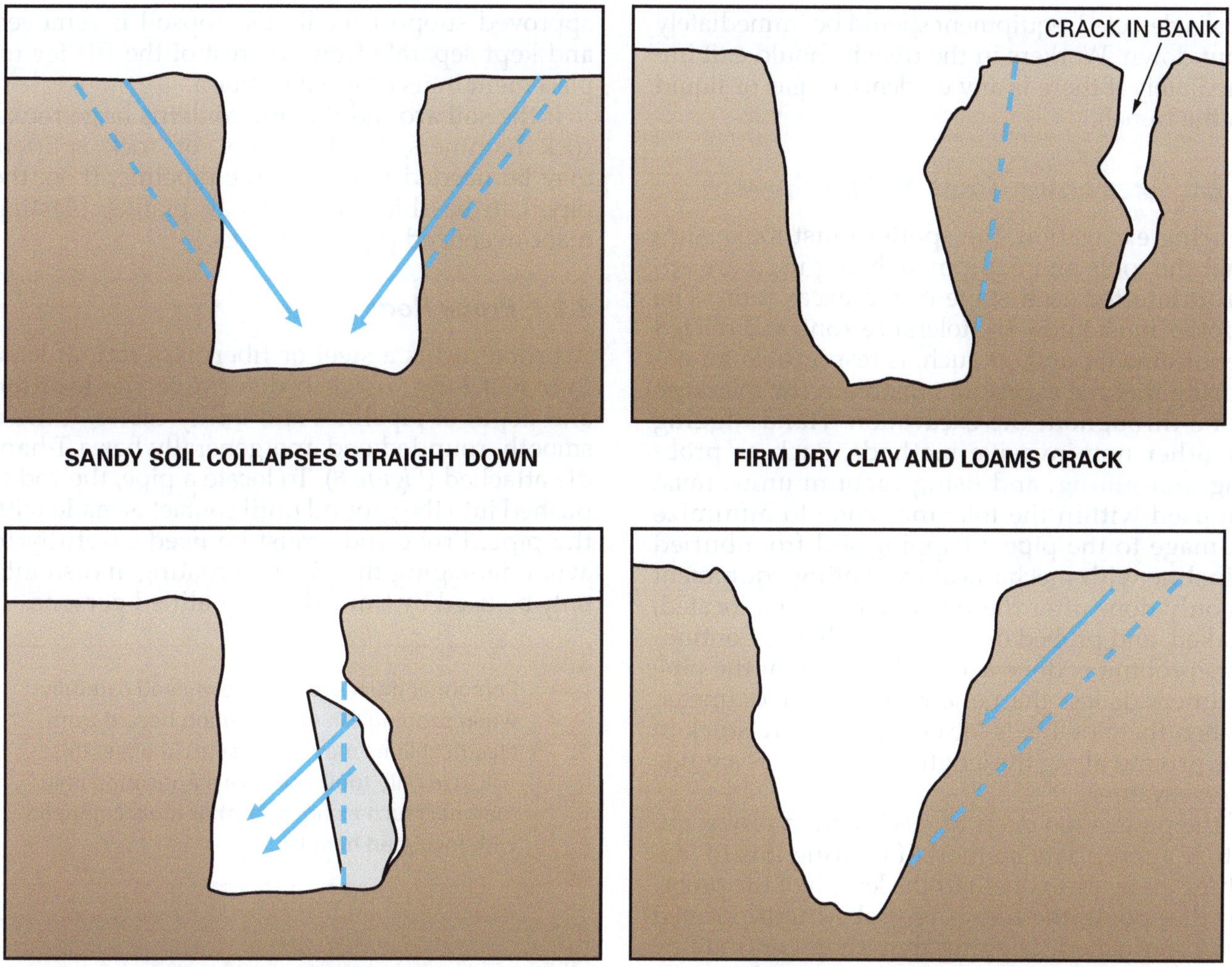

Figure 7 Behavior of clay and other soil types.

2.1.3 Electrical Shock

The danger of electrical shock exists during any excavation work. For example, the insulation of an electrical cable could be pierced during the excavation. In addition, when overhead lines are near the excavation, the excavator could come in contact with them, knocking them into the excavation or electrifying the boom. If the boom is electrified, the pipe it is lowering into the excavation will be electrified as well and could impart a deadly shock.

2.1.4 Falling Objects

Anything around the edges of the excavation could become a dangerous falling object if dropped or knocked over the edge. This includes piles of soil, tools, chunks of dirt from the excavator bucket, parts of the bucket, and other objects. Workers on the surface must keep such objects away from the edge of the excavation, and must keep any excavated dirt at least two feet away. Those in the trench should have on PPE at all times.

> **WARNING!**
>
> Never park vehicles next to the trench. In addition to the potential for the vehicle to fall into the trench, vehicle exhaust could fill a trench if a vehicle is parked too close to it, resulting in carbon monoxide poisoning. Always park vehicles far away from the trench and chock the wheels.

2.1.5 Fire or Explosion

Gases and flammable liquids can accumulate in a trench from natural sources or from the pipeline being repaired, creating a fire and/or explosion hazard. The trench may have been checked and cleared of such substances before workers entered the trench, but the buildup may occur later. In situations where gas may enter the trench, a constant monitoring system may be needed. When gas is detected, equipment should be immediately shut down. Workers in the trench should exit immediately if there is any evidence of gas or liquid in the trench.

2.2.0 Excavation Tools and Equipment

During excavation, the spotter must make sure that the tools and equipment being used are appropriate for each stage of the excavation. The spotter must know the tolerance zone and related requirements, and, as such, is responsible for ensuring that the excavators maintain the tolerance zone throughout the excavation. Hand digging or other non-invasive methods, such as probing, pot-holing, and using vacuum units, must be used within the tolerance zone to minimize damage to the pipe. Stripping soil from buried pipeline with mechanical excavating equipment is only done after the pipeline has been located, staked, and probed or exposed by hand. Continuous probing is done across the width of the pipe at intervals less than one-half the pipe diameter. When the pipeline is located, probes are stuck in the ground along the length of the pipe as a guide to excavators.

If the pipe is too deep to reach with a probe, another appropriate method of locating it is to machine excavate to one-half the length of the probe. At that point, the line is re-probed until located and confirmed. Digging may then continue by hand or with a backhoe running parallel to the line. After the line is exposed, excavation must proceed parallel to the line so that the operator can keep it in view. The backhoe bucket should never be closer than 18 inches to the pipeline. Note that the regulations regarding the distance from the pipeline vary from state to state, so always check local regulations.

Final stripping is done by hand. The cover is removed carefully to avoid damaging the pipe, and spaces are dug at specified intervals under the pipe for the insertion of dirt, plugs, skids, or approved support mediums. Topsoil is removed and kept separate from the rest of the fill for replacement in its original position.

If the soil around the line contains large rocks, rock hammers, jackhammers, or rock splitters may be needed to expose the pipeline. If so, the pipe can be shielded with oak planks, blasting mats, oversized pipe, or sleeves.

2.2.1 Probe Rods

A probe rod is a steel or fiberglass rod, at least four feet long, used to determine the location and depth of pipelines and utility cables. It has a smooth, rounded end and generally has a T-handle attached (*Figure 8*). To locate a pipe, the rod is pushed into the ground until contact is made with the pipe. Probe rods must be used carefully to avoid damaging the pipeline coating and should only be used by trained and qualified persons.

WARNING!

Personnel must proceed slowly and carefully when probing. An electrocution hazard from electrical lines may be present. To offset this risk, use only tools made of a nonconductive material (such as fiberglass) or tools equipped with insulated handles.

NOTE

Some companies do not allow probing at all. Always follow company-specific policies.

2.2.2 Hand digging and Pot-Holing

Hand digging and pot-holing are methods that may be used when excavating within the tolerance zone. Pot-holing is a method of physically locating the pipeline by hand digging small holes in the ground. Pot-holing can be more hazardous than probing because tools with sharp edges could po-

Cave-In Caused by Water

Four employees were laying a lateral sewer line at a building site. The foreman and a laborer were laying an 8"-diameter, 20'-long plastic sewer pipe in the bottom of a trench that was 36" wide, 9' deep, and approximately 50' long. The trench was neither sloped nor shored, and there was water entering it along a rock seam near the bottom. The west side of the trench caved in near the bottom, burying one employee to his chest and completely covering the other. The rescue operation took two hours for the first man and four hours for the second. Both men died.

The Bottom Line: Working in an excavation can be fatal. Always ensure your trench is safe.

Source: The Occupational Safety and Health Administration (OSHA)

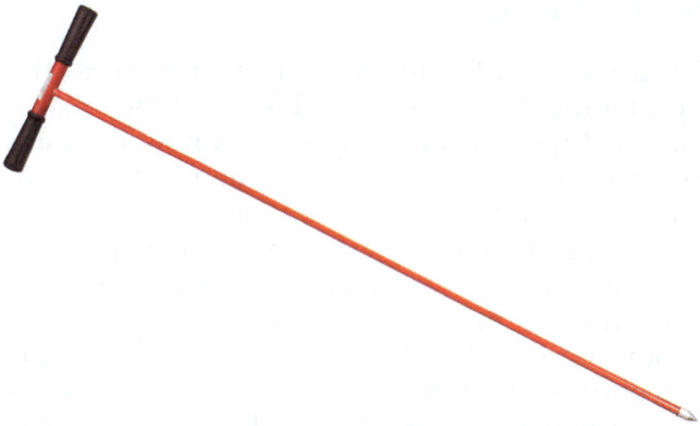

Figure 8 Typical probe rod.

tentially damage the pipeline or other utility lines. When pot-holing, individuals should proceed with caution until the pipeline is exposed.

> **WARNING!**
>
> When hand digging or pot-holing, personnel must use extreme caution and proceed slowly because electrical lines could be present. It is critical that personnel only use tools made of nonconductive materials or tools equipped with electrically insulated handles to minimize the risk of electrocution.

2.2.3 Vacuum-Powered Tools

The advantage of using vacuum-powered tools is that they offer the least risk of damaging the pipeline. Shown in *Figure 9*, a **hydro-vac** unit (also known as a *water-vac*) is an effective tool that hydraulically bores and air-vacuums out a hole over a pipeline. It opens a clean, smooth hole with no damage to the pipeline or cable. Water-vac units minimize disruption of other work on the site.

Air-vac units operate on the same basic principle as water-vac equipment. Air-vac trucks bore holes with compressed air rather than water and vacuum the debris from the hole over the pipeline.

> **WARNING!**
>
> Operating air-vac or water-vac equipment can be dangerous; always use extreme caution. Be sure to stay clear of the vacuum's path as the vacuum force can be extremely strong. It is also imperative that personnel wear appropriate personal protective equipment (PPE) for eye protection since dirt and rock particles are often propelled into the air.

Both types of vacuum units are designed to reduce the risk associated with using a pick, shovel, post hole digger, or other sharp metal tools and probes to expose the pipeline.

2.2.4 Backhoes

A backhoe is a large, powerful piece of construction equipment that can be safely used to quickly expose underground pipe and cables by moving large amounts of cover. The size and power of the back-hoe can significantly reduce the time needed to perform the job. Only personnel qualified to operate a backhoe may do so, and the operator must be competent at working around below-grade facilities.

All mechanical excavating equipment must have a bar installed across the teeth of the bucket so that there are no exposed teeth that could puncture the pipe. It is never acceptable to excavate using a backhoe when personnel are in a trench.

Figure 9 Hydro-vac truck.

CAUTION

Personnel must exercise care while digging around known pipeline and cable locations. Digging too deep or off center from the target area could cause damage to pipes and cables before the backhoe operator is able to reverse the direction of the equipment. The initial excavation often causes the temporary line markings to move. To help the backhoe operator, the line should be periodically relocated in the hole as the hole deepens and the pipeline is exposed.

WARNING!

It is never acceptable to excavate using a backhoe when employees are in a trench.

As a reminder, before resorting to heavy equipment to expose a pipeline or utility service, personnel must make sure that other critical preliminary steps have been completed. These could include any or all of the following:

- Consult available alignment sheets/strip maps to obtain the pipe location and depth information.
- Verify that locating equipment items are operational.
- Physically locate the line.
- Ensure that appropriate supervisors are informed.
- Use alternate methods (hand digging or water-vacuuming) to perform the work, if possible.
- Proceed with caution.

2.2.5 Well Points

During an excavation project, groundwater may be encountered. When working in waterlogged soils, the excavator may want to install **well points** to dewater the ditch line. Well points are devices that can effectively dry out extensive areas along the ditch line by continuously removing underground water. Dewatering is important because material that contains too much moisture may result in trench failure, leaving the project site unprepared for building or compaction.

Each well point functions like a submersible pump. A hollow steel rod with wire mesh on the bottom end is driven into the ground. A flexible hose is attached to its top end and connected to a portable pump. The pump carries the water away from the work site. Many such rods are installed along the line, and the hoses connect to the pump through a pipe along the ground. Dewatering may, in some cases, require additional permits.

2.3.0 Sloping

To understand safe excavation practices, the spotter must be knowledgeable about the excavation technique called sloping and how it relates to the types of soil that may be encountered. Many excavations are started with a vertical cut. Although some soils will stand to considerable depths when cut vertically, most will not. They will fall to a more stable angle, dumping large amounts of material into the trench. Sloping is a technique that involves cutting the side walls of a trench at an angle that makes it more stable. The angle should start at the top of the trench wall and continue to the floor of the trench. The slope is cut at the **angle of repose** for the particular soils. The angle of repose is the greatest angle above the horizontal plane at which the soil will rest without sliding. Refer to the *Appendix* at the end of this module for more information about the properties of soil that may be encountered during an excavation project.

2.3.1 Soil Types

OSHA classifies soils into four categories and determines the maximum allowable slope for each. The classifications are based on stability and range from stable rock at the high end to Type C as least stable. The maximum angles for each soil type are shown in *Figure 10.*

The four OSHA categorized soil types are as follows:

- *Stable rock* – Natural solid material that remains intact when exposed. Stable rock includes solid rock, shale, and cemented sands or gravel.
- *Type A soil* – Cohesive soil, with a compressive strength of at least 1.5 tons per square foot. These soils include clay, silty clay, sandy clay, and clay loam. No soil can be Type A if the soil is fissured, subject to vibration, has been previously disturbed, or has any factors that decrease stability.
- *Type B soil* – Cohesive soil, with a compressive strength of 0.5 to 1.5 tons per square foot. These soils include granular soils, such as angular gravel, crushed rock, silt, sandy loam, unstable rock, and any unstable Type A soils.
- *Type C soil* – Unstable, cohesive soil with a compressive strength less than 0.5 tons per square foot. These soils include gravel, loamy soil, sand, submerged soil, soil with water seepage, and unstable submerged rock. Previously excavated soil is classified as Type C soil.

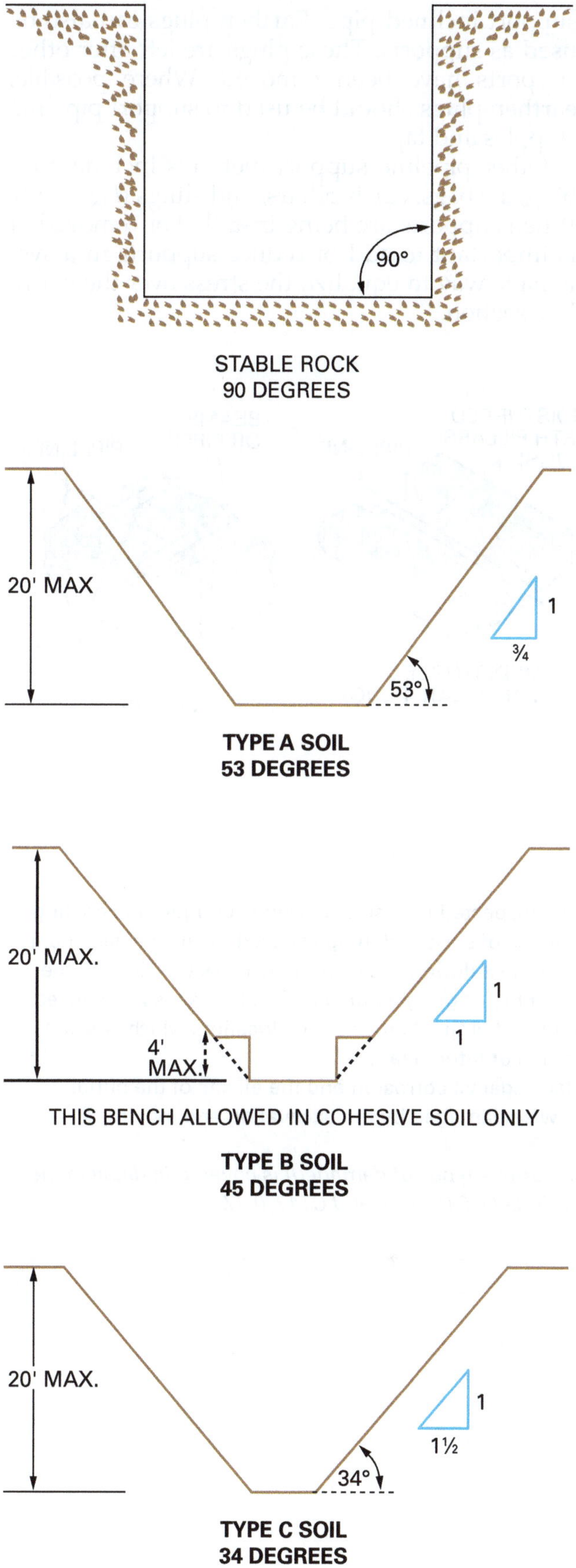

Figure 10 Maximum angles for soil types.

2.3.2 Sloping Requirements

Sloping requirements for an excavation vary based on a variety of factors. It is critical for personnel involved in excavations to understand the following sloping requirements:

- The angle of repose must be flattened when water is present or when there are silty conditions, loose boulders, and areas where erosion and deep frost are possible.
- The excavated soil, or spoil, must be banked at least two feet away from the edge of the trench, as must any tools and other materials.
- All excavations deeper than 20 feet must also have some type of pre-engineered, certified support system installed.
- Special precautions must be taken in sloping the sides of an excavation adjacent to a previously filled excavation or fill.
- End plates, sheet piling, or other methods of retaining soil should be used at the ditch ends if they cannot be sloped properly.

Benching is allowed in cohesive soils. A **bench** is an excavation of the walls in steps, which must rise on the approximate allowable angle of repose. Formed with two or more sections that look like stairs in cross section, it is also called a step-back excavation. The same rules apply for benching as for slopes. The dimensions of a simple bench are shown in *Figure 11*.

2.4.0 Protecting Pipe During Excavation

It is important that the pipeline is protected during excavation as well as during **backfilling**. To avoid damaging the pipe, personnel must handle the pipe carefully and support it properly.

> **NOTE**
> Backfilling a trench after excavation is a separate covered task. It is covered in detail in NCCER Module CT39_0-17, *Backfilling a Trench Following Maintenance.*

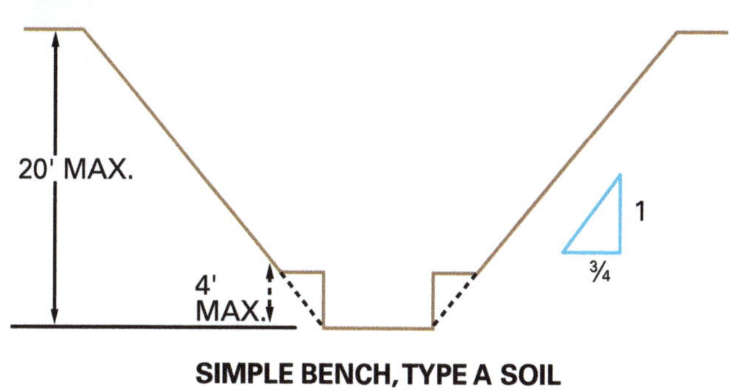

Figure 11 Simple bench.

The pipe must be properly secured and supported to reduce the stress on the pipe and the welds. When the pipe is first exposed and as supporting material is removed from under the pipe, skids or other supports must be placed beneath the pipe about every 20 feet. The maximum span between the locations of the supports should be determined by an engineer, based on the pipe specifications and other factors.

Skids provide support across the ditch, and skids, blocks, sandbags, airbags, and other materials are used on the bottom of the ditch to support inclined pipe. Earthen plugs can also be used as supports. These plugs are left after other supports have been removed. Where possible, earthen plugs should be used to support pipeline stopples and taps.

Other pipeline support methods include cribbing, air bags, earth pillars, and slings (*Figure 12*). When supports are being installed or removed, it is important to add or reduce supports in a systematic way to equalize the stress over the entire line section.

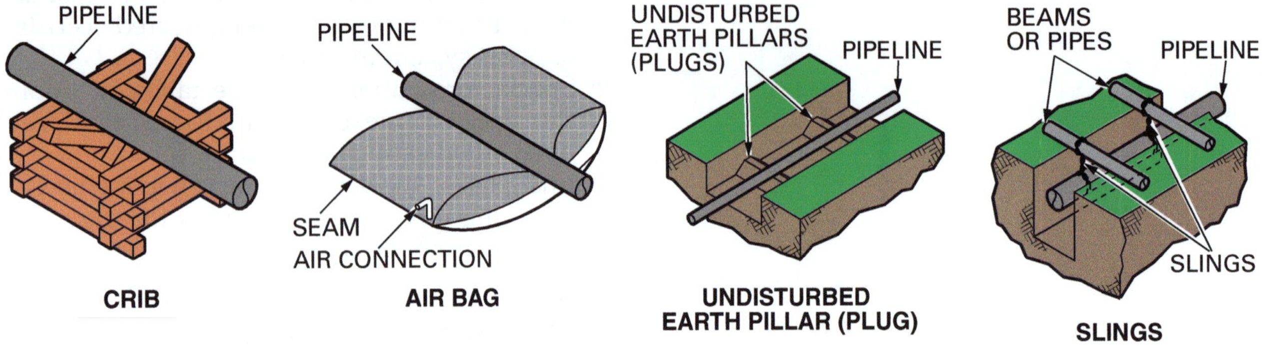

Figure 12 Pipeline support methods.

Inspecting Pipe for Damage

Whenever pipe is exposed and before it is backfilled, it must be inspected for damage. Dents, gouges, and coating damage are types of damage often encountered. Gouges are areas of surface damage caused by the displacement or removal of metal. Dents and gouges are usually caused by contact during excavation, by rocks too close to the pipe, or by objects falling into the excavation. When a pipe is dented or gouged, the pipe wall thickness is reduced at the point of impact. As a result, stress on the pipe is concentrated at the dent or gouge location, which weakens the pipe. The weakened condition of the pipe wall can be a source of future leaks.

The coating on the pipe is essential to protect the pipe surface against corrosion and the effects of the burial environment. The coating may easily be cut, scraped, or otherwise damaged by the excavation and moving equipment, or by rocks in the fill.

Note: Inspection of a pipeline after it has been exposed (for various types of damaged) is covered in detail in the following separate NCCER Covered Task Modules: CT5_1-17, CT5_2-17, CT5_3-17, and CT12_0-17.

Additional Resources

It is assumed that you will learn to perform the purging procedure under the direct supervision of your crew leader, foreman, or instructor. For independent study, refer to the recommended Additional Resources at the back of this module.

2.0.0 Section Review

1. Any time an excavation is left unattended, _____.
   a. the pipe location must be verified again
   b. permits must be renewed
   c. the site must be re-inspected
   d. One-Call must be notified

2. Hand digging is the only type of excavation permitted within a _____.
   a. tolerance zone
   b. buffer zone
   c. critical zone
   d. delimited zone

3. Sloping requirements dictate that an excavation must have a pre-engineered support system, such as trenching, shoring, etc., installed if it is deeper than _____.
   a. 5 feet
   b. 10 feet
   c. 15 feet
   d. 20 feet

4. An example of a support method used to protect pipe during an excavation is _____.
   a. cribbing
   b. probing
   c. stripping
   d. pot-holing

SECTION THREE

3.0.0 MEASURING PIPE CLEARANCE AND OBSERVING AN EXCAVATION

Objective

Describe the procedures for measuring pipe clearance and observing an excavation (CT32_0-17).

a. Identify the prerequisites, competencies, and abnormal operating conditions associated with observation of excavation activities.
b. Describe the general procedures for measuring pipe clearance.
c. Describe the general procedures for observing excavation activities.

Performance Tasks

1. Observe excavation activities (CT32_0-17).
   - Identify potential abnormal operating conditions that may occur during performance of this CT, and know the appropriate actions to take in response to them.
   - Utilize the appropriate personal protective equipment according to relevant company procedures.
   - Notify the control center of the beginning and end of work as required by company procedures.
   - Identify and mark hazards related to the pipeline that surround the excavation site (including underground hazards such as appurtenances to the pipeline).
   - Determine the method of installation for the pipe or foreign structure. If the directional drilling or boring method was used, determine the depths of the pipelines or other structures and compare the depths to determine clearance.
   - Ensure that the spoil pile is placed in a location that does not affect the integrity of the pipeline.
   - Require proper preparation of mechanical equipment for safe digging around the pipeline.
   - Ensure that soil excavated above, below, or on either side of the pipeline meets the company-specified tolerance zone.
   - Require hand or vacuum excavation of the soil nearest the pipeline.
   - Ensure that the pipe is supported as necessary while soil is being removed.
   - Keep area safe from encroachment.
   - Verify the excavation is safe to enter.
   - If structures are in an open trench, measure the distance between the pipeline and nearest foreign structure.
   - Approve the excavation for backfilling by notifying a supervisor, local crew leader, or other appropriate company personnel.
   - Complete appropriate documentation and notifications as required by operator's procedures.

3.1.0 Prerequisites, Competencies, and Abnormal Operating Conditions

Personnel must possess a variety of skills in order to observe excavation activities. Personnel must also be aware of the particular AOCs that may occur when during an excavation and know the appropriate actions to take in response to them.

3.1.1 Prerequisites and Competencies

Personnel responsible for observing an excavation must have an understanding of what the task entails. They must also be able to communicate effectively with the both the control center and local personnel throughout the process.

Personnel observing an excavation must fully understand and follow all applicable company policies and procedures, including damage prevention programs, safety procedures, and required documentation. They must be aware of the marked area to be excavated, as well as applicable digging criteria.

When observing an excavation, personnel must have an understanding of tools that are appropriate for the excavation, including (but not limited to) backhoes, jackhammers, hydro-vacs, vacuum excavators, shovels, and other hand tools. They must also be familiar with the One-Call System, and be able to consult a current One-Call Ticket when required.

3.1.2 Abnormal Operating Conditions

Personnel performing an observation of excavation activities must be able to identify potential abnormal operating conditions (AOCs) and know the proper response to each, should they occur. The following are some potential AOCs, and the appropriate action personnel should take in response to them:

- *An unintentional hit to the pipeline with no release of product occurs* – Personnel should stop the excavation. Notify appropriate local operations personnel and call 911 if appropriate. Request a line shutdown, if necessary.
- *An unintentional hit to the pipeline with release of product occurs* – Personnel should stop the excavation. Follow operator's emergency response procedures for notification and call 911 if appropriate. Request a line shutdown, if necessary.
- *An unintentional or pre-existing release of hazardous liquid or gas occurs* – Personnel should stop the excavation. Follow operator's emergency response plan.
- *The unintended ignition of hazardous liquid or gas causes a fire or explosion* – Evacuate the area, notify the appropriate personnel to shut down the system and isolate affected piping, and notify the fire department.
- *Discovery of unexpected physical damage to the pipeline (such as scraping, dents, or breaks in the coating of a pipeline) occurs* – Personnel should stop the excavation. Report damage per company policy. Resume excavation as appropriate, per company policy.
- *Discovery of an undocumented pipeline in the area of excavation occurs* – Personnel should stop the excavation. Make notifications per company policy, including to the appropriate One-Call Center. Resume excavation as appropriate, per company policy.
- *Improper support for the pipeline during excavation occurs* – Personnel should stop the excavation. Ensure proper support before resuming operations.

3.2.0 Measuring Pipe Clearance

There are two common types of methods used for measuring pipe clearance: directional drilling or boring methods, and open trench methods. This section provides general procedures for both.

3.2.1 Directional Drilling or Boring Method

The following steps outline a very general procedure for measuring pipe clearance using a directional drilling or boring method:

WARNING!

The steps in this process can vary widely by circumstances, company, location, and other factors. Always follow specific company and/or operator procedures for observing an excavation, including any applicable safety procedures.

Step 1 Identify potential abnormal operating conditions (AOCs) that may occur, and know the appropriate actions to take in response to them.

Step 2 Utilize the appropriate personal protective equipment according to relevant company procedures.

Step 3 Determine the depth of the company pipeline at the location in question. This may be done by one of the following methods:
- Search records and maps for the pipeline
- Probe the pipeline depth
- Perform spot excavations
- Use a depth-finding pipeline locator

Step 4 Determine the depth of the foreign pipeline or structure using one or more of the following methods or a suitable alternative:
- As-built records or maps for the foreign pipeline owner, such as directional drilling records
- Depth-finding pipeline locator

CAUTION

Depth-finding locators may not be accurate to the degree required. Allow additional clearance to avoid damaging the pipeline.

Step 5 Compare the depths of the two structures to determine clearance. This may done by:
- Calculation
- Using graphics, such as plotting a scale profile drawing of the two structures and noting the clearance. In some cases, this may be a simple as subtracting the depths to determine the clearance; however, other situations may call for plotting out the results accurately.

Step 6 Approve the excavation for backfilling by notifying your supervisor, local crew leader, or other appropriate personnel. Excavations should not be backfilled until approved by a company representative.

Step 7 Complete the exposed pipe report, foreign line crossing report, or other appropriate documentation and notifications as require by operator's procedures. Records are critical to ensure safety and pipeline integrity.

3.2.2 Open Trench Method

The following steps outline a very general procedure for measuring pipe clearance using an open trench method:

> **WARNING!**
> The steps in this process can vary widely by circumstances, company, location, and other factors. Always follow specific company and/or operator procedures for observing an excavation, including any applicable safety procedures.

Step 1 Identify potential AOCs that may occur, and know the appropriate actions to take in response to them.

Step 2 Utilize the appropriate personal protective equipment according to relevant company procedures.

Step 3 Verify the excavation is safe to enter.

> **WARNING!**
> Excavation poses cave-in hazards and other confined space issues, such as a hazardous atmosphere.

Step 4 Measure the distance between the pipeline and the nearest foreign structure.

> **CAUTION**
> Failure to maintain proper clearance can result in cathodic protection interference and potential stress from foreign structures contacting pipe due to soil settlement.

Step 5 Approve the excavation for backfilling by notifying your supervisor, local crew leader, or other appropriate personnel. Excavations should not be backfilled until approved by a company representative.

Step 6 Complete the exposed pipe report, foreign line crossing report, or other appropriate documentation and notifications as require by operator's procedures. Records are critical to ensure safety and pipeline integrity.

3.3.0 Observing Excavation Activities

The following steps outline a very general procedure for observing excavation activities:

> **WARNING!**
> The steps in this process can vary widely by circumstances, company, location, and other factors. Always follow specific company and/or operator procedures for observing an excavation, including any applicable safety procedures.

Step 1 Identify potential AOCs that may occur, and know the appropriate actions to take in response to them.

Step 2 Utilize the appropriate personal protective equipment according to relevant company procedures.

Step 3 Notify the control center of the beginning and end of work as required by company procedures.

Step 4 Identify and mark hazards related to the pipeline that surround the excavation site (including underground hazards such as appurtenances to the pipeline).

Step 5 Ensure that the spoil pile is placed in a location that does not affect the integrity of the pipeline.

Step 6 Require proper preparation of mechanical equipment for safe digging around the pipeline.

Step 7 Ensure that soil excavated above, below, or on either side of the pipeline meets the company-specified clearance distance.

Step 8 Require hand or vacuum excavation of the soil nearest the pipeline.

Step 9 Ensure that the pipe is supported as necessary while soil is being removed.

Step 10 Keep area safe from encroachment.

Step 11 Complete appropriate documentation and notifications as required by operator's procedures.

Additional Resources

It is assumed that you will learn to perform the purging procedure under the direct supervision of your crew leader, foreman, or instructor. For independent study, refer to the recommended Additional Resources at the back of this module.

3.0.0 Section Review

1. If a pipeline is unintentionally hit, resulting in no release of product, the excavation should be stopped. True or false?
   a. True
   b. False

Summary

Excavating is one of the most hazardous activities performed in pipeline maintenance. Companies designate a specific worker to observe excavation activities in order to protect the pipeline at all times during the excavation project. This designated observer is often called the spotter. The spotter has specific tasks he or she is responsible for both prior to and during excavation.

Before excavation can begin, the spotter must confirm that all required permits and notifications have been made and that the line has been located and properly marked. To safeguard personnel at the excavation site, the spotter must also have a strong understanding of the Occupational Safety and Health Administration (OSHA) regulations that pertain to an excavation site. Although the spotter is not necessarily the competent person as defined by OSHA, the spotter must always take all necessary safety precautions to ensure the safety of personnel at the excavation site.

The spotter must determine the required tolerance zone and communicate that information to the excavator(s). He or she must also ensure that the tolerance zone is maintained during the excavation itself, and require that workers use the appropriate tools and equipment to protect the pipe at all times. Ensuring the proper support of the pipe itself is also the responsibility of the spotter.

Without a doubt, performing the role of an excavation observer requires a strong understanding of the dangers involved in excavation and the proper sloping techniques to make sure trenches are safe. Finally, recognizing the abnormal operation conditions that may arise and the proper reaction to each is another critical responsibility.

Review Questions

1. The purpose of the One-Call System is for contractors and the general public to notify utilities of their intention to _____.
   a. establish an aerial marker
   b. establish a tolerance zone
   c. locate underground cables by hand digging
   d. use equipment for excavating, demolition, or other work

2. Before excavating pipeline systems, permits may be required from _____.
   a. APWA
   b. PHMSA
   c. GPS
   d. ROW

3. Before pipeline or cable excavation commences, a qualified individual must verify the _____.
   a. depth of cover
   b. pipeline material
   c. APWA Uniform Color Code
   d. pipeline pressure

4. After the pipeline is precisely located, personnel install temporary _____.
   a. well points
   b. probe rods
   c. line markers
   d. water-vac units

5. At an excavation site, the person specifically defined by OSHA as a competent person is focused on _____.
   a. preventing pipeline damage
   b. ensuring the safety of personnel
   c. preventing hazards to the environment
   d. providing proper control center notifications

6. Ramps in an excavation must be designed by a (an)_____.
   a. engineer
   b. designated observer/spotter
   c. inspector
   d. competent person

7. Trench boxes should be installed _____.
   a. before personnel enter the trench
   b. by two workers inside the trench
   c. after excavation work is complete
   d. just before backfilling begins

8. Excavated soil should be placed at least _____.
   a. 1 foot away from the edge of an excavation
   b. 2 feet away from the edge of an excavation
   c. 12 feet away from the edge of an excavation
   d. 18 feet away from the edge of an excavation

9. One advantage of using vacuum-powered tools to excavate is that they _____.
   a. eliminate hazards to personnel
   b. remove soil cover more quickly
   c. reduce the potential for pipe damage
   d. cost less than other excavation tools

10. When working in waterlogged soil, the excavator may dewater the ditch line with _____.
    a. well points
    b. sprinklers
    c. drains
    d. hoses

11. OSHA has four classifications for soil, which are based on _____.
    a. stability
    b. cohesion
    c. vibration
    d. granularity

12. An excavation of the wall in steps, designed to prevent cave-ins, is called _____.
    a. shoring
    b. benching
    c. trenching
    d. sloping

13. When excavating a pipeline, the various support methods used include slings, cribbing, air bags, and _____.
    a. soil columns
    b. earth pillars
    c. step backs
    d. stress supports

14. When protecting the pipeline, the maximum span between the locations of the supports should be determined by a(n) _____.
    a. inspector
    b. engineer
    c. contractor
    d. excavator

15. If an AOC occurs during an excavation, the first response must be to _____.
    a. call 911
    b. complete required reporting
    c. stop the excavation
    d. shut down the pipeline segment

Trade Terms Quiz

Fill in the blank with the correct term that you learned from your study of this module.

1. The technique for covering a completed pipeline so that adequate fill material is underneath and over the pipe to protect it from loose rocks, abrasion, shifting, and washouts is called _________.
2. The amount of material needed to bring existing ground to desired grade is the _________.
3. A technique used to cut the side walls of a trench at an angle that will be more stable is called _________.
4. The area surrounding underground pipeline or cables within which no power digging for excavation is allowed is called the _________.
5. A highly visible device used to communicate the location of the pipeline is a(n) _________.
6. A procedure used to locate underground installations by digging a small hole in the ground is _________.
7. Devices installed in waterlogged soil to dry out areas along the pipe ditch line are _________.
8. The greatest angle above the horizontal plane at which the soil will rest without sliding is called the _________.
9. An excavation formed with two or more sections that look like stairs in cross section is called a(n) _________.
10. Methods used to ensure the stability of adjoining buildings, walls, or other structures endangered by an excavation are known as _________.
11. An organization that excavators may contact to determine the presence of underground structures within the area of their proposed excavation activities is the _________.
12. Soil removed during an excavation is called _________.
13. A tool used to hydraulically remove soil and expose a pipeline is called a _________.
14. Conditions within a pipeline system, identified by the operator, that may result in a hazard to persons, property, or the environment are called _________.

Trade Terms

Abnormal operating conditions (AOCs)
Angle of repose
Backfilling
Bench
Fill
Hydro-vac
Line marker
One-Call System
Pot-holing
Sloping
Spoil
Support systems
Tolerance zone
Well Points

Soil Properties

The properties of the soil in which excavations are to be made are very important when planning trenching and shoring techniques. Properties of soil include the grain size, the soil gradation, and the grain shape. Grains are soil particles that have been produced from rocks by weathering and chemical processes.

Grain sizes are determined by measuring with calibrated sieves or screens. The sieves are graded by the size of the holes in their mesh. The grains of soil are poured into a series of different size sieves until most of the grains will not pass through the holes in the mesh. The grains caught by each size or grade sieve are classified by that size (*Figure A01*).

In most soils, all of the grains are not the same size. When passing the samples through a series of sieves, some proportion will be caught in different size meshes. The proportion of each size will identify the soil's gradation.

The soil is first classified as coarse-grained or fine-grained. The coarse-grained soil is further classified for gradation according to the following criteria:

- *Well-graded* – Shows a good representation of all particle sizes from largest to smallest.
- *Poorly-graded* – Has an excess or deficiency of a certain particle size.
- *Uniform gradation* – Most particles are about the same size.
- *Gap or skip gradation* – A mixture of two or more uniform soils.

The first two classifications may be determined in the field with experience; the others may need to be done in the lab.

The grain shape has an important effect on the properties of the soil. The shape can determine how the grains will stick together, support one another, or slide over one another in an unsupported wall. They will also affect the compaction requirements when backfilling the trench.

The most common shapes are bulky grains, flaky grains, and elongated grains and fibers. Bulky grains are rounded, subrounded, subangular, and angular and usually make up most of the coarse-grained particles. Flaky grains are present in many fine-grained soils. These particles slide against one another and can be compacted well to form a tight mass.

These properties determine the classifications of types of soils that may be encountered in excavating. The average soil found on most sites is a mixture of sand, silt, clay, and gravel. The amount of each component in the mixture, along with any water, air, gases, chemicals, or other materials in the soil, affects its workability and stability (*Figure A02*).

Sand is not very stable and does not hold water unless some other material is present. As a result, sand excavations require more support than those in other soils.

Silt has fine particles that do hold water, and it is soft when wet. Dry silt is easily broken and powdery. It is not very stable unless it has been compressed and hardened.

Clay is also made up of fine particles, but is plastic when wet. Dried clay is hard and difficult to break up. Excavations in clay can stand up well unless water is present, which can cause landslides.

Gravel is rock with a range of sizes and, if compacted well, is stable and provides water drainage. The presence of water is a critical factor in any type of soil.

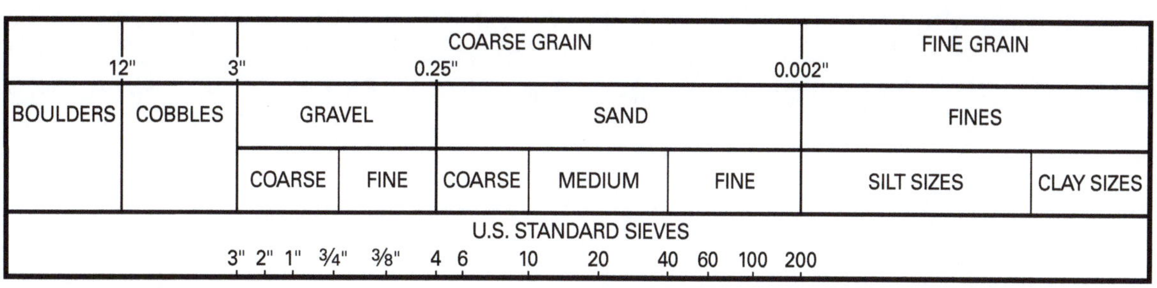

Figure A01 Grain classifications.

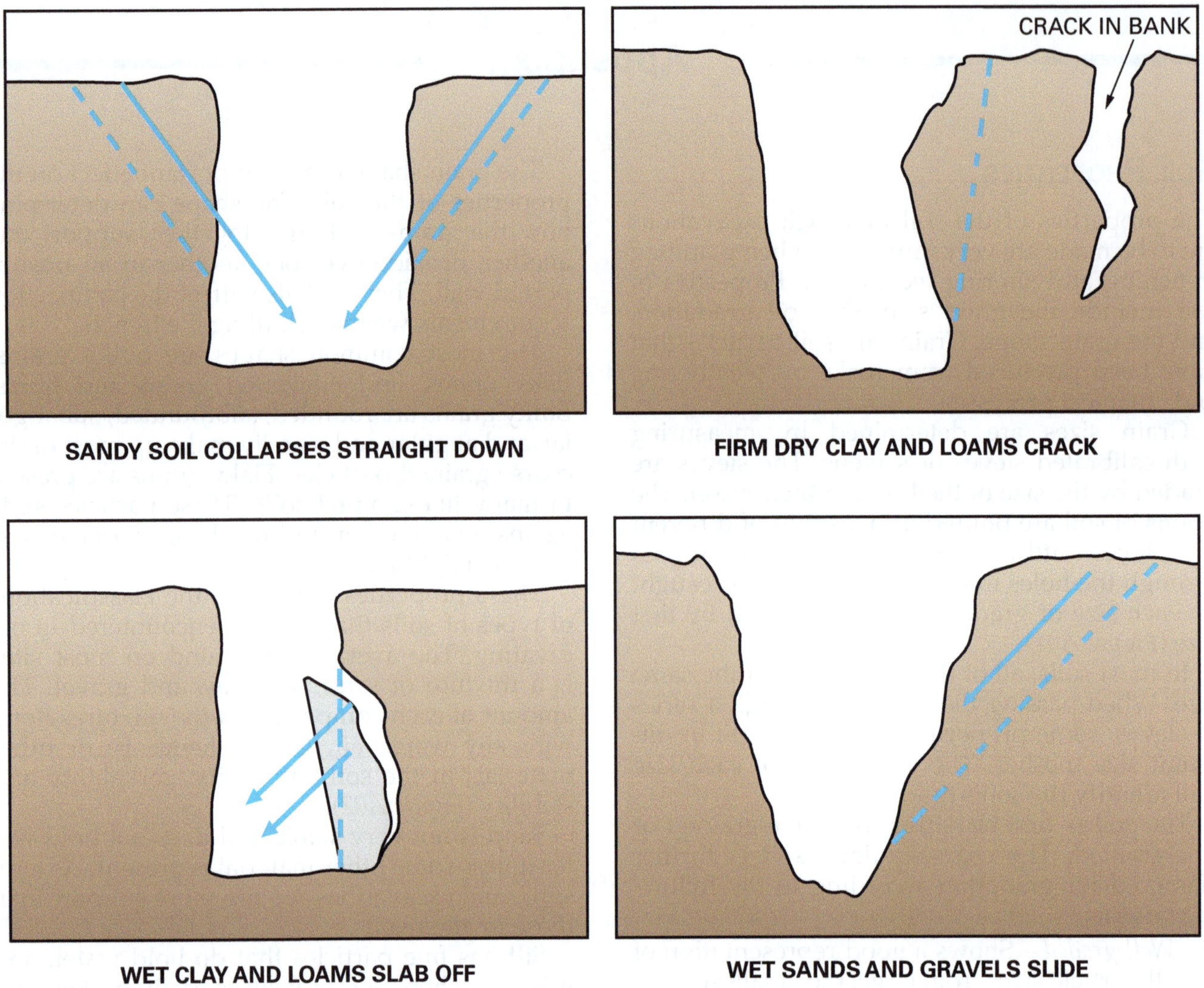

Figure A02 Soil behaviors.

Trade Terms Introduced in This Module

Abnormal operating conditions (AOCs): Conditions within a pipeline system, identified by the operator, that may indicate (1) a malfunction of a component or (2) a deviation from normal operations that may indicate a condition exceeding design limits or result in a hazard to persons, property, or the environment.

Angle of repose: The greatest angle above the horizontal plane at which the soil will rest without sliding. This angle varies depending on factors such as soil type and weather conditions, and is usually between 0 and 90 degrees.

Backfilling: The technique for covering a completed pipeline so that adequate fill material is underneath and over the pipe to protect it from loose rocks, abrasion, shifting, and washouts.

Bench: Excavation formed with two or more sections that look like stairs in cross section; used to prevent cave-ins.

Fill: Amount of material needed to bring existing ground to a desired grade.

Hydro-vac: A tool used to hydraulically remove soil and expose a pipeline, minimizing the risk associated with other sharp metal tools. (Also called a *water-vac*.)

Line marker: A highly visible device used to communicate the location of the pipeline.

One-Call System: An organization that excavators may contact (required in some states) to determine the presence of underground substructures within the area of their proposed excavation activities, and to advise operators of such substructures of the pending activity that may impact their facility.

Pot-holing: A procedure used to locate underground installations by digging a small hole in the ground.

Sloping: A technique used to cut the side walls of a trench at an angle that will be more stable; the angle should start at the top of the trench wall and continue to the floor of the trench.

Spoil: Soil removed during an excavation.

Support systems: Methods used to ensure the stability of adjoining buildings, walls, or other structures endangered by an excavation. (Examples of support systems include shoring, bracing, or underpinning.)

Tolerance zone: Area surrounding underground pipeline or cables within which no power digging for excavation is allowed; personnel must hand dig only.

Well points: Devices installed in waterlogged soil to dry out areas along the pipe ditch line. Hundreds may be needed in some areas.

Additional Resources

The following recommended resources can provide additional helpful information related to the requirements for this covered task:

API Recommended Practice 1161, Third Edition, © 2014 American Petroleum Institute. API Publishing Services, 1220 L Street, NW, Washington, DC 20005. Distributed by Thompson Reuters (Scientific) LLC, **www.techstreet.com**

NCCER Module AOCCC-17, *Abnormal Operating Conditions Control Center.*

NCCER Module AOCFG-17, *Abnormal Operating Conditions Field and Gas.*

NCCER Module CT5_1-17, *Examine for Mechanical Damage on Buried or Submerged Pipe.*

NCCER Module CT5_2-17, *Examine for External Corrosion on Buried or Submerged Pipe.*

NCCER Module CT5_3-17, *Inspect the Condition of External Coating on Buried or Submerged Pipe.*

NCCER Module CT39_0-17, *Backfilling a Trench Following Maintenance.*

49 *CFR* 192, **www.ecfr.gov**

49 *CFR* 195, **www.ecfr.gov**

American Public Works Association (APWA), **www.apwa.net**

Code of Federal Regulations 29, Part 1926. **www.osha.gov**

Code of Federal Regulations 49, Parts 192 and 195. **www.ecfr.gov**

Occupational Safety and Health Administration (OSHA), 200 Constitution Avenue, NW, Washington, DC 20210. **www.osha.gov**

Figure Credits

Copyright © 2013 TrenchSafety and Supply, Inc. Used with permission., Module Opener, Figure 6

ONEOK North System, L.L.C., Figure 1

Courtesy of S4Carlisle Publishing Services, Figure 4 (right)

Courtesy of Humboldt Mfg. Co., Figure 8 (top)

Gradall Industries, Inc., Figure 9

Section Review Answer Key

Answer	Section Reference	Objective
Section One		
1. d	1.1.0	1a
2. a	1.2.0	1b
3. c	1.4.0	1c
Section Two		
1. c	2.1.0	2a
2. a	2.2.0	2b
3. d	2.3.2	2c
4. a	2.4.0	2d
Section Three		
1. a	3.1.2	3a

NCCER CURRICULA — USER UPDATE

NCCER makes every effort to keep its textbooks up-to-date and free of technical errors. We appreciate your help in this process. If you find an error, a typographical mistake, or an inaccuracy in NCCER's curricula, please fill out this form (or a photocopy), or complete the online form at **www.nccer.org/olf**. Be sure to include the exact module ID number, page number, a detailed description, and your recommended correction. Your input will be brought to the attention of the Authoring Team. Thank you for your assistance.

Instructors – If you have an idea for improving this textbook, or have found that additional materials were necessary to teach this module effectively, please let us know so that we may present your suggestions to the Authoring Team.

NCCER Product Development and Revision

13614 Progress Blvd., Alachua, FL 32615

Email: curriculum@nccer.org
Online: www.nccer.org/olf

❑ Trainee Guide ❑ Lesson Plans ❑ Exam ❑ PowerPoints Other ______________

Craft / Level: Copyright Date:

Module ID Number / Title:

Section Number(s):

Description:

Recommended Correction:

Your Name:

Address:

Email: Phone:

CT34_0-17

Inspect Existing Pipe Following Movement

Overview

A visual inspection of the worksite area and all pipes and pipe components after movement activities are complete helps to ensure that the integrity of the pipeline system has not been compromised. An inspection of the coating must be made over the segment's entire length, and repairs to the coating must be made as necessary. The visual inspection must ensure that the pipeline meets strength and serviceability requirements and that there is no visible damage before and during the backfilling of the excavation. A visual inspection of the worksite is also conducted as part of the return-to-service wrap-up activities.

Trainees with successful module completions may be eligible for credentialing through the NCCER Registry. To learn more, go to **www.nccer.org** or contact us at 1.888.622.3720. Our website has information on the latest product releases and training, as well as online versions of our *Cornerstone* magazine and Pearson's product catalog.

Your feedback is welcome. You may email your comments to **curriculum@nccer.org**, send general comments and inquiries to **info@nccer.org**, or fill in the User Update form at the back of this module.

This information is general in nature and intended for training purposes only. Actual performance of activities described in this manual requires compliance with all applicable operating, service, maintenance, and safety procedures under the direction of qualified personnel. References in this manual to patented or proprietary devices do not constitute a recommendation of their use.

CT34_0-17
Inspect Existing Pipe Following Movement

Objectives

When you have completed this module, you will be able to do the following:

1. Explain the process for inspecting existing pipe following movement.
   a. Explain the purpose and procedures for inspecting existing pipe.
2. Describe the procedure used to inspect pipe following movement (CT34_0-17).
   a. Identify the prerequisites, competencies, and abnormal operating conditions associated with pipe inspection following movement.
   b. Describe the procedure used to inspect pipe following movement.

Performance Task

Under the supervision of the instructor, you should be able to do the following:

1. Inspect existing pipe following movement (CT34_0-17).
   - Identify potential abnormal operating conditions that may occur during performance of this CT, and know the appropriate actions to take in response to them.
   - Utilize the appropriate personal protective equipment according to relevant company procedures.
   - Visually inspect entire length of exposed pipeline to ensure there are no wrinkles, gouges, or other apparent physical damage.
   - Visually inspect entire length of exposed pipeline to ensure the coating has not been damaged during the move.
   - Inspect ditch and backfill spoil to ensure materials capable of damaging the coating are not present.
   - Complete appropriate documentation as required by operator's procedures.

Trade Terms

Abnormal operating conditions (AOCs)
Backfilling
Corrosion
Dent
Existing longitudinal stress
Good working order
Gouge
Highly volatile liquid (HVL)
In-service pipe
Maximum allowable operating pressure (MAOP)
Maximum operating pressure (MOP)
Out-of-service pipe
Pipeline movement plan
Pipeline profile

Industry Recognized Credentials

If you are training through an NCCER-accredited sponsor, you may be eligible for credentials from NCCER's Registry. The ID number for this module is CT34_0-17. Note that this module may have been used in other NCCER curricula and may apply to other level completions. Contact NCCER's Registry at 888.622.3720 or go to **www.nccer.org** for more information.

Contents

Figures

SECTION ONE

1.0.0 PIPE INSPECTION FOLLOWING A MOVEMENT

Objectives

Explain the process for inspecting existing pipe following movement.

a. Explain the purpose and procedures for inspecting existing pipe.

Trade Terms

Abnormal operating conditions (AOCs): Conditions within a pipeline system, identified by the operator, that may indicate (1) a malfunction of a component or (2) a deviation from normal operations that may indicate a condition exceeding design limits or result in a hazard to persons, property, or the environment.

Backfilling: The process of replacing soil after an excavation to fill the ditch and safely cover the pipeline.

Corrosion: The deterioration of a pipe surface (usually metal) that results from a reaction its environment.

Dent: A depression in the surface of a pipe that affects the curvature of the pipe without reducing the pipe wall thickness.

Existing longitudinal stress: The longitudinal stress associated with a pipeline at rest, excluding any residual stress in girth welds or over-bends.

Good working order: All valves are required by 49 CFR 195.420(a) to be in good working order, which can be interpreted to mean that a valve's performance, based on in-service criteria, meets all of the necessary functions of the valve.

Gouge: Damage to the surface of a pipe that displaces or removes metal and reduces the pipe wall thickness.

Highly volatile liquid (HVL): A hazardous liquid that has a vapor pressure exceeding 40 psi absolute at 100°F. It will form a vapor cloud when released into the atmosphere.

In-service pipe: A system or part of a system that contains liquid hydrocarbons or other hazardous liquids under pressure. In-service lines are used on a routine basis to move these liquids from one point to another in the system.

Maximum allowable operating pressure (MAOP): The maximum internal pressure permitted during the operation of a gas pipeline.

Maximum operating pressure (MOP): The maximum internal pressure permitted during the operation of a hazardous liquid or carbon dioxide pipeline.

Out-of-service pipe: Pipeline that has been removed from service, drained, purged, and sealed to ensure that there is no danger to people, property, or the environment. Out-of-service pipe includes those segments of a line that have been abandoned or idled until a further need for their use arises.

Pipeline movement plan: A plan that must be developed and approved before a pipeline can be moved. It typically includes information about the pipeline, specifications for excavation and moving, emergency procedures, safety measures, notification requirements, and project communications information.

Pipeline profile: A drawing that maps the location of the pipeline and its position on or under the ground. The pipeline profile details the location and depth of cover for the target line.

Following any project for pipeline relocation, the pipe must be visually inspected once the move is complete to ensure the movement activities have not created any potential issues or pipeline damage. The pipeline movement plan and the pipeline profile need to be reviewed and confirmed to ensure that the line has been relocated as required.

The visual inspection of pipes and their components is necessary so that any damage or abnormality is detected and appropriate corrective actions are taken. The visual inspection process includes re-inspection of the pipes and pipe components after the move is complete. Additional inspections are also conducted during subsequent phases of the move. The types of inspections and their focus will depend on the specific move requirements and circumstances for the in-service pipe move. Inspection requirements for out-of-service pipe may vary as well.

1.1.0 Required Activities After the Movement

Movement of a pipeline can subject the pipe and its components to additional stress and contact that could result in damage or unsafe conditions. Any damage or abnormalities must be corrected before the pipeline can be returned to service. The excavation must be backfilled properly to ensure the pipe is not damaged and stress on the pipe is minimized. Once all required activities after the move are completed, the pipeline is placed back into service according to company-specific procedures. This section discusses the following related topics:

- Minimizing stress
- Inspecting pipe and coating
- Backfilling of excavation
- Return to service requirements

1.1.1 Minimizing Stress

The physical movement of the pipe can create secondary stress during lifting, jacking, supporting, excavating, and any other external forces exerted on the pipe as part of movement activities. It is important to limit and minimize secondary stress during the move and after the move to avoid any subsequent damage and potential pipe failure. The failure mode of a secondary stress is generally a small crack leading to leakage. Secondary stresses on a buried pipe are typically due to cyclic thermal expansion and contraction. Proper supporting and **backfilling** techniques need to be followed to minimize secondary stress that could potentially damage the pipe. Improper backfilling, inadequate soil compaction, and poor fill quality can cause future settlement issues that could create unwanted movement and stress on the pipe.

Pay special attention to the areas at each end of the pipe segment that may have been damaged by longitudinal stresses during the movement. Proper lifting methods and move techniques need to be followed to minimize secondary stress on the pipe. The imposed deflections and resulting stresses may be temporary or permanent, depending on the circumstances. The pipeline may have been lowered, raised, or moved laterally. The longitudinal stresses associated with the displaced pipe alignment will develop and remain present as long as the pipe remains in the displaced configuration. Even though the physical displacement of the pipe is minimal, the newly positioned pipe will be subjected to new longitudinal stresses during and after its move. These new stresses are in addition to the **existing longitudinal stress** on the pipe section being moved. Take appropriate safety measures and consider the cause and effect consequences related to the movement activities.

Bending stresses associated with the spanning effect between lift or support points along the pipeline may also develop. If the pipe is being lowered to a new trench profile, these spanning stresses disappear once the pipe is resting on the new trench bottom. If the pipe is being raised, the spanning stresses will remain in place along with bending and extensional stress until the pipe is lowered back into its original configuration. Always follow company procedures and move plan details to ensure that the in-service move is done in a safe manner. Check the appropriate documentation for pipe stress calculations based on API *RP 1117, Movement of In-Service Pipelines.*

1.1.2 Inspect Pipe and Coating

Pipe must be inspected for damage whenever it is exposed, and before it is backfilled. After a move, check the exposed pipe and the pipe coating for evidence of mechanical damage or external **corrosion**. The coating is essential for protecting the pipe surface against corrosion and the effects of the burial environment. The coating may easily be cut, scraped, or otherwise damaged by the excavation and moving equipment, or by rocks in the fill. Repair all coating damage before the pipeline is reburied. If any corrosion is found, investigate further to determine its extent and the necessary corrective actions.

A **dent** or **gouge** may be encountered after a move. Dents and gouges are usually caused by contact during excavation, by rocks too close to the pipe, or by objects falling into the excavation. When a pipe is dented or gouged, the pipe wall thickness is reduced at the point of impact. As a result, stress on the pipe is concentrated at the dent or gouge location, which weakens the pipe. The weakened condition of the pipe wall can be a source of future leaks.

After a move, the condition and operability of valves also needs to be assessed to ensure that they are still in **good working order**. Evaluate all mechanical damage and repair it as necessary, in accordance with company repair procedures. Document and report inspections and results, and maintain reports for the life of the pipeline.

Corrosion, mechanical damage, coating inspections, and valve maintenance and inspections are covered in detail in other NCCER Covered Task Modules. The following separate tasks can be referenced as needed for additional information:

- Measuring pit depth with a pit gauge is covered in NCCER Module CT8_1-17, *Measure Pit Depth with Pit Gauge.*
- Measuring wall thickness with a handheld ultrasonic meter is covered in NCCER Module CT8_2-17, *Measure Wall Thickness with Handheld Ultrasonic Meter.*
- Examining for mechanical damage on buried or submerged pipe is covered in NCCER Module CT5_1-17, *Examine for Mechanical Damage on Buried or Submerged Pipe.*

- Examining for external corrosion on buried or submerged pipe is covered in NCCER Module CT5_2-17, *Examining for External Corrosion on Buried or Submerged Pipe.*
- Inspecting the condition of external coating on buried or submerged pipe is covered in NCCER Module CT5_3-17, *Inspecting the Condition of External Coating on Buried or Submerged Pipe.*
- Measuring corroded area is covered in NCCER Module CT8_3-17, *Measure Corroded Area.*
- Other tasks related to corrosion are covered in the following NCCER Modules:
  - CT7_1-17, *Visual Inspection of Atmospheric Coatings*
  - CT7_2-17, *Prepare Surface for Coating Using Hand and Power Tools*
  - CT7_3-17, *Prepare Surface for Coating by Abrasive Water Blasting*
  - CT7_4-17, *Prepare Surface for Coating by Abrasive Blasting Media Other than Water*
  - CT7_5-17, *Apply Coating Using Hand Application Methods*
  - CT7_6-17, *Apply Coating Using Spray Application*
  - CT7_7-17, *Perform Coating Inspection*
- Other tasks related to valve inspection and maintenance are covered in the following NCCER Modules:
  - CT19_2-17, *Valve Lubrication*
  - CT19_3-17, *Valve Seat Sealing*
  - CT19_4-17, *Valve Stem Packing Maintenance*
  - CT19_5-17, *Adjust Actuator/Operator, Electric*
  - CT19_6-17, *Adjust Actuator/Operator, Pneumatic*
  - CT19_7-17, *Adjust Actuator/Operator, Hydraulic*
  - CT20_0-17, *Inspect Main-Line Valves*
  - CT21_3-17, *Internal Inspection of Valves and Their Components*
  - CT22_2-17, *Inspect, Test, and Calibrate HVL Tank Pressure Relief Valves*
  - CT23_1-17, *Maintain and Repair Relief Valves*
  - CT23_2-17, *Inspect, Test, and Calibrate Relief Valves*

1.1.3 Backfilling the Excavation

After maintenance or repairs, backfilling is required, and the soil must be replaced in the ditch and over the pipeline. Backfilling and bedding must be provided in a manner that will offer firm support for the pipeline. Backfill material must be free of rocks, roots, and any other foreign material that could damage the pipe coating or cause uneven loading of the pipe.

Pipe and pipe coating must be inspected to be certain they were not damaged during excavation or maintenance activities, and the inspection must continue during the backfilling operation. Pipe and pipe coating must be continuously monitored during backfilling to ensure that they are not damaged.

NOTE

Only qualified workers familiar with the movement plan should supervise and monitor excavation activities. Personnel monitoring an excavation may delegate more than one task to a single worker. Any worker completing one or more excavation tasks must be qualified to do so. These tasks may include directing the equipment operator, site supervising, operating equipment, shoveling by hand, and providing assistance.

The pipe must be properly supported after any temporary supports are removed. The pipe must then be provided with adequate cover. Cover must meet the specifications outlined in the pipeline movement plan. Allow extra compaction at road crossings or other locations where heavy loads may cross a pipeline. Follow regulations and company policies for required depths of cover.

The right-of-way (ROW) and surrounding property must be restored as closely as possible to its original state and to the satisfaction of the landowner and any regulatory agency. Erosion prevention measures may be required.

The general procedure for backfilling an excavation is outlined in the following steps:

Step 1 Utilize the appropriate personal protective equipment according to relevant company procedures.

Step 2 Perform a visual inspection of the backfill material to identify any foreign objects that could cause damage to the pipeline system.

Step 3 Take appropriate action to remove any foreign material or objects.

Step 4 Determine if the backfill material is suitable for backfill around the pipeline:

- Soft material should be used as bedding near the coating.
- Excessive rocks may make replacement of backfill necessary.
- Excessive roots, brush, and other material may make replacement of backfill necessary.

NOTE

Shielding (rock shields) may be required to protect the pipeline from foreign objects.

Step 5 Use suitable materials to fill within 6 inches of the pipe or to the level specified by company specifications. Drawings of how the backfill material should be put in place are shown in *Figure 1*.

Step 6 Fill equally along both sides of the pipe until the pipe is covered. Use the excavator to backfill until the pipe is completely covered with 2 feet of soil. Use a dozer or loader to finish backfilling, making sure to operate the heavy equipment in a position that is perpendicular to the pipe.

NOTE

Extra compaction is required at road crossings and other locations where heavy loads may cross the pipeline.

Step 7 Compact the soil under and around the pipe for proper pipe support, using tamping tools, backhoes, and other tools and equipment as needed.

Step 8 Cover the pipe to the cover depth required by specifications.

Step 9 Compact the cover material.

Step 10 Crown the backfill according to company specifications, to allow for settlement.

Step 11 Complete appropriate documentation as required by operator's procedures.

NOTE

The proper cover for buried pipelines is defined in 49 *CFR* 195.248(a). All pipe must be buried so that it is below the level of cultivation. The required depth of cover varies by location and type of soil and is specifically outlined in the code. The pipe must be installed so that the cover between the top of the pipe and the ground level, road bed, river bottom, or underwater natural bottom complies with this code.

NOTE

Provide extra cover at road crossings, ditches, and streams. These areas face an increased risk from third-party damage, and additional cover provides some extra protection. In special circumstances like these, reference procedural manuals, such as your company manual and 49 *CFR* 195.

LAYING CONDITION	DESCRIPTION
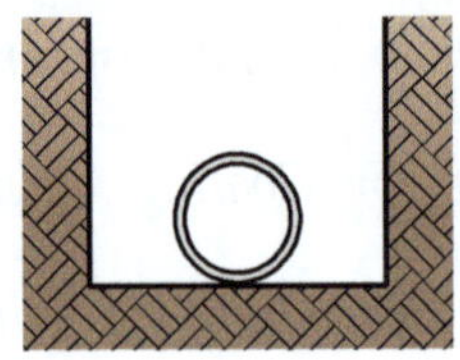 TYPE 1	FLAT-BOTTOM TRENCH. LOOSE BACKFILL.
 TYPE 2	FLAT-BOTTOM TRENCH. BACKFILL LIGHTLY CONSOLIDATED TO CENTER LINE OF PIPE.
 TYPE 3	PIPE BEDDED IN 4" MINIMUM LOOSE SOIL OR BEDDING MATERIAL. BACKFILL LIGHTLY CONSOLIDATED TO TOP OF PIPE.
 TYPE 4	PIPE BEDDED IN SAND, GRAVEL, OR CRUSHED STONE TO DEPTH OF ⅛ PIPE DIAMETER, 4" MINIMUM. BACKFILL COMPACTED TO TOP OF PIPE.
 TYPE 5	PIPE BEDDED TO ITS CENTER LINE IN COMPACTED GRANULAR MATERIAL, 4" MINIMUM UNDER PIPE. COMPACTED GRANULAR OR SELECT MATERIAL TO TOP OF PIPE.

NOTE: THESE OPTIONS VARY WITH JOB SPECIFICATIONS AND LOCATIONS.

Figure 1 Proper placement of backfill material.

1.1.4 Return-to-Service Requirements

After all move-related activities are complete, a hand-off or return to service process must occur. Always follow approved company policies and processes regarding return-to-service procedures and protocols. Communicate with personnel before resuming normal operations. Proper communication includes verbal and written methods to all appropriate personnel such as control center management, field management, and key stakeholders, as required by the company policy for its return-to-service. The following information is provided as general reference for establishing return-to-service protocols:

- Establish communications with company-designated responders between control center and field operations as needed.
- Validate line integrity including, but not limited to, the following:
  - Visual inspection of the general area where the line move occurred.
  - Verify the status of all valves in the work site area.
  - Verify that line pressures are consistent with normal operation.
  - Verify all telemetered equipment back to SCADA if applicable.
  - Investigate all valleys, peaks, places near rivers/streams, and any other high consequence areas (HCAs) that may have subjected the pipeline that was moved to potential damage.
  - Verify that all work on the related pipeline has been completed successfully and documented.
  - Verify final line pressure for the section of line that was moved.
- Investigate **abnormal operating conditions (AOCs)**, verify pipeline system integrity, and identify root-cause before start-up. Communicate with field engineers, corrosion technicians, pipeline integrity technicians, and other field personnel as needed.
- Verify and document (either through email or properly executed checklist) that the incident has been repaired and has integrity for start-up.
- Communicate with the control center; restart requires approval by appropriate management (i.e., field management plus control center management) and review with key stakeholders.
- Document and maintain all related information as required.

Conventional liquid, gas, and **highly volatile liquid (HVL)** all have different pressure requirements. It is important to reconfirm the type of product in the section of pipe that was moved so that the operating pressure can be reestablished. The pressure in the line would have been reduced based on the requirements for determining the allowable line pressure. Recall the general guidelines that were used to determine the allowable line pressure of the in-service pipeline that was moved:

- If conventional liquids are in the section, the pressure was lowered to 50 percent or less of the **maximum operating pressure (MOP)**.
- If gases are in the section, the pressure was lowered to 50 percent or less of the **maximum allowable operating pressure (MAOP)**.
- If HVLs are in the section, the pressure was lowered to a minimum of 50 psi over the vapor pressure of the product.

Check the actual pressure in the pipeline so that the pressure can be adjusted accordingly. Communicate with appropriate field and control room personnel as needed. Confirm the line pressure in the section of pipe that was moved and restore the pressure to normal operating levels in accordance with company policies and operating procedures for normal operations.

Additional Resources

It is assumed that you will learn to perform these procedures under the direct supervision of your crew leader, foreman, or instructor. For independent study, refer to the recommended Additional Resources at the back of this module.

1.0.0 Section Review

1. Secondary stress on a pipeline can be minimized by _____.
   a. communicating with the control center
   b. reestablishing operating pressure
   c. proper supporting and backfilling techniques
   d. inspecting the coating for damage

2. Ensuring that valves are in good working order is part of inspecting a pipe after movement.
   a. True
   b. False

3. Cover must meet the specifications outlined in the _____.
   a. profile plan
   b. excavation plan
   c. pipeline movement plan
   d. design plan

SECTION TWO

2.0.0 INSPECTING PIPE FOLLOWING MOVEMENT

Objective

Describe the procedure used to inspect pipe following movement (CT34_0-17).

a. Identify the prerequisites, competencies, and abnormal operating conditions associated with pipe inspection following movement.

b. Describe the procedure used to inspect pipe following movement.

Performance Task

1. Inspect existing pipe following movement (CT34_0-17).
   - Identify potential abnormal operating conditions that may occur, and know the appropriate actions to take in response to them.
   - Utilize the appropriate personal protective equipment according to relevant company procedures.
   - Visually inspect entire length of exposed pipeline to ensure there are no wrinkles, gouges, or other apparent physical damage.
   - Visually inspect entire length of exposed pipeline to ensure the coating has not been damaged during the move.
   - Inspect ditch and backfill spoil to ensure materials capable of damaging the coating are not present.
   - Complete appropriate documentation as required by operator's procedures.

This section provides general procedures for inspecting pipe following movement. It also provides the prerequisites and competencies required for the procedures, as well as the AOCs that may be encountered and the proper responses to them.

2.1.0 Prerequisites, Competencies, and Abnormal Operating Conditions

Personnel must possess a variety of skills in order to inspect pipe after an in-service line move. A general working knowledge of the types of pipes and components is required to ensure that personnel are able to recognize any issues that may compromise the operation of the pipeline system.

Individuals performing visual inspections of pipes and pipe components should have some basic knowledge and skills related to the following:

- Knowledge of the specifications covered in API *standard 5L* for seamless and welded steel line pipe. This standard specifies requirements for the manufacture of two product specification levels (PSL 1 and PSL 2) of seamless and welded steel pipes for use in pipeline transportation systems in the petroleum and natural gas industries. This standard is not applicable to cast pipe.
- Knowledge of pipeline transportation systems for liquids and slurries, covered in ASME *B31.4*.
- Knowledge of standards for welding of pipelines and related facilities, covered in API *standard 1104*.
- OSHA confined spaces, hot work permit, and excavation standards.
- Lockout/tagout procedures.

Personnel must also be aware of the particular AOCs that may occur when visually inspecting pipe and pipe components and know the proper actions to take in responses to them.

2.1.1 Prerequisites and Competencies

Personnel responsible for inspecting the pipe after the move must have an understanding of what the task entails. The task typically begins by checking for any over-stress conditions and damage to the pipe, pipe components, and coating. Any issues should be addressed and corrected as needed. The excavation is backfilled after successful completion of all activities and related inspections. The backfilling process is monitored to ensure that pipes and components are not damaged by equipment, rocks, or debris. The task ends when all activities are complete and the pipe is returned to service.

Proper demonstration of use for inspection equipment and personal protective equipment is required. Personnel must have an understanding of pipe and pipe component identification. Personnel must have the ability to recognize unacceptable damage, corrosion, or any condition that would make the return to service unsafe. Proper interpretation of relevant pipeline documentation is required to ensure that all pipes and components meet design specifications and the structural integrity has not been compromised.

Personnel must know the regulations associated with the movement of an in-service pipeline.

In addition, personnel must also ensure design calculations have been completed and the allowable line pressure is confirmed. Note that allowable pressure calculations for HVLs are based on the vapor pressure of the product in transport.

> **NOTE**
>
> The allowable line pressure for an in-service move will be significantly lower (50 percent or less) than the maximum allowable operating pressure (MAOP) or the maximum operating pressures (MOP). Make sure you understand that MAOP values are for gases and MOP values are for conventional liquids.

2.1.2 Abnormal Operating Conditions

Personnel inspecting pipe after a move must be able to identify abnormal operating conditions (AOCs) and know the proper response to each, should they occur. The following are some potential AOCs, and the appropriate action the controller should take in response to them:

- *Pipeline system damage or corrosion is discovered, possibly leading to an unexpected release of product* – Personnel must be aware of the corrosion requirements. Perform initial assessment and notify appropriate personnel. In all cases, personnel must first determine the extent of the damage, and then make the appropriate notifications so that the appropriate repairs can be made. Unexplained pressure deviation (increase, decrease, or lack of pressure) needs to be investigated and the situation needs to be addressed so that appropriate corrective action is taken. Appropriate notifications must be made to all impacted stakeholders.
- *Maps, survey records, and drawings that are not up-to-date provide false or inaccurate information* – Make sure you are familiar with the source of the information, the documentation systems, and the various types of documentation so that you can recognize potential out of date or superseded information. Proper interpretation of the maps, survey records, and drawings is a critical requirement for locating lines and ensuring lines are effectively located.
- *Personnel notice damage from third-party construction on the right-of-way, or unsupervised third-party construction in progress* – If the contractor performing the unsupervised work is available, personnel should discuss the danger of the situation with him or her. Proper notification of the damage or unsupervised work must be made as soon as it is practical to do so.
- *Hazardous liquid or carbon dioxide is encountered during the process of moving the line* – Personnel trained to do so should try to identify and isolate the source of the leak. Proper notifications should be made as soon as possible, including calling 911 if necessary. If there is evidence of a leak, stop task activities, move to a safe distance, and notify appropriate pipeline personnel. Evidence of a leak may include odor or hazardous gas or liquids, dead vegetation, audible cues, sheen on water, or bubbles.
- *The ignition of flammable vapors causes a fire or explosion* – Personnel should limit access to the area and make the proper notifications, including calling 911 if necessary.

2.2.0 Procedure

After the physical movement of a pipeline section is complete, a visual inspection of the exposed pipe must be conducted. Upon satisfactory inspection, the excavation must be backfilled. The exposed pipe should be checked for wrinkles, dents, gouges, or other damage. If damage is found, it must be repaired before the system can be placed back into normal operation. The inspection also includes checking the coating. The entire segment must be inspected and repairs to coating made as necessary. All repairs must be completed *before* backfilling and covering the pipe can occur. All tasks must be completed according to company-specific policies and procedures.

Using a pipeline inspection form to keep track of inspection elements and procedures may be useful. A pipeline inspection form is a checklist that provides a general guideline for inspection. An example of a pipeline inspection form is shown in *Figure 2A* and *Figure 2B*. Note that personnel must always follow company-specific procedures for inspecting pipeline that has been relocated or lowered-in-place.

The general procedure for inspecting a pipeline following movement is outlined in the following steps:

Step 1 Identify potential abnormal operating conditions that may occur, and know the appropriate actions to take in response to them.

Step 2 Utilize the appropriate personal protective equipment according to relevant company procedures.

Step 3 Visually inspect the entire length of the pipeline section to ensure that there are no wrinkles, gouges, or other apparent physical damage.

PIPELINE INSPECTION REPORT

Frequency: At Time of Inspection

PART A — LOCATION

Reason For Inspection ____________

1. Location: Division ____________ Country ____________ State ______ Date of Inspection ____________

A) Main Line
Alignment Sheet ________ Pipeline Size/Segment ________
Beg. M.P. ____________ End M.P. ____________
Beg. Sta. No. ____________ End Sta. No. ____________
Note: Station numbers shall be measured from a known landmark (Road Crossing etc.)

B) Company Facility: Pump Station, Tank Farm, etc.
Location Name ____________
Physical Location of Pipe Inspected ____________
Milepost ____________

PART B — MATERIAL/CONDITIONS

2. Piping:
Pipe Size ____________ In. O.D. Norm. W.T. ____________ In. Actual W.T. ____________ Grade ________

3. External Pipe Coating.

A) Coating
☐ Bare
☐ LAC 450
☐ Taped
☐ Insulated
☐ Other ________

B) Application
☐ Mill Coated
☐ Yard Coated
☐ Field Coated
☐ Other

C) Coating Thickness
Factory Applied ____ mile
Field Applied ____ mile

D) Condition of Coating
☐ Good ☐ Fair
☐ Poor

E) Bond
☐ Good
☐ Fair
☐ Poor

F) Cause of Coating Failure
☐ Damage
☐ Disbondment
☐ Defective Application
☐ Decomposition
☐ Other ________

4. Environment
☐ Clay ☐ Loan ☐ Sand ☐ Gumbo ☐ Caliche ☐ Rock ☐ Other ________
☐ Wet ☐ Dry ☐ Moist Depth of Cover (Inches) ____________
Soil Resistivity Measurement ____________ (ohm-cm) Measurement Method ____________

5. Coating Repaired ☐ No ☐ Yes
Repair System ____________

6. Pipe-to-Soil Potential at Inspected Location (At Grade) ____________ (volts) (At Pipe/Soil Interface) ____________ (volts)

7. Were any Nondestructive Tests Completed? ☐ No ☐ Yes Type ____________

8. Evidence of: ☐ Corrosion ☐ Defect ☐ Damage ☐ Stress Corrosion Cracking (S.C.C.) ☐ Caicareous Deposits

9. Test Station Added ☐ No ☐ Yes I.D. No. ____________

NOTES:
1) If S.C.C. or Corrosion is Noted, Complete Below and Back of Report (External Only, See Note 4)
2) If Defect or Damage is Noted, Complete Back of Report
3) If Repairs are Required, AAPL Form 502 "REPAIR REPORT FOR LEAK DAMAGE, CORROSION OR TEST FAILURE" Must be Attached
4) AAPL Form 508 "INTERNAL CORROSION INSPECTION REPORT" Should be Completed When Internal Corrosion Is Detected.

PART C — EXTERNAL CORROSION

10. General Corrosion Information:

A. DESCRIPTION
☐ Less than 10% of Norm. Wall Thickness, Complete Front of Report
☐ Greater than 10% of Norm. W.T., Complete Front and Back of Report
☐ Pitting ☐ General ☐ Localized
Max Depth ________ (In.) Longitudinal Length ________ (In.)

B. STATUS
☐ Active
☐ Inactive
☐ Unknown

C. CAUSE
☐ Galvanic
☐ Stray Current
☐ Unknown
☐ Other ________

11. Cathodic Protection: ☐ Impressed ☐ Galvanic

12. Pipe-to-Soil Potential:
a) Last Pipe-to-Soil Measurement at Nearest Point Each Side of Corroded Area
Upstream ____________ (volts) Downstream ____________ (volts)
b) Distance from Corroded Area to Measurement Points:
Upstream ____________ (feet) Downstream ____________ (feet)
c) Date of Measurements: Upstream ____________ Downstream ____________

13. pH of Soil Near Corroded Area ____________ pH of Moisture Under Coating ____________

14. Pipe Surface Temperature if Fluid is Heated ____________ (deg F)

Remarks ____________

Distribution: (Original) Corr. Ctrl/Eng (Copy) Division Office (Copy) Other ________

____________ ___/___/___ ____________ ___/___/___
Technician or Inspector *Date* *Supervisor* *Date*

Figure 2A Pipeline inspection form (Sheet 1 of 2).

PIPELINE INSPECTION REPORT

Frequency: At Time of Inspection

PART D — DEFECT, DAMAGE, OR CORROSION INVESTIGATION

Notes:
1) Tabulate all Dents, Gouges, Slivers, Corrosion, S.C.C., etc.
2) List Reference Point – Valve, Tee, Center Road, etc., as shown on Alignment Sheet with appropriate Station Number, if available.
3) Indicate Location of Longitudinal Seam. If Pipe is Vertical Substitute Compass direction for o'clock Position. For Spiral Weld Pipe, Note as Such.
4) Defects Such as Dents, Slivers, Gouges, etc., shall be tabulated individually. Corrosion Cracks and/or Corrosion Pits Closely Grouped May be Tabulated as One Defect.
5) Defect Width and Length Will be Measured Along the Girth and Longitudinal Axis, Respectively.

Reference Point: ☐ Upstream End (Station No.) ____________ Description ____________

☐ Downstream End (Station No.) ____________ Description ____________

Upstream End **Downstream End**

SKETCH DEPICTING PIPE CUT AT BOTTOM AND LAID FLAT

NOTE: Indicate Girth Weld Location If Known

________ O'Clock Position →

← Length of Inspected Area →

PART E — DEFECT, DAMAGE, OR CORROSION MEASUREMENTS AND LOCATION

No.	Type	Distance from Girth Weld	Distance from Long Seam	Width (Circumferencial)	Length (Longitudinal)	Maximum Depth	Number of Anomalies
1.							
2.							
3.							
4.							
5.							
6.							
7.							
8.							
9.							
10.							
11.							
12.							
13.							
14.							

If Repair is Required, send to Division Manager for Approval.

Repair Required ☐ No ☐ Yes Type of Repair ____________

Approved by ____________ Title ____________

Remarks: ____________

Figure 2B Pipeline inspection form (Sheet 2 of 2).

> **NOTE**
> Pay special attention to areas at each end of the pipe segment that may have been damaged by longitudinal stresses during the movement.

Step 4 If any damage is observed along the pipe, notify the appropriate personnel according to company policies and procedures.

Step 5 Visually inspect the entire length of the pipeline section to ensure that the coating has not been damaged.

Step 6 If coating damage is noted, take necessary steps to repair the coating.

Step 7 Assess the condition and operability of valves to ensure that they are still in good working order.

Step 8 Inspect the ditch and the backfill material to ensure that no materials capable of damaging the pipeline coating are present.

Step 9 Complete appropriate documentation as required by operator's procedures.

Additional Resources

It is assumed that you will learn to perform these procedures under the direct supervision of your crew leader, foreman, or instructor. For independent study, refer to the recommended Additional Resources at the back of this module.

2.0.0 Section Review

1. The standard the covers pipeline transportation system for liquids and slurries is _____.
   a. *ASME B31.4*
   b. *OSHA 395*
   c. *API 61L*
   d. *API 1455*

2. Longitudinal stress during the move typically occurs _____.
   a. 4 feet from the lift point
   b. at each end of the pipe segment
   c. in the center of the pipe segment
   d. 4 feet from the support point

SUMMARY

Whenever pipe has been moved, it must be visually inspected after the move to ensure the structural integrity of the pipe and pipe components is maintained. If any damage is found, the damage must be repaired before the system can be placed back in service. An inspection of the coating must also be made over the entire length of the exposed pipe, and repairs to coating must be made as necessary. Upon completion of the required inspection, the ditch is backfilled and covered according to specifications and to the satisfaction of the landowner. The work-site location activities are finalized so that the pipe can be returned to normal operations.

Review Questions

1. It is important to limit and minimize secondary stress during the move and after the move to avoid any subsequent damage and _____.
   a. potential pipe failure
   b. temperature fluctuation
   c. line purge issues
   d. pressure deviation

2. If a pipeline containing gases were being moved, how much would the line pressure be lowered?
   a. 50 percent or less of the maximum operating pressure
   b. 50 percent or less of the vapor pressure
   c. 50 percent or less of the minimum operating pressure
   d. 50 percent or less of the maximum allowable operating pressure

3. When a pipe is dented or gouged, the pipe wall thickness is _____.
   a. subjected to shear forces
   b. reduced at the point of impact
   c. increased at the point of impact
   d. protected by the coating

4. When validating line integrity after the line move, you should verify the final _____.
   a. line temperature
   b. soil compaction
   c. line pressure
   d. soil temperature

5. After successful completion of all activities and related inspections, the excavation is _____.
   a. backfilled
   b. hydro tested
   c. backflushed
   d. reinforced

6. Hard backfill material should be used near the coating.
   a. True
   b. False

Trade Terms Quiz

Fill in the blank with the correct term that you learned from your study of this module.

1. Conditions within a pipeline system, identified by the operator, that may indicate (1) a malfunction of a component or (2) a deviation from normal operations that may indicate a condition exceeding design limits or result in a hazard to persons, property, or the environment are called ________.
2. The maximum internal pressure permitted during the operation of a hazardous liquid or carbon dioxide pipeline is called a(n) ________.
3. A drawing that maps the location of the pipeline and its position on/under the ground is called a(n) ________.
4. Pipeline that has been removed from service, drained, purged, and sealed to ensure there is no danger to people, property, or the environment is called a(n) ________.
5. A plan that must be developed and approved before a pipeline can be moved is called a(n) ________.
6. The deterioration of a pipe surface (usually metal) that results from a reaction its environment is called ________.
7. A hazardous liquid that has a vapor pressure exceeding 40 pounds per square inch absolute at 100°F and that will form a vapor cloud when released into the atmosphere is called a(n) ________.
8. A system or part of a system that contains liquid hydrocarbons or other hazardous liquids under pressure and are used on a routine basis to move these liquids from one point to another in the system is called a(n) ________.
9. The maximum internal pressure permitted during the operation of a gas pipeline is called ________.
10. The stress associated with a pipeline at rest, excluding any residual stress in girth welds or over-bends, is called ________.
11. The process of replacing soil after an excavation is called ________.
12. When a valve's performance, based on in-service criteria, meets all of the necessary functions of the valve it is in ________.
13. Damage to the surface of a pipe that displaces or removes metal and reduces the pipe wall thickness is a(n) ________.
14. A depression in the surface of a pipe that affects the curvature of the pipe without reducing the pipe wall thickness is a(n) ________.

Trade Terms

Abnormal operating conditions (AOCs)
Backfilling
Corrosion
Dent
Existing longitudinal stress
Good working order
Gouge
Highly volatile liquid (HVL)
In-service pipe
Maximum allowable operating pressure (MAOP)
Maximum operating pressure (MOP)
Out-of-service pipe
Pipeline movement plan
Pipeline profile

Trade Terms Introduced in This Module

Abnormal operating conditions (AOCs): Conditions within a pipeline system, identified by the operator, that may indicate (1) a malfunction of a component or (2) a deviation from normal operations that may indicate a condition exceeding design limits or result in a hazard to persons, property, or the environment.

Backfilling: The process of replacing soil after an excavation to fill the ditch and safely cover the pipeline.

Dent: A depression in the surface of a pipe that affects the curvature of the pipe without reducing the pipe wall thickness.

Corrosion: The deterioration of a pipe surface (usually metal) that results from a reaction its environment.

Existing longitudinal stress: The longitudinal stress associated with a pipeline at rest, excluding any residual stress in girth welds or over-bends.

Good working order: All valves are required by 49 CFR 195.420(a) to be in good working order, which can be interpreted to mean that a valve's performance, based on in-service criteria, meets all of the necessary functions of the valve.

Gouge: Damage to the surface of a pipe that displaces or removes metal and reduces the pipe wall thickness.

Highly volatile liquid (HVL): A hazardous liquid that has a vapor pressure exceeding 40 psi absolute at 100°F. It will form a vapor cloud when released into the atmosphere.

In-service pipe: A system or part of a system that contains liquid hydrocarbons or other hazardous liquids under pressure. In-service lines are used on a routine basis to move these liquids from one point to another in the system.

Maximum allowable operating pressure (MAOP): The maximum internal pressure permitted during the operation of a gas pipeline.

Maximum operating pressure (MOP): The maximum internal pressure permitted during the operation of a hazardous liquid or carbon dioxide pipeline.

Out-of-service pipe: Pipeline that has been removed from service, drained, purged, and sealed to ensure there is no danger to people, property, or the environment. Out-of-service pipe includes those segments of a line that have been abandoned or idled until a further need for their use arises.

Pipeline movement plan: A plan that must be developed and approved before a pipeline can be moved. Typically includes information about the pipeline, specifications for excavation and moving, emergency procedures, safety measures, notification requirements, and project communications information.

Pipeline profile: A drawing that maps the location of the pipeline and its position on/under the ground. The pipeline profile details the location and depth of cover for the target line.

Additional Resources

The following recommended resources can provide additional helpful information related to the requirements for this covered task:

API Recommended Practice 1161, Third Edition, © 2014 American Petroleum Institute. API Publishing Services, 1220 L Street, NW, Washington, DC 20005. Distributed by Thompson Reuters (Scientific) LLC, **www.techstreet.com**.

49 CFR 192, **www.ecfr.gov**

49 CFR 195, **www.ecfr.gov**

NCCER Module AOCFG-17, *Abnormal Operating Conditions Field and Gas.*

NCCER Module AOCCC-17, *Abnormal Operating Conditions Control Center.*

NCCER Module 67102-02, *Basic Pipeline Pneumatics and Equipment.*

NCCER Module CT32_0-17, *Observation of Excavation Activities.*

NCCER Module CT39_0-17, *Backfilling a Trench Following Maintenance.*

NCCER Module 75120-13, *Confined Spaces and Excavations.*

NCCER Module CT14_1-17, *Locate Line.*

NCCER Module CT33_1-17, *Determining Allowable Line Pressure in Section to be Moved.*

NCCER Module CT33_2-17, *Preparation for Movement Activities*.

NCCER Module CT33_3-17, *Moving In-Service Pipeline.*

NCCER Module 67103-02, *Pipeline Communications.*

NCCER Module 62105-14, *Introduction to Pipeline Documents.*

The following websites provide resources for products and training:

Occupational Safety and Health Administration (OSHA), **www.osha.gov**

Pipeline and Hazardous Materials Safety Administration (PHMSA), **www.phmsa.dot.gov**

Figure Credits

Copyright © 2013 Trench Safety and Supply, Inc. Used with permission. Module Opener

Section Review Answer Key

Answer	Section Reference	Objective
Section One		
1. c	1.1.1	1a
2. a	1.1.2	1a
3. c	1.1.3	1a
Section Two		
1. a	2.1.0	2a
2. b	2.2.0	2b

NCCER CURRICULA — USER UPDATE

NCCER makes every effort to keep its textbooks up-to-date and free of technical errors. We appreciate your help in this process. If you find an error, a typographical mistake, or an inaccuracy in NCCER's curricula, please fill out this form (or a photocopy), or complete the online form at **www.nccer.org/olf**. Be sure to include the exact module ID number, page number, a detailed description, and your recommended correction. Your input will be brought to the attention of the Authoring Team. Thank you for your assistance.

Instructors – If you have an idea for improving this textbook, or have found that additional materials were necessary to teach this module effectively, please let us know so that we may present your suggestions to the Authoring Team.

NCCER Product Development and Revision

13614 Progress Blvd., Alachua, FL 32615

Email: curriculum@nccer.org
Online: www.nccer.org/olf

❑ Trainee Guide ❑ Lesson Plans ❑ Exam ❑ PowerPoints Other ______________

Craft / Level: Copyright Date:

Module ID Number / Title:

Section Number(s):

Description:

Recommended Correction:

Your Name:

Address:

Email: Phone:

CT37_0-17

Install or Repair Support Structures on Existing Aboveground Components

Overview

Pipelines must often cross highways, railroads, ditches, or bodies of water. Structures are required to support the pipe at these crossings. This module covers the procedures for repairing or installing pipeline support structures on aboveground pipelines.

Trainees with successful module completions may be eligible for credentialing through the NCCER Registry. To learn more, go to **www.nccer.org** or contact us at 1.888.622.3720. Our website has information on the latest product releases and training, as well as online versions of our *Cornerstone* magazine and Pearson's product catalog.

Your feedback is welcome. You may email your comments to **curriculum@nccer.org**, send general comments and inquiries to **info@nccer.org**, or fill in the User Update form at the back of this module.

This information is general in nature and intended for training purposes only. Actual performance of activities described in this manual requires compliance with all applicable operating, service, maintenance, and safety procedures under the direction of qualified personnel. References in this manual to patented or proprietary devices do not constitute a recommendation of their use.

CT34_0-17
Install or Repair Support Structures on Existing Aboveground Components

Objectives

When you have completed this module, you will be able to do the following:

1. Describe pipeline repair criteria.
   a. Identify pipeline repair standards.
   b. Identify pipe support structures and components.
2. Describe how to repair or install support structures (CT37_0-17).
   a. Identify the prerequisites, competencies, and abnormal operating conditions associated with repairing or installing support structures.
   b. Describe the procedure for repairing or installing support structures.

Performance Tasks

Under the supervision of the instructor, you should be able to do the following:

1. Repair or install support structures on existing aboveground components (CT37_0-17).
   - Identify potential abnormal operating conditions that may occur during performance of this CT, and know the appropriate actions to take in response to them.
   - Utilize the appropriate personal protective equipment according to relevant company procedures.
   - Determine proper construction, spacing, and design of pipe support, including support structure foundation in accordance with company practice, standards or engineering design.
   - Obtain proper tools and equipment to complete the installation.
   - Install temporary support (if applicable).
   - Lower internal pipe pressure to 50% of maximum operating pressure if pipe movement is required.
   - Lift pipe with a jack using pipe protective device (pipe saddles) and hoist with proper rigging techniques or another appropriate lift technique (if applicable).
   - Remove existing pipe support (if applicable).
   - Perform a pipe wall inspection, and measure pipe wall thickness, if applicable.
   - Repair pipe support in accordance with company practice, standards, or engineering design (if applicable).
   - Install repaired or replacement support and insulator.
   - Lower pipe to desired position and remove temporary pipe support (if applicable). Adjust support height to insure a level pipe run (if using adjustable support).
   - Complete appropriate documentation as required by operator's procedures.

Trade Terms

Abnormal operating conditions (AOCs)
Flare
Hoist
Hot tie-in
Insulator
Jack
Pipe saddles

Industry Recognized Credentials

If you are training through an NCCER-accredited sponsor, you may be eligible for credentials from NCCER's Registry. The ID number for this module is CT37_0-17. Note that this module may have been used in other NCCER curricula and may apply to other level completions. Contact NCCER's Registry at 888.622.3720 or go to **www.nccer.org** for more information.

Contents

Figures

SECTION ONE

1.0.0 PIPELINE REPAIR

Objective

Describe pipeline repair criteria.

a. Identify pipeline repair standards
b. Identify pipe support structures and components

Trade Terms

Flare: A burner or other device used to burn hydrocarbon vapors off a pipeline for the purpose of avoiding a dangerous buildup of the vapors.

Hot tie-in: Any weld completed on an in-service pipeline.

The need for pipeline repair can arise from an emergency, routine pipeline maintenance, or pipeline modifications. When repairs are made in an emergency, the primary goal is to end the emergency and minimize damage. This is often a hazardous situation, so company emergency and safety procedures become critically important. There is no time to plan and prepare for the repair work. When repairs can be planned ahead of time, try to minimize hazards, schedule necessary interruptions, and avoid damage.

1.1.0 Pipeline Repair

Pipeline repair may require that the pipeline be shut down, or a temporary bypass system may be used. Special fittings and machines can be used to temporarily bypass a section of pipeline without interrupting service. This may require cutting and welding on a pipeline operating at reduced pressure but still containing product. Any weld made on an in-service pipeline is known as a hot tie-in.

Working on pipelines that currently have, or have had, product inside can be hazardous. An empty pipeline section that has been filled with air can ignite and explode. Nitrogen or other inert-gas purging is usually done to make these vapors noncombustible. A safer condition for electric welding on liquid petroleum pipelines is for the pipe to be completely filled with moving liquid.

Another technique used by the natural gas pipeline industry is a flare, a procedure in which the gas in the pipeline is purposely ignited to avoid an explosive ignition by a chance spark from welding. In all cases of pipeline repair, it is critical that company-specific procedures be followed closely to avoid the related hazards

1.1.1 Repair Standards

Before performing pipeline repairs of any kind, personnel must understand general repair standards in order to minimize risk to people, property, and the environment. These standards include the following:

- Estimating the extent of work to be done and how long it will take to complete
- Notifying the appropriate people of the repair work in order for work to be organized and scheduled
- Planning for limiting pipeline pressure during repairs
- Planning for excavation activities, including pipeline locating and marking, One-Call notification, and provisions for pipeline and excavation support and excavating procedures
- Following excavation procedures
- Checking for and possibly grounding of induced voltage
- Arranging for qualified persons to evaluate the coating, pipe integrity, and corrosion when the pipe is exposed
- Cleaning and coating the pipe according to specifications after repairs
- Restoring pipeline cover according to specifications
- Returning surrounding areas to their original condition as much as possible
- Completing repair reports to document the repair — these should include any reports from inspections and tests performed relative to the repair. An example of a permanent repair report is shown in *Figure 1*.

1.2.0 Support Structures for Aboveground Pipe

Overhead pipe installations require support structures that are able to withstand climate and load stresses. These structures must be maintained and repaired as needed to provide the stable support needed for the pipeline. In certain climates, pipe that is fully underground or transitional (partly aboveground and partly underground) may require an insulating system. This system must also be maintained and repaired as necessary.

Permanent Repair Report

LPL Report Number: ____________________

Report Date: ____________________ Report Type: ____________________

Location

Reporting Office Superior	Date Discovered	Date Repaired	Report Date	Safety-Related Condition Report Required?
Pipeline Location - Pump Station or Milepost	Line No.	Line Size	Special Cost Number	
Specific Location - Location from a Known Stationing on a Route Sheet			Route Sheet Number	

Quarter	Section	Township	Range	Tract No.	Meridian/County	Province/State

Property Owner's Name and Address Phone:
Property Tenant's Name and Address Phone:
Full Description of Property Damage and Any Communications with Owner or Tenant (Attach a sketch if necessary)

Leak

Break or Leak Classification				
Describe Specific Nature and Cause of Break or Leak				
Oil Out of Line - Barrels	Oil Recovered - Barrels	Net Loss - Barrels 0	Crude Type	Batch Number
Disposition of Lost Oil				
Name and Address of Person Reporting Leak			Was a Report Fee Paid? No - $	

Repair

Nature of Repair & Fitting added to Mainline (Attach Sketch - Identify Upstream & Pipe Orientation)		
Nature and Type of Soil	General Condition of Line See Corrosion Inspection Report	Type of Existing Coating
Condition of Trench (dry, wet, etc.)	Cathodic Protection Potential See Corrosion Inspection Report	Type and Length of Repair Costing
LPL Personnel Must Also Complete a Corrosion Inspection Report		

Other

Was There any Stock Pipe Installed? (if yes, attach hydrotest records)		Hydrotest Number	
Remarks:			
Government Agency Notified	Individual Contacted	Date	Time
Report Completed by and Signature		Date Entered on System Drawings by Tech. Services	

Figure 1 Example of a permanent repair report.

When underground crossings are not practical, overhead crossings may be built. With overhead crossings, consideration must be given to problems involving the strength and flexibility of pipe and supporting structures not ordinarily encountered in pipeline construction. Strength requirements are variable because of wind and ice loads. Flexibility must be provided to accommodate dimensional changes in pipe and structures from temperature and load variations. Vibrations that cannot be completely eliminated provide other stresses on the pipeline. Maintenance of these crossings requires monitoring and reducing the effects of these stresses. A major part of this is the installation, replacement, and repair of support structures.

1.2.1 Checking Pipe Supports and Support Structures

Vertical support members (VSMs) can be tilted by jacking or by settling of permafrost or other ground movement. Tilting can occur without being a major problem requiring immediate attention, but it may be an indication of a problem that might develop into a more serious condition as time passes. *Figure 2* shows a run of aboveground pipeline with anchor supports and intermediate supports.

> **NOTE**
> Always follow company-specific policies and procedures when inspecting tilting of any pipeline support member.

Aboveground pipeline support structures (bents) are designed and built to be strong and durable. Aboveground construction is often employed as the pipeline crosses areas where the soil is unstable because it is exposed to a freeze-thaw cycle. If underground construction were used in these areas, the expansion and contraction of the pipe could be contained by the surrounding fill material. *Figure 3* shows potential crossbeam movement in a support structure.

Aboveground pipe expands and contracts in a more predictable manner. The aboveground design accommodates pipe motion by allowing lateral movement on the intermediate supports between secured anchor points on some supports (*Figure 4*).

1.2.2 Damage Prevention Sliders

Aboveground pipes often have damage-prevention slider assemblies designed to limit deflection of the mainline pipe from thermal loads and seismic forces. The slider components must be visually inspected to ensure that they are not damaged. The slide assembly is composed of three elements: pedestal, baseplate, and slide rods.

The pedestal provides a connection between the anchor clamp and the platform. It is usually fabricated from pipe with a ¾-inch wall thickness (*Figure 5*).

The baseplate supports the pedestal. This plate transmits the load of the pipe and its contents to the platform and provides the low-friction surfaces that allow the pipe to be displaced. The baseplate is welded to the bottom of the pedestal and is parallel to the main-line pipe. The slide rods are attached to the baseplate.

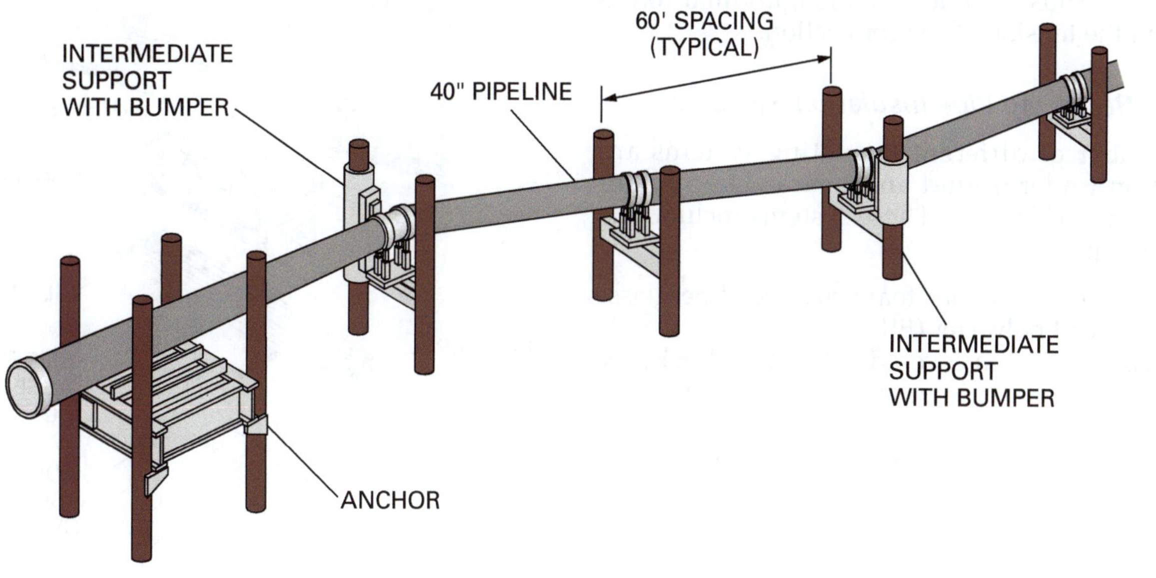

Figure 2 Aboveground pipeline structure.

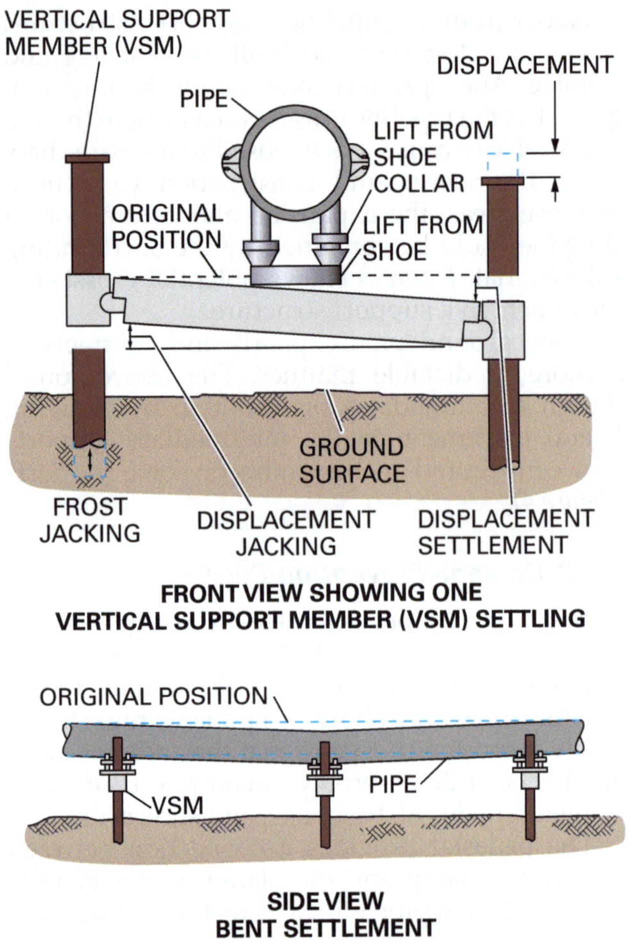

Figure 3 Pipeline-support movement.

Slide rods are fabricated from angle iron and steel plate, and are installed so the plate is sandwiched between the angle pieces. The rod assemblies are attached to the plates with 1-inch-diameter bolts. These bolts, installed through holes in the angle pieces and a long slot in the plates, allow the rods to slide when longitudinal forces exceed the tension of the connection.

1.2.3 Repairing Pipe Insulation Systems

A number of different insulating systems are used on underground and/or transition mainline piping (*Figure 6*). These systems include the following:

- Insulated spray-on foam-covered fiberglass-reinforced polyester (FRP)
- Premolded foam covered with an FRP jacket or individual FRP pieces glued onto the pipe
- On-site applied foam
- Annular insulation, polystyrene, foamglass, or polyvinyl chloride (PVC) foam on underground pipe

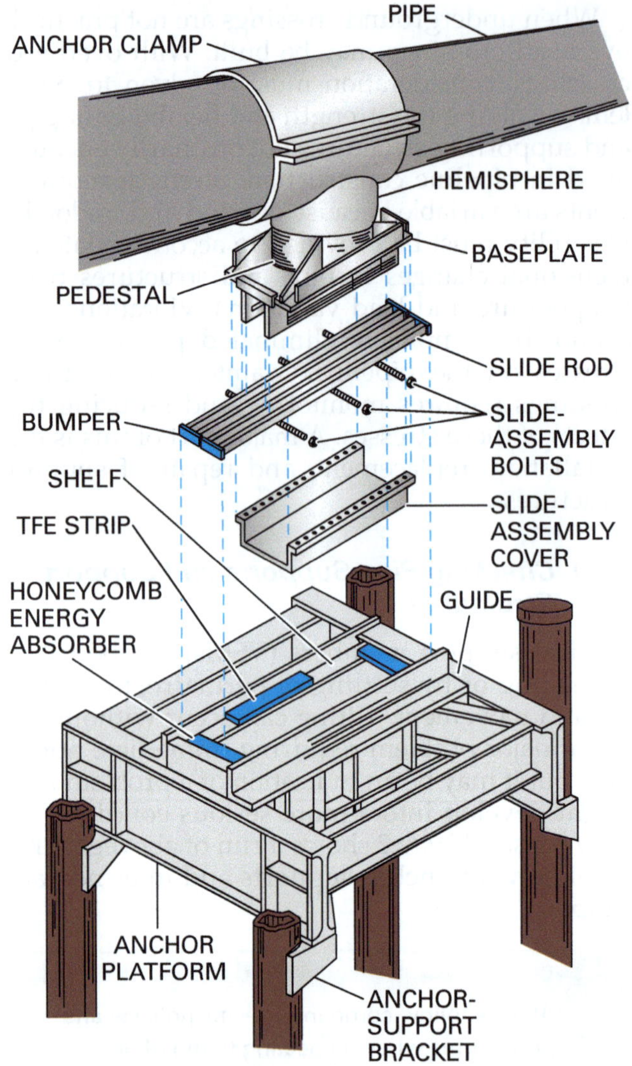

Figure 4 Pipeline anchor support.

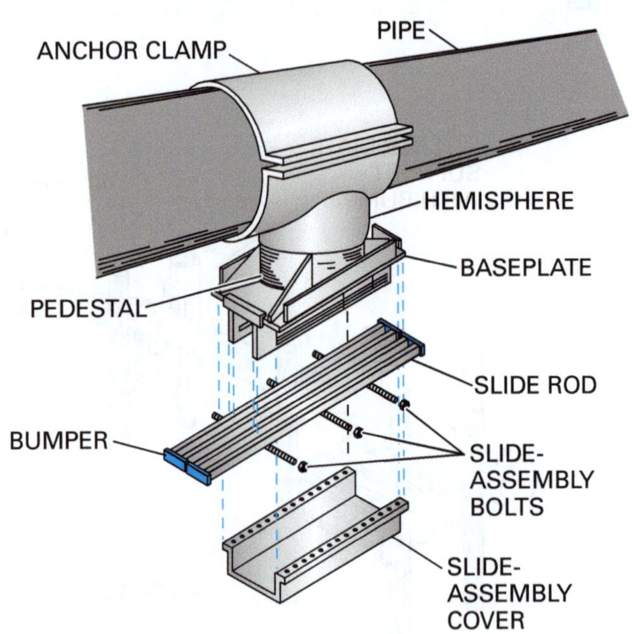

Figure 5 Pedestal.

Figure 6 Transition pipeline.

The following steps outline the procedure for general repair of foam insulation:

> **NOTE**
> The steps presented here are general in nature, and may not always apply to your particular situation. Always follow company-specific policies and procedures.

Step 1 Clean the area around the damage with solvent or with water mixed with a small amount of detergent.

Step 2 Apply a bonding agent approximately 4 inches beyond and around the damaged area.

> **NOTE**
> Small holes and cracks can be repaired with construction-grade silicone sealant.

Additional Resources

It is assumed that you will learn to perform the purging procedure under the direct supervision of your crew leader, foreman, or instructor. For independent study, refer to the recommended Additional Resources at the back of this module.

1.0.0 Section Review

1. Any weld made on an in-service pipeline is known as a _____.
   a. groove
   b. joint
   c. hot tie-in
   d. job

2. The components of a damage-prevention slide assembly are the pedestal, baseplate, and _____.
   a. bents
   b. jacks
   c. bonding agents
   d. slide rods

SECTION TWO

2.0.0 REPAIRING OR INSTALLING SUPPORT STRUCTURES

Objective

Describe how to repair or install support structures (CT37_0-17).

a. Identify the prerequisites, competencies, and abnormal operating conditions associated with repairing or installing support structures.
b. Describe the procedure for repairing or installing support structures.

Performance Task

1. Repair or install support structures on existing aboveground components (CT37_0-17).
   - Identify potential abnormal operating conditions (AOCs) that may occur during performance of this CT, and know the appropriate actions to take in response to them.
   - Utilize the appropriate personal protective equipment according to relevant company procedures.
   - Determine proper construction, spacing, and design of pipe support, including support structure foundation in accordance with company practice, standards or engineering design.
   - Obtain proper tools and equipment to complete the installation.
   - Install temporary support (if applicable).
   - Lower internal pipe pressure to 50% of maximum operating pressure if pipe movement is required.
   - Lift pipe with a jack using pipe protective device (pipe saddles) and hoist with proper rigging techniques or another appropriate lift technique (if applicable).
   - Remove existing pipe support (if applicable).
   - Perform a pipe wall inspection, and measure pipe wall thickness, if applicable.
   - Repair pipe support in accordance with company practice, standards, or engineering design (if applicable).
   - Install repaired or replacement support and insulator.
   - Lower pipe to desired position and remove temporary pipe support (if applicable). Adjust support height to insure a level pipe run (if using adjustable support).
   - Complete appropriate documentation as required by operator's procedures.

Trade Terms

Abnormal operating conditions (AOCs): Conditions within a pipeline system, identified by the operator, that may indicate (1) a malfunction of a component or (2) a deviation from normal operations that may indicate a condition exceeding design limits or result in a hazard to persons, property, or the environment.

Hoist: A hydraulically, electrically, or manually operated piece of equipment used to lift pipe with straps or chains from above the pipe.

Insulator: Non-conductive material placed between the pipe and the support to prevent metal-to-metal contact.

Jack: A hydraulically or manually operated piece of equipment used to lift pipe from underneath.

Pipe saddles: Curved fitted pieces of steel placed between the pipe exterior and a force-exerting piece of equipment. Pipe saddles are used for pipe protection during lifting operations.

This section provides the procedures for repairing, replacing, and installing aboveground support structures. It also provides the prerequisites and competencies required for the procedures, as well as the **abnormal operating conditions (AOCs)** that may be encountered.

2.1.0 Prerequisites, Competencies, and Abnormal Operating Conditions

Personnel must possess a variety of skills in order to install support structures. The various operations require a series of steps, which a pipeline technician must be competent to perform. Personnel must also be aware of the particular abnormal operating conditions (AOCs) that may occur during the procedure and know the proper responses to them.

2.1.1 Prerequisites and Competencies

Personnel assigned to repair or replace a support structure must be able to determine the proper materials and construction techniques needed for a given application. They must also be able to understand and follow company practices for the procedure. Personnel must be able to ensure that the procedure does not over stress or damage the pipe being supported.

Personnel must understand pipe rigging and lifting techniques, as well as the proper usage of the hand and power tools and additional installation equipment needed for the task. They must also be able to read and interpret facilities drawings, protective and control device drawings, and pipe system drawings.

It may be required that personnel performing the task be qualified in support structure welding practices and techniques, as well as measuring pipe wall thickness. More information on this task can be found in NCCER Module CT8_2-17, *Measure Wall Thickness with Ultrasonic Meter.*

Personnel must understand and be able to follow all company safety policies and procedures, including excavation safety and the proper use of PPE.

2.1.2 Abnormal Operating Conditions

Personnel installing support structures must be able to identify potential AOCs and know the proper response to each, should they occur. The following are some potential AOCs, and the proper action personnel should take in response to them:

- *A component failure (such as a valve, weld/seam, flange, or joint failure) occurs* – Isolate the component or shut down the system and depressurize the affected piping and components.
- *Damage to the pipeline system such as wrinkling, buckling, gouging, dents, or an over-stress is discovered* – Perform an initial damage assessment, document the damage as required, and notify the appropriate personnel of the nature and location of the damage.
- *An unexpected hazardous liquid or gas release is encountered during the procedure* – Eliminate any ignition sources, and then notify the control center to isolate the leak or shut down the line, whichever is necessary. If the leak is excessive and cannot be controlled, evacuate the area and only take defensive action using the proper PPE.
- *The unintended ignition of hazardous liquid or gas causes a fire or explosion* – Evacuate the area, notify the appropriate personnel to shut down the system, isolate affected piping, and notify the fire department.

2.2.0 Procedure for Repairing or Installing Support Structures

This task is normally performed or supervised by pipeline maintenance foremen, project engineers, or designated inspectors. Contract pipeline construction companies and contract pipeline maintenance personnel also commonly carry out this task. Support for each aboveground structure must provide protection from the forces exerted by the anticipated loads. The following checklist is presented as a general guideline; however, always follow the company's written and approved procedures for performing this task.

The following steps outline the general procedure for installing, replacing, or repairing support structures:

Step 1 Identify potential abnormal operating conditions that may occur during performance of this CT, and know the appropriate actions to take in response to them.

Step 2 Utilize the appropriate personal protective equipment according to relevant company procedures.

Step 3 Determine the proper construction, spacing, and design of the pipe support including the supporting structure foundation per company practice, or engineering design standards. Ensure that the structure will support a fully loaded pipe facility.

Step 4 Obtain the proper tools and equipment required to safely complete the installation.

Step 5 Install a temporary support, if applicable. Ensure that the pipe is adequately supported during the removal or repair of an existing support, and during the installation of a new support and **insulator**.

Step 6 Lower the internal pressure to 50 percent of the maximum operating pressure (MOP) if pipe movement is required. This step aids in the safe movement of the pipe, and ensures compliance with 49 *CFR* 195.424(a).

Step 7 Lift the pipe with a **jack** using pipe protective devices, such as **pipe saddles**, or **hoist** the pipe with proper rigging techniques, if applicable. Allow room to remove or repair the existing support, or to install a new support and insulator.

Step 8 Remove existing pipe support, if applicable. Removal of an existing support may be required in order to eliminate an obstacle that might interfere with the repair of the pipe, inspection of the pipe wall, or the installation of insulators.

Step 9 Perform a pipe wall inspection, and measure pipe wall thickness, if applicable. This procedure is to ensure that there is adequate wall thickness for the required pressure. It will be performed by a qualified individual in accordance with CT8_2-17.

Step 10 Repair the pipe support in accordance with company practice and standards, or with acceptable engineering design. Correct any repairable defect in the existing supports.

Step 11 Install the repaired, new, or replacement support and insulator. Ensure that the pipe is adequately supported. Protect the pipe surface to prevent metal loss.

Step 12 Lower the pipe to desired position for normal operations.

Step 13 Remove the temporary pipe support, if applicable.

Step 14 Adjust the support height to ensure a level pipe, if using an adjustable support. For this operation, the goal is to minimize the stress on the pipe while maximizing the aesthetics.

Step 15 Complete appropriate documentation as required by operator's procedures.

Additional Resources

It is assumed that you will learn to perform the purging procedure under the direct supervision of your crew leader, foreman, or instructor. For independent study, refer to the recommended Additional Resources at the back of this module.

2.0.0 Section Review

1. If an unexpected hazardous liquid or gas release is encountered during the procedure, the first step would be to _____.
   a. notify the control center
   b. eliminate any ignition sources
   c. evacuate the area
   d. take defensive action using proper PPE

2. Prior to moving pipe, the internal pressure should be lowered to what percent of the maximum operating pressure?
   a. 25 percent
   b. 40 percent
   c. 50 percent
   d. 75 percent

SUMMARY

Pipeline repairs are often necessary because of an emergency situation, which can be extremely hazardous and result in damage to the environment. If potential repairs are identified and performed ahead of time, these hazards and damage can be minimized. Repair procedures which may be needed include installing sleeves, clock springs or other composite repair, tap installation, and cutting out and replacing pipe.

Overhead pipe installations require support structures, which are subject to climate and load stresses. These structures must be maintained and often repaired to provide the stable support needed for the pipeline.

Review Questions

1. Hot tie-ins are made on pipes that are _____.
   a. out of service
   b. in service
   c. completely drained
   d. newly installed

2. In certain climates, pipe that is fully underground or transitional may require _____.
   a. bents
   b. collars
   c. a pedestal
   d. insulating

3. Should the unintended ignition of hazardous liquid or gas cause a fire or explosion, the proper initial response is to _____.
   a. call 911
   b. notify the control center
   c. evacuate the area
   d. contain the event, if practical

4. When lifting pipe, use a protective device such as a _____.
   a. bent
   b. saddle
   c. slide rod
   d. hoist

Trade Terms Quiz

Fill in the blank with the correct term that you learned from your study of this module.

1. Any weld completed on an in-service pipeline is called a(n) ________.
2. A hydraulically or manually operated piece of equipment used to lift pipe from underneath is called a(n) ________.
3. Curved fitted pieces of steel placed between the pipe exterior and a force-exerting piece of equipment are called ________.
4. Conditions within a pipeline system, identified by the operator, that may indicate (1) a malfunction of a component or (2) a deviation from normal operations that may indicate a condition exceeding design limits or result in a hazard to persons, property, or the environment are called ________.
5. A hydraulically, electrically, or manually operated piece of equipment used to lift pipe with straps or chains from above the pipe is called a(n) ________.
6. Non-conductive material placed between the pipe and the support to prevent metal-to-metal contact is called a(n) ________.
7. A burner or other device used to burn hydrocarbon vapors off a pipeline for the purpose of avoiding a dangerous buildup of the vapors is called a(n) ________.

Trade Terms

Abnormal operating conditions (AOCs)
Flare
Hot tie-in
Hoist
Insulator
Jack
Pipe saddles

Trade Terms Introduced in this Module

Abnormal operating conditions (AOCs): Conditions within a pipeline system, identified by the operator, that may indicate (1) a malfunction of a component or (2) a deviation from normal operations that may indicate a condition exceeding design limits or result in a hazard to persons, property, or the environment.

Flare: A burner or other device used to burn hydrocarbon vapors off a pipeline for the purpose of avoiding a dangerous buildup of the vapors.

Hoist: A hydraulically, electrically, or manually operated piece of equipment used to lift pipe with straps or chains from above the pipe.

Hot tie-in: Any weld completed on an in-service pipeline.

Insulator: Non-conductive material placed between the pipe and the support to prevent metal-to-metal contact.

Jack: A hydraulically or manually operated piece of equipment used to lift pipe from underneath.

Pipe saddles: Curved fitted pieces of steel placed between the pipe exterior and a force-exerting piece of equipment. Pipe saddles are used for pipe protection during lifting operations.

Additional Resources

The following recommended resources can provide additional helpful information related to the requirements for this covered task:

API Recommended Practice 1161, Third Edition, © 2014 American Petroleum Institute. API Publishing Services, 1220 L Street, NW, Washington, DC 20005. Distributed by Thompson Reuters (Scientific) LLC, **www.techstreet.com**.

49 *CFR* 192, **www.ecfr.gov**

49 *CFR* 195, **www.ecfr.gov**

NCCER Module AOCCC-17, *Abnormal Operating Conditions Control Center.*

NCCER Module AOCFG-17, *Abnormal Operating Conditions Field and Gas.*

NCCER Module 66101-02, *Introduction to the Pipeline Industry.*

NCCER Module 62105-14, *Introduction to Pipeline Documents.*

NCCER Module 60102-02, *Basic Pipeline Hydraulics and Equipment.*

NCCER Module 63104-02, *Piping and Mechanical Blueprint Reading.*

NCCER Module CT8_2-17, *Measure Wall Thickness with Ultrasonic Meter*

NCCER Module CT40_1-17, *Fit Full Encirclement Welded Split Sleeve.*

NCCER Module CT40_3-17, *Apply Composite Sleeve.*

NCCER Module CT40_4-17, *Install Mechanical Bolt-On Split Repair Sleeve.*

NCCER Module CT40_5-17, *Install Weldable Compression Coupling.*

NCCER Module CT40_6-17, *Install and Remove Plugging Machine.*

NCCER Module CT40_7-17, *Install a Tap 2 Inches and Under on a Pipeline System.*

NCCER Module CT40_8-17, *Install a Tap Larger than 2 Inches on a Pipeline System.*

NCCER Module CT40_9-17, *Install and Remove Completion Plug on Pipelines Larger Than 2 Inches.*

The following websites offer resources for products and training:

Occupational Safety and Health Administration (OSHA), **www.osha.gov**

Pipeline and Hazardous Materials Safety Administration (PHMSA), **www.phmsa.dot.gov**

Figure Credits

Courtesy of PLIDCO, Module Opener

Courtesy of Herbert E Lindberg, Figure 6

Section Review Answer Key

Answer	Section Reference	Objective
Section One		
1. c	1.1.0	1a
2. d	1.2.2	1b
Section Two		
1. b	2.1.2	2a
2. c	2.2.0	2b

NCCER CURRICULA — USER UPDATE

NCCER makes every effort to keep its textbooks up-to-date and free of technical errors. We appreciate your help in this process. If you find an error, a typographical mistake, or an inaccuracy in NCCER's curricula, please fill out this form (or a photocopy), or complete the online form at **www.nccer.org/olf**. Be sure to include the exact module ID number, page number, a detailed description, and your recommended correction. Your input will be brought to the attention of the Authoring Team. Thank you for your assistance.

Instructors – If you have an idea for improving this textbook, or have found that additional materials were necessary to teach this module effectively, please let us know so that we may present your suggestions to the Authoring Team.

NCCER Product Development and Revision

13614 Progress Blvd., Alachua, FL 32615

Email: curriculum@nccer.org

Online: www.nccer.org/olf

❑ Trainee Guide ❑ Lesson Plans ❑ Exam ❑ PowerPoints Other ______________

Craft / Level: Copyright Date:

Module ID Number / Title:

Section Number(s):

Description:

Recommended Correction:

Your Name:

Address:

Email: Phone:

CT38_1-17

Visually Inspect Pipe and Pipe Components Prior to Installation

Overview

The safe operation of a pipeline system relies on the proper installation of pipes and pipe components. Prior to installation, all pipes and components must be inspected to ensure they meet system pressure and specified design requirements. A visual inspection of pipes and pipe components prior to installation helps to ensure that the integrity of the pipeline system meets strength and serviceability requirements. This module explains the procedures for visually inspecting pipes and pipe components prior to installation. Additional types of visual inspections associated with installation are also discussed.

Trainees with successful module completions may be eligible for credentialing through the NCCER Registry. To learn more, go to **www.nccer.org** or contact us at 1.888.622.3720. Our website has information on the latest product releases and training, as well as online versions of our *Cornerstone* magazine and Pearson's product catalog.

Your feedback is welcome. You may email your comments to **curriculum@nccer.org**, send general comments and inquiries to **info@nccer.org**, or fill in the User Update form at the back of this module.

This information is general in nature and intended for training purposes only. Actual performance of activities described in this manual requires compliance with all applicable operating, service, maintenance, and safety procedures under the direction of qualified personnel. References in this manual to patented or proprietary devices do not constitute a recommendation of their use.

CT38_1-17

Visually Inspect Pipe and Pipe Components Prior to Installation

Objectives

When you have completed this module, you will be able to do the following:

1. Explain the purpose for visually inspecting pipes and pipe components and identify types of visual inspections.
   a. Explain the purpose for visually inspecting pipe and pipe components.
   b. Identify and describe the types of visual inspections for pipe and pipe components.
2. Describe the procedures used to visually inspect pipes and pipe components prior to installation (CT38_1-17).
   a. Identify the prerequisites, competencies, and abnormal operating conditions (AOCs) associated with visually inspecting pipes and pipe components.
   b. Describe the general procedures used to visually inspect pipes and pipe components.

Performance Task

Under the supervision of the instructor, you should be able to do the following:

1. Visually inspect pipes and pipe components prior to installation (CT38_1-17).
   - Identify potential abnormal operating conditions (AOCs) that may occur during performance of this CT, and know the appropriate actions to take in response to them.
   - Utilize the appropriate personal protective equipment according to relevant company procedures.
   - Inspect pipe and components for corrosion; mechanical defects such as cracks, grooves, gouges, dents, or out-of-round pipe; or coating damage such as cuts, scratches, or holidays.
   - Inspect bends for buckles and/or wrinkles.
   - Ensure that components are rated for the intended service through visual verification of the markings on the pipe and components.
   - Make appropriate notifications to resolve any unsatisfactory conditions.
   - Complete appropriate documentation as required by operator's procedures.

Trade Terms

Abnormal operating conditions (AOCs)
Components
Corrosion
Defects
Maximum operating pressure (MOP)
Undercut

Industry Recognized Credentials

If you are training through an NCCER-accredited sponsor, you may be eligible for credentials from NCCER's Registry. The ID number for this module is CT38_1-17. Note that this module may have been used in other NCCER curricula and may apply to other level completions. Contact NCCER's Registry at 888.622.3720 or go to **www.nccer.org** for more information.

Contents

Figures

SECTION ONE

1.0.0 VISUALLY INSPECT PIPES AND PIPE COMPONENTS

Objective

Explain the purpose for visually inspecting pipes and pipe components and identify types of visual inspections.

a. Explain the purpose for visually inspecting pipe and pipe components.
b. Identify and describe the types of visual inspections for pipe and pipe components.

Trade Terms

Components: Parts of a pipeline that may be subjected to pump pressure, including, but not limited to, pipe, valves, elbows, tees, flanges, and closures.

Defects: Discontinuities that render a part or product unable to meet minimum applicable acceptance standards or specifications.

The visual inspection of pipes and pipe components is necessary so that any damage or abnormality of components or material connecting to an existing pipeline are detected and appropriate corrective actions are taken. The visual inspection process includes an initial inspection of the pipes and pipe components and may include different types of inspections during subsequent phases of the installation. The type of inspection will depend on the specific item(s) being inspected and installed. Upon receipt, all pipes and components need to be visually inspected externally as well as internally.

The initial visual inspection helps to ensure that the proper materials are selected and there is no visible damage prior to installation. Additional inspections are also conducted for the prefit-up, fit-up, and weld root passes, and for the completed welds. A satisfactory outcome of the visual inspections ensures that each phase of work is completed successfully and installation requirements are satisfied.

1.1.0 Inspect Pipes and Pipe Components Prior to Installation

In order to properly inspect pipes and pipe components, personnel must be able to identify and recognize the different types of pipes and components that are being installed. A review of the appropriate documentation is required to ensure that the specified pipes and pipe components have been selected for the given job.

1.1.1 Initial Visual Inspection of Pipes and Pipe Components

An initial visual inspection involves physically looking for obvious mechanical defects and other issues that can affect the integrity of the pipes and pipe components. A general knowledge of the installation requirements is needed so that the visual inspection will confirm that the pipes and pipe components are appropriate for the job. Pipe size, materials, ratings, and pipe component compatibility need to be taken into consideration. Look for visible markings to confirm each item being inspected. Check the available documentation for relevant information and confirm the pipes and pipe components are correct as specified.

An inspector must have knowledge of the following inspection criteria:

- General appearance of different pipe coatings and potential coating defects such as cuts, scratches, or other defects characterized by visually determined loss of coating (holiday).
- Each length of pipe with a nominal outside diameter of 4 ½ inches (11.43 cm) or more must be marked on the pipe or pipe coating with the specification to which it was made, the specified minimum yield strength or grade, and the pipe size. The marking must be applied in a manner that does not damage the pipe or pipe coating and must remain visible until the pipe is installed.
- Each valve must be marked on the body of the nameplate with the following information, at a minimum:
  - Manufacturer's name or trademark
  - Class designation or the maximum working pressure to which the valve may be subjected
  - Body material designation (the end connection material, if more than one type is used)
  - Nominal valve size

> **NOTE**
> A butt-weld fitting must meet the marking and end preparation required by the operator's specification.

- The appropriate supporting design documentation for the pipeline system installation must also be reviewed so that the pipes and pipe components required for the installation are known. Inspectors must check relevant documentation and confirm that the design specifications match the pipes and pipe components selected for the given installation.

1.2.0 Types of Visual Inspections for Installation of Pipes and Pipe Components

After the initial visual inspection has been completed and the proper pipes and pipe components have been confirmed, the following types of inspections will need to be performed as the installation continues:

- Prefit-up or hold point inspections
- Visual inspection of fit-ups using Hi-Lo gauges
- Visual inspection using borescopes

1.2.1 Prefit-Up or Hold Point Inspections

A hold point is an area that must be inspected before proceeding with work activities such as code welding. A prefit-up inspection involves the following tasks:

- Ensuring that the base metal is clean and free of all foreign material.
- Checking the bevel angle on the pipe to ensure that it is within tolerance.
- Checking the weld area to ensure that it is free of damage, such as nicks and gouges.

1.2.2 Visual Inspection of Fit-Ups

During the installation process, the pipes and pipe components may need to be inspected for proper fit-up prior to welding. A Hi-Lo gauge can be used to check the fit-up for accurate alignment so that the installation process can continue. They are typically used to provide a good visual indication of any major mismatches between the pipes and components being installed, and can also help to determine what corrective actions should be taken if pipes and pipe components need to be repositioned or realigned.

> **NOTE**
> A taper gauge can be used to accurately establish the root gap for the weld to ensure that the pipes and pipe components are properly aligned.

To check for internal misalignment of pipe, Hi-Lo gauges have two prongs, or alignment stops, that are pulled tightly against the inside diameter of the joint so that one stop is flush with each side of the joint. The variation between the two stops is read on a scale marked on the gauge. *Figure 1* shows checking internal misalignment with a Hi-Lo gauge.

To measure pipe misalignment with a Hi-Lo gauge, insert the prongs of the gauge into the root opening. Pull up on the gauge until the prongs are snug against both inside surfaces, and read the misalignment indicated on the scale.

1.2.3 Visual Inspection Using Borescopes

Borescopes are used for visual inspection to detect surface flaws inside pipes, valves, rotating equipment, and completed welds in areas that are restricted from direct visual inspection. *Figure 2* shows three typical types of borescopes.

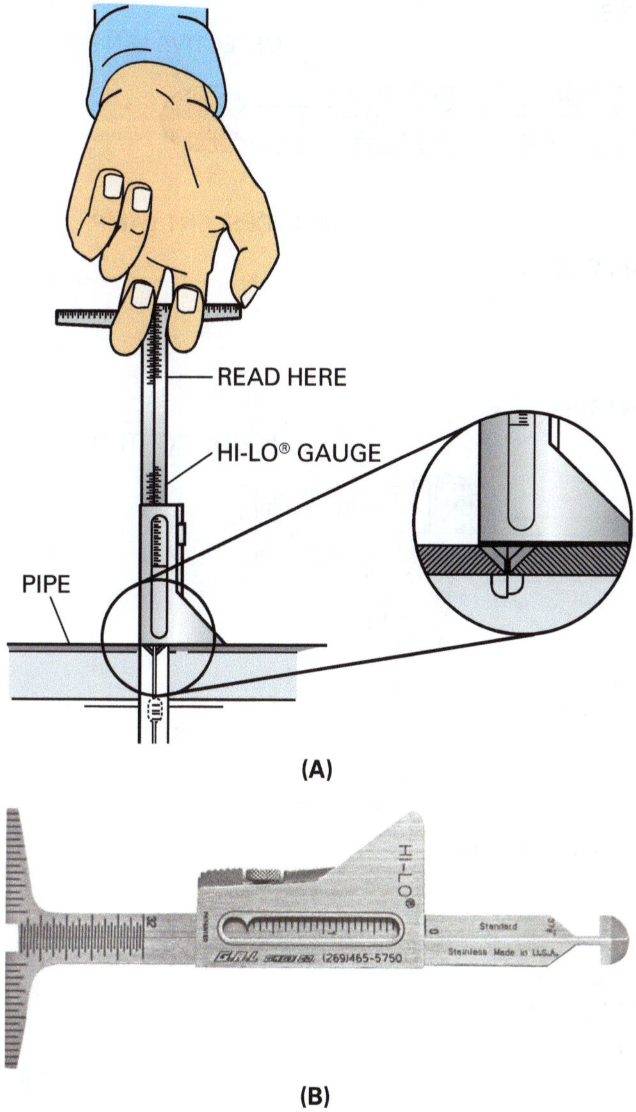

Figure 1 Using a Hi-Lo gauge to check internal misalignment.

Flexible borescopes are most commonly used by pipeline workers. Two of the most common types of flexible borescopes are:

- *Flexible Fiberscopes* – A fiberscope consists of a light guide bundle, an image guide bundle, an objective lens, interchangeable viewing heads, and remote controls to control the rotation of the tip. Fiberscopes are available in varying diameters and come in lengths up to 40 feet. The interchangeable viewing heads provide various directions and fields of view on a single fiberscope. Fields of view are typically 40 to 60 degrees; special applications can range from 10 to 120 degrees. Most fiberscopes provide adjustable focusing of the objective lens.

CAUTION

Specialized training is required to use a flexible fiberscope.

- *Videoscopes with CCD Probes* – Videoscopes transmit color or black-and-white images to a video monitor. The end of the electronic videoscope contains a CCD chip, which consists of thousands of light-sensitive elements in a pattern of rows and columns. Videoscopes with CCD probes produce images similar to borescopes, but they normally produce a higher resolution. The most significant difference is that videoscopes with CCD probes have a longer working length than fiberscopes. Videoscopes with CCD probes are operated in the same manner as flexible fiberscopes. A typical videoscope is shown in *Figure 3*.

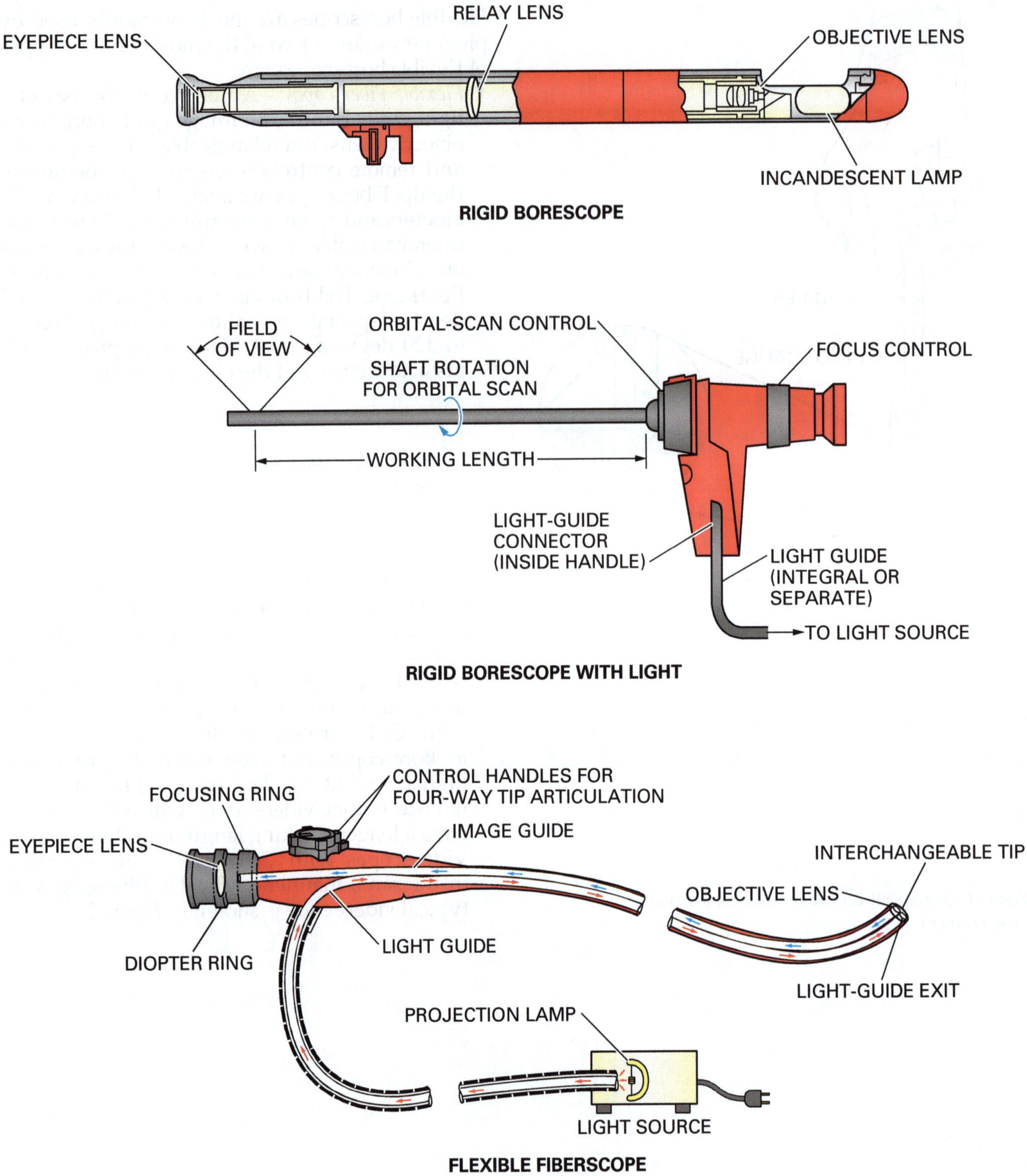

Figure 2 Typical borescopes.

Figure 3 Typical videoscope.

Additional Resources

It is assumed that you will learn to perform these procedures under the direct supervision of your crew leader, foreman, or instructor. For independent study, refer to the recommended Additional Resources at the back of this module.

1.0.0 Section Review

1. Each length of pipe with a nominal outside diameter of 4 ½ inches (11.43 cm) or more must be marked on the pipe or pipe coating with the specification to which it was made, the specified minimum yield strength or grade, and the _____.
   a. date of manufacture
   b. country of origin
   c. pipe size
   d. pressure profile

2. Two of the most common types of flexible borescopes are flexible fiberscopes and _____.
   a. videoscopes with CCD probes
   b. light tubes with XYZ probes
   c. digital imaging probes
   d. light scopes with fiber probes

Section Two

2.0.0 Visually Inspect Pipes and Pipe Components Prior to Installation

Objective

Describe the procedures used to visually inspect pipes and pipe components prior to installation (CT38_1-17).

a. Identify the prerequisites, competencies, and abnormal operating conditions (AOCs) associated with visually inspecting pipes and pipe components.
b. Describe the general procedures used to visually inspect pipes and pipe components.

Performance Task

Visually inspect pipes and pipe components prior to installation (CT38_1-17).

- Identify potential abnormal operating conditions (AOCs) that may occur during performance of this CT, and know the appropriate actions to take in response to them.
- Utilize the appropriate personal protective equipment according to relevant company procedures.
- Inspect pipe and components for corrosion; defects such as cracks, grooves, gouges, dents, or out-of-round pipe; or coating damage such as cuts, scratches, or holidays.
- Inspect bends for buckles and/or wrinkles.
- Ensure that components are rated for the intended service through visual verification of the markings on the pipe and components.
- Make appropriate notifications to resolve any unsatisfactory conditions.
- Complete appropriate documentation as required by operator's procedures.

Trade Terms

Abnormal operating conditions (AOCs): Conditions within a pipeline system, identified by the operator, that may indicate (1) a malfunction of a component or (2) a deviation from normal operations that may indicate a condition exceeding design limits or result in a hazard to persons, property, or the environment.

Corrosion: Surface rust and pitting that may be identified during a visual inspection of pipes and pipe components.

Maximum operating pressure (MOP): The maximum pressure at which a pipeline or segment of a pipeline or a component may be normally operated.

Undercut: A groove melted into the base metal adjacent to the toe or root of a weld and left unfilled by weld metal.

This section describes the procedures for visually inspecting pipes and pipe components prior to installation. It also provides the prerequisites and competencies required for the procedures, and discusses the **abnormal operating conditions (AOCs)** that may be encountered.

2.1.0 Prerequisites, Competencies, and Abnormal Operating Conditions

Personnel must possess a variety of skills to perform a visual inspection of pipes and pipe components. A general working knowledge of the types of pipes and components is required to ensure that personnel are able to recognize any issues that may compromise the installation of the pipeline system. Personnel must also be aware of the particular abnormal operating conditions (AOCs) that may occur when visually inspecting pipes and pipe components and know the proper responses to them.

2.1.1 Prerequisites and Competencies

Personnel responsible for visual inspection of pipes and pipe components must understand that the task typically involves the on-site visual inspection of pipes and pipe components that are to be installed in the pipeline system. Personnel performing visual inspections of pipes and pipe components should have basic knowledge and skills related to the following:

- Knowledge of the design **maximum operating pressure (MOP)** for the given pipeline system.
- Ability to ensure through visual verification of the markings on the pipes and components that all items are compatible with the MOP.
- Knowledge of the proper use of personal protective equipment required for this task.
- Familiarity with pipe and pipe component identification.
- Experience with inspection equipment.
- Ability to recognize coating defects, damage, and corrosion.
- Ability to interpret relevant pipeline documentation to ensure all pipes and components meet design specifications.

2.1.2 Abnormal Operating Conditions

Personnel performing visual inspections of pipes and pipe components must be able to identify AOCs and know the proper response to each, should they occur. The following are some potential AOCs, and the appropriate action personnel should take in response to them:

- *Pipeline system damage (wrinkle, buckle, overstress, gouge, dent) is discovered* – Document as required and make appropriate notifications.
- *Component failure (valve, weld/seam, flange, joint) occurs* – Document as required and make appropriate notifications.
- *Abnormal facility conditions are discovered or materials do not meet specifications* – Document as required and make appropriate notifications.
- *Unexpected hazardous liquid or carbon dioxide is encountered or a failure of an isolation device is discovered* – Eliminate ignition sources and notify appropriate personnel.
- *Fire or explosion occurs* – Evacuate the area and notify the fire department.

2.2.0 Procedures for Visual Inspection of Pipes and Pipe Components

The purpose of a visual inspection is to ensure that the pipes and components have not been damaged in a manner that could impair their strength or reduce their serviceability, and to ensure that the pipes and components are rated for their intended service. It is important to visually inspect all pipes and pipe components that are to be installed in the pipeline system to ensure that they are not damaged and that they meet design specification requirements.

This task does not include the assessment of damage or any determination of the measures that should be taken to mitigate the damage found during an inspection.

2.2.1 Perform Initial Visual Inspection of Pipes and Pipe Components

Always follow company policy and procedures when performing visual inspections. In general, use the following steps to visually inspect pipes and components prior to installation:

Step 1 Identify potential abnormal operating conditions (AOCs) that may occur, and know the appropriate actions to take in response to them.

Step 2 Utilize the appropriate personal protective equipment according to relevant company procedures.

Step 3 Clean the pipe and components so that defects that might impair the strength of the pipe can be easily detected.

Step 4 Visually inspect pipe and pipe components for the following:

NOTE

If the materials do not meet the required specifications, document as required, and notify the proper personnel.

- Complete/proper identification markings
- Corrosion
- Roundness
- Buckling or wrinkling
- Coating integrity (scratches, cuts, holidays)
- Surface defects (cracks, grooves, gouges, dents)

Step 5 Document the inspection according to operator's procedures.

2.2.2 Visually Inspect Pipes and Pipe Component Fit-Ups

Use the following general steps to perform a typical visual inspection of fit-ups. Always follow company policy and procedures.

Step 1 Identify potential AOCs that may occur, and know the appropriate actions to take in response to them.

Step 2 Utilize the appropriate personal protective equipment according to relevant company procedures.

Step 3 Check the pipes and components for corrosion, dents, and proper beveling.

Step 4 Check the classification and grade of the pipes and components.

Step 5 Check the fit-up for misalignment or mismatch, using a Hi-Lo gauge.

Step 6 Check the fit-up for the proper root gap with a taper gauge. *Figure 4* shows how to use a taper gauge to measure a root gap.

Step 7 Inspect tack welds for any weld defects such as cracks, crater cracks, inclusions, poor fusion, undercut, inadequate penetration, and burn-through.

Step 8 Document the inspection according to operator's procedures.

2.2.3 Visually Inspect Pipes and Pipe Component Welds Using a Borescope

Borescopes are optical devices used for performing visual inspections of areas that are difficult to view directly. The following steps are a general procedure for using a borescope to inspect welds on pipes and pipe components:

NOTE

Always follow company policy and procedures.

Step 1 Identify potential AOCs that may occur, and know the appropriate actions to take in response to them.

Step 2 Utilize the appropriate personal protective equipment according to relevant company procedures.

Step 3 Select the appropriate borescope for the job; consider access to the area to be inspected. If the path is straight, use a rigid borescope; if it is indirect or curved, choose a flexible borescope.

NOTE

Videoscopes offer similar flexibility but house a very small CCD camera chip near the tip of the scope and relay the image electronically to a monitor rather than an eyepiece.

Step 4 Review the manufacturer's manual for operation and maintenance.

Step 5 Prepare the item for inspection, observing all required safety precautions.

Step 6 Insert the tip of the borescope into the area to be viewed.

CAUTION

Do not crush or bend the insertion tube. Avoid impact with the tip, and avoid sharp edges. Do not insert it into areas with chemicals or excessive heat; check the manufacturer's manual for the heat rating.

Step 7 For flexible borescopes, use the articulating tip to steer the scope while performing the inspection. Adjust the position of the borescope optics to view the area as needed.

Step 8 Store and secure the borescope after use to prevent damage.

Step 9 Document the inspection according to operator's procedures.

> **NOTE**
>
> For additional information regarding welding inspections, see NCCER module CT38_3-17, *Visually Inspect that Welds Meet DOT Requirements.*

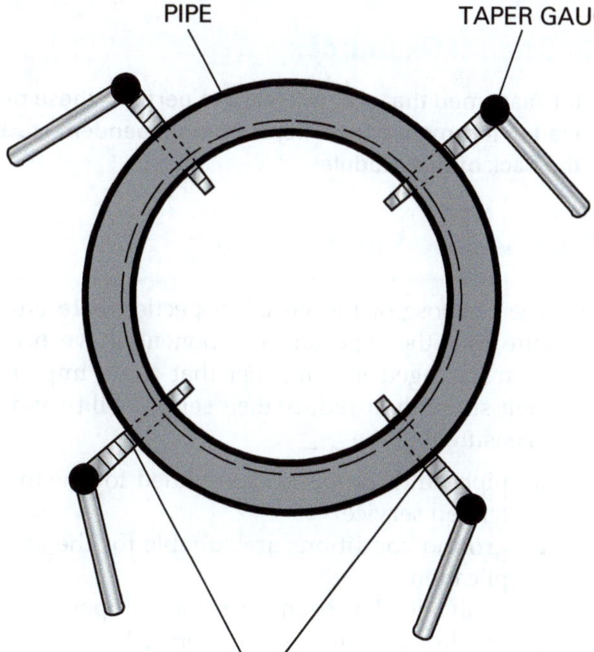

Figure 4 Measuring the root gap with a taper gauge.

Additional Resources

It is assumed that you will learn to perform these procedures under the direct supervision of your crew leader, foreman, or instructor. For independent study, refer to the recommended Additional Resources at the back of this module.

2.0.0 Section Review

1. The purpose of the visual inspection is to ensure that the pipe and components have not been damaged in a manner that could impair their strength or reduce their serviceability and to ensure that the _____.
   a. pipes and components are rated for the intended service
   b. ground conditions are suitable for the application
   c. excavation has been performed properly
   d. pipeline design was done correctly

2. When visually inspecting pipes and pipe components, the final step is to _____.
   a. confirm that the pipe matches the component
   b. request any needed repairs
   c. document the inspection according to operator's procedures
   d. check the root gap with a taper gauge

Summary

Pipeline system safety relies on the proper installation of the pipes and pipe components used in the design of the system. A visual inspection of the pipes and components must be performed before installation. In addition, other inspections may be needed during the different phases of the installation. Visual inspections help to ensure that the specified pipe and pipe components have no visible damage prior to installation and meet the rating requirement for the intended service. A satisfactory outcome for each visual inspection will help to ensure that the installation is completed properly.

Review Questions

1. An initial visual inspection involves physically looking for obvious mechanical defects and other issues concerning the _____.
   a. integrity of the pipes and pipe components
   b. setup requirements and work plan
   c. weather conditions and environmental factors
   d. work load and scheduling

2. The markings on a valve nameplate must include, at a minimum, _____.
   a. the date of manufacture
   b. the yield strength or grade
   c. the nominal valve size
   d. the name and address of the manufacturer

3. The type of visual inspection for which a Hi-Lo gauge is typically used is the _____.
   a. root penetration
   b. fit-up
   c. hold point
   d. prefit-up

4. This covered task includes the assessment and mitigation of damage found during a visual inspection.
   a. True
   b. False
   c. It depends on the damage.
   d. Only if required by company policy and procedures.

5. If personnel encounter mechanical damage during an inspection, the appropriate response is to _____.
   a. document the damage as required and make appropriate notifications
   b. determine the cause and attempt to repair it
   c. evacuate the area and call 911
   d. alert the control center immediately

6. When visually inspecting pipes and pipe components fit-ups, use a taper gauge to check the _____.
   a. base gap
   b. root gap
   c. gap depth
   d. penetration gap

Trade Terms Quiz

Fill in the blank with the correct term that you learned from your study of this module.

1. Parts of a pipeline which may be subjected to pump pressure including, but not limited to, pipe, valves, elbows, tees, flanges, and closures are called ________.
2. A groove melted into the base metal adjacent to the toe or root of a weld and left unfilled by weld metal is called a(n) ________.
3. Conditions within a pipeline system, identified by the operator, that may indicate (1) a malfunction of a component or (2) a deviation from normal operations that may indicate a condition exceeding design limits or result in a hazard to persons, property, or the environment are called ________.
4. Surface rust or pitting that may be identified during a visual inspection of pipes and pipe components is called ________.
5. Discontinuities that render a part or product unable to meet minimum applicable acceptance standards or specifications are called ________.
6. The maximum pressure at which a pipeline or segment of a pipeline or a component may be normally operated is called the ________.

Trade Terms

Abnormal operating conditions (AOCs)
Components
Corrosion
Defects
Maximum operating pressure (MOP)
Undercut

Trade Terms Introduced in This Module

Abnormal operating conditions (AOCs): Conditions within a pipeline system, identified by the operator, that may indicate (1) a malfunction of a component or (2) a deviation from normal operations that may indicate a condition exceeding design limits or result in a hazard to persons, property, or the environment.

Components: Parts of a pipeline which may be subjected to pump pressure including, but not limited to, pipe, valves, elbows, tees, flanges, and closures.

Corrosion: Surface rust or pitting that may be identified during a visual inspection of pipes and pipe components.

Defects: Discontinuities that render a part or product unable to meet minimum applicable acceptance standards or specifications.

Maximum operating pressure (MOP): The maximum pressure at which a pipeline or segment of a pipeline or a component may be normally operated.

Undercut: A groove melted into the base metal adjacent to the toe or root of a weld and left unfilled by weld metal.

Additional Resources

The following recommended resources can provide additional helpful information related to the requirements for this covered task:

API Recommended Practice 1161, Third Edition, © 2014 American Petroleum Institute. API Publishing Services, 1220 L Street NW, Washington, DC 20005. Distributed by Thompson Reuters (Scientific) LLC, **www.techstreet.com**

49 *CFR* 192, **www.ecfr.gov**.

49 *CFR* 195, **www.ecfr.gov**.

NCCER Module AOCCC-17, *Abnormal Operating Conditions Control Center*

NCCER Module AOCFG-17, *Abnormal Operating Conditions Field and Gas*

NCCER Module 62105-14, *Introduction to Pipeline Documents*

NCCER Module 63104-02, *Piping and Mechanical Blueprint Reading*

NCCER Module 67102-02, *Basic Pipeline Pneumatics and Equipment*

NCCER Module 67103-02, *Pipeline Communications*

NCCCER Module CT7_7-17, *Perform Coating Inspection*

NCCER Module CT38_3-17, *Visually Inspect that Welds Meet DOT Requirements*

Figure Credits

Courtesy of Olympus, Module Opener

G.A.L. Gage Co., Figure 1B

DEWALT, Figure 3

Section Review Answer Key

Answer	Section Reference	Objective
Section One		
1. c	1.1.1	1a
2. a	1.2.3	1b
Section Two		
1. a	2.2.0	2a
2. c	2.2.1	2b

NCCER CURRICULA — USER UPDATE

NCCER makes every effort to keep its textbooks up-to-date and free of technical errors. We appreciate your help in this process. If you find an error, a typographical mistake, or an inaccuracy in NCCER's curricula, please fill out this form (or a photocopy), or complete the online form at **www.nccer.org/olf**. Be sure to include the exact module ID number, page number, a detailed description, and your recommended correction. Your input will be brought to the attention of the Authoring Team. Thank you for your assistance.

Instructors – If you have an idea for improving this textbook, or have found that additional materials were necessary to teach this module effectively, please let us know so that we may present your suggestions to the Authoring Team.

NCCER Product Development and Revision

13614 Progress Blvd., Alachua, FL 32615

Email: curriculum@nccer.org
Online: www.nccer.org/olf

❑ Trainee Guide ❑ Lesson Plans ❑ Exam ❑ PowerPoints Other ____________

Craft / Level: Copyright Date:

Module ID Number / Title:

Section Number(s):

Description:

Recommended Correction:

Your Name:

Address:

Email: Phone:

CT39_0-17
Backfilling a Trench Following Maintenance

Overview

After maintenance has been completed on an excavated pipeline, the trench must be backfilled. Backfilling a trench is hazardous, and protecting the pipe from damage during backfilling operations is essential. This module describes the tasks involved in properly backfilling a trench after maintenance has been completed on a pipeline.

Trainees with successful module completions may be eligible for credentialing through the NCCER Registry. To learn more, go to **www.nccer.org** or contact us at 1.888.622.3720. Our website has information on the latest product releases and training, as well as online versions of our *Cornerstone* magazine and Pearson's product catalog.

Your feedback is welcome. You may email your comments to **curriculum@nccer.org**, send general comments and inquiries to **info@nccer.org**, or fill in the User Update form at the back of this module.

This information is general in nature and intended for training purposes only. Actual performance of activities described in this manual requires compliance with all applicable operating, service, maintenance, and safety procedures under the direction of qualified personnel. References in this manual to patented or proprietary devices do not constitute a recommendation of their use.

CT39_0-17

Backfilling a Trench Following Maintenance

Objectives

When you have completed this module, you will be able to do the following:

1. Identify and describe the safety considerations and tasks involved in backfilling a trench following maintenance.
    a. Identify the potential hazards associated with excavation sites and the associated Occupational Safety and Health Administration (OSHA) regulations.
    b. Explain the visual inspection and preparation of fill material.
    c. Explain methods used to support pipe during backfilling.
    d. Explain the process of placing bedding.
    e. Explain the process of compacting soil.
2. Describe the requirements and procedures for safely backfilling a trench following maintenance (CT39_0-17).
    a. Identify the prerequisites, competencies, and abnormal operating conditions associated with safely backfilling a trench following maintenance.
    b. Describe the general procedure used to safely backfill a trench following maintenance.

Performance Task

Under the supervision of your instructor, you should be able to do the following:

1. Backfill a trench following maintenance (CT39_0-17).
    - Identify potential abnormal operating conditions (AOCs) that may occur during performance of this CT, and know the appropriate actions to take in response to them.
    - Utilize the appropriate personal protective equipment (PPE) according to relevant company procedures.
    - Identify foreign objects that could damage the pipeline system.
    - Perform a visual inspection of backfill material, using a shovel or other tool to spot check backfill material.
    - Take appropriate action to remove foreign material or objects. In regard to safety, beware of trenching and excavation hazards.
    - Identify unsuitable backfill material around the pipeline. Soft material must be used near the coating. Replace if de-rocking is not feasible, or if backfill contains too much brush, roots, or other foreign material to remove cost-effectively.
    - Place suitable backfill material within six inches of pipe or within the distance specified by company policy, whichever is greater.
    - Demonstrate ability to compact soil for proper pipe support, size of support, and spacing of support. Use tamping tool, backhoe, or any other required tools to compact soil.
    - Complete appropriate documentation as required by operator's procedures.

Trade Terms

Abnormal operating conditions (AOCs)
Backfilling
Competent person
Crown
Fill
Gradation
Padding
Spoil
Subsidence
Well points

Industry Recognized Credentials

If you're training through an NCCER-accredited sponsor, you may be eligible for credentials from NCCER's Registry. The ID number for this module is CT39_0-17. Note that this module may have been used in other NCCER curricula and may apply to other level completions. Contact NCCER's Registry at 1.888.622.3720 or go to **www.nccer.org** for more information.

Contents

Figures and Tables

Section One

1.0.0 Backfilling Tasks and Safety Considerations

Objective

Identify and describe the safety considerations and tasks involved in backfilling a trench following maintenance.

a. Identify the potential hazards associated with excavation sites and the associated Occupational Safety and Health Administration (OSHA) regulations.
b. Explain the visual inspection and preparation of fill material.
c. Explain methods used to support pipe during backfilling.
d. Explain the process of placing bedding.
e. Explain the process of compacting soil.

Trade Terms

Backfilling: The technique for covering a completed pipeline so that adequate fill material is underneath and over the pipe to protect it from loose rocks, abrasion, shifting, and washouts.

Competent person: An individual who is capable of identifying existing and predictable hazards in the surroundings, or working conditions that are unsanitary, hazardous, or dangerous to personnel, and who has authorization to take prompt corrective measures to eliminate them.

Crown: To create a slightly built-up section of backfill material, providing a slope so that water will drain to either side. Crowning is performed to allow for settlement or rainwater erosion.

Fill: Amount of material needed to bring existing ground to a desired grade.

Gradation: The distribution (in percent by weight) of individual particle sizes. Materials such as clay, silt, and sand can all have varying gradations.

Padding: Screened or sifted dirt, clean gravel, or foam placed in a ditch to protect pipe from damage caused by sharp rocks or rough soil.

Spoil: The soil that is removed from the ground during an excavation. The spoil pile should be kept a minimum of two feet from the edge of an excavation.

Subsidence: A depression in the earth that is caused by unbalanced stresses in the soil surrounding an excavation.

Well points: Devices installed in waterlogged soil to dry out areas along the pipe ditch line. Hundreds may be needed in some areas.

After maintenance or repair work is completed on a pipe that has been excavated, the trench must be backfilled with soil. **Backfilling** a trench following pipeline maintenance is a delicate operation with the goal of protecting the pipeline from damage. The **fill** material must be inspected to ensure that it will not harm the pipeline, and proper backfilling procedures must be followed to provide adequate support under the pipe as well as to properly cover the pipe.

When backfilling is completed, the surface should be restored as closely as possible to its original condition, including the replacement of topsoil. A designated observer/spotter, sometimes also called the swamper, is responsible for observing progress and watching for potential dangers in order to prevent damage to the pipeline, including the pipe coating.

Excavators, or trackhoes, and dozers are commonly used to move fill into a trench. Operators of backfilling equipment must be competent at working around below-grade facilities. The backfill is laid around the pipe in such a manner that it supports the pipe evenly, reducing stresses that could later cause pipe damage. In order to properly protect the pipe, the bedding, side fill, and cover should provide a uniform, even blanket around the pipe.

For the safety of personnel as well as the pipeline, designated personnel must make certain that if an excavation site is left unattended at any time before it is completely backfilled, it is properly secured. Temporary fencing is typically used, but the appropriate methods for securing the site vary and equipment operators must follow company-specific policies and procedures.

1.1.0 Trench Dangers

Working in or near a trench is one of the most hazardous pipeline maintenance tasks. Something as simple as slipping, tripping, or falling can be deadly if it happens in or near a trench. In excavations that exceed 4 feet in depth, access/egress must be provided within 25 feet of each worker. Trenches must be inspected daily before work begins, after rainstorms and other weather events, and when there is any sign of change in stability. Any time the excavation is left unattended, it must be reinspected. Appropriate personal protective equipment (PPE) must be used at all times.

There are several preventative measures to ensure the safety of personnel working near trenches. For example, some company procedures require pressure reduction in the pipeline before excavation begins. Personnel may need to evacuate excess water from a trench. When working in waterlogged soils, personnel may install **well points** to dewater the ditch line. Well points are devices that can effectively dry out extensive areas along the ditch line by continuously removing underground water. Dewatering is important because material that contains too much moisture may result in trench failure. Each well point functions like a submersible pump. A hollow steel rod with wire mesh on the bottom end is driven into the ground. A flexible hose is attached to its top end and connected to a portable pump. The pump carries the water away from the work site. Many such rods are installed along the line, and the hoses connect to the pump through a pipe along the ground. Dewatering may, in some cases, require additional permits.

It is critical for each worker to be aware of the hazards involved in excavation work and how to avoid them. This section provides information about OSHA regulations for working at excavation sites, as well as specific information about the following hazards:

- Cave-ins
- Gases
- Electricity
- Falling objects
- Fire or explosion

1.1.1 Cave-Ins

Cave-ins are the most obvious dangers associated with excavations, and are the single most hazardous part of excavation work. If a worker is buried by collapsing trench walls, the weight alone can be fatal. Being buried up to the waist can provide enough pressure to result in death. If a worker's face is buried and the weight does not result in death, the lack of air may result in suffocation. One cubic yard of earth can weigh 3,000 pounds. Observe *Figure 1*, and try to visualize just how small an amount of earth that is compared to its weight.

To prevent cave-ins, the walls of earth that are left after digging must be adequately supported. Excavations of 5 feet or deeper must use shielding (*Figure 2*) and shoring devices or sloping to prevent cave-ins. A shoring or shielding system provides a structure such as a metal hydraulic, mechanical, or timber system that supports the sides of an excavation. This structure may be sloping, vertical, or horizontal. Acceptable trench protective devices and how they are to be used are defined in 29 *CFR* 1926. In general, the trench is no wider than it is deep. Any excavation 20 feet or deeper requires a professional engineer to design an appropriate system.

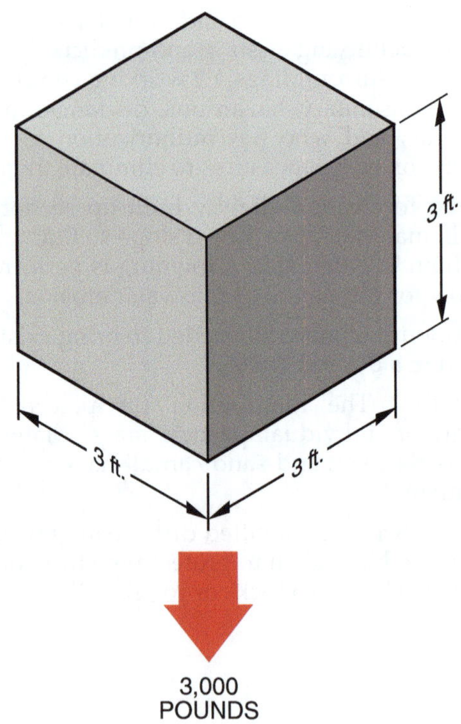

Figure 1 Dimensions of 1 cubic yard of earth.

Figure 2 Examples of shielding (shoring) devices.

To protect themselves from potential cave-ins, personnel must be mindful of the following hazards and associated safety rules:

- An excavation is not safe to enter unless the proper support structure or sloping system is provided.
- It is not safe to assume that frozen earth is stable; frozen ground is not an alternative to shoring.
- Saturated soil increases the possibility of a cave-in. The increased pressure of the water exerted on the soil can be the factor that causes the walls to collapse. Personnel should not be allowed in a trench that contains water because it could cause erosion of the trench.
- Excavations in or near backfilled or previously excavated ground are especially dangerous because the soil will be loose and will not support itself.
- Clay can be extremely dangerous if dried by the sun. Large chunks of material can break off a trench wall after having been stable and solid for a long period of time (*Figure 3*).
- Shoring must be adequate to overcome additional pressures from piles of excavated material, adjoining structures, vehicular traffic, and nearby equipment.
- Shoring uprights and cross braces should be checked to ensure they are properly installed.
- Ensure that an exit is available (within 25 feet) where any work is to be done. Ladders should have rails that extend 3 feet above the landing.
- Never enter an excavation to install shoring or trench boxes. These must be in place before personnel are allowed to enter the trench.
- Never enter an excavation behind the trench box after it has been moved; the excavation should be backfilled as soon as the box is moved.

Personnel must also keep in mind that an open cut or trench can become more unstable as work progresses. Stresses and weaknesses can occur in an open trench or excavation for a number of reasons. For example, any change in moisture content can affect the stability of a trench or excavation. *Figure 4* illustrates the most frequently encountered signs of unstable trenches; refer to it as you read through each indication.

Tension cracks usually form one-quarter to one-half of the way down from the top of a trench, and may result in sliding or slipping. Tension cracks can also cause toppling. Toppling occurs when the trench's vertical face shears along the tension crack line and topples into the excavation. An unsupported excavation can create an unbalanced stress in the soil, which in turn causes **subsidence** at the surface and bulging of the vertical face of the trench. If uncorrected, this

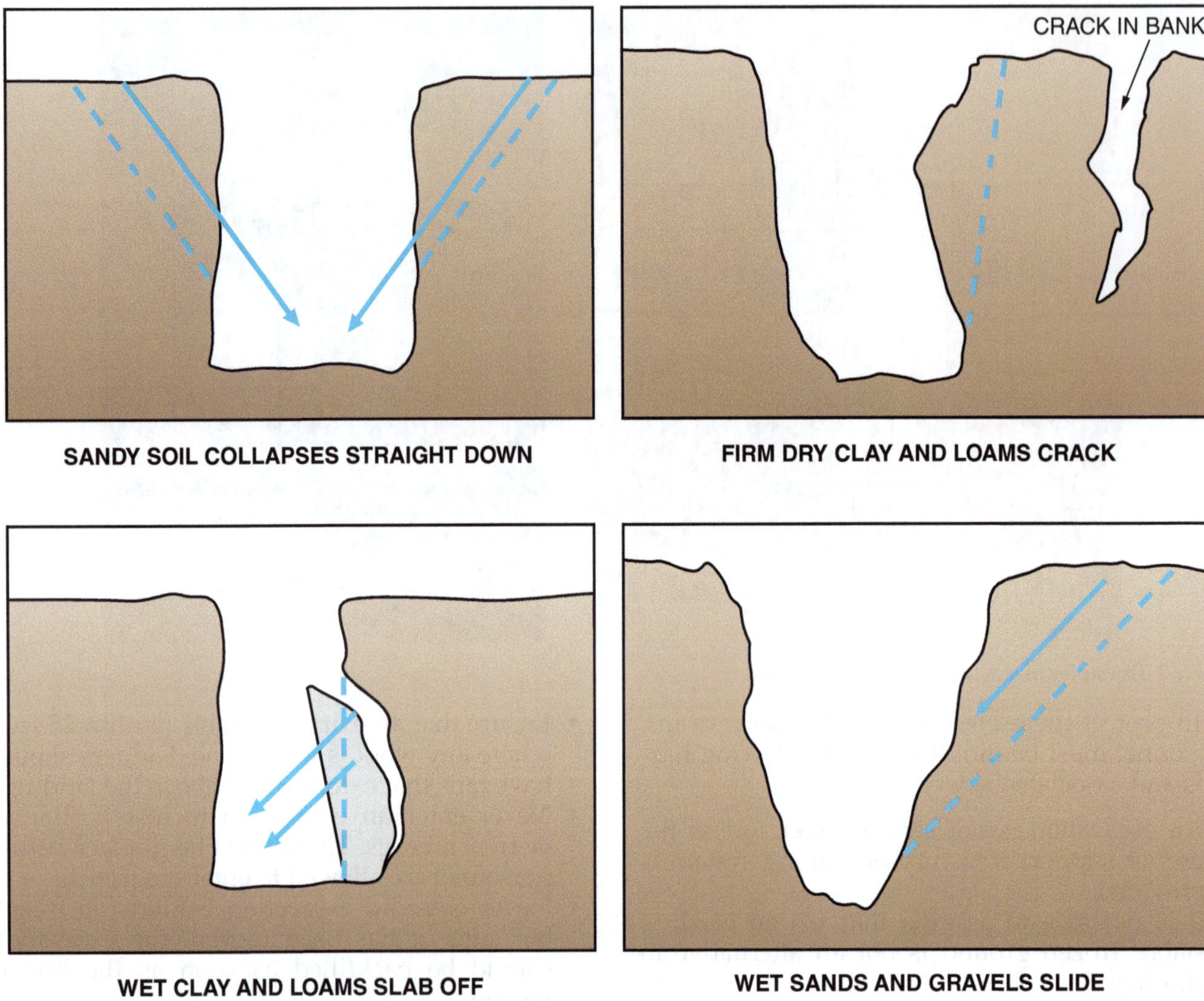

Figure 3 Behavior of clay and other soil types.

condition can cause wall failure and trap workers in the trench, or greatly stress the protective system. Bottom heaving is caused by downward pressure created by the weight of adjoining soil. This pressure causes a bulge in the bottom of the cut. Heaving and squeezing can occur even when shoring and shielding are properly installed.

CAUTION

Protective systems are designed for even loads of earth. Heaving and squeezing can place uneven loads on the shielding and may stress particular parts of the protective system.

Another indication of an unstable trench is boiling. Boiling is when water flows upward into the bottom of the cut. A high water table is one of the causes of boiling. Boiling can happen quickly and can occur even when shoring or trench boxes are used.

WARNING!

Boiling is a sign of an unstable trench, and can happen very quickly. If boiling starts, stop what you are doing and leave the trench immediately.

Signs of an unstable trench must be reported immediately. Work must be stopped until corrective measures are taken. This is necessary even if a support system is being used. Refer to OSHA regulations regarding the safety requirements for trench support systems.

1.1.2 Gases

In some situations, an excavation should be considered a confined space, and appropriate evaluation and controls must be undertaken to ensure against a contaminated atmosphere or toxic substances. Excavations can fill with dangerous gases, or oxygen can be displaced from the bot-

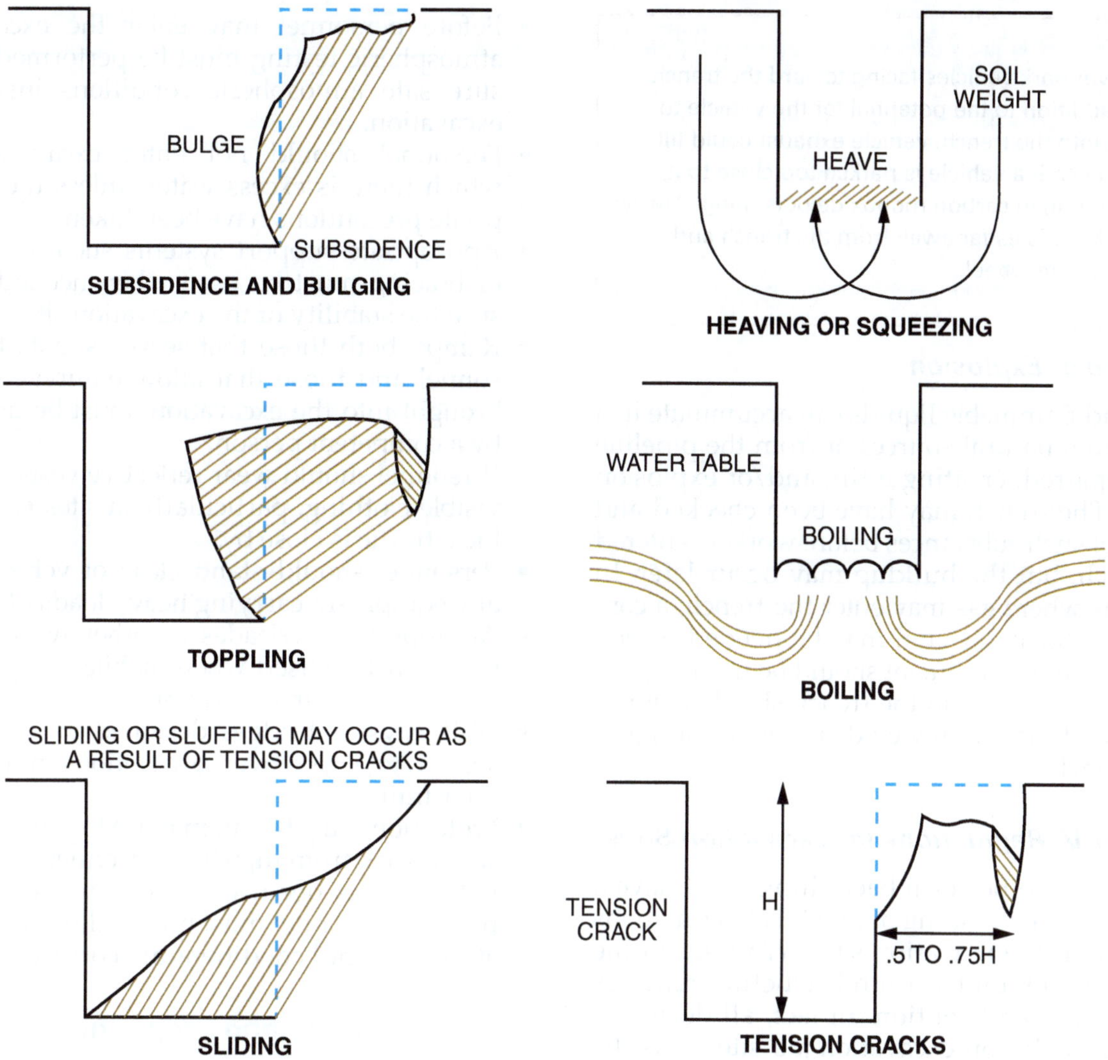

Figure 4 Indications of unstable trenches.

tom of the excavation if located near facilities that store large volumes of chemicals. The liquids or gases, even exhaust fumes from vehicles, could flow or leak into the excavation. When excavating a pipeline for repairs, the liquids or gases could have already leaked into the ground around the pipe. An excavation should never be entered until the atmosphere is determined to be safe by authorized persons and methods.

1.1.3 Electrical Shock

The danger of electrical shock exists during any excavation work. For example, the insulation of an electrical cable could be pierced. In addition, when overhead lines are near the excavation, the excavator could come in contact with them, knocking them into the excavation or electrifying the boom. If the boom is electrified, the pipe it is lowering into the excavation will be electrified as well and could impart a deadly shock.

1.1.4 Falling Objects

Anything around the edges of the excavation could become a dangerous falling object if dropped or knocked over the edge. This includes piles of soil, tools, chunks of dirt from the excavator bucket, parts of the bucket, and other objects. Workers on the surface must keep such objects away from the edge of the excavation, and must keep any excavated dirt at least 2 feet away from the edge. Those in the trench should be wearing PPE at all times.

Do not park vehicles facing toward the trench. Always park vehicles far away from the trench and chock the wheels.

WARNING!

Never park vehicles facing toward the trench. In addition to the potential for the vehicle to fall into the trench, vehicle exhaust could fill a trench if a vehicle is parked too close to it, resulting in carbon monoxide poisoning. Always park vehicles far away from the trench and chock the wheels.

1.1.5 Fire or Explosion

Gases and flammable liquids can accumulate in a trench from natural sources or from the pipeline being repaired, creating a fire and/or explosion hazard. The trench may have been checked and cleared of such substances before workers entered the trench, but the buildup may occur later. In situations where gas may enter the trench, a constant monitoring system may be needed. When gas is detected, equipment should be immediately shut down. Workers in the trench should exit immediately if there is any evidence of gas or liquid in the trench.

1.1.6 OSHA Regulations for Excavation Sites

Personnel who perform backfilling must have a thorough understanding of the OSHA regulations that pertain to excavation sites. Working in the vicinity of any underground structure requires extreme care and caution. In fact, all decisions affecting safety on excavation job sites must be made by a person specifically defined by OSHA as a **competent person**. According to OSHA regulations, a competent person is an individual who is capable of identifying existing and predictable hazards in the surroundings, or working conditions that are unsanitary, hazardous, or dangerous to personnel, and who has authorization to take prompt corrective measures to eliminate them. An excavation site must be inspected by a competent person before any personnel may enter the site. The competent person must also inspect the excavation site daily or when hazards or site conditions change.

OSHA has strict and specific requirements that apply to any open excavation work, including trenches. The purpose of these requirements is to ensure the safety of excavation workers. These requirements are explained in detail in 29 *CFR* 1926. The following is a summary of the general requirements:

- Any obstacles near the excavation site that pose a potential hazard to personnel must be removed.
- Before personnel may enter the excavation, atmospheric testing must be performed to ensure safe atmospheric conditions inside the excavation.
- Personnel should not enter excavations in which there is excess water unless the appropriate precautions have been taken.
- Appropriate support systems such as shoring or bracing must be used when needed to ensure the stability of the excavation site.
- Ramps, both those that serve as exits for personnel and those that allow equipment to be brought into the excavation, must be designed by a competent person.
- Personnel should wear reflective vests or other visible clothing, particularly at sites where vehicle traffic is close by.
- Personnel should stand clear of vehicles and any equipment carrying heavy loads of soil.
- Appropriate barricades or other warning devices must be used when mobile equipment is operated near an excavation.
- Walkways or bridges designed to allow personnel to cross over excavations must have guardrails.
- Protection must be put in place to prevent loose soil or rock from injuring personnel.
- If the site is deemed unsafe by a competent person, personnel must not be allowed to work at the site until conditions are corrected.

1.2.0 Inspecting and Preparing Fill

The soil that is removed from the ground during excavation is called **spoil**. The spoil pile is to be kept a minimum of two feet from the edge of an excavation. The spoil pile, also called the *spoil bank*, is the main source for backfill material, but loss through excavation may require more material.

The backfill material placed around the pipe must not damage the pipe or coating. Any rocks and foreign objects left in the ditch or in the backfill material can potentially damage the pipe and its coating. A visual inspection must be performed in order to ensure that the fill is free of any items that could affect compaction, cathodic protection systems, coating, or cause corrosion. These include the following:

- Cans, tools, clamps, or scrap metal
- Roots or brush
- Rocks or large chunks of hard-packed dirt
- Soil that contains hydrocarbons
- Acids or other caustic substances

Fill dirt, particularly that which is used for bedding and is close to the pipe, may be screened and sifted to remove rocks, large chunks, and other debris. The spoil from a new ditch made through rock cannot be used as fill because of the amount of rock debris. New fill dirt must be brought in to pad and fill most of the ditch. Sometimes rocky soil may be used in the upper part of the ditch, as long as the rocks are small, well away from the pipe, and only used up to about one foot from the surface. The final layer must be clean fill dirt or topsoil. If excavated soil has been contaminated with product from a leak before or during repair, it should not be used as fill.

Spoil from ditches dug in frozen ground can be as dangerous as rocky soil. The frozen bits of earth are irregular and hard as rock. They may be used if they can be broken up sufficiently to not pose a hazard to the pipe. However, frozen soil is difficult to distribute evenly and, when it thaws, may result in areas of unsupported pipe or uneven fill. In some areas, soil that was unfrozen when excavated can freeze at the surface, particularly overnight. Using this fill would cause the same problems.

An understanding of soil types is required to properly backfill a trench. OSHA classifies soils into four categories. The classifications are based on stability and range from stable rock at the high end to Type C as least stable. *Table 1* describes the characteristics of each soil type.

1.3.0 Supporting the Pipe

It is important to protect the pipeline during backfilling. To avoid damaging the pipe, personnel must handle the pipe carefully and support it properly. The pipe must be properly secured and supported to reduce the stress on the pipe and the welds. The maximum span between the locations of the supports should be determined by an engineer, based on the pipe specifications and other factors.

Skids provide support across the ditch, and skids, blocks, sandbags, airbags, and other materials are used on the bottom of the ditch to support inclined pipe. Earthen plugs can also be used as supports. These plugs are left after other supports have been removed. Where possible, earthen plugs should be used to support pipeline stopples and taps.

Other pipeline support methods include cribbing, air bags, earth pillars, and slings (*Figure 5*). When supports are being installed or removed, it is important to add or reduce supports in a systematic way to equalize the stress over the entire line section.

1.4.0 Placing Bedding

The material placed in the bottom of the trench, on which the pipe is laid, is called bedding (noun). The act of placing this material in the trench is also referred to as bedding (verb). The purpose of

Table 1. Soil types and their characteristics.

Soil type	Description	Examples
Stable rock	Natural solid material that remains intact when exposed.	Solid rock, shale, and cemented sands or gravel.
Type A	Cohesive soil, with a compressive strength of at least 1.5 tons per square foot. No soil can be Type A if the soil is fissured, subject to vibration, has been previously disturbed, or has any factors that decrease stability.	Clay, silty clay, sandy clay, and clay loam.
Type B	Cohesive soil, with a compressive strength of 0.5 to 1.5 tons per square foot.	Granular soils, such as angular gravel, crushed rock, silt, sandy loam, and unstable rock, and any unstable Type A soils.
Type C	Unstable, cohesive soil with a compressive strength less than 0.5 tons per square foot.	Gravel, loamy soil, sand, submerged soil, soil with water seepage, unstable submerged rock, and previously excavated soil.

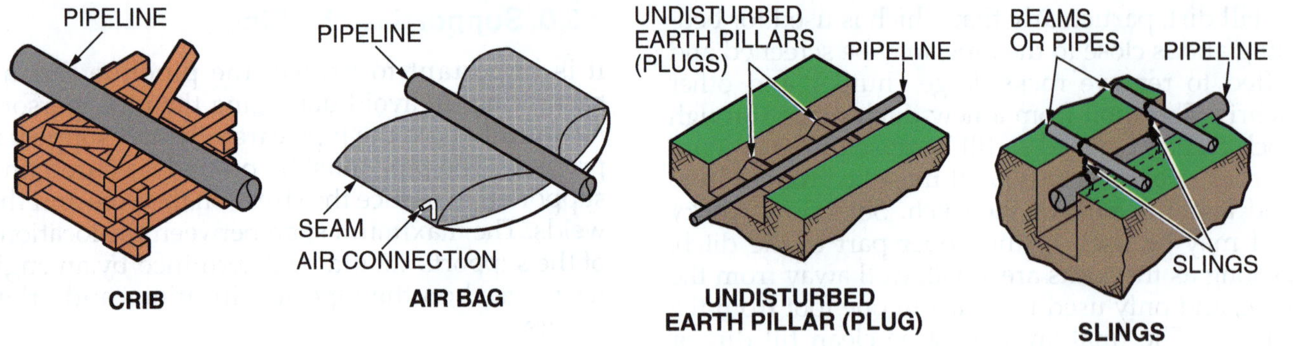

Figure 5 Pipeline support methods.

bedding is to protect the pipe. Before bedding is placed, skids must be removed. This is only done as other support is provided under the pipe. In rocky, rough soils, padding of earth, sand bags, or other approved supports may be laid under the pipe. Synthetic materials are also often used to provide a padded surface on the bottom of the ditch.

A sound pipe installation begins with a stable foundation. This foundation may include a layer of stabilization material to support the bedding, a layer of bedding material to cushion and support the pipe, and initial backfill to blanket the pipe. These three elements can be critical to the load-bearing ability of the pipeline. Pipeline specifications should always be followed when preparing the ditch for replacement of the pipe. A trench bedding arrangement is shown in *Figure 6*.

1.4.1 Stabilization

Some soil conditions may require a stabilization layer to support the pipe bedding. Unless the pipe is moved to a different trench, the need for such a layer should be evident in the existing trench. If so, the stabilization layer under the exposed pipe may need to be repaired after excavation.

If the pipeline is laid in a new trench, the bottom should be inspected to determine the need for stabilization. A worker can walk in the trench, shifting weight from one foot to the other to determine if the soil underneath moves, shifts, or gives. If it does not feel solid, stabilization may be needed.

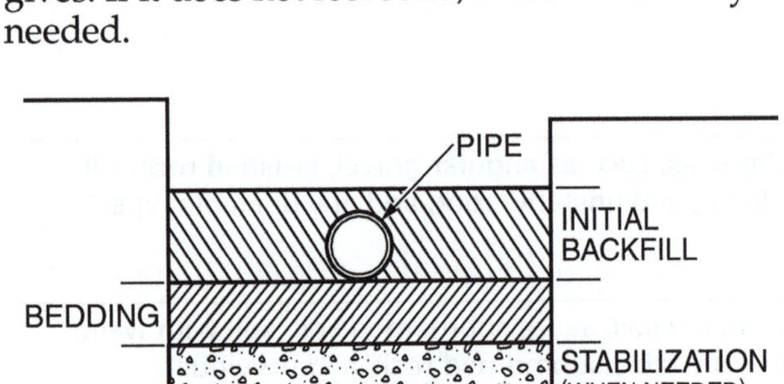

Figure 6 Trench bedding arrangement.

Crushed stone is often used, in sizes ranging from less than an inch to large riprap (loose pieces of rock). The material is poured into the trench and spread to a uniform depth.

1.4.2 Bedding

Bedding should be sufficient to provide uniform, firm support for the pipe and to maintain pipe grade. The type of bedding support is dictated by engineer-determined specifications. The load-bearing capacity of the pipeline is determined from the strength of the pipe itself and the type of bedding.

Bedding extends from the top of the stabilization layer, if present, to the bottom of the pipe. Some materials used include native soil, sand, stone, or concrete. Specifications commonly recommend 4 to 6 inches of bedding material. The material is poured into the trench and leveled to grade along the full length of the trench. Lasers are often used to check grade.

The most commonly-used bedding method is referred to as Class B. Using this method, the pipe is placed on a granular material bed that is carefully shaped to fit the bottom of the pipe. The width of the bed is equal to at least 60 percent of the pipe diameter. Backfill is placed by hand in 6-inch layers, and compacted for a depth of at least 12 inches over the top of the pipe. The bedding is then tamped thoroughly to ensure good compaction around and under the pipe.

1.4.3 Initial Backfill

After laying the pipe on the bedding, the initial backfill is added. This is material placed from the bedding or the bottom of the pipe to some specified level above the pipe. The purpose of this initial backfill is to protect the pipe. Use only clean fill material free of rocks, clumps, foreign objects, or organic materials. Topsoil, clay, or sand are examples of acceptable initial backfill material.

Initial backfill may include any of the following activities:

- Compacting, tamping, or filling to the center of the pipe
- Tamping or filling to the top of the pipe
- Tamping or filling to some point above the pipe

1.5.0 Performing Soil Compaction

Fill that is not properly compacted will settle naturally over time, giving the effect of shrinking. Loose, uncompacted earth is subject to erosion by rainwater. Clay-bearing soils that are improperly compacted may absorb moisture and swell. Compaction is the deliberate application of pressure or vibration to eliminate air pockets and increase the density of soil. By consolidating the soil particles in the fill, compaction also strengthens the pipeline cover and increases the protection of the blanket of fill around the pipe.

Compaction can be accomplished by driving a heavy crawler-type tractor over the backfilled ditch. Provide extra compaction at road crossings, ditches, streams, and other locations where heavy loads are expected to cross the pipeline.

Personnel will normally **crown** backfill after compaction to allow for some settlement or rainwater erosion. Crowning involves creating a slightly built up section of backfill material to provide a slope so that water will drain to either side of the area. Ditch breaking may be required on the slope of a ditch, diversion level, or surface to prevent backfill from washing away.

Cave-In Caused by Water

Four employees were laying a lateral sewer line at a building site. The foreman and a laborer were laying an 8"-diameter, 20'-long plastic sewer pipe in the bottom of a trench that was 36" wide, 9' deep, and approximately 50' long. The trench was neither sloped nor shored, and there was water entering it along a rock seam near the bottom. The west side of the trench caved in near the bottom, burying one employee to his chest and completely covering the other. The rescue operation took two hours for the first man and four hours for the second. Both men died.

The Bottom Line: Working in an excavation can be fatal. Always ensure your trench is safe.

Source: The Occupational Safety and Health Administration (OSHA)

1.5.1 Moisture Content

Moisture content is the single most important variable in soil compaction. Moisture acts as a lubricant, allowing soil particles to easily slide across one another and fill in air gaps. In soil that contains too little moisture, the particles resist sliding. In soil that contains too much moisture, the particles float and permit the gaps to be filled with water. Water cannot be compressed and when it drains, it will leave air gaps. Since both conditions are unacceptable, a great deal of effort goes into determining the optimum moisture content.

Soil is commonly sprayed with water as it is being placed in an excavation so it may be compacted to the specified density, but native soil often contains too much moisture to be compacted properly. In this case, heavy equipment may be required to turn the soil to allow it to dry out. Then, when the moisture content is acceptable, the fill can be placed and compacted to specification.

NOTE

One method of determining if the soil has sufficient moisture to be properly compacted is called the *hand test.* Pick up a handful of soil and squeeze it in your hand. The soil should easily form a ball. Then drop it. It should break into a couple of pieces. If the soil is powdery and does not retain its shape, it is too dry. If the soil shatters when dropped, it is too dry. If the soil leaves traces of moisture on your fingers when squeezed, it is too wet for compaction. If it stays in one piece when dropped, it is also too wet.

1.5.2 Compacting Soil by Layers

Fill is typically placed in layers known as *lifts*, with the specifications determining the maximum thickness of each lift. As each lift is placed, it is compacted to the desired density, and another lift is added until the specified depth is reached. Common lift thicknesses are 4", 6", 8", and 10".

Compacting equipment is designed for particular soil types and lift thicknesses, so it is important to check the manufacturer's specifications when selecting equipment. For example, a very heavy rubber-tired roller works well for coarse rock fills, but not on cohesive clay soil, which needs a sheepsfoot roller.

1.5.3 Checking Quality

Each compacted layer must meet design specifications to ensure that the final product will stand up under projected loads. It must also meet smoothness requirements. (All quality control is determined according to specs provided by engineers.) Each stabilized layer must be checked for the following, using various types of meters and screens:

- Compaction or density
- Thickness
- Mix uniformity
- Gradation
- Loadbearing capacity
- Moisture content
- Binder content

The amount of compaction needed to bring the soil to the engineers' requirements is just a measure of the density of the soil. There are tests that will determine if the compaction performed meets the specifications. One method of checking compaction is the *proctor test,* which measures how many blows are required to compact the soil. Densometers can also be used to check compaction.

Moisture content can be tested using a moisture meter, by simply attaching a probe and inserting it to the desired depth with every lift. Shakers are used to separate rock and other debris from spoil prior to backfilling, typically starting with a 2" screen. A number of different screens can be applied to achieve the desired specs.

Inspecting Pipe for Damage

Whenever pipe is exposed and before it is backfilled, it must be inspected for damage. Dents, gouges, and coating damage are types of damage often encountered. Gouges are areas of surface damage caused by the displacement or removal of metal. Dents and gouges are usually caused by contact during excavation, by rocks too close to the pipe, or by objects falling into the excavation. When a pipe is dented or gouged, the pipe wall thickness is reduced at the point of impact. As a result, stress on the pipe is concentrated at the dent or gouge location, which weakens the pipe. The weakened condition of the pipe wall can be a source of future leaks.

The coating on the pipe is essential to protect the pipe surface against corrosion and the effects of the burial environment. The coating may easily be cut, scraped, or otherwise damaged by the excavation and moving equipment, or by rocks in the fill.

Additional Resources

It is assumed that you will learn to perform these procedures under the direct supervision of your crew leader, foreman, or instructor. For independent study, refer to the recommended Additional Resources at the back of this module.

1.0.0 Section Review

1. Trench boxes should be installed _____.
   a. by two workers inside the trench
   b. just before backfilling begins
   c. before personnel enter the trench
   d. after excavation work is complete

2. OSHA has four classifications for soil, which are based on _____.
   a. stability
   b. cohesion
   c. vibration
   d. granularity

3. When excavating a pipeline, the various support methods used include slings, cribbing, air bags, and _____.
   a. soil columns
   b. earth pillars
   c. step backs
   d. stress supports

4. There are three elements critical to the load-bearing ability of the pipeline: stabilization material, initial backfill, and a layer of _____.
   a. bedding material
   b. fill dirt
   c. rocks
   d. frozen soil

5. In order to avoid problems due to settlement or rainwater erosion, replaced soil is _____.
   a. crowned
   b. sloped
   c. covered
   d. compacted

SECTION TWO

2.0.0 BACKFILL A TRENCH FOLLOWING MAINTENANCE

Objective

Describe the requirements and procedures for safely backfilling a trench following maintenance (CT39_0-17).

a. Identify the prerequisites, competencies, and abnormal operating conditions associated with safely backfilling a trench following maintenance.
b. Describe the general procedure used to safely backfill a trench following maintenance.

Performance Task

Backfill a trench following maintenance (CT39_0-17).

- Identify potential abnormal operating conditions (AOCs) that may occur during performance of this CT, and know the appropriate actions to take in response to them.
- Utilize the appropriate personal protective equipment (PPE) according to relevant company procedures.
- Identify foreign objects that could damage the pipeline system.
- Perform a visual inspection of backfill material, using a shovel or other tool to spot check backfill material.
- Take appropriate action to remove foreign material or objects. In regard to safety, beware of trenching and excavation hazards.
- Identify unsuitable backfill material around the pipeline. Soft material must be used near the coating. Replace if de-rocking is not feasible, or if backfill contains too much brush, roots, or other foreign material to remove cost-effectively.
- Place suitable backfill material within six inches of pipe or within the distance specified by company policy, whichever is greater.
- Demonstrate ability to compact soil for proper pipe support, size of support, and spacing of support. Use tamping tool, backhoe, or any other required tools to compact soil.
- Complete appropriate documentation as required by operator's procedures

Trade Terms

Abnormal operating conditions (AOCs): Conditions within a pipeline system, identified by the operator, that may indicate (1) a malfunction of a component or (2) a deviation from normal operations that may indicate a condition exceeding design limits or result in a hazard to persons, property, or the environment.

This section provides general procedures for backfilling a trench following maintenance. It also provides the prerequisites and competencies required for the procedures, as well as the abnormal operating conditions (AOCs) that may be encountered.

2.1.0 Prerequisites, Competencies, and Abnormal Operating Conditions

Personnel must use a variety of skills to backfill a trench following maintenance. Personnel must be able to perform backfilling or replacing of soil back into the ditch and over the pipeline after maintenance or repairs. Backfill material must be free of any foreign material that could damage the pipe coating or cause uneven loading of the pipe. Pipe and pipe coating must be inspected to be certain it was not damaged during excavation or maintenance activities, and the inspection continues during the backfilling operation. Pipe must be properly supported after any temporary supports are removed. Finally, the pipe must be provided with adequate cover.

The safe and proper use of tools, required machinery, and personal protective equipment (PPE) is required. Personnel must also be aware of the particular abnormal operating conditions (AOCs) that may occur when backfilling a trench and know the proper responses to them.

2.1.1 Prerequisites and Competencies

Personnel responsible for backfilling a trench following maintenance must have an understanding of what the task entails. Familiarity with types of soils and soil compaction methods is necessary, along with a thorough knowledge of excavation hazards and the safety guidelines that apply to excavations, pipeline facilities, and backfilling operations. Personnel should have knowledge of different types of pipeline stress and pipe coatings. They should have experience with common hand tools and heavy equipment operations, and an understanding of the specifications for general backfill, bedding, and padding materials.

2.1.2 Abnormal Operating Conditions

Personnel backfilling a trench following maintenance must be able to identify abnormal operating conditions (AOCs) and know the proper response to each, should they occur. The following are some potential AOCs, and the appropriate action personnel should take in response to them:

- Unexpected physical damage to the pipeline coating, caused by objects or debris in the backfill material, occurs – Personnel should remove the items from the backfill material. Make notifications per company policy to ensure the coating is repaired as needed. Complete the required reporting.
- Unexpected physical damage to the pipeline, such as scraping, dents, or gouges, caused by equipment accidentally striking the pipe, occurs – Personnel should inspect the damage. Make notifications per company policy to ensure the coating is repaired as needed. Complete the required reporting.
- Stress to the pipeline, such as pipe sag, caused by pipe movement, occurs – Personnel should make notifications per company policy. Level and support the pipe using appropriate methods. Complete the required reporting.
- Fire or explosion due to pipe failure, stress, movement of pipe, or equipment damage occurs – Personnel should make notifications per company policy. Take appropriate action to protect people and the environment. Complete the required reporting.

2.2.0 Procedure

The following steps outline the general procedure for properly backfilling a trench:

Step 1 Identify potential abnormal operating conditions (AOCs) that may occur, and know the appropriate actions to take in response to them.

Step 2 Utilize the appropriate personal protective equipment (PPE) according to relevant company procedures.

Step 3 Identify soil types and their suitability to support the pipe. Some soil types compact better than others, making them more suitable to support the pipeline.

Step 4 Perform a visual inspection of backfill material to identify foreign objects that might damage the pipeline system. For example, use a shovel to spot check for large objects or foreign material that might be mixed into the backfill material. If any foreign material or objects are found, make sure that they are removed.

Step 5 Determine when backfill material is too coarse to be used directly against the pipeline. Remember that soft material must be used near the coating. The backfill may need to be replaced if there are too many rocks or roots, or too much brush or other foreign material. To avoid coating damage, use suitable backfill material within six inches of the pipe or according to company policy.

NOTE

Shielding (rock shields) are sometimes required to protect the pipeline from foreign objects.

Step 6 Fill equally along both sides of the pipe until the pipe is covered. This will help prevent pipe movement and provide additional support for the pipe. Backfill in small layers or lifts to ensure proper support of the pipeline.

Step 7 Compact the soil under and around the pipe for proper pipe support. Tamping is necessary to compact the soil. Use tamping tools, backhoes, and other tools and equipment as needed.

Step 8 Cover the pipe to the cover depth required by specifications.

> **NOTE**
>
> The proper cover for buried pipelines is defined in 49 *CFR* 195.248(a). All pipe must be buried so that it is below the level of cultivation. The required depth of cover varies by location and type of soil and is specifically outlined in the code. The pipe must be installed so that the cover between the top of the pipe and the ground level, roadbed, river bottom, or underwater natural bottom complies with this code.

> **NOTE**
>
> Provide extra cover at road crossings, ditches, and streams. These areas face an increased risk from third-party damage, and additional cover provides some extra protection. In special circumstances like these, reference 49 *CFR* 195 and follow company-specific procedures.

Step 9 Compact the cover material.

Step 10 Crown the backfill according to company specifications to allow for settlement or rainwater erosion.

Step 11 Complete appropriate documentation as required by operator's procedures.

Additional Resources

It is assumed that you will learn to perform these procedures under the direct supervision of your crew leader, foreman, or instructor. For independent study, refer to the recommended Additional Resources at the back of this module.

2.0.0 Section Review

1. During backfilling, if workers discover stress to the pipeline caused by pipe movement, it is appropriate to proceed with leveling and supporting the pipe before continuing to backfill. True or False?
   a. True
   b. False

SUMMARY

After maintenance has been completed on an excavated pipeline, the trench must be backfilled. Protecting the pipe from damage during backfilling operations is essential. The fill material must be inspected to ensure that it is free of foreign objects that could harm the pipe. Proper backfilling procedures must be followed to provide adequate support under the pipe as well as to properly cover the pipe. A designated observer/spotter observes continuously in order to prevent damage to the pipeline, including the pipe coating.

Specific procedures for preparing the trench with stabilization material, bedding, and initial backfill must be followed. The soil is compacted and crowned to prevent settlement. For the safety of workers, it is important to ensure that trenches are safe places in which to do the necessary work. It is also important to know how to deal with hazardous conditions when encountered. Using heavy equipment to dig and backfill can be dangerous to workers and to the pipeline itself. Specific procedures must be followed to avoid accidents and damage from excavators and backhoes.

Dents and gouges, damage to pipe coating, or pipe stress caused by pipe movement are examples of abnormal operating conditions (AOCs) that may be discovered before exposed pipe is backfilled. Personnel involved in backfilling must be able to recognize and react to the AOCs that they may encounter.

Review Questions

1. After pipeline maintenance, the purpose of backfilling a trench is to _____.
   a. compact the soil
   b. break the ditch
   c. tamp the pipeline
   d. protect the pipeline

2. The worker who is responsible for observing progress and watching for potential dangers in order to prevent damage to the pipeline during backfilling is the _____.
   a. designated spotter
   b. competent person
   c. safety inspector
   d. equipment operator

3. After compacting cover material, the next step is to _____.
   a. remove any foreign objects
   b. fill equally until the pipe is covered
   c. crown the backfill according to company procedures
   d. perform a visual inspection of backfill material

4. Any time an excavation is left unattended, _____.
   a. the pipe location must be verified again
   b. permits must be renewed
   c. the site must be reinspected
   d. One-Call must be notified

5. If left in a ditch, foreign materials such as scrap metal, tools, or clamps can damage the pipe or diminish _____.
   a. cathodic protection
   b. the stabilization layer
   c. protective coating
   d. vacuum force

6. As defined by OSHA, the least stable soil type is _____.
   a. stable rock
   b. Type A
   c. Type B
   d. Type C

7. The type of supports for a pipeline that can be left after other supports have been removed are the _____.
   a. skids
   b. earthen plugs
   c. blocks
   d. sandbags

8. To determine if the soil underneath moves, shifts, or gives, personnel can _____.
   a. perform a vibratory analysis
   b. walk in the trench
   c. use a seismograph
   d. probe the soil

9. Compaction is the deliberate application of _____.
   a. pressure
   b. water
   c. evaporation methods
   d. backfill

10. If unexpected physical damage to the pipeline is discovered during backfilling, the proper initial reaction is to _____.
    a. proceed with backfilling
    b. inspect the damage
    c. isolate and/or shut down the pipeline segment or facility
    d. call 911

Trade Terms Quiz

Fill in the blank with the correct term that you learned from your study of this module.

1. The technique for covering a completed pipeline so that adequate fill material is underneath and over the pipe to protect it from loose rocks, abrasion, shifting, and washouts is called ________.
2. The amount of material needed to bring existing ground to desired grade is the ________.
3. The soil that is removed from the ground during an excavation and which must be kept in a pile a minimum of two feet from the edge of an excavation is called the ________.
4. Screened or sifted dirt, clean gravel, or foam placed in a ditch to protect pipe from damage caused by sharp rocks or rough soil is called ________.
5. Devices installed in waterlogged soil to dry out areas along the pipe ditch line are ________.
6. Conditions within a pipeline system, identified by the operator, that may indicate (1) a malfunction of a component or (2) a deviation from normal operations that may indicate a condition exceeding design limits or result in a hazard to persons, property, or the environment are called ________.
7. To create a slightly built up section of soil, providing a slope so that water will drain to either side, means to ________.
8. A depression in the earth that is caused by unbalanced stresses in the soil surrounding an excavation is called ________.
9. An individual who is capable of identifying existing and predictable hazards in the surroundings is called a(n) ________.
10. The distribution (in percent by weight) of individual particle sizes is called ________.

Trade Terms

Abnormal operating conditions (AOCs)
Backfilling
Competent person
Crown
Fill
Gradation
Padding
Spoil
Subsidence
Well points

Trade Terms Introduced in This Module

Abnormal operating conditions (AOCs): Conditions within a pipeline system, identified by the operator, that may indicate (1) a malfunction of a component or (2) a deviation from normal operations that may indicate a condition exceeding design limits or result in a hazard to persons, property, or the environment.

Backfilling: The technique for covering a completed pipeline so that adequate fill material is underneath and over the pipe to protect it from loose rocks, abrasion, shifting, and washouts.

Competent person: An individual who is capable of identifying existing and predictable hazards in the surroundings, or working conditions that are unsanitary, hazardous, or dangerous to personnel, and who has authorization to take prompt corrective measures to eliminate them.

Crown: To create a slightly built up section of backfill material, providing a slope so that water will drain to either side. Crowning is performed to allow for settlement or rainwater erosion.

Fill: Amount of material needed to bring existing ground to a desired grade.

Gradation: The distribution (in percent by weight) of individual particle sizes. Materials such as clay, silt, and sand can all have varying gradations.

Padding: Screened or sifted dirt, clean gravel, or foam placed in a ditch to protect pipe from damage caused by sharp rocks or rough soil.

Spoil: The soil that is removed from the ground during an excavation. The spoil pile should be kept a minimum of two feet from the edge of an excavation.

Subsidence: A depression in the earth that is caused by unbalanced stresses in the soil surrounding an excavation.

Well points: Devices installed in waterlogged soil to dry out areas along the pipe ditch line. Hundreds may be needed in some areas.

Additional Resources

The following recommended resources can provide additional helpful information related to the requirements for this covered task:

API Recommended Practice 1161, Third Edition, © 2014 American Petroleum Institute. API Publishing Services, 1220 L Street, NW, Washington, DC 20005. Distributed by Thompson Reuters (Scientific) LLC, **www.techstreet.com**.

49 *CFR* 192, **www.ecfr.gov**.

49 *CFR* 195, **www.ecfr.gov**.

26*CFR* 1926, **www.osha.gov**

NCCER Module AOCCC-17, *Abnormal Operating Conditions Control Center*

NCCER Module AOCFG-17, *Abnormal Operating Conditions Field and Gas*

The following websites provide resources for products and training:

American Public Works Association (APWA), **www.apwa.net**

Occupational Safety and Health Administration (OSHA), 200 Constitution Avenue, NW, Washington, DC 20210. **www.osha.gov**

Figure Credits

Copyright © 2013 TrenchSafety and Supply, Inc. Used with permission., Module Opener, Figure 2

Section Review Answer Key

Answer	Section Reference	Objective
Section One		
1. c	1.1.1	1a
2. a	1.2.0	1b
3. b	1.3.0	1c
4. a	1.4.0	1d
5. a	1.5.0	1e
Section Two		
1. a	2.1.2	2a

NCCER CURRICULA — USER UPDATE

NCCER makes every effort to keep its textbooks up-to-date and free of technical errors. We appreciate your help in this process. If you find an error, a typographical mistake, or an inaccuracy in NCCER's curricula, please fill out this form (or a photocopy), or complete the online form at **www.nccer.org/olf**. Be sure to include the exact module ID number, page number, a detailed description, and your recommended correction. Your input will be brought to the attention of the Authoring Team. Thank you for your assistance.

Instructors – If you have an idea for improving this textbook, or have found that additional materials were necessary to teach this module effectively, please let us know so that we may present your suggestions to the Authoring Team.

NCCER Product Development and Revision

13614 Progress Blvd., Alachua, FL 32615

Email: curriculum@nccer.org

Online: www.nccer.org/olf

❑ Trainee Guide ❑ Lesson Plans ❑ Exam ❑ PowerPoints Other ____________________

Craft / Level: Copyright Date:

Module ID Number / Title:

Section Number(s):

Description:

Recommended Correction:

Your Name:

Address:

Email: Phone:

CT52_1-17

Conduct Vegetation Survey

Overview

Pipeline personnel must sometimes perform leak surveys to confirm the safety of the pipeline. These must always be performed after any blasting has taken place in the proximity of the pipeline. In addition, gas pipelines must be surveyed at specific intervals as required by federal regulations. This module explains the procedures for conducting a leakage survey on gas pipelines. The condition of vegetation in the immediate area of a pipeline is observed as a means of detecting a natural gas leak.

Trainees with successful module completions may be eligible for credentialing through the NCCER Registry. To learn more, go to **www.nccer.org** or contact us at 1.888.622.3720. Our website has information on the latest product releases and training, as well as online versions of our *Cornerstone* magazine and Pearson's product catalog.

Your feedback is welcome. You may email your comments to **curriculum@nccer.org**, send general comments and inquiries to **info@nccer.org**, or fill in the User Update form at the back of this module.

This information is general in nature and intended for training purposes only. Actual performance of activities described in this manual requires compliance with all applicable operating, service, maintenance, and safety procedures under the direction of qualified personnel. References in this manual to patented or proprietary devices do not constitute a recommendation of their use.

CT52_1-17
Conduct Vegetation Survey

Objectives

When you have completed this module, you will be able to do the following:

1. Describe the factors that impact leakage survey frequency.
   a. Describe the factors that impact leakage survey frequency.
2. Describe procedures for conducting a vegetation survey on gas pipelines (CT52_1-17).
   a. Describe the prerequisites, competencies, and abnormal operating conditions associated with conducting a vegetation survey.
   b. Describe the procedures for conducting a vegetation survey.

Performance Task

Under the supervision of your instructor, you should be able to do the following:

1. Conduct a vegetation survey (CT52_1-17).
   - Identify potential abnormal operating conditions that may occur, and know the appropriate actions to take in response to them.
   - Utilize the appropriate personal protective equipment (PPE) according to relevant company procedures.
   - Identify the portion of the pipeline system to be surveyed to ensure that the intended system is the one that will be surveyed.
   - Ensure that you understand the growth characteristics of the local plants so that you can correctly identify a plant problem that is caused by a natural gas leak.
   - Locate the centerline of the pipeline being surveyed to ensure that the area immediately above the pipe is surveyed. Use a calibrated line locator if necessary.
   - Visually survey the surface conditions over the pipeline, looking for signs of abnormal changes in vegetation growth to identify possible points of natural gas leakage. Also look for changes in the soil condition, unusual sounds, or concentrations of insects, which may indicate the presence of a leak.
   - If a leak is detected, attempt to pinpoint the source of the leak to further determine the severity.
   - If the appropriate leakage detection equipment is available, calibrate the instrument before using. Classify the leak based on the percentage of gas, volume, and location per company leak grading criteria. If the leak classification is serious, follow company procedures.
   - Document the results of the survey and make the appropriate notifications.

Trade Terms

Abnormal operating conditions (AOCs)
Combustible gas detector
Explosive limits
Flame ionization gas detector
Lower explosive limit (LEL)
Upper explosive limit (UEL)
Vegetation survey

Industry-Recognized Credentials

If you are training through an NCCER-accredited sponsor, you may be eligible for credentials from NCCER's Registry. The ID number for this module is CT52_1-17. Note that this module may have been used in other NCCER curricula and may apply to other level completions. Contact NCCER's Registry at 1.888.622.3720 or go to **www.nccer.org** for more information.

Contents

SECTION ONE

1.0.0 CONDUCTING A LEAKAGE SURVEY

Objective

Describe the factors that impact leakage survey frequency.

a. Describe the factors that impact leakage survey frequency.

Trade Terms

Combustible gas detector: An instrument that is capable of detecting product or crude at very slight concentrations, generally within 7 to 10 seconds; pulls a volume of the atmosphere being sampled through itself to analyze for product or crude. Two examples of combustible gas detectors include a percent scale, which reads the percentage of gas in air, and an LEL scale, which reads the percentage of the lower explosive level of gas in air.

Explosive limits: The range that will support combustion of a flammable gas; also called flammable limits. Each gas has its unique range, but typical natural gas limits are 5 to 15 percent gas in air.

Flame ionization gas detector: More sensitive than any other combustible gas detector, this instrument may read down to one part per million and generally responds within two seconds. The atmosphere being sampled is pulled into the unit by a pump. These units may be mounted on vehicles to allow for a mobile survey of the pipeline.

Lower explosive limit (LEL): The lowest concentration of gas in air that will support combustion. Readings below this limit are considered to be too lean in mixture to support combustion.

Upper explosive limit (UEL): The highest concentration of gas in air that will support combustion. Readings above this limit are considered to be too rich in mixture to support combustion.

Vegetation survey: Observing the growth of normal vegetation over the pipeline in order to detect any changes in the vegetation that may indicate a natural gas leak.

Federal regulations require periodic leakage surveys of gas pipelines. The intervals between surveys vary based on several factors, including the class location of the line, whether the line goes through a business district, and the type of line.

1.1.0 Leakage Survey Frequency

Class location designation is based on population density. More specifically, the class location of a line is defined by the number of dwellings intended for human occupancy that exist within 220 yards on either side of the centerline of any continuous one-mile length of pipeline. Note that each separate dwelling unit in a multiple dwelling unit building is counted as a separate building intended for human occupancy. The four class locations are defined as follows:

- *Class 1 location* – An offshore area or an area containing 10 or fewer buildings intended for human occupancy.
- *Class 2 location* – An area containing between 10 and 46 buildings intended for human occupancy.
- *Class 3 location* – An area containing 46 or more buildings intended for human occupancy, an area where the pipeline lies within 100 yards of a building, or a small, well-defined outside area that is occupied by 20 or more persons at least 5 days a week for 10 weeks in any 12-month period.
- *Class 4 location* – An area where buildings with four or more stories above ground are prevalent.

NOTE

Class locations may be adjusted by other factors. Refer to 49 *CFR* 192.5 for further detail about class locations.

The required intervals for leakage surveys also depend on whether the pipeline goes through a business district and whether it is a distribution or a transmission line. According to 49 *CFR* 192.723, a leakage survey must be conducted in business districts at intervals not exceeding 15 months but at least once each calendar year. A leakage survey must be conducted outside business districts at least once every 5 calendar years at intervals not exceeding 63 months. However, for distribution lines that lack cathodic protection for which electrical surveys for corrosion are impractical, a leakage survey must be conducted at least once every 3 calendar years at intervals not exceeding 39 months.

According to 49 *CFR* 192.706, leakage surveys of a transmission line must be conducted at intervals not exceeding 15 months, but at least once each calendar year. However, in the case of a transmission line that transports gas without an odor or odorant, in conformity with code 49 *CFR* 192.625, leakage surveys using leak detector equipment must be conducted more frequently. In Class 3 locations, these transmission lines must be surveyed at intervals not exceeding 7 ½ months, but at least twice each calendar year. In Class 4 locations, they must be surveyed at intervals not exceeding 4 ½ months but at least 4 times each calendar year.

A leakage survey is performed by means of a visual inspection or by use of leak detection instruments that detect the presence of natural gas, propane, or other gases. If a leak does exist, the person performing the inspection must be alert to an unexpected gas release or the possibility of fire or explosion.

Preparation for the task includes being familiar with company procedures for emergencies, wearing appropriate personal protective equipment (PPE), and understanding company criteria for classifying leaks. It is also necessary to understand the characteristics of the gas in the pipeline. An important characteristic is the **explosive limits**, or the range that will support combustion of a flammable gas. Each gas has its own range, typically 5 to 15 percent gas in air. The **lower explosive limit (LEL)** is the lowest percentage of gas in air that will support combustion or explosion. The **upper explosive limit (UEL)** is the highest concentration of a gas that will support combustion. Elements of conducting a leakage survey include conducting a **vegetation survey**, using a **combustible gas detector**, and using a **flame ionization gas detector**.

Additional Resources

It is assumed that you will learn to perform these procedures under the direct supervision of your crew leader, foreman, or instructor. For independent study, refer to the recommended Additional Resources at the back of this module.

1.0.0 Section Review

1. Leak surveys in Class 3 locations must be conducted at least _____.
   a. once a year
   b. twice a year
   c. three times a year
   d. four times a year

2. Inspection intervals vary by class location of the line, whether the line goes through a business district, and the _____.
   a. line rating
   b. product being transported
   c. length of the line
   d. type of line

SECTION TWO

2.0.0 CONDUCTING A VEGETATION SURVEY

Objective

Describe the procedures for conducting a vegetation survey on gas pipelines (CT52_1-17).

a. Describe the prerequisites, competencies, and abnormal operating conditions associated with conducting a vegetation survey.
b. Describe the procedures for conducting a vegetation survey.

Performance Task

Conduct a vegetation survey (CT52_1-17).

- Identify potential abnormal operating conditions that may occur, and know the appropriate actions to take in response to them.
- Utilize the appropriate personal protective equipment (PPE) according to relevant company procedures.
- Identify the portion of the pipeline system to be surveyed to ensure that the intended system is the one that will be surveyed.
- Ensure that you understand the growth characteristics of the local plants so that you can correctly identify a plant problem that is caused by a natural gas leak.
- Locate the centerline of the pipeline being surveyed to ensure that the area immediately above the pipe is surveyed. Use a calibrated line locator if necessary.
- Visually survey the surface conditions over the pipeline, looking for signs of abnormal changes in vegetation growth to identify possible points of natural gas leakage. Also look for changes in the soil condition, unusual sounds, or concentrations of insects, which may indicate the presence of a leak.
- If a leak is detected, attempt to pinpoint the source of the leak to further determine the severity.
- If the appropriate leakage detection equipment is available, calibrate the instrument before using. Classify the leak based on the percentage of gas, volume, and location per company leak grading criteria. If the leak classification is serious, follow company procedures.
- Document the results of the survey and make the appropriate notifications.

Trade Terms

Abnormal operating conditions (AOCs): Those conditions in which operating design limits are exceeded, or when an operating condition or event occurs that could present a hazard to persons, property, or the environment.

This section provides the procedures for conducting a vegetation survey on a pipeline system. It also provides the prerequisites and competencies required for the procedures, as well as the abnormal operating conditions (AOCs) that may be encountered.

2.1.0 Prerequisites, Competencies, and Abnormal Operating Conditions

Personnel must possess a variety of skills in order to conduct pipeline surveys. The various operations require a series of steps, which a pipeline technician must be competent to perform. Personnel must also be aware of the particular AOCs that may occur during the procedure and know the proper responses to them.

2.1.1 Prerequisites and Competencies

Personnel performing vegetation surveys must understand the characteristics of natural gas and other gases being transported by pipeline. They must able to locate the section of pipeline to be surveyed using maps or other documents, and then use a pipe locator to determine the centerline of the buried pipe.

Personnel must be familiar with the growth characteristics of the plants in the area to be surveyed, and understand how these characteristics are affected by the presence of natural gas. They must be able to visually inspect and analyze the condition of the vegetation on the right-of-way (ROW), looking for changes in the soil, as well as plant deterioration or thawing, due to a gas leak in the immediate vicinity.

Personnel must understand the hazards of the gas in the line being surveyed, know how to use the appropriate devices to detect leaks, and be able to properly grade leaks according to the company leak-classifying criteria. They must also be aware of and able to report any changes on or near the ROW that may affect the classification of a section of pipeline, which could impact future leakage survey frequency.

Personnel must know and be able to follow the company's policies and procedures for responding to emergencies, and for the proper use of personal protective equipment (PPE).

2.1.2 Abnormal Operating Conditions

An unexpected gas release may be encountered during the survey. Depending upon the location and the size of the leak, a hazardous condition may exist that could result in an explosion and/or fire. Determine the extent of the leak and act accordingly. Appropriate actions may include evacuation of affected buildings or the immediate area, calling 911, and/or contacting the company emergency center.

A fire or explosion may occur due to a hazardous gas leak. Should this occur, evacuate the area immediately and call 911 and the company emergency center.

2.2.0 Conducting a Vegetation Survey

Vegetation surveys are performed as a part of routine right-of-way (ROW) inspections and as part of the job whenever maintenance personnel are working on the pipeline route. A vegetation survey is often the first alert to problems. Vegetation and other surface conditions will be affected by the escape of product or crude into the soil surrounding the pipeline.

A vegetation survey may include the following steps:

Step 1 Identify potential abnormal operating conditions that may occur, and know the appropriate actions to take in response to them.

Step 2 Utilize the appropriate personal protective equipment (PPE) according to relevant company procedures.

Step 3 Identify the portion of the pipeline system to be surveyed to ensure that the intended system is the one that will be surveyed.

Step 4 Ensure that you understand the growth characteristics of the local plants so that you can correctly identify a plant problem that is caused by a natural gas leak.

Step 5 Locate the centerline of the pipeline being surveyed to ensure that the area immediately above the pipe is surveyed. Use a calibrated line locator if necessary.

Step 6 Visually survey the surface conditions over the pipeline, looking for signs of abnormal changes in vegetation growth to identify possible points of natural gas leakage. Also look for changes in the soil condition, unusual sounds, or concentrations of insects, which may indicate the presence of a leak.

Step 7 If a leak is detected, attempt to pinpoint the source of the leak to further determine the severity.

Step 8 If the appropriate leakage detection equipment is available, calibrate the instrument before using. Classify the leak based on the percentage of gas, volume, and location per company leak grading criteria. If the leak classification is serious, follow company procedures.

Step 9 Document the results of the survey and make the appropriate notifications.

WARNING!

Depending on the location and size of the leak, a hazardous condition may exist that could result in an explosion and/or fire. Determine the extent of the leak and act accordingly. Appropriate actions may include evacuating affected buildings or the immediate area, calling local emergency services, and contacting the company emergency center.

Additional Resources

It is assumed that you will learn to perform these procedures under the direct supervision of your crew leader, foreman, or instructor. For independent study, refer to the recommended Additional Resources at the back of this module.

2.0.0 Section Review

1. A gas leak may be detected by changes in the soil, as well as plant deterioration or _____.
   a. thawing
   b. shape
   c. type
   d. brittleness

2. Vegetation surveys are performed during _____.
   a. routine ROW inspections only
   b. leakage surveys only
   c. post-blast leakage surveys only
   d. routine ROW inspections and leakage surveys

SUMMARY

Federal regulations require periodic leakage surveys of gas pipelines. The intervals required vary based on the location of the line and the specific type of line. As part of routine right-of-way (ROW) inspections or when working on the pipeline route, personnel perform vegetation surveys for any signs of a release. The condition of vegetation in the immediate area of a pipeline is one of the means of detecting a natural gas leak. More in-depth surveys may also be conducted with a combustible gas detector or a flame ionization gas detector. When performing any type of leak survey, personnel must remain aware of the possibility of an unexpected gas release. The utmost caution must always be used to protect people, property, and the environment.

Review Questions

1. The highest concentration of a gas that will support combustion is the _____.
   a. maximum explosive limit
   b. explosive limit
   c. lower explosive limit (LEL)
   d. upper explosive limit (UEL)

2. Leakage surveys of a transmission line must be conducted at intervals not exceeding 15 months, but at least _____.
   a. once per calendar year
   b. twice per calendar year
   c. three times per calendar year
   d. four times per calendar year

3. Regulations dictate the intervals of leakage surveys based on _____.
   a. class location
   b. proximity to a business district
   c. type of line
   d. class location, proximity to a business district, and type of line

4. A vegetation survey is _____.
   a. often unnecessary
   b. often the first alert to problems
   c. a bi-monthly requirement
   d. only performed in the spring and summer

Trade Terms Quiz

Fill in the blank with the correct term that you learned from your study of this module.

1. The range that will support combustion of a flammable gas, typically between 5 and 15 percent gas in air, is called the ________.
2. Conditions within a pipeline system, identified by the operator, that may indicate (1) a malfunction of a component or (2) a deviation from normal operations that may indicate a condition exceeding design limits or result in a hazard to persons, property, or the environment are called ________.
3. An instrument that is capable of detecting product or crude at very slight concentrations by pulling a volume of the atmosphere being sampled through itself to analyze for product or crude is a(n) ________.
4. The highest concentration of gas in air that will support combustion is the ________.
5. The observation of the growth of normal vegetation over the pipeline in order to detect any changes in the vegetation that may indicate a natural gas leak is a(n) ________.
6. More sensitive than any other combustible gas detector, the instrument that may be read down to one part per million and generally responds within two seconds is called a(n) ________.
7. The lowest concentration of gas in air that will support combustion is the ________.

Trade Terms

Abnormal operating conditions (AOCs)
Combustible gas detector
Explosive limits
Flame ionization gas detector
Lower explosive limit (LEL)
Upper explosive limit (UEL)
Vegetation survey

Trade Terms Introduced in This Module

Abnormal operating conditions (AOCs): Those conditions in which operating design limits are exceeded, or when an operating condition or event occurs that could present a hazard to persons, property, or the environment.

Combustible gas detector: An instrument that is capable of detecting product or crude at very slight concentrations, generally within 7 to 10 seconds; pulls a volume of the atmosphere being sampled through itself to analyze for product or crude. Two examples of combustible gas detectors include a percent scale, which reads the percent of gas in air, and an LEL scale, which reads the percent of the lower explosive level of gas in air.

Explosive limits: The range that will support combustion of a flammable gas; also called flammable limits. Each gas has its unique range, but typical natural gas limits are 5 to 15 percent gas in air.

Flame ionization gas detector: More sensitive than any other combustible gas detector, this instrument may read down to one part per million and generally responds within two seconds. The atmosphere being sampled is pulled into the unit by a pump. These units may be mounted on vehicles to allow for a mobile survey of the pipeline.

Lower explosive limit (LEL): The lowest concentration of gas in air that will support combustion. Readings below this limit are considered to be too lean in mixture to support combustion.

Upper explosive limit (UEL): The highest concentration of gas in air that will support combustion. Readings above this limit are considered to be too rich in mixture to support combustion.

Vegetation survey: Observing the growth of normal vegetation over the pipeline in order to detect any changes in the vegetation that may indicate a natural gas leak.

Additional Resources

The following recommended resources can provide additional helpful information related to the requirements for this covered task:

API Recommended Practice 1161, Third Edition, © 2014 American Petroleum Institute. API Publishing Services, 1220 L Street, NW, Washington, DC 20005. Distributed by Thompson Reuters (Scientific) LLC, **www.techstreet.com**.

49 *CFR* 192, **www.ecfr.gov**.

49 *CFR* 195, **www.ecfr.gov**.

NCCER Module AOCCC-17, *Abnormal Operating Conditions Control Center*

NCCER Module AOCFG-17, *Abnormal Operating Conditions Field and Gas*

NCCER Module 66101-14, *Introduction to the Pipeline Industry*

NCCER Module 62105-14, *Introduction to Pipeline Documents*

NCCER Module 63104-02, *Piping and Mechanical Blueprint Reading*

NCCER Module CT52_2-17, *Conduct a Leak Survey with a Combustible Gas Detector*

NCCER Module CT52_3-17, *Conduct a Leak Survey with a Flame Ionization Unit*

Section Review Answer Key

Answer	Section Reference	Objective
Section One		
1. b	1.0.0	1a
2. d	1.0.0	1a
Section Two		
1. a	2.1.1	2a
2. d	2.2.0	2b

NCCER CURRICULA — USER UPDATE

NCCER makes every effort to keep its textbooks up-to-date and free of technical errors. We appreciate your help in this process. If you find an error, a typographical mistake, or an inaccuracy in NCCER's curricula, please fill out this form (or a photocopy), or complete the online form at **www.nccer.org/olf**. Be sure to include the exact module ID number, page number, a detailed description, and your recommended correction. Your input will be brought to the attention of the Authoring Team. Thank you for your assistance.

Instructors – If you have an idea for improving this textbook, or have found that additional materials were necessary to teach this module effectively, please let us know so that we may present your suggestions to the Authoring Team.

NCCER Product Development and Revision

13614 Progress Blvd., Alachua, FL 32615

Email: curriculum@nccer.org
Online: www.nccer.org/olf

❑ Trainee Guide ❑ Lesson Plans ❑ Exam ❑ PowerPoints Other ____________

Craft / Level: Copyright Date:

Module ID Number / Title:

Section Number(s):

Description:

Recommended Correction:

Your Name:

Address:

Email: Phone:

CT52_2-17

Conduct a Leak Survey with a CGD

Overview

Pipeline personnel must sometimes perform leak surveys to confirm the safety of the pipeline. These must always be performed after any blasting has taken place in the proximity of the pipeline. In addition, gas pipelines must be surveyed at specific intervals as required by federal regulations. This module explains parameters for performing post-blasting leak surveillance, as well as procedures for conducting a leak survey using a combustible gas detector (CGD) on a gas pipeline.

Trainees with successful module completions may be eligible for credentialing through the NCCER Registry. To learn more, go to **www.nccer.org** or contact us at 1.888.622.3720. Our website has information on the latest product releases and training, as well as online versions of our *Cornerstone* magazine and Pearson's product catalog.

Your feedback is welcome. You may email your comments to **curriculum@nccer.org**, send general comments and inquiries to **info@nccer.org**, or fill in the User Update form at the back of this module.

This information is general in nature and intended for training purposes only. Actual performance of activities described in this manual requires compliance with all applicable operating, service, maintenance, and safety procedures under the direction of qualified personnel. References in this manual to patented or proprietary devices do not constitute a recommendation of their use.

CT52_2-17
Conduct a Leak Survey with a CGD

Objectives

When you have completed this module, you will be able to do the following:

1. Describe the purpose and procedures for conducting a leak survey on gas pipelines.
2. Describe procedures for conducting a leak survey with a combustible gas detector (CGD) on a gas pipeline.
   a. Identify prerequisites and competencies associated with surveying with a CGD.
   b. Identify abnormal operating conditions (AOCs) associated with surveying with a CGD.
   c. Explain the procedures for surveying with a CGD.

Performance Task

Under the supervision of your instructor, you should be able to do the following:

1. Conduct a leak survey with a CGD (CT52_2-17).
   - Identify potential abnormal operating conditions that may occur during performance of this CT, and know the appropriate actions to take in response to them.
   - Utilize the appropriate personal protective equipment (PPE) according to relevant company procedures.
   - Identify the portion of the pipeline system to be surveyed.
   - Locate the centerline of the pipeline to be surveyed.
   - Determine what equipment is needed for the survey, and turn on and calibrate any instruments to be used.
   - Survey in available openings along pipeline ROW and pinpoint leaks when leak indications are found.
   - Classify the grade of the leak following company criteria.
   - If leak is classified as the most serious, follow company response procedures.
   - Complete appropriate documentation as required by operator's procedures.

Trade Terms

Abnormal operating conditions (AOCs)
Bar holes
Combustible gas detector
Explosive limit
Flame ionization gas detector
Lower explosive limit (LEL)
Probe bar
Upper explosive limit (UEL)
Vegetation survey

Industry-Recognized Credentials

If you're training through an NCCER-accredited sponsor, you may be eligible for credentials from NCCER's Registry. The ID number for this module is CT52_2-17. Note that this module may have been used in other NCCER curricula and may apply to other level completions. Contact NCCER's Registry at 1.888.622.3720 or go to **www.nccer.org** for more information.

Contents

SECTION ONE

1.0.0 CONDUCTING A LEAK SURVEY

Objective

Describe the purpose and procedures for conducting a leak survey on gas pipelines.

Trade Terms

Combustible gas detector (CGD): An instrument that is capable of detecting product or crude at very slight concentrations, generally within 7 to 10 seconds; pulls a volume of the atmosphere being sampled through itself to analyze for product or crude. Two examples of combustible gas detectors include a percent scale, which reads the percentage of gas in air, and an LEL scale, which reads the percentage of the lower explosive level of gas in air.

Explosive limit: The percentage range of a flammable gas in the air that will support combustion of the gas. Each gas has a unique range, but typical natural gas limits are 5 to 15 percent gas in air. (Also called flammable limits.)

Flame ionization gas detector: More sensitive than any other combustible gas detector, this instrument may read down to one part per million and generally responds within two seconds. The atmosphere being sampled is pulled into the unit by a pump. These units may be mounted on vehicles to allow for a mobile survey of the pipeline.

Lower explosive limit (LEL): The lowest concentration of gas in air that will support combustion. Readings below this limit are considered to be too lean in mixture to support combustion.

Upper explosive limit (UEL): The highest concentration of gas in air that will support combustion. Readings above this limit are considered to be too rich in mixture to support combustion.

Vegetation survey: Observation of the growth of normal vegetation over the pipeline in order to detect any changes in the vegetation that may indicate a natural gas leak.

After blasting operations have taken place near any section of pipeline, a leak survey must be performed. A leak survey is a visual inspection of the pipeline right-of-way (ROW) directly affected by a blasting operation. Additional segments of pipe both upstream and downstream of the blast area must also be surveyed. The survey serves to check for evidence of leaks that indicate damage to the pipeline. Pipeline maintenance personnel must be able to perform a leak survey in accordance with their company policy after any blasting has taken place in proximity to the pipeline.

> **WARNING!**
>
> Depending on the location and size of the leak, a hazardous condition may exist that could result in an explosion and/or fire. Determine the extent of the leak and act accordingly. Appropriate actions may include evacuating affected buildings or the immediate area, calling local emergency services, and contacting the company emergency center.

Federal regulations require periodic leak surveys of gas pipelines. The intervals between surveys vary based on several factors, including the class location of the line, whether the line goes through a business district, and the type of line.

Class location designation is based on population density. More specifically, the class location of a line is defined by the number of dwellings intended for human occupancy that exist within 220 yards on either side of the centerline of any continuous one-mile length of pipeline. Note that each separate dwelling unit in a multiple dwelling unit building is counted as a separate building intended for human occupancy. The four class locations are defined as follows:

- *Class 1 location* – An offshore area or an area containing 10 or fewer buildings intended for human occupancy.
- *Class 2 location* – An area containing between 10 and 46 buildings intended for human occupancy.

- *Class 3 location* – An area containing 46 or more buildings intended for human occupancy, an area where the pipeline lies within 100 yards of a building, or a small, well-defined outside area that is occupied by 20 or more persons at least 5 days a week for 10 weeks in any 12-month period.
- *Class 4 location* – An area where buildings with four or more stories above ground are prevalent.

NOTE

Class locations may be adjusted by other factors. Refer to 49 *CFR* 192.5 for further detail about class locations.

The required intervals for leak surveys also depend on whether the pipeline goes through a business district and whether it is a distribution or a transmission line. According to 49 *CFR* 192.723, a leak survey must be conducted in business districts at intervals not exceeding 15 months but at least once each calendar year. A leak survey must be conducted outside business districts at least once every 5 calendar years at intervals not exceeding 63 months. However, for cathodically unprotected distribution lines for which electrical surveys for corrosion are impractical, a leak survey must be conducted at least once every 3 calendar years at intervals not exceeding 39 months.

According to 49 *CFR* 192.706, leak surveys of a transmission line must be conducted at intervals not exceeding 15 months but at least once each calendar year. However, in the case of a transmission line that transports gas without an odor or odorant, in conformity with code 49 *CFR* 192.625, leak surveys using leak detector equipment must be conducted more frequently. In Class 3 locations, these transmission lines must be surveyed at intervals not exceeding 7 $^1/_2$ months but at least twice each calendar year. In Class 4 locations, they must be surveyed at intervals not exceeding 4 $^1/_2$ months but at least 4 times each calendar year.

A leak survey is performed by means of a visual inspection or by use of leak detection instruments that detect the presence of natural gas, propane, or other gases. If a leak does exist, the person performing the inspection must be alert to an unexpected gas release or the possibility of fire or explosion.

Preparation for the task includes being familiar with company procedures for emergencies, wearing appropriate personal protective equipment (PPE), and understanding company criteria for classifying leaks. It is also necessary to understand the characteristics natural gas and other gasses in the pipeline, including their effect on plant growth. An important characteristic is the **explosive limit**, or the range that will support combustion of a flammable gas. Each gas has its own range, typically 5 to 15 percent gas in air. The **lower explosive limit (LEL)** is the lowest percentage of gas in air that will support combustion or explosion. The **upper explosive limit (UEL)** is the highest percentage of a gas that will support combustion. Elements of conducting a leak survey include conducting a **vegetation survey**, using a **combustible gas detector (CGD)**, and using a **flame ionization gas detector**.

Additional Resources

It is assumed that you will learn to perform these procedures under the direct supervision of your crew leader, foreman, or instructor. For independent study, refer to the recommended Additional Resources at the back of this module.

1.0.0 Section Review

1. Leak surveys in Class 3 locations must be conducted at least _____.
   a. once a year
   b. twice a year
   c. three times a year
   d. four times a year

2. Inspection intervals vary by class location of the line, whether the line goes through a business district, and the _____.
   a. line rating
   b. product being transported
   c. length of the line
   d. type of line

Section Two

2.0.0 Conducting a Leak Survey with a CGD

Objective

Describe procedures for conducting a leak survey with a combustible gas detector (CGD) on a gas pipeline.

a. Identify prerequisites and competencies associated with surveying with a CGD.
b. Identify abnormal operating conditions (AOCs) associated with surveying with a CGD.
c. Explain the procedures for surveying with a CGD.

Performance Task

Conduct a leak survey with a CGD (CT52_2-17).

- Identify potential abnormal operating conditions that may occur during performance of this CT, and know the appropriate actions to take in response to them.
- Utilize the appropriate personal protective equipment (PPE) according to relevant company procedures.
- Identify the portion of the pipeline system to be surveyed.
- Locate the centerline of the pipeline to be surveyed.
- Determine what equipment is needed for the survey, and turn on and calibrate any instruments to be used.
- Survey in available openings along pipeline ROW and pinpoint leaks when leak indications are found.
- Classify the grade of the leak following company criteria.
- If leak is classified as the most serious, follow company response procedures.
- Complete appropriate documentation as required by operator's procedures.

Trade Terms

Abnormal operating conditions (AOCs): Conditions within a pipeline system, identified by the operator, that may indicate (1) a malfunction of a component or (2) a deviation from normal operations that may indicate a condition exceeding design limits or result in a hazard to persons, property, or the environment.

Bar holes: Holes punched in the ground by use of a probe bar (see probe bar); allow a sampling hose and probe of a combustible gas detector (CGD) to be inserted to detect gas.

Probe bar: A device used to bore or probe a cylindrical hole in the earth's surface that is large enough to insert the suction probe of gas detection equipment.

This section provides the procedures for conducting a leak survey with a combustible gas detector on a pipeline system. It also provides the prerequisites and competencies required for these procedures, as well as the abnormal operating conditions (AOCs) that may be encountered.

2.1.0 Prerequisites and Competencies

Personnel performing pipeline surveys using a CGD must understand the characteristics of natural gas and other gases being transported by pipeline, including their effect on plant growth. They must be able to locate the section of pipeline to be surveyed using maps or other documents, and then use a pipe locator to determine the centerline of the buried pipe.

Personnel must know which instrument is best suited for the type of leak survey being conducted. They must also know how to properly calibrate and then operate the instrument in order to perform the survey.

Personnel must understand the hazards of the gas in the line being surveyed, know how to use the appropriate devices to detect leaks, and be able to properly grade leaks according to the company leak-classifying criteria. They must also be aware of and able to report any changes on or near the right-of-way (ROW) that may affect the classification of a section of pipeline, which could affect the frequency of future surveys.

Personnel must know and be able to follow the company's policies and procedures for responding to emergencies, and for the proper use of PPE.

2.2.0 Abnormal Operating Conditions

Personnel must also be aware of the particular AOCs that may occur during a leak survey and be prepared to respond to them with the appropriate action from memory. The following are some potential AOCs, and the appropriate action personnel should take in response to them:

- *An unexpected gas release occurs* – Personnel should determine the extent of the leak and act accordingly. Depending on the location and the size of the leak, a hazardous condition may exist. Appropriate actions may include evacuation of affected buildings or the immediate area, calling 911, and/or contacting the company emergency center.
- *A fire or explosion occurs* – Personnel should evacuate the area immediately and call 911 and the company emergency center.

2.3.0 Procedure

A combustible gas detector is an instrument that is capable of detecting product or crude at very slight concentrations. This instrument pulls in a volume of the atmosphere being sampled to analyze for the presence of product or crude. The sample may be obtained manually or by automatic means. Follow these steps to use a combustible gas detector:

Step 1 Identify potential abnormal operating conditions that may occur, and know the appropriate actions to take in response to them.

Step 2 Utilize the appropriate personal protective equipment according to relevant company procedures.

Step 3 Identify the portion of the pipeline system to be surveyed.

Step 4 Locate the centerline of the pipe being surveyed to ensure that the area immediately above the pipe is surveyed. Use a calibrated pipe locator if necessary.

Step 5 Turn on the properly calibrated combustible gas detector.

Step 6 Survey in available openings along the right-of-way (ROW) to check locations where leaking gases might migrate to or places where gases might accumulate.

Step 7 Pinpoint the leak sources if leak indications are found. As an example, bar holing may be used to pinpoint the leak. **Bar holes** are holes punched in the ground by use of a **probe bar** to allow the sampling hose and probe of a combustible gas detector to be inserted to detect gas.

Step 8 Using company criteria, classify the grade of the leak based on percentage of gas, volume, and location. This classification helps determine the severity of the leak.

Step 9 If the leak classification is critical, follow company procedures to prevent harm to people, property, and the environment.

Step 10 Complete appropriate documentation as required by operator's procedures.

Additional Resources

It is assumed that you will learn to perform these procedures under the direct supervision of your crew leader, foreman, or instructor. For independent study, refer to the recommended Additional Resources at the back of this module.

2.0.0 Section Review

1. Personnel performing leak surveys must be able to properly grade leaks according to _____.
   a. the instrument being used
   b. federal regulations
   c. company criteria
   d. the type of leak survey

2. If an unexpected gas release is encountered, personnel must first _____.
   a. determine the extent of the leak
   b. contact the operator
   c. perform a vegetation survey
   d. call 911

3. When surveying with a combustible gas detector, classify the grade of a leak based on _____.
   a. percentage of gas, volume, and ionization
   b. percentage of gas, volume, and location
   c. product flow, volume, and ionization
   d. ionization and product flow

SUMMARY

After blasting operations have taken place near any segment of pipeline, designated pipeline personnel must perform a leak survey. A leak survey includes visually inspecting the pipeline ROW directly affected by the blasting, surveying additional segments of the pipeline, and checking for leaks indicative of damage to the pipeline.

Federal regulations require periodic leak surveys of gas pipelines. The intervals required vary based on the location of the line and the specific type of line. As part of routine ROW inspections or when working on the pipeline route, personnel perform vegetation surveys for any signs of a release. More in-depth surveys may also be conducted with a combustible gas detector (CGD) or a flame ionization gas detector. When performing any type of leak survey, personnel must remain aware of the possibility of an unexpected gas release. The utmost caution must always be used to protect people, property, and the environment.

Review Questions

1. The highest concentration of a gas that will support combustion is the _____.
   a. pressure
   b. explosive limit
   c. lower explosive limit (LEL)
   d. upper explosive limit (UEL)

2. Leak surveys of a transmission line must be conducted at intervals not exceeding 15 months but at least _____.
   a. four times per calendar year
   b. three times per calendar year
   c. twice per calendar year
   d. once per calendar year

3. Regulations dictate the intervals of leak surveys based upon _____.
   a. class location
   b. proximity to a business district
   c. type of line
   d. class location, proximity to a business district, and type of line

4. An area containing between 10 and 46 buildings intended for human occupancy is categorized as a _____.
   a. Class 1 location
   b. Class 2 location
   c. Class 3 location
   d. Class 4 location

5. When transmission lines do not have an odor or odorant, leak surveys must be performed _____.
   a. more frequently
   b. less frequently
   c. at the same intervals
   d. at least 4 times each year

6. Unlike some other detectors, combustible gas detectors can detect product or crude at _____.
   a. very slight concentrations
   b. moderate concentrations
   c. high concentrations
   d. extreme concentrations

7. Bar-holing is one technique to _____.
   a. measure the lower explosive limit (LEL)
   b. conduct a vegetation survey
   c. pinpoint leaks
   d. monitor for pressure

Trade Terms Quiz

Fill in the blank with the correct term that you learned from your study of this module.

1. The percentage range of a flammable gas in the air that will support combustion is called the _________.

2. An instrument that is very sensitive to gas that pulls a volume of the atmosphere being sampled by a pump to detect the gas is a(n) _________.

3. A device used to bore or probe a cylindrical hole in the earth's surface for insertion of gas detection equipment is called a _________.

4. The highest concentration of gas in air that will support combustion is the _________.

5. The observation of the growth of normal vegetation over the pipeline in order to detect any changes in the vegetation that may indicate a natural gas leak is called a(n) _________.

6. Conditions within a pipeline system, identified by the operator, that may indicate (1) a malfunction of a component or (2) a deviation from normal operations that may indicate a condition exceeding design limits or result in a hazard to persons, property, or the environment are called _________.

7. An instrument that is capable of detecting product or crude at very slight concentrations by pulling a volume of the atmosphere being sampled through itself to analyze for product or crude is a(n) _________.

8. To allow a sampling hose and probe of a CGD to be inserted, _________ are punched in the ground with a probe bar.

9. The lowest concentration of gas in air that will support combustion is the _________.

Trade Terms

Abnormal operating conditions (AOCs)
Bar holes
Combustible gas detector (CGD)
Explosive limit
Flame ionization gas detector
Lower explosive limit (LEL)
Probe bar
Upper explosive limit (UEL)
Vegetation survey

Trade Terms Introduced in This Module

Abnormal operating conditions (AOCs): Conditions within a pipeline system, identified by the operator, that may indicate (1) a malfunction of a component or (2) a deviation from normal operations that may indicate a condition exceeding design limits or result in a hazard to persons, property, or the environment.

Bar holes: Holes punched in the ground by use of a probe bar (see probe bar); allow a sampling hose and probe of a combustible gas detector (CGD) to be inserted to detect gas.

Combustible gas detector (CGD): An instrument that is capable of detecting product or crude at very slight concentrations, generally within 7 to 10 seconds; pulls a volume of the atmosphere being sampled through itself to analyze for product or crude. Two examples of combustible gas detectors include a percent scale, which reads the percentage of gas in air, and an LEL scale, which reads the percentage of the lower explosive level of gas in air.

Explosive limit: The percentage range of a flammable gas in the air that will support combustion of the gas. Each gas has a unique range, but typical natural gas limits are 5 to 15 percent gas in air. (Also called flammable limits.)

Flame ionization gas detector: More sensitive than any other combustible gas detector, this instrument may read down to one part per million and generally responds within two seconds. The atmosphere being sampled is pulled into the unit by a pump. These units may be mounted on vehicles to allow for a mobile survey of the pipeline.

Lower explosive limit (LEL): The lowest concentration of gas in air that will support combustion. Readings below this limit are considered to be too lean in mixture to support combustion.

Probe bar: A device used to bore or probe a cylindrical hole in the earth's surface that is large enough to insert the suction probe of gas detection equipment.

Upper explosive limit (UEL): The highest concentration of gas in air that will support combustion. Readings above this limit are considered to be too rich in mixture to support combustion.

Vegetation survey: Observation of the growth of normal vegetation over the pipeline in order to detect any changes in the vegetation that may indicate a natural gas leak.

Additional Resources

The following recommended resources can provide additional helpful information related to the requirements for this covered task:

API Recommended Practice 1161, Third Edition, © 2014 American Petroleum Institute. API Publishing Services, 1220 L Street, NW, Washington, DC 20005. Distributed by Thompson Reuters (Scientific) LLC, **www.techstreet.com**.

49 *CFR* 192, **www.ecfr.gov**.

49 *CFR* 195, **www.ecfr.gov**.

NCCER Module 66101-14, *Introduction to the Pipeline Industry*.

NCCER Module AOCFG-17, *Abnormal Operating Conditions Field and Gas*.

NCCER Module AOCCC-17, *Abnormal Operating Conditions Control Center*.

NCCER Module 62105-14, *Introduction to Pipeline Documents*.

NCCER Module 63104-02, *Piping and Mechanical Blueprint Reading*.

NCCER Module CT52_1-17, *Conduct Vegetation Survey*.

NCCER Module CT52_3-17, *Conduct a Leak Survey with a Combustible-Gas Detector*.

Section Review Answer Key

Answer	Section Reference	Objective
Section One		
1. b	1.0.0	1
2. d	1.0.0	1
Section Two		
1. c	2.1.0	2a
2. a	2.2.0	2b
3. b	2.3.0	2c

NCCER CURRICULA — USER UPDATE

NCCER makes every effort to keep its textbooks up-to-date and free of technical errors. We appreciate your help in this process. If you find an error, a typographical mistake, or an inaccuracy in NCCER's curricula, please fill out this form (or a photocopy), or complete the online form at **www.nccer.org/olf**. Be sure to include the exact module ID number, page number, a detailed description, and your recommended correction. Your input will be brought to the attention of the Authoring Team. Thank you for your assistance.

Instructors – If you have an idea for improving this textbook, or have found that additional materials were necessary to teach this module effectively, please let us know so that we may present your suggestions to the Authoring Team.

NCCER Product Development and Revision

13614 Progress Blvd., Alachua, FL 32615

Email: curriculum@nccer.org

Online: www.nccer.org/olf

❑ Trainee Guide ❑ Lesson Plans ❑ Exam ❑ PowerPoints Other ____________

Craft / Level: Copyright Date:

Module ID Number / Title:

Section Number(s):

Description:

Recommended Correction:

Your Name:

Address:

Email: Phone:

CT52_3-17

Conduct a Leak Survey with a Flame Ionization Unit

Overview

Pipeline personnel must sometimes perform leak surveys to ensure the safety of the pipeline. After any blasting has taken place in proximity to the pipeline, designated personnel must perform a leak survey. In addition, gas pipelines must be surveyed at specific intervals as required by federal regulations. This module explains the procedures for conducting a leak survey on gas pipelines using a flame ionization unit.

Trainees with successful module completions may be eligible for credentialing through the NCCER Registry. To learn more, go to **www.nccer.org** or contact us at 1.888.622.3720. Our website has information on the latest product releases and training, as well as online versions of our *Cornerstone* magazine and Pearson's product catalog.

Your feedback is welcome. You may email your comments to **curriculum@nccer.org**, send general comments and inquiries to **info@nccer.org**, or fill in the User Update form at the back of this module.

This information is general in nature and intended for training purposes only. Actual performance of activities described in this manual requires compliance with all applicable operating, service, maintenance, and safety procedures under the direction of qualified personnel. References in this manual to patented or proprietary devices do not constitute a recommendation of their use.

CT52_3-17
Conduct a Leak Survey with a Flame Ionization Unit

Objectives

When you have completed this module, you will be able to do the following:

1. Describe the factors that impact leak survey frequency.
2. Describe procedures for conducting a leak survey with a flame ionization unit (CT52_3-17).
   a. Describe the prerequisites, competencies, and abnormal operating conditions associated with conducting a leak survey with a flame ionization unit.
   b. Describe the flame ionization unit leak survey procedure.

Performance Tasks

Under the supervision of your instructor, you should be able to do the following:

1. Conduct a leak survey with a flame ionization unit (CT52_3-17).
   - Identify potential abnormal operating conditions that may occur during performance of this CT, and know the appropriate actions to take in response to them.
   - Utilize the appropriate personal protective equipment according to relevant company procedures.
   - Identify the portion of the pipeline system to be surveyed and locate the centerline.
   - Determine what equipment is to be used with the survey.
   - Turn on and calibrate the instrument to be used, if necessary.
   - For a walking survey, check in available openings along pipeline right-of-way (ROW), including bar holes.
   - For a driving survey, drive the survey vehicle at the speed recommended by company procedures.
   - Pinpoint leaks when leak indications are found and classify the grade of the leak following company criteria.
   - If leak is classified as the most serious, follow company response procedures.
   - Complete appropriate documentation as required by operator's procedures.

Trade Terms

Abnormal operating conditions (AOCs)
Bar holes
Combustible gas detector
Explosive limits
Flame ionization gas detector
Lower explosive limit (LEL)
Probe bar
Upper explosive limit (UEL)
Vegetation survey

Industry-Recognized Credentials

If you are training through an NCCER-accredited sponsor, you may be eligible for credentials from NCCER's Registry. The ID number for this module is CT52_3-17. Note that this module may have been used in other NCCER curricula and may apply to other level completions. Contact NCCER's Registry at 1.888.622.3720 or go to **www.nccer.org** for more information.

Contents

SECTION ONE

1.0.0 CONDUCTING A LEAK SURVEY

Objective

Describe the factors that impact leak survey frequency.

Trade Terms

Combustible gas detector: An instrument that is capable of detecting product or crude at very slight concentrations, generally within 7 to 10 seconds; pulls a volume of the atmosphere being sampled through itself to analyze for product or crude. Two examples of combustible gas detectors include a percent scale, which reads the percentage of gas in air, and an LEL scale, which reads the percentage of the lower explosive level of gas in air.

Explosive limits: The range that will support combustion of a flammable gas; also called flammable limits. Each gas has its unique range, but typical natural gas limits are 5 to 15 percent gas in air.

Flame ionization gas detector: More sensitive than any other combustible gas detector, this instrument may read down to one part per million and generally responds within two seconds. The atmosphere being sampled is pulled into the unit by a pump. These units may be mounted on vehicles to allow for a mobile survey of the pipeline.

Lower explosive limit (LEL): The lowest concentration of gas in air that will support combustion. Readings below this limit are considered to be too lean in mixture to support combustion.

Upper explosive limit (UEL): The highest concentration of gas in air that will support combustion. Readings above this limit are considered to be too rich in mixture to support combustion.

Vegetation survey: Observing the growth of normal vegetation over the pipeline in order to detect any changes in the vegetation that may indicate a natural gas leak.

Federal regulations require periodic leak surveys of gas pipelines. The intervals vary based on several factors, including the class location of the line, whether the line goes through a business district, and the type of line.

Class location designation is based on population density. More specifically, the class location of a line is defined by the number of dwellings intended for human occupancy that exist within 220 yards on either side of the centerline of any continuous one-mile length of pipeline. Note that each separate dwelling unit in a multiple dwelling unit building is counted as a separate building intended for human occupancy. The four class locations are defined as follows:

- *Class 1 location* – An offshore area or an area containing 10 or fewer buildings intended for human occupancy.
- *Class 2 location* – An area containing between 10 and 46 buildings intended for human occupancy.
- *Class 3 location* – An area containing 46 or more buildings intended for human occupancy, an area where the pipeline lies within 100 yards of a building, or a small, well-defined outside area that is occupied by 20 or more persons at least 5 days a week for 10 weeks in any 12-month period.
- *Class 4 location* – An area where buildings with four or more stories above ground are prevalent.

> **NOTE**
> Class locations may be adjusted by other factors. Refer to *49CFR192.5* for further detail about class locations.

The required intervals for leak surveys also depend on whether the pipeline goes through a business district and whether it is a distribution or a transmission line. According to 49 *CFR 192.723*, a leak survey must be conducted in business districts at intervals not exceeding 15 months but at least once each calendar year. A leak survey must be conducted outside business districts at least once every 5 calendar years at intervals not exceeding 63 months. However, for cathodically unprotected distribution lines for which electrical surveys for corrosion are impractical, a leak survey must be conducted at least once every 3 calendar years at intervals not exceeding 39 months.

According to 49 *CFR 192.706*, leak surveys of a transmission line must be conducted at intervals not exceeding 15 months but at least once each calendar year. However, in the case of a transmission line that transports gas without an odor or odorant, in conformity with code 49 *CFR 192.625*, leak surveys using leak detector equipment must be conducted more frequently. In Class 3 locations, these transmission lines must be surveyed at intervals not exceeding 7 $^{1}/_{2}$ months but at least twice each calendar year. In Class 4 locations, they must be surveyed at intervals not exceeding 4 $^{1}/_{2}$ months but at least 4 times each calendar year.

A leak survey is performed by means of a visual inspection or by use of leak detection instruments that detect the presence of natural gas, propane, or other gases. If a leak does exist, the person performing the inspection must be alert to an unexpected gas release or the possibility of fire or explosion.

Preparation for the task includes being familiar with company procedures for emergencies, wearing appropriate personal protective equipment (PPE), and understanding company criteria for classifying leaks. It is also necessary to understand the characteristics of the gas in the pipeline. An important characteristic is the **explosive limits**, or the range that will support combustion of a flammable gas. Each gas has its own range, typically 5 to 15 percent gas in air. The **lower explosive limit (LEL)** is the lowest percentage of gas in air that will support combustion or explosion. The **upper explosive limit (UEL)** is the highest percentage of a gas that will support combustion. Elements of conducting a leak survey include conducting a **vegetation survey**, using a **combustible gas detector**, and using a **flame ionization gas detector**.

Additional Resources

It is assumed that you will learn to perform these procedures under the direct supervision of your crew leader, foreman, or instructor. For independent study, refer to the recommended Additional Resources at the back of this module.

1.0.0 Section Review

1. Leak surveys in Class 3 locations must be conducted at least _____.
   a. once a year
   b. twice a year
   c. three times a year
   d. four times a year

2. Inspection intervals vary by class location of the line, whether the line goes through a business district, and the _____.
   a. line rating
   b. product being transported
   c. length of the line
   d. type of line

SECTION TWO

2.0.0 CONDUCTING A LEAK SURVEY WITH A FLAME IONIZATION UNIT

Objective

Describe procedures for conducting a leak survey with a flame ionization unit.

a. Describe the prerequisites, competencies, and abnormal operating conditions associated with conducting a leak survey with a flame ionization unit.
b. Describe the flame ionization unit leak survey procedure.

Performance Tasks

Conduct a leak survey with a flame ionization unit (52_3-17).

- Identify potential abnormal operating conditions that may occur during performance of this CT, and know the appropriate actions to take in response to them.
- Utilize the appropriate personal protective equipment according to relevant company procedures.
- Identify the portion of the pipeline system to be surveyed and locate the centerline.
- Determine what equipment is to be used with the survey.
- Turn on and calibrate the instrument to be used, if necessary.
- For a walking survey, check in available openings along pipeline right-of-way (ROW), including bar holes.
- For a driving survey, drive the survey vehicle at the speed recommended by company procedures.
- Pinpoint leaks when leak indications are found and classify the grade of the leak following company criteria.
- If leak is classified as the most serious, follow company response procedures.
- Complete appropriate documentation as required by operator's procedures.

Trade Terms

Abnormal operating conditions (AOCs): Conditions within a pipeline system, identified by the operator, that may indicate (1) a malfunction of a component or (2) a deviation from normal operations that may indicate a condition exceeding design limits or result in a hazard to persons, property, or the environment.

Bar holes: Holes punched in the ground by use of a probe bar; allow a sampling hose and probe of a combustible gas detector to be inserted to detect gas.

Probe bar: A device used to bore or probe a cylindrical hole in the earth's surface that is large enough to insert the suction probe of gas detection equipment.

This section provides the procedures for conducting a leak survey on a pipeline system. It also provides the prerequisites and competencies required for the procedures, as well as the **abnormal operating conditions (AOCs)** that may be encountered.

> **WARNING!**
> A leak survey utilizing a flame ionization unit is not recommended for a pipeline system transporting propane in a gaseous state.

2.1.0 Prerequisites, Competencies, and Abnormal Operating Conditions

Personnel must possess a variety of skills in order to conduct pipeline surveys. The various operations require a series of steps, which a pipeline technician must be competent to perform. Personnel must also be aware of the particular AOCs that may occur during the procedure and know the proper responses to them.

2.1.1 Prerequisites and Competencies

Personnel performing pipeline surveys using a flame ionization detector must understand the characteristics of natural gas and other gases being transported by pipeline. They must be able to locate the section of pipeline to be surveyed using maps or other documents, and then use a pipe locator to determine the centerline of the buried pipe.

Personnel must know which instrument is best suited for the type of leak survey being conducted. They must also know how to properly calibrate and then operate the instrument in order to perform the survey.

Personnel must understand the hazards of the gas in the line being surveyed, know how to use the appropriate devices to detect leaks, and be able to properly grade leaks according to the company leak-classifying criteria. They must also be aware of and able to report any changes on or near the right-of-way (ROW) that may affect the classification of a section of pipeline, which could affect the frequency of future surveys.

Personnel must know and be able to follow the company's policies and procedures for responding to emergencies, and for the proper use of PPE.

2.1.2 Abnormal Operating Conditions

Personnel conducting a leak survey with a flame ionization unit must be able to identify abnormal operating conditions (AOCs) and know the proper response to each, should they occur. The following are some potential AOCs, and the appropriate action personnel should take in response to them:

- *An unexpected gas release occurs* – Personnel should determine the extent of the leak and act accordingly. Depending on the location and the size of the leak, a hazardous condition may exist that could result in an explosion and/or fire. Appropriate actions may include evacuation of affected buildings or the immediate area, calling 911, and/or contacting the company emergency center.
- *A fire or explosion occurs* – Personnel should evacuate the area immediately and call 911 and the company emergency center.

2.2.0 Conducting a Leak Survey with a Flame Ionization Unit

A flame ionization gas detector is capable of detecting leaking natural gas and other flammable gases. Leaking gas is drawn into the unit through a suction device and then burned in a hydrogen flame. The burning gas gives off ions, which are collected on an electrically charged plate. The current across the collector plate is proportional to the rate of ionization. This ionization rate corresponds to the concentration of gas in the air being sampled. While more conventional gas detectors have a response time of one to two seconds, the flame ionization gas detector operates in milliseconds.

The flame ionization leak survey may be performed by walking the pipeline route with the unit or by driving over the route with the unit mounted on the vehicle. This instrument is more sensitive than any other detector and may read down to one part per million. This increased sensitivity allows the unit to respond more quickly than other instruments, increasing its ability to detect leaks at a normal walking speed or moderate driving speed.

Follow these steps to conduct a leak survey using a flame ionization gas detector:

Step 1 Identify potential abnormal operating conditions that may occur, and know the appropriate actions to take in response to them.

Step 2 Utilize the appropriate personal protective equipment according to relevant company procedures.

Step 3 Identify the portion of the pipeline system to be surveyed.

Step 4 Locate the centerline of the pipe being surveyed to ensure that the area immediately above the pipe is surveyed. Use a calibrated pipe locator if necessary.

Step 5 Turn on and calibrate the flame ionization gas detector. Refer to the instruction manual as necessary in order to ensure proper operation of the device.

NOTE

If the survey is a walking survey, proceed to Step 6. If it is a driving survey, proceed to Step 7.

Step 6 Survey in available openings along the right-of-way (ROW) to check locations where leaking gases might migrate to or places where gases might accumulate.

Step 7 When performing a driving survey, drive the survey vehicle at the speed recommended by company procedures. Ensure that the flame ionization unit is operating properly.

Step 8 Pinpoint leak sources if leak indications are found. A **probe bar** may be used to create **bar holes** to aid in pinpointing underground leakage.

Step 9 Using company criteria, classify the grade of the leak based on percentage of gas, volume, and location. This classification determines the severity of the leak.

Step 10 If the leak classification is critical, follow company procedures to prevent harm to people, property, and the environment.

Step 11 Complete appropriate documentation as required by operator's procedures.

WARNING!

Depending on the location and size of the leak, a hazardous condition may exist that could result in an explosion and/or fire. Determine the extent of the leak and act accordingly. Appropriate actions may include evacuating affected buildings or the immediate area, calling local emergency services, and contacting the company emergency center.

Additional Resources

It is assumed that you will learn to perform these procedures under the direct supervision of your crew leader, foreman, or instructor. For independent study, refer to the recommended Additional Resources at the back of this module.

2.0.0 Section Review

1. Personnel performing a leak survey must understand the hazards of the gas in the line being surveyed, know how to use the appropriate devices to detect leaks, and be able to properly grade leaks according to_____.
   a. the company's leak-classification criteria
   b. the criteria specified in the instrument's instruction manual
   c. *49CFR192.706*
   d. the EPA's leak-classification criteria

2. Bar-holing may be used to _____.
   a. excavate the pipe
   b. withdraw a sample of gas or oil
   c. remove vegetation
   d. pinpoint a leak

SUMMARY

Federal regulations require periodic leak surveys of gas pipelines. The intervals required vary based on the location of the line and the specific type of line. As part of routine ROW inspections or when working on the pipeline route, personnel perform vegetation surveys for any signs of a release. More in-depth surveys may also be conducted with a combustible gas detector or a flame ionization gas detector. When performing any type of leak survey, personnel must remain aware of the possibility of an unexpected gas release. The utmost caution must always be used to protect people, property, and the environment.

Review Questions

1. The highest concentration of a gas that will support combustion is the _____.
   a. maximum explosive limit
   b. explosive limit
   c. lower explosive limit (LEL)
   d. upper explosive limit (UEL)

2. Leak surveys of a transmission line must be conducted at intervals not exceeding 15 months but at least _____.
   a. once per calendar year
   b. twice per calendar year
   c. three times per calendar year
   d. four times per calendar year

3. Regulations dictate the intervals of leak surveys based on _____.
   a. class location
   b. proximity to a business district
   c. type of line
   d. class location, proximity to a business district, and type of line

4. The sensitivity of a flame ionization detector allows for _____.
   a. more outside air to be sampled
   b. a faster leak detection response time
   c. a more thorough leak detection survey
   d. increased accuracy in pinpointing leaks

Trade Terms Quiz

Fill in the blank with the correct term that you learned from your study of this module.

1. The range that will support combustion of a flammable gas, typically between 5 and 15 percent gas in air, is called the ________.
2. Conditions within a pipeline system, identified by the operator, that may indicate (1) a malfunction of a component or (2) a deviation from normal operations that may indicate a condition exceeding design limits or result in a hazard to persons, property, or the environment are called ________.
3. An instrument that is capable of detecting product or crude at very slight concentrations by pulling a volume of the atmosphere being sampled through itself to analyze for product or crude is a(n) ________.
4. The highest concentration of gas in air that will support combustion is the ________.
5. The observation of the growth of normal vegetation over the pipeline in order to detect any changes in the vegetation that may indicate a natural gas leak is a(n) ________.
6. More sensitive than any other combustible gas detector, the instrument that may be read down to one part per million and generally responds within two seconds is called a(n) ________.
7. The lowest concentration of gas in air that will support combustion is the ________.
8. Holes punched in the ground by use of a probe bar; allow a sampling hose and probe of a combustible gas detector to be inserted to detect gas are called ________.
9. A device used to bore or probe a cylindrical hole in the earth's surface that is large enough to insert the suction probe of gas detection equipment is called a(n) ________.

Trade Terms

Abnormal operating conditions (AOCs)
Bar holes
Combustible gas detector
Explosive limits
Flame ionization gas detector
Lower explosive limit (LEL)
Probe bar
Upper explosive limit (UEL)
Vegetation survey

Trade Terms Introduced in This Module

Abnormal operating conditions (AOCs): Conditions within a pipeline system, identified by the operator, that may indicate (1) a malfunction of a component or (2) a deviation from normal operations that may indicate a condition exceeding design limits or result in a hazard to persons, property, or the environment.

Bar holes: Holes punched in the ground by use of a probe bar; allow a sampling hose and probe of a combustible gas detector to be inserted to detect gas.

Combustible gas detector: An instrument that is capable of detecting product or crude at very slight concentrations, generally within 7 to 10 seconds; pulls a volume of the atmosphere being sampled through itself to analyze for product or crude. Two examples of combustible gas detectors include a percent scale, which reads the percent of gas in air, and an LEL scale, which reads the percent of the lower explosive level of gas in air.

Explosive limits: The range that will support combustion of a flammable gas; also called flammable limits. Each gas has its unique range, but typical natural gas limits are 5 to 15 percent gas in air.

Flame ionization gas detector: More sensitive than any other combustible gas detector, this instrument may read down to one part per million and generally responds within two seconds. The atmosphere being sampled is pulled into the unit by a pump. These units may be mounted on vehicles to allow for a mobile survey of the pipeline.

Lower explosive limit (LEL): The lowest concentration of gas in air that will support combustion. Readings below this limit are considered to be too lean in mixture to support combustion.

Probe bar: A device used to bore or probe a cylindrical hole in the earth's surface that is large enough to insert the suction probe of gas detection equipment.

Upper explosive limit (UEL): The highest concentration of gas in air that will support combustion. Readings above this limit are considered to be too rich in mixture to support combustion.

Vegetation survey: Observing the growth of normal vegetation over the pipeline in order to detect any changes in the vegetation that may indicate a natural gas leak.

Additional Resources

The following recommended resources can provide additional helpful information related to the requirements for this covered task.

API Recommended Practice 1161, Third Edition, © 2014 American Petroleum Institute. API Publishing Services, 1220 L Street, NW, Washington, DC 20005. Distributed by Thompson Reuters (Scientific) LLC, **www.techstreet.com**.

49 *CFR* 192, **www.ecfr.gov**.

49 *CFR* 195, **www.ecfr.gov**.

NCCER Module 66101-14, *Introduction to the Pipeline Industry*.

NCCER Module AOCFG-17, *Abnormal Operating Conditions Field & Gas*.

NCCER Module 62105-14, *Introduction to Pipeline Documents*.

NCCER Module 63104-02, *Piping and Mechanical Blueprint Reading*.

NCCER Module CT52_1-17, *Conduct Vegetation Survey*.

NCCER Module CT52_2-17, *Conduct a Leak Survey with a Combustible gas detector*.

Section Review Answer Key

Answer	Section Reference	Objective
Section One		
1. b	1.0.0	1
2. d	1.0.0	1
Section Two		
1. a	2.1.1	2a
2. d	2.2.0	2b

NCCER CURRICULA — USER UPDATE

NCCER makes every effort to keep its textbooks up-to-date and free of technical errors. We appreciate your help in this process. If you find an error, a typographical mistake, or an inaccuracy in NCCER's curricula, please fill out this form (or a photocopy), or complete the online form at **www.nccer.org/olf**. Be sure to include the exact module ID number, page number, a detailed description, and your recommended correction. Your input will be brought to the attention of the Authoring Team. Thank you for your assistance.

Instructors – If you have an idea for improving this textbook, or have found that additional materials were necessary to teach this module effectively, please let us know so that we may present your suggestions to the Authoring Team.

NCCER Product Development and Revision

13614 Progress Blvd., Alachua, FL 32615

Email: curriculum@nccer.org

Online: www.nccer.org/olf

❑ Trainee Guide ❑ Lesson Plans ❑ Exam ❑ PowerPoints Other ________________

Craft / Level: Copyright Date:

Module ID Number / Title:

Section Number(s):

Description:

Recommended Correction:

Your Name:

Address:

Email: Phone:

CT59_0-17

Vault Maintenance

Overview

Designated pipeline personnel are often required to enter confined spaces on the pipeline to perform routine pipeline tasks such as inspections, maintenance, and repairs. Entry into confined spaces requires a detailed understanding of the hazards involved. This module describes confined-space entry hazards and the associated safety requirements. It also describes the tasks involved in one type of inspection that requires adherence to confined-space entry regulations: the maintenance of vaults.

Trainees with successful module completions may be eligible for credentialing through the NCCER Registry. To learn more, go to **www.nccer.org** or contact us at 1.888.622.3720. Our website has information on the latest product releases and training, as well as online versions of our *Cornerstone* magazine and Pearson's product catalog.

Your feedback is welcome. You may email your comments to **curriculum@nccer.org**, send general comments and inquiries to **info@nccer.org**, or fill in the User Update form at the back of this module.

This information is general in nature and intended for training purposes only. Actual performance of activities described in this manual requires compliance with all applicable operating, service, maintenance, and safety procedures under the direction of qualified personnel. References in this manual to patented or proprietary devices do not constitute a recommendation of their use.

CT59_0-17
Vault Maintenance

Objectives

When you have completed this module, you will be able to do the following:

1. Explain the hazards associated with confined-space entry.
   a. Identify the types of confined spaces.
   b. Explain the use of ventilation in confined spaces.
   c. Explain the hazards associated with flammable atmospheres.
   d. Explain the hazards associated with toxic atmospheres.
   e. Explain the hazards associated with irritant atmospheres.
   f. Explain the hazards associated with asphyxiating atmospheres.
   g. Explain other general/physical hazards encountered during confined-space entry.
2. Explain the requirements for confined-space entry.
   a. Explain the permits required for confined-space entry.
   b. Explain the safety requirements for confined-space entry.
   c. Explain the duties and responsibilities of personnel required for confined-space entry.
   d. Explain the emergency preparation required for confined-space entry.
3. Explain how to perform vault maintenance (CT59_0-17).
   a. Explain safety requirements for vault inspections.
   b. Explain vault maintenance procedures.

Performance Task

Under the supervision of your instructor, you should be able to do the following:

1. Perform vault maintenance (CT59_0-17).
   - Identify potential abnormal operating conditions (AOCs) that may occur during performance of this CT, and know the appropriate actions to take in response to them.
   - Utilize the appropriate personal protective equipment (PPE) according to relevant company procedures.
   - Test the atmosphere around the vault for the presence of a flammable or toxic atmosphere.
   - Before entry, test the atmosphere at various locations inside the vault for a flammable or toxic atmosphere before removing the vault cover.
   - Enter the vault and inspect for any physical damage to the piping leaving or entering the vault.
   - If applicable, inspect the vault ventilation equipment for proper operation.
   - Inspect the vault for proper ventilation.
   - Inspect the vault cover for adequate load support.
   - Complete appropriate documentation as required by operator's procedures.

Trade Terms

Abnormal operating condition (AOC)
Combustible gas indicator (CGI)
Confined space
Hazardous atmosphere
Hydrogen sulfide (H_2S)
Occupational Safety and Health Administration (OSHA)
Oxygen-deficient atmosphere
Safety data sheet (SDS)
Vault

Industry Recognized Credentials

If you're training through an NCCER-accredited sponsor, you may be eligible for credentials from NCCER's Registry. The ID number for this module is CT59_0-17. Note that this module may have been used in other NCCER curricula and may apply to other level completions. Contact NCCER's Registry at 1.888.622.3720 or go to **www.nccer.org** for more information.

Contents

Figures and Tables

SECTION ONE

1.0.0 CONFINED-SPACE ENTRY HAZARDS

Objective

Explain the hazards associated with confined-space entry.

a. Identify the types of confined spaces.
b. Explain the use of ventilation in confined spaces.
c. Explain the hazards associated with flammable atmospheres.
d. Explain the hazards associated with toxic atmospheres.
e. Explain the hazards associated with irritant atmospheres.
f. Explain the hazards associated with asphyxiating atmospheres.
g. Explain other general/physical hazards encountered during confined-space entry.

Trade Terms

Combustible gas indicator (CGI): An instrument used to measure the amount of combustible gas in air.

Confined space: Defined by OSHA as a space that is not intended for human occupancy; has limited means of entry and exit; may be difficult to evacuate quickly due to size, shape, or contents; and may present other hazards to anyone who enters.

Hydrogen sulfide (H_2S) : A colorless gas that is heavier than air and very poisonous, corrosive, and flammable.

Occupational Safety and Health Administration (OSHA): An agency of the US Department of Labor that regulates workplace safety and health. OSHA sets standards and provides training, education, and outreach.

Oxygen-deficient atmosphere: Atmosphere that contains less than 19.5 percent oxygen and is therefore unsafe for workers without a respirator.

Vault: An underground structure having a volumetric internal content of 200 cubic feet or more that contains pressure-regulating and pressure-limiting equipment.

A **confined space** refers to a relatively small or restricted space, such as a pit, storage tank, sewer, casing, or silo. Trenches can also be confined spaces. A **vault** is an underground installation on the pipeline where valves, meters, and other equipment attached to the pipeline are accessed; vaults generally meet the definition for confined spaces. The size of the vault can vary from closet to room size. As with all pipeline installations, the vault and the equipment housed within it require routine inspections and maintenance.

The definitions for confined spaces are outlined by the **Occupational Safety and Health Administration (OSHA)**. OSHA's standard defines a confined space as a space that:

- Is large enough and so configured that an employee can bodily enter and perform assigned work
- Has a limited or restricted means of entry or exit, for example, a tank, vessel, silo, storage bin, hopper, vault, or pit
- Is not designed for continuous employee occupancy

OSHA further defines a permit-required confined space as a space that meets any one of the following criteria:

- Contains or has the potential to contain a hazardous atmosphere
- Contains a material that has the potential for engulfing an entrant
- Has an internal configuration that might cause an entrant to be trapped or asphyxiated by inwardly converging walls or by a floor that slopes downward and tapers to a smaller cross section
- Contains any other recognized serious safety or health hazard

Working in confined spaces can be extremely dangerous if the proper precautions and safety procedures are not followed. Confined-space hazards include explosion, poisoning, and asphyxiation. Personnel must be prepared to encounter any associated hazards, and must follow all ventilation requirements and other safety requirements.

1.1.0 Confined-Space Types and Hazards

Confined spaces are categorized generally as spaces with open tops and a depth that will restrict the natural movement of air, as well as enclosed spaces with very limited openings for entry. In either case, the space may contain mechanical equipment with moving parts.

Trenches, degreasers, pits, and certain types of storage tanks, which may be classified as open-topped confined spaces, usually contain no moving parts. Although the risk of explosion is reduced when a confined space contains no moving parts, gases that are heavier than air, such as butane, propane, and other hydrocarbons, will remain in depressions and will flow to low points where they are difficult to remove, so asphyxiation is possible.

Other hazards may develop due to the work performed in the confined space or because of corrosive residues that accelerate the decomposition of scaffolding supports and electrical components. Confined spaces such as sewers, casings, tanks, silos, vaults, and compartments of ships usually have limited access, which increases the risk of injury.

Hazards specific to a confined space are dictated by the following:

- The material stored or used in the confined space, for example, damp activated carbon in a filtration tank will absorb oxygen, thus creating an oxygen-deficient atmosphere
- The activity carried out, such as fermentation of molasses that creates ethyl-alcohol vapors and decreases the oxygen content of the atmosphere
- The external environment, as in the case of sewer systems that may be affected by high tides, heavier-than-air gases, or flash floods

The most hazardous kind of confined space is that in which there is limited access and the presence of mechanical devices. Digesters and boilers usually contain power-driven equipment that may be inadvertently activated after entry unless properly isolated.

1.2.0 Ventilation

Before forced ventilation is initiated, information regarding restricted areas within the confined space, the existence of voids, the nature of contaminants present, and the size of the space needs to be obtained. The type of work to be done and the number of persons involved also needs to be determined.

1.2.1 Ventilation Air

Ventilation air should not create an additional hazard due to recirculation of contaminants, improper arrangement of the inlet duct, or by the substitution of anything but normal fresh air. Normal fresh air is defined as consisting of approximately:

- 20.9 percent oxygen
- 78.1 percent nitrogen
- 1 percent argon
- Small amount of other gases

OSHA defines atmosphere with normal levels of oxygen as atmosphere containing between 19.5 percent and 23.5 percent oxygen.

The terms *air* and *oxygen* are sometimes considered to be synonymous. However, this is a dangerous assumption. The substitution of oxygen for fresh air for ventilation in a confined space will expand the limits of flammability and increase the hazards of fire and explosion.

1.2.2 Ventilation Means

A positive means of ventilation must be established and operated the entire time the work is in progress. The accepted practice and recommended level of ventilation is one complete change of air every three minutes. The ventilation system must be positioned upwind so that only clean, fresh air is drawn into a vessel or space, and must be arranged to allow for complete purging of the space. The exhausted air must be positioned so that it will not be a hazard to other personnel or equipment. A simple ventilation system is shown in *Figure 1*.

It may be necessary to provide local exhausting in addition to forced ventilation when the work may generate fumes or vapors. Welding is one example of the need for local exhausting. This exhaust should also be arranged to avoid creating a hazard to others. *Figure 2* shows a local exhaust arrangement.

Natural ventilation may or may not be sufficient to achieve adequate exchange of fresh air in a space. There could possibly be dead zones in the circulation. When using natural ventilation, continuous air monitoring is recommended.

CAUTION

When using natural ventilation methods, always perform continuous air monitoring.

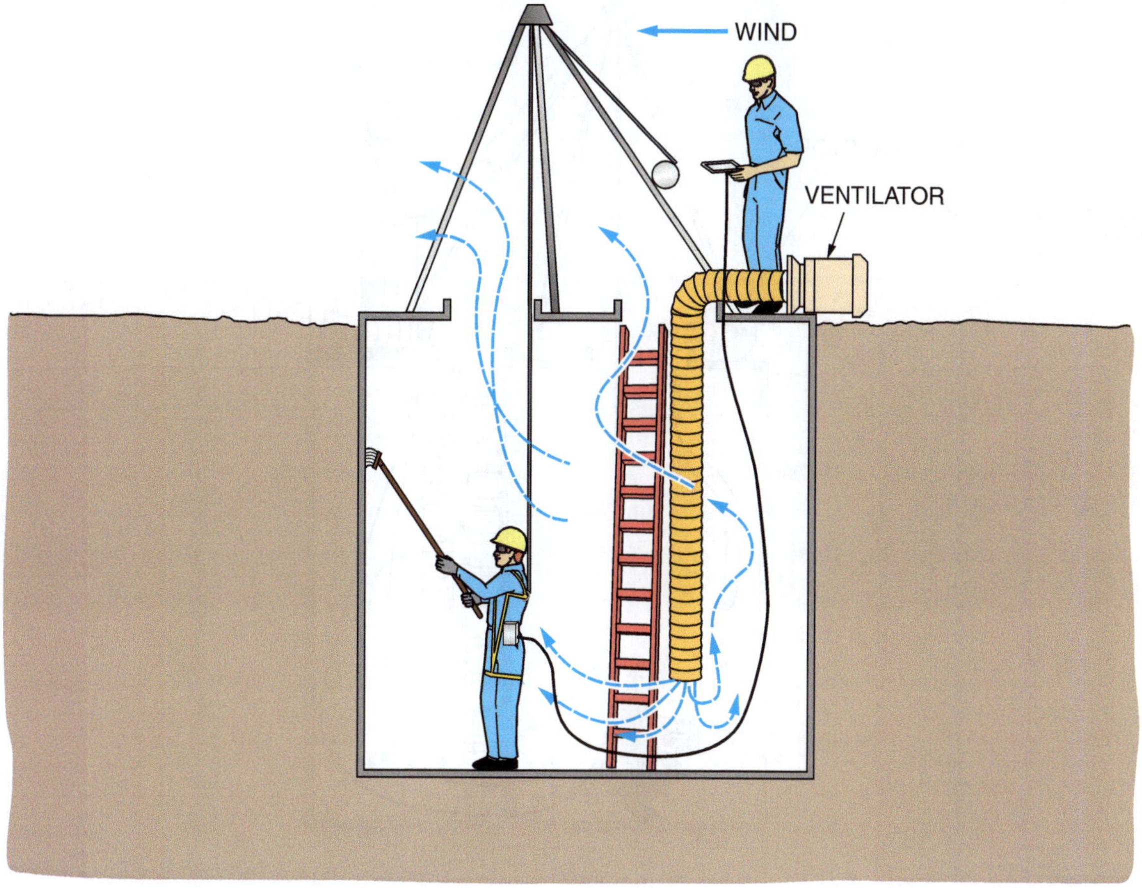

Figure 1 Simple ventilation system.

Air movers should be placed at the top of the vessel or space when vapors that are lighter than air may be present. More than one air mover and fresh air inlet should be used when possible, to ensure that adequate ventilation is occurring. Upward forced ventilation is illustrated in *Figure 3*.

Air movers should be placed at the bottom of a vessel or space when vapors heavier than air may be present. When this system is used, be certain the exhausted air does not become a hazard for other area personnel. Downward forced ventilation is illustrated in *Figure 4*.

An air horn, a type of air mover, should not be positioned to blow into a vessel or space. The use of an air hose to ventilate a vessel or space is prohibited. Equipment is available for ventilation systems that minimize the restriction that may occur when the worker must enter the opening used for ventilation. A manhole would be an example. This equipment would also provide for access in emergencies. *Figure 5* shows retrieval of a worker from a space containing ventilating equipment.

Retrieval equipment should be used whenever rescue may be difficult, there is the possibility of a fall, or the entrant is using self-contained or supplied breathing air.

1.3.0 Flammable Atmospheres

Flammable atmospheres generally arise from enriched oxygen atmospheres, vaporization of flammable liquids, byproducts of work, chemical reactions, concentrations of combustible dusts, and desorption of chemicals from the inner surfaces of the confined space.

Flammable gases such as acetylene, butane, propane, hydrogen, methane, and natural or manufactured gases or vapors from liquid hydrocarbons can be trapped in confined spaces. Since many gases are heavier than air, these will seek lower levels as in pits, sewers, and various types of storage tanks and vessels.

In a closed-top tank, lighter-than-air gases will rise and collect in the top to produce a flammable concentration if trapped above the opening. The

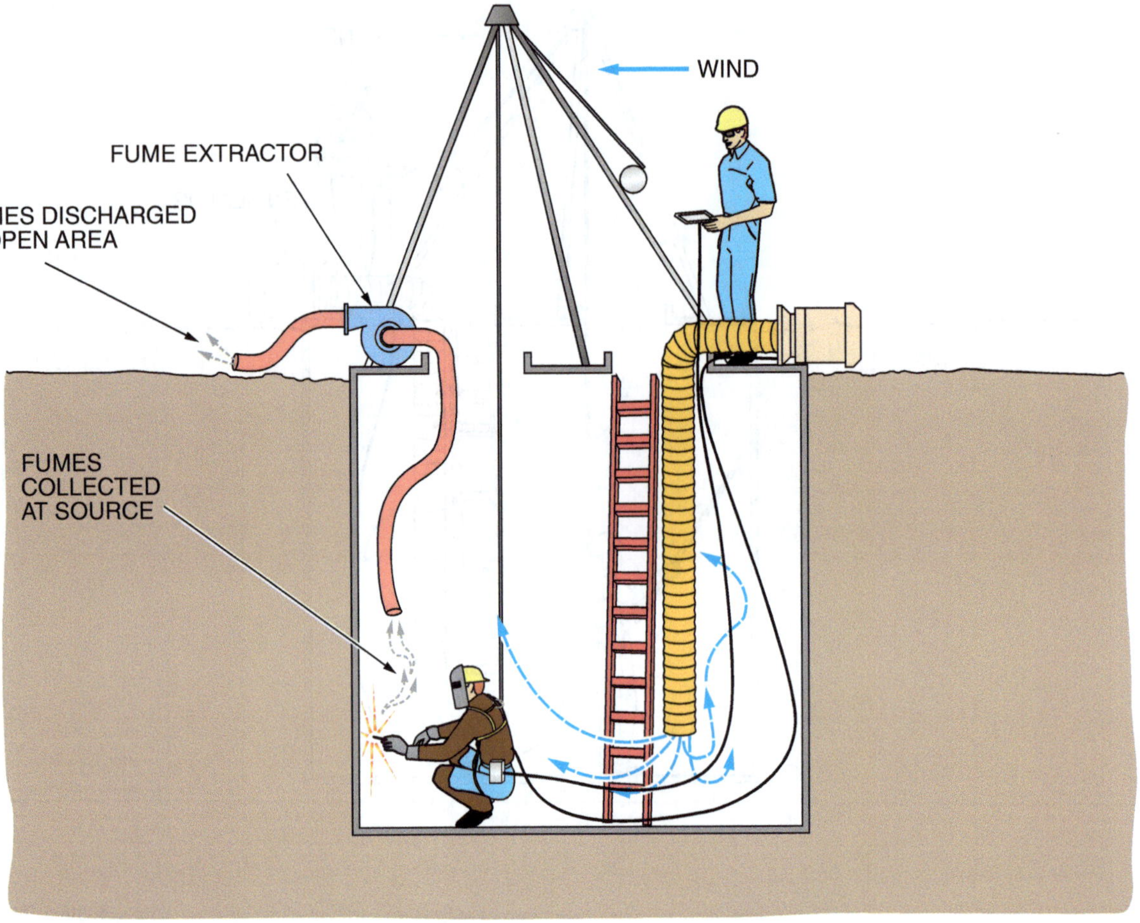

Figure 2 Local exhaust arrangement.

byproducts of work procedures such as spray painting can generate flammable or explosive conditions within a confined space. Welding in a confined space is a major cause of explosions in areas that contain combustible gases.

Combustible dust concentrations are usually found during the processes of loading, unloading, and conveying grain products, nitrated fertilizers, finely ground chemical products, and any other combustible material.

Warning:

Atmosphere with an oxygen level above 23.5 percent is considered enriched, and can result in easy ignition of flammable material such as clothes.

1.4.0 Toxic Atmospheres

The sources of toxic atmospheres that may be encountered in confined spaces include the following:

- *The manufacturing process* – For example, in producing polyvinyl chloride, hydrogen chloride is used as well as vinyl chloride monomer, which is carcinogenic.
- *The product stored* – Removing decomposed organic material from a tank can release toxic substances such as **hydrogen sulfide (H_2S)**.
- *The operation performed in the confined space* – This may include welding or brazing with metals capable of producing toxic fumes.

Mechanical or human error during loading, unloading, and production may also be a source of toxic gases that are not part of the planned operation.

Carbon monoxide (CO) is a major hazard in confined spaces. It is a highly toxic gas that can build up in such spaces and is odorless, colorless, and not detectable by a **combustible gas indicator (CGI)**, also known as a combustible gas detector (CGD). CO must be tested for specifically.

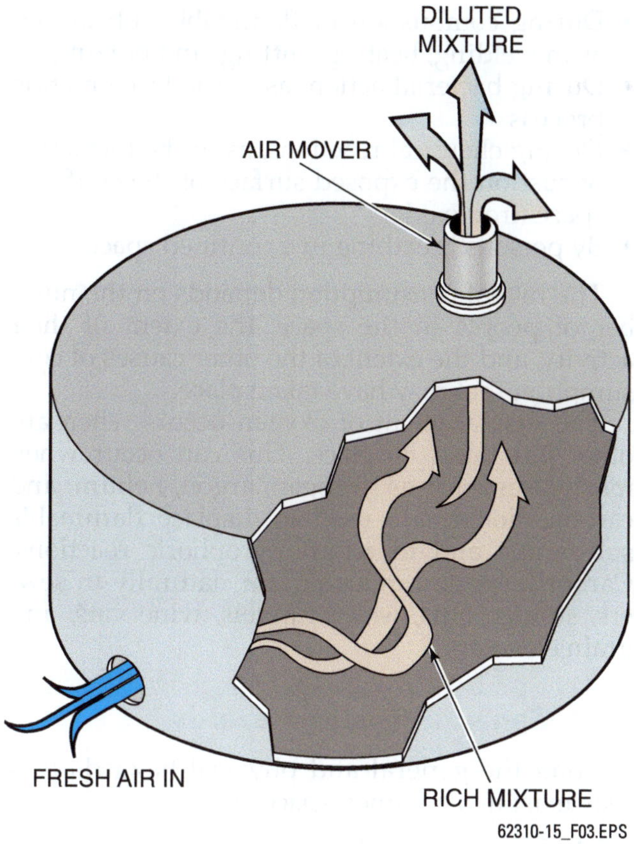

Figure 3 Upward forced ventilation.

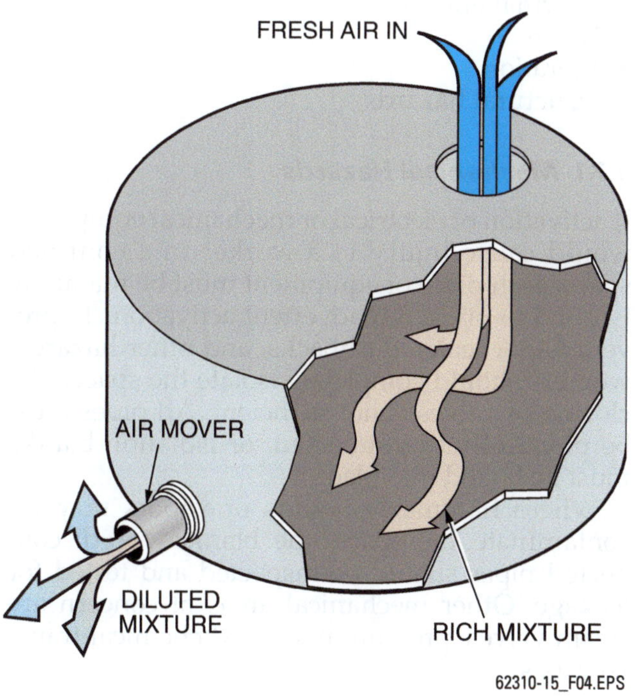

Figure 4 Downward forced ventilation.

Figure 5 Worker retrieval past ventilating equipment.

CO is approximately the same density as air and is formed from incomplete combustion of organic materials such as wood, oil, coal, gas, and gasoline. It can also be formed by microbial decomposition of organic matter in sewers, silos, and fermentation tanks. However, the formation of CO may result from chemical reactions or work activities, and fatalities due to CO poisoning are not confined to particular industries.

Early stages of CO intoxication are nausea and headache. CO may be fatal at 1,000 parts per million (ppm) in air and is considered dangerous even at 200 ppm in air. It forms carboxyhemoglobin in the blood, which prevents the distribution of oxygen in the body.

1.5.0 Irritant Atmospheres

Irritant or corrosive atmospheres can be divided into primary and secondary groups:

- The primary irritants exert no systemic toxic effects (toxic effects on the entire body). Examples include chlorine, ozone, hydrochloric acid, hydrofluoric acid, sulfuric acid, nitrogen dioxide, ammonia, and sulfur dioxide.
- A secondary irritant may produce systemic toxic effects in addition to surface irritation. Examples include benzene, carbon tetrachloride, ethyl chloride, trichloroethane, trichloroethylene, and chlor propene.

Prolonged exposure to irritant or corrosive concentrations in a confined space may produce little or no evidence of irritation, but may result in a general weakening of the defense reflexes from changes in sensitivity. The danger in this situation is that the worker is usually not aware of any increase in exposure to toxic substances.

Irritant gases vary widely among all areas of industrial activity. They can be found in plastics plants, the petroleum industry, tanneries, refrigeration industries, paint manufacturing, mining operations, and others.

1.6.0 Asphyxiating Atmospheres

The normal atmosphere is composed of oxygen, nitrogen, argon, and other gases. The normal range for oxygen levels in the atmosphere is between 19.5 percent and 23.5 percent. *Table 1* shows the various effects of decreased oxygen levels on workers.

A reduction of the oxygen in a confined space may be the result of either consumption or displacement. The consumption of oxygen takes place:

Table 1. Effects of Decreased Oxygen Levels in the Atmosphere

Oxygen Level	Effects
19.5%–23.5%	Normal
17%	Deterioration of night vision, increased breathing volume, accelerated heartbeat
14%–16%	Very poor muscular coordination, rapid fatigue, intermittent respiration
6%–10%	Nausea, vomiting, inability to perform, unconsciousness
< 6%	Spasmodic breathing, convulsive movements, and death in minutes

- During combustion of flammable substances, as in welding, heating, cutting, and brazing
- During bacterial action, as in the fermentation process
- During chemical reactions, as in the formation of rust on the exposed surface of the confined space (iron oxide)
- By persons breathing in a confined space

The rate of consumption depends on the number of people in the space, the extent of their activity, and the extent of the other causes of consumption that may have taken place.

The displacement of oxygen occurs when another gas takes its place. This can occur when inert gases such as nitrogen, argon, helium, and carbon dioxide are used to displace flammable substances and to retard pyrophoric reactions. Carbon dioxide can also occur naturally in sewers, storage bins, wells, tunnels, wine vats, and grain elevators.

1.7.0 General/Physical Hazards

Among the general and physical hazards that may exist in a confined space are:

- Mechanical hazards
- Communication problems
- Entry and exit time
- Thermal effects
- Noise
- Vibration
- Structural hazards

1.7.1 Mechanical Hazards

If activation of electrical or mechanical equipment would cause injury to a worker in a confined space, each piece of equipment must be manually isolated to prevent inadvertent activation. To prevent vapor leaks, flashbacks, and other hazards, workers should completely isolate the space. The closing of valves is not sufficient. All pipes must be physically disconnected, or isolation blanks must be bolted into place.

Where flammable liquids or vapors may recontaminate the space, the blanked or disconnected pipes should be inspected and tested for leakage. Other mechanical areas of concern are steam valves, pressure lines, and chemical-transfer pipes.

1.7.2 Communication Problems

Communication between the worker inside the confined space and the standby person is of the utmost importance. When visual monitoring of the worker is not possible because of the design of the confined space or location of the entry hatch, a voice- or alarm-activated explosion-proof type of communication system is necessary. Suitable, approved illumination that provides sufficient visibility for work is also required.

1.7.3 Entry and Exit Time

Entry and exit time is of major significance as a physical limitation, and is directly related to the potential hazard of the confined space. The extent of precautions taken and the standby equipment needed to maintain a safe work area will be determined by the means of access and rescue. The following factors are considered:

- Type of confined space to be entered
- Access to the entrance
- Number and size of openings
- Barriers within the space
- Occupancy load
- Time requirement for exiting in event of fire or vapor incursion
- Time required to rescue injured workers

1.7.4 Thermal Effects

The thermal effects in a confined space may be either from heat or from cold. When a body temperature of approximately 102°F (38.9°C) is exceeded, workers are less efficient and are prone to heat exhaustion, heat cramps, and heat stroke.

In cold environments, workers may be subject to frostbite, trench foot, and general hypothermia. Protective insulated clothing for both hot and cold environments will add additional bulk to the worker, and must be considered in allowing for movement in the confined space and for exit time.

1.7.5 Noise

Noise problems are usually intensified in confined spaces because the interior tends to cause sound to reverberate and thus to expose the worker to higher sound levels than those found in an open environment.

This intensified noise increases the risk of hearing damage to workers, which could result in temporary or permanent loss of hearing. Noise in a confined space that may not be intense enough to cause hearing damage may still disrupt verbal communication with the emergency standby person on the exterior of the confined space.

1.7.6 Vibration

Whole-body vibration may affect multiple body parts and organs, depending upon the vibration characteristics. Segmental vibration, unlike whole-body vibration, appears to be more localized in creating injury to the fingers and hands of workers using tools such as pneumatic hammers, rotary grinders, and other hand tools that cause vibration.

1.7.7 Structural Hazards

Other hazards in confined spaces may include scaffolding, surface residue, and other structures that might not be able to be eliminated because of the nature of the confined space or the work to be performed. The use of scaffolding in confined spaces has contributed to many accidents caused by workers or materials falling, improper use of guardrails, and lack of maintenance to ensure worker safety.

Surface residues in confined spaces can increase the already-hazardous possibility of electrical shock, reaction of incompatible materials, release of toxic substances, and bodily injury due to slips and falls. Without protective clothing, additional health hazards may arise due to surface residues.

Structural hazards within a confined space, such as baffles in horizontal tanks, trays in vertical towers, bends in tunnels, or overhead structural members, constitute physical hazards that are intensified because of the physical surroundings.

Rescue procedures may require withdrawal of an injured or unconscious person. Careful planning must be given to the relationship between the internal structure, the exit opening, and the worker.

Additional Resources

It is assumed that you will learn to perform these procedures under the direct supervision of your crew leader, foreman, or instructor. For independent study, refer to the recommended Additional Resources at the back of this module.

1.0.0 Section Review

1. Confined spaces are generally categorized as spaces with open tops and a depth that will restrict the natural movement of _____.

   a. air
   b. personnel
   c. equipment
   d. tools

2. The terms *air* and *oxygen* mean the same thing. True or false?

   a. True
   b. False

3. In a confined space that contains combustible gases, a major source of explosions is _____.

   a. welding
   b. site layout
   c. dust concentrations
   d. mechanical errors

4. Carbon monoxide is approximately the same density as _____.

   a. water
   b. air
   c. CO_2
   d. oxygen

5. Primary irritants exert no systemic toxic effects. True or false?

   a. True
   b. False

Section Two

2.0.0 Confined-Space Entry Requirements

Objective

Explain the requirements for confined-space entry.

a. Explain the permits required for confined-space entry.
b. Explain the safety requirements for confined-space entry.
c. Explain the duties and responsibilities of personnel required for confined-space entry.
d. Explain the emergency preparation required for confined-space entry.

Trade Terms

Hazardous atmosphere: An atmosphere capable of causing death, illness, or harm by being explosive, flammable, toxic, or oxygen deficient.

Safety data sheet (SDS): A form containing information on a potentially hazardous material, including names, contents, test results, characteristics, potential hazards, and handling procedures.

Confined-space entry requires a detailed understanding of all safety requirements. Because of the potential hazards involved, many confined spaces require a permit and specific safety provisions. In all cases of confined-space entry, the importance of continuous atmospheric monitoring cannot be stressed enough. Always follow company-specific policies and procedures for confined-space entry.

2.1.0 Permits

OSHA defines a permit-required confined space as one that meets the definition of a confined space and has one or more of the following characteristics:

- Contains or has the potential to contain a hazardous atmosphere
- Contains a material that has the potential for engulfing an entrant
- Has an internal configuration that might cause an entrant to be trapped or asphyxiated by inwardly converging walls or by a floor that slopes downward and tapers to a smaller cross section
- Contains any other recognized serious safety or health hazards

Employers must evaluate the workplace to determine if spaces are permit-required confined spaces. A flowchart can be used to assist with this determination.

2.1.1 Permit System

A permit, signed by the entry supervisor and verifying that pre-entry preparations have been completed and the space is safe to enter, must be posted at entrances or otherwise made available to entrants before they enter a permit space. Duration of permits must not exceed the time required to complete an assignment.

The entry supervisor must terminate entry and cancel permits when an assignment has been completed or when new conditions exist. New conditions must be noted on the canceled permit and used in revising the permit-space program. All canceled permits are kept for one year.

2.1.2 Entry Permits

The entry permit must contain the following information:

- Atmospheric test results
- Tester's initials or signature
- Name and signature of supervisor authorizing the entry
- Name of permit space to be entered, authorized entrants, eligible attendants, and individuals authorized to be entry supervisors
- Purpose of entry and known space hazards
- Measures to be taken to isolate permit spaces and to eliminate or control space hazards

> **Emergency Rescue**
>
> Confined-space entry may be required for emergency rescue of personnel. The standby person and all rescue personnel should be aware of the structural design of the space, emergency exit procedures, and life support system required. Always follow company-specific policies and procedures regarding confined-space entry in emergency situations.

- Name and telephone numbers of rescue and emergency services
- Date and authorized duration of entry
- Acceptable entry conditions
- Communications procedures and equipment to maintain contact during entry
- Additional permits, such as for hot work, that have been issued to authorize work in the permit space
- Special equipment and procedures including personal protective equipment (PPE) and alarm systems
- Any other information needed to ensure personnel safety

A sample entry permit is shown in *Figure 6A* and *Figure 6B*. Some companies may require personnel to use a confined-space entry checklist along with the entry permit.

2.2.0 Safety Requirements

If there are permit-required spaces in the workplace, employers must inform personnel who could be exposed of the existence, location, and danger posed by the spaces. This can be accomplished by posting danger signs or by another equally effective means. An example of a confined-space entry warning sign is shown in *Figure 7*.

If personnel are not to enter and work in permit spaces, employers must take effective measures to prevent their personnel from entering the permit spaces. If personnel are to enter permit spaces, the employer must develop a written permit-space program, which shall be made available to personnel or their representatives.

2.2.1 Alternate Procedures

Under certain conditions, the employer may use alternate procedures for worker entry into a permit space. For example, if employers can demonstrate with monitoring and inspection data that the only hazard is an actual or potential hazardous atmosphere, which can be made safe for entry by use of continuous forced-air ventilation and monitoring, they may be exempted from some requirements, such as permits and attendants. Even in such circumstances, however, the internal atmosphere of the space must be tested before any personnel enter, first for oxygen content, second for flammable gases, and third for potential toxic air contaminants.

2.2.2 Written Program

Among other things, the OSHA standard requires the employer's program to:

- Identify and evaluate permit-space hazards before allowing personnel entry
- Test conditions in the permit space before entry operations and monitor the space during entry
- Perform appropriate testing for atmospheric hazards in the following sequence: oxygen content, combustible gases or vapors, and toxic gases or vapors
- Implement necessary measures to prevent unauthorized entry
- Establish and implement the means, procedures, and practices to eliminate or control hazards necessary for safe permit-space entry operations. This may include:

  - Specifying acceptable entry conditions
  - Isolating the permit space
  - Providing barriers
  - Verifying acceptable entry conditions
  - Purging, flushing, or ventilating the permit-space, or making the permit space inert

- Identify personnel job duties
- Provide, maintain, and require, at no cost to personnel, the use of personal protective equipment (PPE) and any other equipment necessary for safe entry, which may include protective clothing, masks, and hard hats, in addition to testing, monitoring procedures, ventilation, communications, and lighting equipment; and barriers, shields, and ladders
- Ensure that at least one attendant is stationed outside the permit space for the duration of entry operations
- Coordinate entry operations when personnel of more than one employer are to be working in the permit space
- Implement appropriate procedures for summoning rescue and emergency services
- Establish, in writing, and implement a system for the preparation, issuance, use, and cancellation of entry permits
- Review established entry operations and annually revise the permit-space entry program
- When an attendant is required to monitor multiple spaces, implement the procedures to be followed during an emergency in one or more of the permit spaces being monitored

1. Work Description

Area_____________________ Equipment Location_____________________

Work to be done:

2. Gas Test	☐ Instrument Check	Results	Recheck	Recheck
Required	☐			
☐ Yes	☐ Oxygen % 20.8 Min.			
☐ No	☐ Combustible % LFL			
		Date/Time/Sig.	Date/Time/Sig.	Date/Time/Sig.

3. Special Instructions: ☐ Check issuer before beginning work ☐ None

4. Hazardous Materials: ☐ None What did the line/equipment last contain?

5. Special Protection Required: ☐ None ☐ Forced Air Ventilation

☐ Avoid Skin Contact ☐ Gloves ______________ ☐ Suit ________________

☐ Goggles or Face Shield ☐ Respirator _________________ ☐ Safety Harness

☐ Self-Contained Breathing Equipment ☐ Hose-line Breathing Equipment

☐ Other, Specify ________ ☐ Standby – Name: ______________________________

6. Fire Protection Required: ☐ None ☐ Portable Fire Extinguisher ☐ Fire Hose and Nozzle

☐ Fire Watch ☐ Other, Specify:

7. Condition of Area and Equipment

Required Yes	No	THESE KEY POINTS MUST BE CHECKED
		a. Lines disconnected & blinded or where disconnecting is not possible, blinds installed? (Includes drains, vents, and instrument leads) and appropriate valves locked out?
		b. Equipment cleaned, washed, purged, ventilated?
		c. Low-voltage or GFCI-protected electrical equipment provided?
		d. Explosion-proof electrical equipment provided?
		e. Lifelines required to be attached to safety harnesses?

Comments

rev.6/10/2015

62310-15_F06A.EPS

Figure 6A First page of a sample entry permit.

2.2.3 Training and Education

Employers must provide proper training for all employees who are required to work in permit spaces before each employee's initial work assignment begins. Upon completing this training, employers must ensure that personnel have acquired the understanding, knowledge, and skills necessary for the safe performance of their duties.

Additional training is required when any of the following are true:

- The job duties change.
- There is a change in the permit-space program or the permit-space operation presents a new hazard.
- A worker's job performance shows deficiencies.

8. Approval	Date	Time	Permit Authorization Area Supv.	Permit Acceptance Maint./Contractor Supv./Engineer	Date	Time
Issued by						
Endorsed by						
Endorsed by						
Endorsed by						

9. Individual Review

I have been instructed on proper Safety Procedures and proper Confined Space Entry Procedures. I have signed in on the appropriate Master Card and have affixed personal locks on energy isolation devices as appropriate.

Signatures of all personnel covered by this permit

Forward to Production Superintendent 7 days after completion of work.

rev.6/10/2015

62310-15_F06B.EPS

Figure 6B Second page of a sample entry permit.

Training is also required for rescue team members, including cardiopulmonary resuscitation (CPR) and first aid training.

Upon completion of training, personnel must receive a certificate that includes the worker's name, signatures or initials of trainers, and the dates of training. Personnel must also be trained in their assigned duties.

2.3.0 Duties and Responsibilities

Every person involved with the permit-space entry has assigned duties and specific responsibilities. These personnel include the following:

- Authorized entrants
- Entry attendant
- Entry supervisor

62310-15_F07.EPS

Figure 7 Confined-space entry warning sign.

2.3.1 Authorized Entrants

The workers who will be entering the space must know the hazards, including information on the mode of exposure, and the signs, symptoms, and consequences of exposure. When using natural ventilation methods, continuous atmospheric monitoring should be used. Workers must be trained in the use of, as well as properly use PPE as required for the particular workspace. An important component of PPE for confined-space entry is a respirator. Some commonly used respirators are shown in *Figure 8*.

> **CAUTION**
>
> When using natural ventilation methods, always perform continuous atmospheric monitoring.

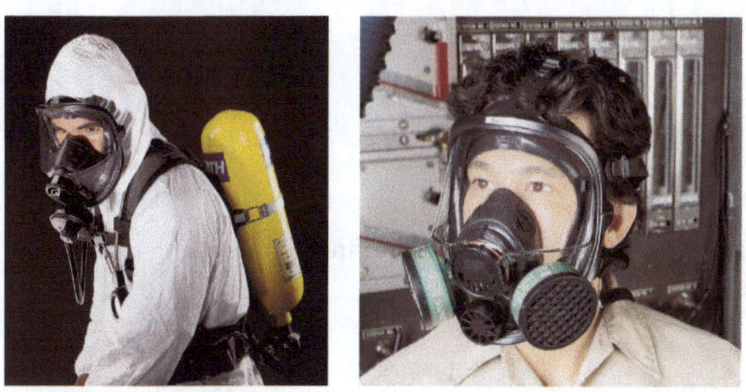

SELF-CONTAINED BREATHING APPARATUS (SCBA)

FULL FACEPIECE MASK

62310-15_F08.EPS

Figure 8 Respirators used for confined-space entry.

Other duties include the following:

- Entrants must maintain communication with the attendants to enable the attendants to monitor the entrants' status as well as to alert the entrants to evacuate, as necessary.
- Entrants must exit from the space as soon as possible when ordered to do so by an authorized person, when the entrant recognizes that the signs or symptoms of exposure exist, when a prohibited condition exists, or when an automatic alarm is activated.
- When an entrant recognizes the existence of a prohibited condition or signs of exposure, the entrant must alert the attendant immediately.

2.3.2 Entry Attendant

The duties of the attendant during permit-space entry operations include the following. The attendant must:

- Remain outside the permit space during operations unless relieved by another authorized attendant
- Perform no-entry rescues when specified by employer's rescue procedures
- Know the existing and potential hazards, including information on the mode of exposure, signs or symptoms of exposure, and the consequences and physiological effects of exposure
- Maintain communication with and keep an accurate account of those workers entering the permit-required space
- Order evacuation of the permit space when a prohibited condition exists, when a worker shows signs of physiological effects of hazardous exposure, when an emergency outside the confined space exists, and when the attendant cannot effectively and safely perform required duties
- Summon rescue and other services during an emergency
- Ensure that unauthorized persons stay away from permit spaces or exit immediately if they have entered the space
- Inform authorized entrants and the entry supervisor of entry by unauthorized persons
- Perform no other duties that interfere with the attendant's primary duties

2.3.3 Entry Supervisor

The entry supervisor, like everyone else, must know the space hazards, existing or potential, including mode of exposure, signs or symptoms of exposure, and the consequences of exposure. The supervisor must also verify emergency plans and specified entry conditions such as permits, tests, procedures, and equipment before allowing entry. In addition, the entry supervisor must:

- Terminate entry and cancel permits when entry operations are completed or if a new condition exists
- Take appropriate measures to remove unauthorized entrants
- Ensure that entry operations remain consistent with the entry permit and that acceptable entry conditions are maintained

2.4.0 Emergency Preparation

To prepare for possible emergencies, the employer must ensure that rescue service personnel are provided with and trained in the proper use of personal protective and rescue equipment, including respirators. They must be trained to perform assigned rescue duties and must also have had authorized-entrant training. All rescuers must be trained in first aid and in CPR, and at a minimum, one team member must be currently certified in first aid and CPR.

The employer must also ensure that practice rescue exercises are performed yearly and that rescue services are provided access to permit spaces so that they can practice rescue operations. Rescuers must be informed of the hazards in permit spaces.

When appropriate, authorized entrants who enter a permit space must wear a chest or full-body harness with a retrieval line attached to the center of their backs near shoulder level or above their heads. Wristlets can be used if the employer can demonstrate that the use of harnesses is infeasible or creates a greater hazard. *Figure 9* shows a worker entering a confined space using an appropriate harness system.

The employer must ensure that the other end of the retrieval line is attached to a mechanical device or to a fixed point outside the permit space. A mechanical device must be available to retrieve personnel from vertical permit spaces more than 5 feet deep.

If an injured entrant is exposed to a substance for which a **safety data sheet (SDS)** (formerly known as material safety data sheet) or other similar written information is required to be kept at the work site, that SDS or other information must be made available to the medical facility treating the exposed entrant.

62310-15_F09.EPS

Figure 9 Worker entering a confined space using a harness system.

Additional Resources

It is assumed that you will learn to perform these procedures under the direct supervision of your crew leader, foreman, or instructor. For independent study, refer to the recommended Additional Resources at the back of this module.

2.0.0 Section Review

1. New site conditions must be noted on a canceled confined-space permit. True or false?
   a. True
   b. False
2. If an employer can demonstrate with monitoring and inspection data that the only hazard in a confined space is a hazardous atmosphere, which can be made safe by continuous forced-air ventilation, the employer may not need to obtain a permit. True or false?
   a. True
   b. False
3. When an authorized entrant to a confined space recognizes the existence of a prohibited condition, that entrant must immediately alert _____.
   a. the supervisor
   b. the attendant
   c. the operator
   d. emergency officials
4. A mechanical device must be available to retrieve personnel from vertical permit spaces more than _____.
   a. 3 feet deep
   b. 5 feet deep
   c. 7 feet deep
   d. 10 feet deep

SECTION THREE

3.0.0 VAULT MAINTENANCE

Objective

Explain how to perform vault maintenance (CT59_0-17).

a. Explain safety requirements for vault inspections.
b. Explain vault maintenance procedures.

Performance Task

Perform vault maintenance (CT59_0-17)

- Identify potential abnormal operating conditions (AOCs) that may occur during performance of this CT, and know the appropriate actions to take in response to them.
- Utilize the appropriate personal protective equipment (PPE) according to relevant company procedures.
- Test the atmosphere around the vault for the presence of a flammable or toxic atmosphere.
- Before entry, test the atmosphere at various locations inside the vault for a flammable or toxic atmosphere before removing the vault cover.
- Enter the vault and inspect for any physical damage to the piping leaving or entering the vault.
- If applicable, inspect the vault ventilation equipment for proper operation.
- Inspect the vault for proper ventilation.
- Inspect the vault cover for adequate load support.
- Complete appropriate documentation as required by operator's procedures.

Trade Term

Abnormal operating condition (AOC): A condition within a pipeline system, identified by the operator that may indicate (1) a malfunction of a component or (2) a deviation from normal operations that may indicate a condition exceeding design limits or result in a hazard to persons, property, or the environment.

A vault is a pipeline structure that is at least partially underground. Vaults generally meet the definition for confined spaces and may often be permit-required because of the nature of the product in the pipeline. According to 49 *CFR* 192.749, a vault is an underground structure having a volumetric internal content of 200 cubic feet (5.6 m^3) or more that contains pressure-regulating and pressure-limiting equipment.

Vaults generally contain pumps, meters, test stations, or other equipment attached to the pipeline, and they must be entered to perform necessary pipeline maintenance and maintenance on the vault itself. Maintenance begins with routine inspections.

3.1.0 Inspection and Safety Requirements for Vault Maintenance

According to 49 *CFR* Part 192.749, each vault must be inspected at intervals not exceeding 15 months, but at least once annually to determine that it is in good physical condition and adequately ventilated. If gas is found in the vault, the equipment in the vault must be inspected for leaks, and any leaks found must be repaired. The ventilating equipment must also be inspected to determine that it is functioning properly. Each vault cover must be inspected to ensure that it does not present a hazard to public safety.

Elements of the inspection task may include:

- Gas detection
- Leak repair
- Pipe replacement
- Coating repair
- Inspection of ventilation equipment and systems
- Public safety hazard assessment
- Confined-space entry

3.1.1 Prerequisites, Competencies, and Abnormal Operating Conditions

Persons authorized to perform this inspection must be familiar with the hazards of any flammable/toxic gases that may be encountered, and they must be familiar with and trained in company procedures for confined-space entries, personal protective equipment (PPE), and responding to emergencies.

An abnormal operating condition (AOC), or multiple AOCs, may be encountered during vault inspections. Personnel conducting vault maintenance must be able to identify potential abnormal operating conditions (AOCs) and know the proper response to each, should they occur. The following are some potential AOCs, and the appropriate action personnel should take in response to them:

- Discovery of a leak occurs – Personnel should determine the extent of the leak and act accordingly. Depending on the location and the size of the leak, a hazardous condition may exist that could result in an explosion and fire. Appropriate actions may include evacuation of the immediate area, calling 911, and/or contacting the company emergency center.
- Vault structural damage occurs – Personnel should determine the extent of the damage and act accordingly. Appropriate actions may include immediate repair of the damage and/or contacting the appropriate company personnel to report damage and schedule repairs.
- Water in vault occurs – Personnel should use pumps or other water-removal equipment to remove water from the vault before entry.
- Inadequate vault ventilation occurs – Personnel should determine the cause for inadequate ventilation and act accordingly. Appropriate actions may include clearing obstructed vents and/or installing additional vents or ventilating equipment.
- Malfunctioning ventilating equipment occurs – Personnel should determine the origin of equipment failure and act accordingly. Appropriate actions may include immediate repair of equipment or contacting the appropriate company personnel to report the equipment failure.
- Unexpected hazardous atmosphere occurs – Personnel should evacuate the area immediately. Notify appropriate company personnel and request assistance. Depending on the type of toxic atmosphere, appropriate actions may include ventilating the vault until toxic atmosphere is dissipated and safe entry can be made and/or putting on appropriate personal protective equipment (PPE) to determine the source of toxic gas and initiate repairs.
- Fire/explosion occurs – Personnel should evacuate the area the immediately, and call 911 and the company emergency center.
- Oxygen-deficient atmosphere occurs – Personnel should not enter the vault. Notify appropriate company personnel, and request assistance. Appropriate actions may include opening vault cover to allow for adequate ventilation and eventual safe entry and/or putting on appropriate personal protective equipment (PPE) to determine the reason for the oxygen deficiency and initiate repairs.

NOTE

A combustible gas indicator (CGI) should be used to check the area around the vault for the presence of a combustible gas mixture before the cover is removed. Before entering the vault, tests with a CGI and an oxygen and toxic-gas monitor are performed to ensure that there are no fires/explosions, oxygen deficiencies, or toxic-gas hazards. Tests are made at various levels that can be reached from the surface.

3.2.0 Vault Maintenance Procedure

The following steps outline the general procedure for conducting vault maintenance:

Step 1 Identify potential abnormal operating conditions (AOCs) that may occur, and know the appropriate actions to take in response to them.

Step 2 Utilize the appropriate personal protective equipment (PPE) according to relevant company procedures.

Step 3 Step 3 Test the atmosphere around the vault for the presence of a flammable or toxic atmosphere.

Step 4 Step 4 Test for flammable or toxic gases in the atmosphere at various locations inside the vault prior to removing the vault cover to allow entry.

CAUTION

Different toxic and flammable gases have different properties that may allow them to collect in different locations of the space.

Step 5 Step 5 Enter the vault, and inspect it for any physical damage to piping; do the same when leaving the vault.

Step 6 Step 6 Inspect the vault for structural damage, such as cracks, collapse, deterioration, etc., to ensure the vault will not fail and damage gas piping and equipment.

Step 7 Step 7 Inspect the vault ventilation equipment for proper operation, if applicable.

Step 8 Step 8 Inspect the vault for proper ventilation.

Step 9 Step 9 Inspect the vault cover for adequate load support.

Step 10 Step 10 Complete appropriate documentation as required by operator's procedures.

Additional Resources

It is assumed that you will learn to perform these procedures under the direct supervision of your crew leader, foreman, or instructor. For independent study, refer to the recommended Additional Resources at the back of this module.

3.0.0 Section Review

1. Before opening the cover of a vault, personnel should use a combustible gas indicator to check the area for the presence of a combustible gas mixture. True or false?
   a. True
   b. False

2. Structural damage to a vault can result in _____.
   a. damage to gas piping
   b. corrosion of equipment
   c. PHMSA-imposed fines
   d. infestation of termites

SUMMARY

Working in confined spaces can be hazardous if the proper precautions and safety procedures are not followed. Confined-space hazards include explosion, poisoning, and asphyxiation. There are many types of confined spaces that pipeline personnel may encounter. These include pits, storage tanks, sewers, casings, silos, and vaults.

A confined space has the potential to contain a hazardous atmosphere or a material that has the potential for engulfing an entrant.

A vault is a pipeline structure that is at least partially underground. Vaults meet the definition for confined spaces and may often be permit-required because of the nature of the product in the pipeline.

Vaults generally contain pumps, meters, test stations, or other equipment attached to the pipeline, and must be entered to perform necessary pipeline maintenance and maintenance on the vault itself. Maintenance begins with routine inspections.

Review Questions

1. Which of the following is *not* an example of a confined space?
   a. A tank
   b. A vault
   c. A pipeline facility
   d. A silo

2. The definitions for confined spaces are outlined by _____.
   a. API
   b. OSHA
   c. EPA
   d. PHMSA

3. The decomposition of such equipment as scaffolding supports and electrical components within a confined space may be accelerated by _____.
   a. asbestos
   b. corrosive residues
   c. flammable atmospheres
   d. acid rain

4. The most hazardous kind of confined space is that in which there is limited access and the presence of _____.
   a. mechanical devices
   b. oil-based paint
   c. about 78 percent nitrogen in the air
   d. irritant gases

5. Ventilation air should not create an additional hazard due to _____.
   a. too much fresh air
   b. introduction of gases
   c. creation of air currents
   d. recirculation of contaminants

6. OSHA defines atmosphere with normal levels of oxygen as atmosphere containing between _____.
   a. 8.5 percent and 12.5 percent oxygen
   b. 13.5 percent and 18 percent oxygen
   c. 19.5 percent and 23.5 percent oxygen
   d. 24.5 percent and 28.5 percent oxygen

7. So only clean, fresh air is drawn into a space, the ventilation system must be positioned _____.
   a. south of the confined space
   b. downwind
   c. upwind
   d. on a mobile truck

8. When vapors lighter than air are present, air movers should be placed _____.
   a. outside the space
   b. at the top of the space
   c. at the bottom of the space
   d. in the center of the space

9. When using natural ventilation, the air should be _____.
   a. stable
   b. free of nitrogen
   c. continuously monitored
   d. free of contaminants

10. A secondary irritant may produce surface irritation as well as _____.
    a. no systemic toxic effect
    b. a combustible atmosphere
    c. systemic toxic effect
    d. dizziness

11. While the normal oxygen level in the air is 20.9 percent, increased breathing volume and an accelerated heartbeat can start occurring with oxygen levels of _____.
    a. 10 percent
    b. 17 percent
    c. 23 percent
    d. 30 percent

12. Consumption of oxygen takes place by persons breathing and during combustion of flammable substances, during chemical reactions, and during _____.
    a. painting processes
    b. cleaning processes
    c. photosynthesis
    d. bacterial actions

13. To prevent vapor leaks, flashbacks, and other hazards, workers should completely _____.
    a. evacuate the space
    b. isolate the space
    c. purge the space
    d. ventilate the space

14. Workers are less efficient and are prone to heat problems at a body temperature in excess of approximately _____.
    a. 99°F (37.2°C)
    b. 100°F (37.8°C)
    c. 102°F (38.9°C)
    d. 105°F (40.6°C)

15. Vibrations that may affect multiple body parts and organs are called _____.
    a. whole-body vibrations
    b. segmental vibrations
    c. local vibrations
    d. overall vibrations

16. To help determine if confined spaces require a permit, operators may use _____.
    a. OSHA tables
    b. common sense
    c. local codes
    d. a flowchart

17. OSHA mandates that a permit-space entry program be revised _____.
    a. once a month
    b. quarterly
    c. twice a year
    d. annually

18. The attendant of a permit space must _____.
    a. count the personnel entering the permit-required space
    b. terminate entry and cancel permits
    c. enter the permit-required space for rescue procedures
    d. perform other tasks as specified by the entry supervisor

19. Tasks routinely performed as part of vault maintenance include gas detection, leak and coating repairs, and _____.
    a. site layout
    b. public safety hazard assessment
    c. surveying and line location
    d. repairs to breakout tanks

20. The final step in a routine vault inspection is to _____.
    a. inspect the ventilation equipment
    b. look for signs of structural damage
    c. complete appropriate documentation
    d. close the vault cover

Trade Terms Quiz

Fill in the blank with the correct term that you learned from your study of this module.

1. An underground structure having a volumetric internal content of 200 cubic feet or more that contains pressure-regulating and pressure-limiting equipment is a(n) ________.
2. An agency of the US Department of Labor that regulates workplace safety and health, sets standards, and provides training, education, and outreach is the ________.
3. As defined by the federal government, a space that is not intended for human occupancy; has limited means of entry and exit; may be difficult to evacuate quickly due to size, shape, or contents; and may present other hazards to anyone who enters is called a(n) ________.
4. An atmosphere capable of causing death, illness, or harm by being explosive, flammable, toxic, or oxygen deficient is considered to be a(n) ________.
5. An instrument used to measure the amount of combustible gas in air is called a(n) ________.
6. A colorless gas that is heavier than air and very poisonous, corrosive, and flammable is ________.
7. A form containing information on a potentially hazardous material, including names, contents, test results, characteristics, potential hazards, and handling procedures is called a(n) ________.
8. A malfunction of a component or a deviation from normal operations that may indicate a condition exceeding design limits or result in a hazard to persons, property, or the environment is called a(n) ________.
9. Atmosphere that contains less than 19.5 percent oxygen and is therefore unsafe for workers without a respirator is considered to be a(n) ________.

Trade Terms

Abnormal operating condition (AOC)
Combustible gas indicator (CGI)
Confined space
Hazardous atmosphere
Hydrogen sulfide (H_2S)
Occupational Safety and Health Administration (OSHA)
Oxygen-deficient atmosphere
Safety data sheet (SDS)
Vault

Trade Terms Introduced in This Module

Abnormal operating condition (AOC): A condition within a pipeline system, identified by the operator that may indicate (1) a malfunction of a component or (2) a deviation from normal operations that may indicate a condition exceeding design limits or result in a hazard to persons, property, or the environment.

Combustible gas indicator (CGI): An instrument used to measure the amount of combustible gas in air.

Confined space: Defined by OSHA as a space that is not intended for human occupancy; has limited means of entry and exit; may be difficult to evacuate quickly due to size, shape, or contents; and may present other hazards to anyone who enters.

Hazardous atmosphere: An atmosphere capable of causing death, illness, or harm by being explosive, flammable, toxic, or oxygen deficient.

Hydrogen sulfide (H_2S): A colorless gas that is heavier than air and very poisonous, corrosive, and flammable.

Occupational Safety and Health Administration (OSHA): An agency of the US Department of Labor that regulates workplace safety and health. OSHA sets standards and provides training, education, and outreach.

Oxygen-deficient atmosphere: Atmosphere that contains less than 19.5 percent oxygen and is therefore unsafe for workers without a respirator.

Safety data sheet (SDS): A form containing information on a potentially hazardous material, including names, contents, test results, characteristics, potential hazards, and handling procedures.

Vault: An underground structure having a volumetric internal content of 200 cubic feet or more that contains pressure-regulating and pressure-limiting equipment.

Additional Resources

The following recommended resources can provide additional helpful information related to the requirements for this covered task:

API Recommended Practice 1161, Third Edition, © 2014 American Petroleum Institute. API Publishing Services, 1220 L Street, NW, Washington, DC 20005. Distributed by Thompson Reuters (Scientific) LLC, **www.techstreet.com**.

NCCER Module AOCCC-17, *Abnormal Operating Conditions Control Center.*

NCCER Module AOCFG-17, *Abnormal Operating Conditions Field and Gas.*

49 *CFR* 192, **www.ecfr.gov.**

49 CFR 195, **www.ecfr.gov.**

The following websites provide resources for products and training:

Occupational Safety and Health Administration (OSHA), **www.osha.gov.**

Pipeline and Hazardous Materials Safety Administration (PHMSA), **www.phmsa.dot.gov.**

Figure Credits

LPR Construction, Module Opener

Honeywell Safety Products, Figure 8

Courtesy of Capital Safety (DBI-SALA and PROTECTA), Figure 9

Section Review Answer Key

Answer	Section Reference	Objective
Section One		
1. a	1.1.0	1a
2. b	1.2.1	1b
3. a	1.3.0	1c
4. b	1.4.0	1d
5. a	1.5.0	1e
Section Two		
1. a	2.1.1	2a
2. a	2.2.1	2b
3. b	2.3.1	2c
4. b	2.4.0	2d
Section Three		
1. a	3.1.1	3a
2. a	3.2.0	3b

NCCER CURRICULA — USER UPDATE

NCCER makes every effort to keep its textbooks up-to-date and free of technical errors. We appreciate your help in this process. If you find an error, a typographical mistake, or an inaccuracy in NCCER's curricula, please fill out this form (or a photocopy), or complete the online form at **www.nccer.org/olf**. Be sure to include the exact module ID number, page number, a detailed description, and your recommended correction. Your input will be brought to the attention of the Authoring Team. Thank you for your assistance.

Instructors – If you have an idea for improving this textbook, or have found that additional materials were necessary to teach this module effectively, please let us know so that we may present your suggestions to the Authoring Team.

NCCER Product Development and Revision

13614 Progress Blvd., Alachua, FL 32615

Email: curriculum@nccer.org
Online: www.nccer.org/olf

❑ Trainee Guide ❑ Lesson Plans ❑ Exam ❑ PowerPoints Other ____________

Craft / Level: Copyright Date:

Module ID Number / Title:

Section Number(s):

Description:

Recommended Correction:

Your Name:

Address:

Email: Phone:

CTCC-17
Cold Cutting

Overview

When repairs require cutting pipe, cold cutting is one technique used on pipelines that carry hazardous products, because it will not produce sparks. This module explains the safety requirements and tools required for cold cutting operations. It also describes the general procedures for draining a pipeline and performing cold cutting.

Trainees with successful module completions may be eligible for credentialing through the NCCER Registry. To learn more, go to **www.nccer.org** or contact us at 1.888.622.3720. Our website has information on the latest product releases and training, as well as online versions of our *Cornerstone* magazine and Pearson's product catalog.

Your feedback is welcome. You may email your comments to **curriculum@nccer.org**, send general comments and inquiries to **info@nccer.org**, or fill in the User Update form at the back of this module.

This information is general in nature and intended for training purposes only. Actual performance of activities described in this manual requires compliance with all applicable operating, service, maintenance, and safety procedures under the direction of qualified personnel. References in this manual to patented or proprietary devices do not constitute a recommendation of their use.

CTCC-17
COLD CUTTING

Objectives

When you have completed this module, you will be able to do the following:

1. Explain the safety requirements and identify tools used for cold cutting.
   a. Explain the safety requirements for cold cutting.
   b. Identify the tools used for cold cutting.
2. Describe how to drain a pipe before performing cold cutting.
   a. Describe how to estimate drain volumes.
   b. Describe the procedure for draining a pipe.
3. Describe the procedure used to perform cold cutting of pipe (CTCC-17).
   a. Identify the prerequisites, competencies, and abnormal operating conditions (AOCs) associated with cold cutting.
   b. Describe the procedure used to perform cold cutting.
   c. Describe the procedure for replacing pipe.
   d. Describe the procedure for line filling.

Performance Task

Under the supervision of your instructor, you should be able to do the following:

1. Cold cut pipe according to your instructor's specifications (CTCC-17).
   - Identify potential abnormal operating conditions (AOCs) that may occur during performance of this CT, and know the appropriate actions to take in response to them.
   - Utilize the appropriate personal protective equipment according to relevant company procedures.
   - Determine where the cold cut(s) will be made on the pipe, allowing extra pipe for hot cuts and fit-up.
   - Remove the pipeline coating, and clean the immediate area.
   - Install bonding cables between the upstream and downstream pipe sections.
   - Mount the cutter on the pipeline and cold cut the pipe. When the cut is complete, remove the cutter.
   - Complete the draining if there is any product remaining in the pipe, and make additional cuts as necessary to remove the pipe section.
   - Roll or lift the pipe out of position, remove the bonding cables, and remove the pipe from the area.
   - File sharp edges of the pipe, or cover with a split rubber hose to prevent lacerations.
   - Complete appropriate documentation as required by operator's procedures.

Trade Terms

Abnormal operating conditions (AOCs)
American Wire Gauge
Bonded
Cold cutting
Cutting liquid
Draining
Lowering-in belt
Roller cradle
Tack weld
Threadolet

Industry-Recognized Credentials

If you're training through an NCCER-accredited sponsor, you may be eligible for credentials from NCCER's Registry. The ID number for this module is CTCC-17. Note that this module may have been used in other NCCER curricula and may apply to other level completions. Contact NCCER's Registry at 888.622.3720 or go to **www.nccer.org** for more information.

Contents

Figures

SECTION ONE

1.0.0 COLD CUTTING SAFETY AND TOOLS

Objective

Explain the safety requirements and identify tools used for cold cutting.

a. Explain the safety requirements for cold cutting.
b. Identify the tools used for cold cutting.

Trade Terms

American Wire Gauge: A measurement standard used in North America to indicate the diameter of solid round electrically conducting wire. The larger the gauge number, the smaller the wire diameter.

Bonded: Protected against electrical sparks or discharges by the use of low-resistance conductors.

Cold cutting: A technique for cutting pipe that does not create a spark that could ignite hazardous pipeline product.

Cutting liquid: A liquid used to cool and lubricate cutting blades when cold cutting pipe.

When a section of pipe in the pipeline must be replaced for any reason, it must be cut out. **Cold cutting** is one technique for cutting pipe. The advantage of cold cutting is that it does not create a spark, which is always a matter of concern when working with hazardous pipeline products. This section describes the safety requirements and tools used for cold cutting.

1.1.0 Cold Cutting Safety

The following safety equipment is used for cold cutting:

- Appropriate personal protective equipment (PPE)
- Atmospheric multigas monitors
- Bonding cables

At a minimum, the appropriate PPE includes gloves, safety glasses, hard hat, and safety boots/rubber boots. Additional PPE may be required. Always follow site and/or company-specific policies and procedures.

> **WARNING!**
> Cutting pipe may require working in trenches or confined spaces. For their personal safety, personnel must always follow applicable guidelines for working in trenches and confined spaces.

1.1.1 Atmospheric Multigas Monitors

Atmospheric multigas monitors (*Figure 1*) are used to sample the air for hazardous atmosphere before cutting begins. These portable devices can monitor as many as five atmospheric conditions simultaneously, including oxygen, flammable gas, and selected toxic gases. Toxic-gas monitoring capability typically covers both carbon monoxide and hydrogen sulfide sensors.

Some of these detectors also are equipped with data loggers and modems that allow them to communicate directly with a base unit that can receive real-time readings, a function that can be very valuable in emergency situations. Selected models include lithium-ion battery packs that enable the monitor to be used for up to 20 hours of continuous operation without requiring a recharge.

1.1.2 Bonding

The pipe segment must be **bonded** to prevent a spark or electrical discharge when the disconnected pipe is removed. Bonding is performed with cables, using a minimum of #4 AWG (**American Wire Gauge**) stranded copper wire, and clamps designed to provide a continuous connection between the pipe sections. An example of an acceptable method is using cables with magnets on each end to connect to the pipe.

Figure 2 shows an example of bonding cables that are commonly used. Always use the appropriate-size cables according to company policies and procedures.

> **NOTE**
> In some cases, such as where piping forms an electrical bond, it may not be necessary to install bonding cables.

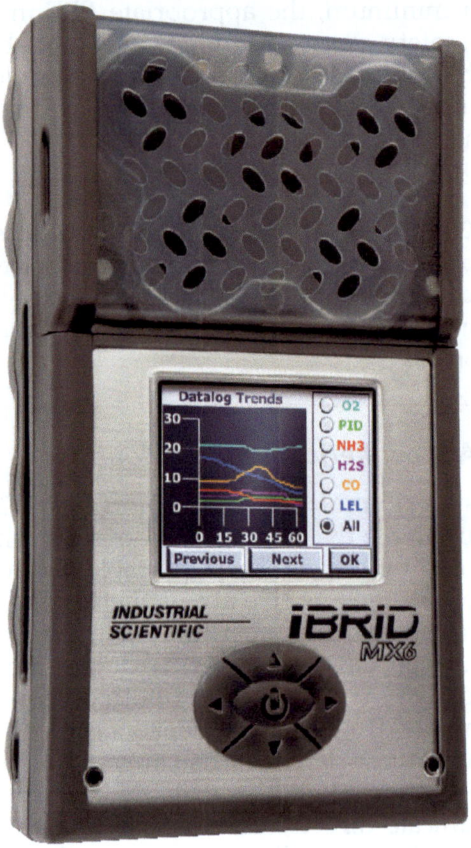

62312-15_F01.EPS

Figure 1 Atmospheric multigas monitor.

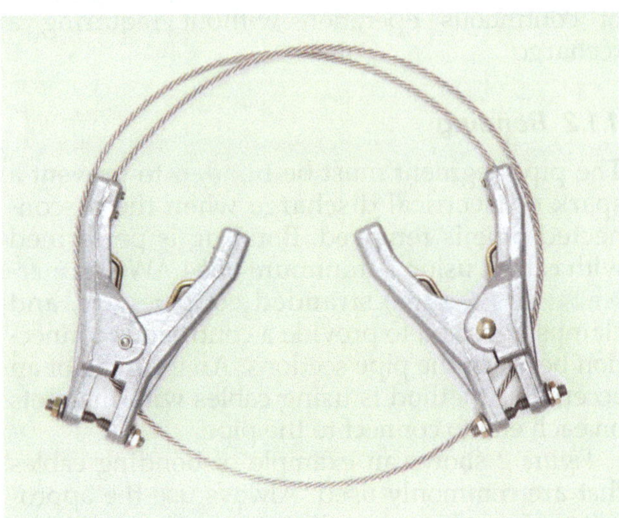

62312-15_F02.EPS

Figure 2 Example of bonding cables.

A general procedure for bonding is described in the following steps. Always follow company-specific procedures for performing this task. Follow these steps to bond the pipe before cutting or separating a pipeline:

Step 1 Install grounding clamps on the pipe; the clamps will remain on both sides of a separation and on the section to be removed. The cable/clamp arrangement is shown in *Figure 3*.

Step 2 Ensure the clamps make contact with clean, bare metal.

Step 3 Select one bonding cable long enough to span the work area, and securely attach each end to the grounding clamps located on either side of the separation.

CAUTION

Once a cable is in place, take care to avoid breaking, cutting, or detaching it as long as a fire hazard exists.

Step 4 Select a second cable long enough to clear the hazardous area when removing a pipe section.

Step 5 Securely attach one end of the second cable to one of the grounding clamps located at either end of the separation.

Step 6 Attach the other end of the second cable to the grounding clamp on the pipe section to be removed.

1.2.0 Tools Used for Cold Cutting

The following tools are used for cold cutting:

- Hand-operated pipe cutters
- Powered wheel cutters
- Cutting liquid

1.2.1 Hand-Operated Pipe Cutters

Hand-operated rotary pipe cutters (*Figure 4*), both manual and pneumatic, are commonly used for cold cutting. Rotary pipe cutters come in a wide range of sizes, and can be used to cut steel, stainless steel, ductile iron, and cast-iron pipe in tight spaces. The tool usually needs only about 4 to 8 inches (≈ 10 to 20 cm) of clearance to operate, and will cut pipe up to 34 inches (86.4 cm) in diameter with wall thicknesses to about $^{1}/_{2}$ inch (1.3 cm). The cutting blades are interchangeable so that blades for a specific material can easily be installed.

> **CAUTION**
>
> Before operating pipe cutters, always check for loose, missing, or broken parts. Check the condition and operation of the cutter wheel and guide wheels/rollers.

Follow these steps to cut pipe with a rotary pipe cutter:

Step 1 Place the cutter around the pipe and tighten it.

Step 2 Using a ratchet action, move the pipe cutter back and forth around the pipe.

Step 3 Feed the blade further into the cut by tightening the tool around the pipe.

Step 4 Repeat Steps 2 and 3 until the pipe is cut.

> **CAUTION**
>
> A pipe may snap off before the cut is all the way through the pipe. To reduce the risk of injury, be sure to stand clear of the pipe in case this happens unexpectedly.

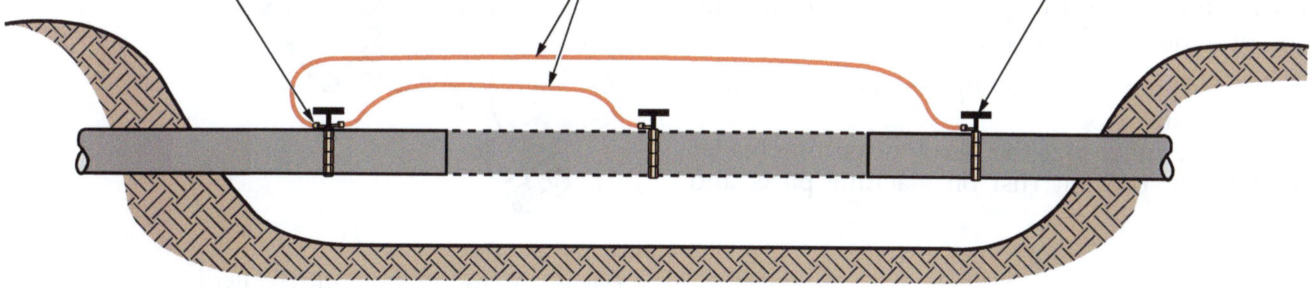

62312-15_F06.EPS

Figure 3 Cable/clamp arrangement.

(A)

(B)

62312-15_F03.EPS

Figure 4 Rotary pipe cutters.

1.2.2 Powered Wheel Cutters

Powered wheel cutters were developed for fast and easy pipe cutting. Different models of wheel cutters come in a variety of sizes capable of cutting pipe ranging from 2 to 36 inches (≈ 5 to 91 cm) in diameter. The Wachs® wheel cutter (*Figure 5*) is used for special pipe-cutting situations where it is important that there is no metal chip contamination. The Wachs® tool uses three cutting wheels that are fed equally into the cut using the Wachs® star-wheel feed system. As the cutting wheels are fed into the cut line, the pipe wall is pierced and displaced by the cutting wheels.

The combination of a parting blade and a beveling blade coupled to the Wachs® LCSF/3 split frame is a more widely used tool in the oil and gas pipeline industry. Using this tool set to bevel the pipe after cutting may still produce some chips, but much of the time the metal is removed in ribbons, making it easier to control chip contamination.

1.2.3 Cutting Liquid

Cutting liquid serves as a coolant and lubricant during the cold cutting process. Using cutting liquid provides the following advantages:

- Keeps the pipe and equipment at a safe, stable temperature
- Helps to maximize the life of the cutting blades
- Helps to prevent rust on machine parts and cutters

NOTE

Cutting liquid must be contained. Always use a drain tray or similar type of equipment to prevent it from falling on the ground.

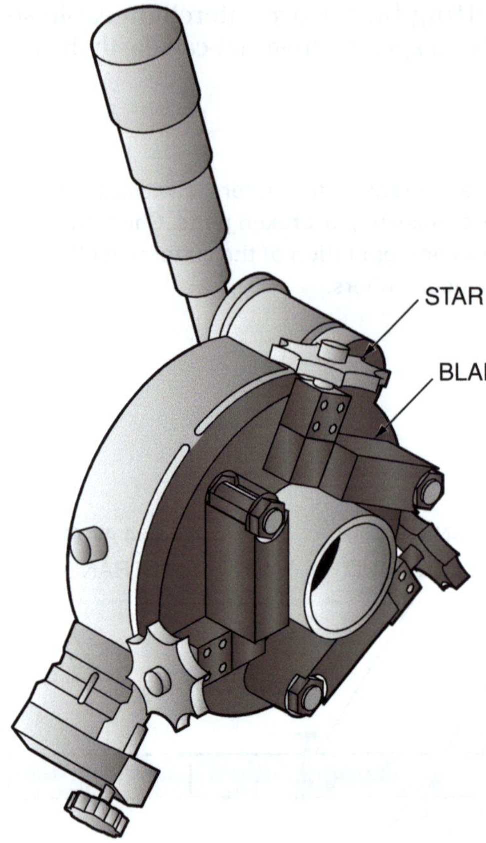

Figure 5 Wachs ® powered wheel cutter.

Additional Resources

It is assumed that you will learn to perform these procedures under the direct supervision of your crew leader, foreman, or instructor. For independent study, refer to the recommended Additional Resources at the back of this module.

1.0.0 Section Review

1. As long as a fire hazard exists, a bonding cable should never be _____.
   a. detached
   b. attached
   c. lengthened
   d. shortened

2. When pipe must be cut in a tight space, use a(n) _____.
   a. powered wheel cutter
   b. rotary pipe cutter
   c. Wachs® LCSF/3 split frame
   d. cutting torch

SECTION TWO

2.0.0 DRAINING A PIPE BEFORE COLD CUTTING

Objective

Describe how to drain a pipe before performing cold cutting.

a. Describe how to estimate drain volumes.
b. Describe the procedure for draining a pipe.

Trade Terms

Draining: The removal of product from a pipeline, typically performed when purging is neither preferable nor possible.

Threadolet: A fitting that provides a threaded connection for an outlet branch.

The removal of product from the pipeline is called draining. This procedure is necessary before replacing a section of main-line pipe. Before the pipe is cut, it must first be drained. Draining is performed when purging is not preferable or possible. Depending on geographic location, draining may also be referred to as *draining up* or *draining down*.

To drain, personnel may use vacuum trucks, threaded O-ring fittings, or valves and/or appropriate fittings. The draining plan should address the potential for contamination associated with the disconnection activities, and it should eliminate any contamination that already exists.

> **WARNING!**
> Always follow company-specific policies and procedures for draining a pipe. The use of a vacuum truck may not be permitted.

The pipeline must be shut down for draining, except when a bypass has been installed. If a shutdown is required, personnel should obtain permission from the pipeline control center. The work must be coordinated and scheduled with the control center. Personnel take the drained product directly into tanker trucks or into appropriate holding containment located a safe distance from the work area. The product is then injected back into the system after work has been completed.

2.1.0 Estimating Drain Volume

The following procedure is a general guideline for estimating drain volume. Always follow company-specific policies and procedures for performing this task.

Step 1 Excavate the site and install pipe supports.

Step 2 Attach tag lines to control the swing of the pipe.

Step 3 Evacuate the line using the nitrogen purge method, or block the line using sectionalizing valves and/or stopples.

Step 4 Determine the distances on both sides of the draining point back to the main-line shutoff valve, stopple, or highest elevation point in the pipe section.

Step 5 Subtract any lengths that dip, or where there are decreases in elevation followed by increases; subtract the total length of the decreases.

Step 6 Multiply the distance by the appropriate line-fill volume for the size of pipe.

> **NOTE**
> Be prepared for greater-than-calculated volumes in case of incorrect calculations, errors in the hydraulic profiles, or malfunctioning shutoff valves.

2.2.0 Draining a Pipe

The following procedure is a general guideline for draining a pipe. Always follow company-specific policies and procedures for performing this task.

> **WARNING!**
> Petroleum vapors or other hazardous materials may be released when draining the line. When draining a pipe, continuously monitor for petroleum vapor using a detector.

Step 1 Isolate the section of pipe to be drained by closing and locking out upstream and downstream valves or by using stopples or a pipeline isolation tool.

Step 2 Connect bonding cable(s) between the pipe and the drain tray or vacuum truck. Install a grounding rod and connect the cable from the rod to the drain tray.

Step 3 Configure the drain connection (*Figure 6*):

- Install a threadolet fitting.
- Install a full-bore valve.
- Mount a tapping machine and drill a hole through the open valve to tap the pipe.
- Install a pressure gauge.

Step 4 Open the drain valve and proceed with draining.

> **WARNING!**
> Ensure a fire watch is standing by while the pipe is being drained.

Step 5 Stop the draining and monitor the pressure gauge:

- If pressure increases, confirm that all main-line valves are closed and locked out.
- If there is no increase in pressure, continue draining.

Step 6 When no more product can be drawn from the connection, open any available vent valves. Insert a small hose or standpipe through the draining nozzle to suck out the remaining product.

Step 7 When draining is complete:

- Dip the pipe to ensure that it is empty.
- Disconnect the drain hoses/piping.

> **NOTE**
> Long lengths of exposed pipe may continue to drain due to liquid expansion from solar heat or higher ambient temperature.

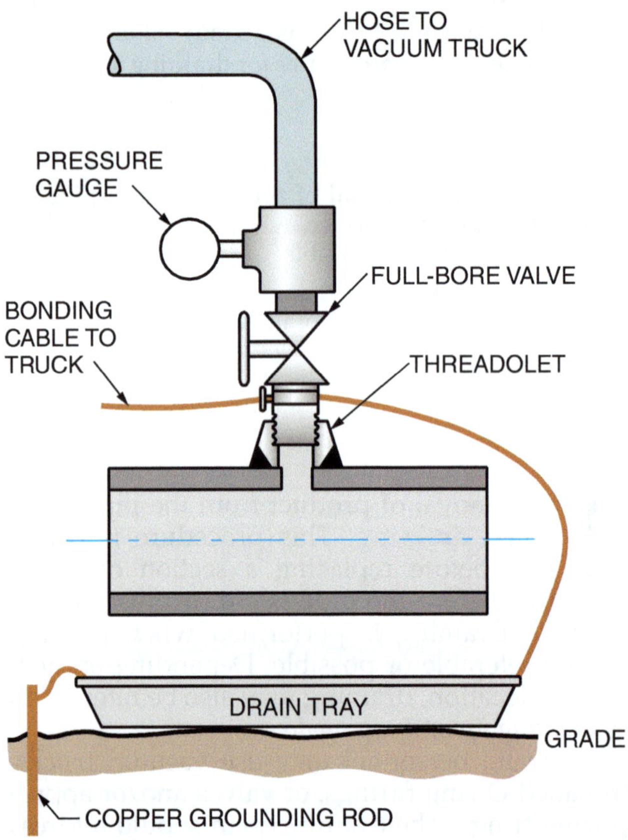

Figure 6 Drain connection.

Additional Resources

It is assumed that you will learn to perform these procedures under the direct supervision of your crew leader, foreman, or instructor. For independent study, refer to the recommended Additional Resources at the back of this module.

2.0.0 Section Review

1. Before draining a pipe, _____.
   a. it must be purged
   b. it must be vacuumed
   c. the drain volume must be estimated
   d. decontamination must occur

2. When draining a pipe, ensure that _____.
   a. emergency personnel have been notified
   b. a fire watch is standing by
   c. bonding cables have been removed
   d. vent valves have been plugged

SECTION THREE

3.0.0 COLD CUTTING PIPE

Objective

Describe the procedure used to perform cold cutting of pipe (CTCC-17).

a. Identify the prerequisites, competencies, and abnormal operating conditions (AOCs) associated with cold cutting.
b. Describe the procedure used to perform cold cutting.
c. Describe the procedure for replacing pipe.
d. Describe the procedure for line filling.

Performance Task

Cold cut pipe according to your instructor's specifications (CTCC-17).

- Identify potential abnormal operating conditions (AOCs) that may occur during performance of this CT, and know the appropriate actions to take in response to them.
- Utilize the appropriate personal protective equipment according to relevant company procedures.
- Determine where the cold cut(s) will be made on the pipe, allowing extra pipe for hot cuts and fit up.
- Remove the pipeline coating, and clean the immediate area.
- Install bonding cables between the upstream and downstream pipe sections.
- Mount the cutter on the pipeline and cold cut the pipe. Once the cut is complete remove the cutter.
- Complete the draining if there is any product remaining in the pipe and make additional cuts as necessary to remove the pipe section.
- Roll or lift the pipe out of position, remove the bonding cables, and remove the pipe from the area.
- File sharp edges of the pipe or cover with a split rubber hose, to prevent lacerations.
- Complete appropriate documentation as required by operator's procedures.

Trade Terms

Abnormal operating conditions (AOCs): Conditions within a pipeline system, identified by the operator, that may indicate (1) a malfunction of a component or (2) a deviation from normal operations that may indicate a condition exceeding design limits or result in a hazard to persons, property, or the environment.

Lowering-in belt: A device consisting of a steel lifting iron and a multiband fabric sling that is used to lift and lower pipe segments into place.

Roller cradle: A pipe cradle that uses rollers attached to the slings to allow the pipe to easily slide in and out of the cradle; also called wheel cradle.

Tack weld: A temporary weld used to hold two pieces of metal in position.

Pipelines may need to be cut for repair or modification. A mechanical cutter may be used on small- or large-diameter lines; however, large-diameter lines are usually cut with a pneumatic cutter. This section describes the procedures used for preparing the pipe for cutting, cutting the pipe, replacing the pipe, and filling the pipe after replacement.

WARNING!

Lines containing hazardous products may not be cut with a cutting torch.

3.1.0 Prerequisites, Competencies, and Abnormal Operating Conditions

Personnel must possess a variety of skills in order to successfully cold cut pipe. The various operations require a series of steps, which a pipeline technician must be competent to perform. Personnel must also be aware of the particular abnormal operating conditions (AOCs) that may occur during the procedure and know the proper responses to them.

3.1.1 Prerequisites and Competencies

Personnel assigned to cold cut pipe must have an understanding of what the task entails. Personnel must understand how to coordinate pipeline startup and shutdown with control center personnel. They need to be familiar with procedures for purging, draining, and refilling the pipe. Personnel must have knowledge of proper valve operation. They must understand how to remove coatings and how to safely install grounding and bonding prior to cutting. Personnel should also be familiar with welding tasks and weld inspections.

If excavation is required, personnel must be aware of company as well as regulatory agency confined space requirements for safe work in an excavation. They must also be aware of the requirements to notify the local One-Call organization if excavation is necessary.

> **WARNING!**
> Always obtain and use the appropriate personal protective equipment for each task. Stay alert to the possibility of unexpected leaks from the exposed pipeline such as hazardous liquid or gas. Safely eliminate all ignition sources and immediately notify the appropriate personnel if a leak is observed.

3.1.2 Abnormal Operating Conditions

Personnel repairing or replacing test leads must be able to identify the following AOCs and know the proper response to each, should they occur. The following are some potential AOCs, and the appropriate action personnel should take in response to them:

- *Pipeline cracking, breakthrough, or other mechanical damage is found* – Anticipate the proper response for unexpected release of product and/or fire or explosion likely to occur as a result of the damage. Determine the extent of the damage, and make proper company notifications so that appropriate repairs can be made. Document as required and notify personnel responsible for correcting the damage
- *Unexpected hazardous liquid or carbon dioxide leak is encountered* – Eliminate ignition source and notify appropriate personnel for corrective action.
- *Fire or explosion occurs* – Limit access to the area and make the proper notifications, including calling 911 if necessary.

3.2.0 Procedure

The following steps outline the general procedure for cold cutting pipe:

> **NOTE**
> The steps presented are general in nature and provided as a guideline. Always follow company-specific policies and procedures for performing this task.

Step 1 Identify potential abnormal operating conditions (AOCs) that may occur, and know the appropriate actions to take in response to them.

Step 2 Utilize the appropriate personal protective equipment according to relevant company procedures.

Step 3 Prepare to make the cut(s):

- Determine where the cold cut(s) will be made on the pipe, allowing extra pipe for hot cuts and fit-up.
- Remove the pipeline coating from the immediate area and ensure the cutting area is clean.
- Install bonding cables between the upstream and downstream pipe sections.
- Mount the cutter on the pipe.

> **WARNING!**
> Ensure that fire extinguishers are ready and the fire watch is on alert.

Step 4 Cold cut the pipe.

> **WARNING!**
> The pipe may spring suddenly as the cut completes. Watch both sections of pipe for signs of movement or binding, and add or change supports as necessary to prevent the pipe from moving.

> **CAUTION**
> Do not use cutters at speeds that create pipe surface temperatures over 250°F.

Step 5 Complete the cut and remove the cutter.

> **NOTE**
>
> If the main-line pipe is subject to compressive forces, the section may continue to be securely bound through frictional forces with the main-line pipe. Additional cuts may be necessary to obtain the necessary clearance for removal.

Step 6 Complete the draining if there is any product remaining in the pipe.

Step 7 Make additional cuts as necessary to remove the pipe section.

Step 8 Roll or lift the pipe out of position.

> **WARNING!**
>
> Use caution when removing pipe because compression forces between the cut section and the original line can cause the section to be wedged in. More cold cuts may be needed to relieve this situation.

Step 9 Remove the bonding cables and remove the pipe from the area.

Step 10 File sharp edges of the pipe, or cover them with a split rubber hose to prevent lacerations.

Step 11 Complete appropriate documentation as required by operator's procedures.

3.3.0 Replacing Pipe

After the section of existing pipe is cut and removed, the new section can be cut to size and welded in place. The welders will cut the new pipe to the correct length and bevel the edges for welding using an oxyacetylene torch. To make a sound weld, the bevel must be cut to a specified angle.

The welder will **tack weld** the new section to the existing pipe. An external line-up clamp is used to hold the two ends together in alignment for this operation.

The final pass, also called a *cap weld*, is applied and the weld is tested by radiography. When the welds are approved, the new section and the ends of the existing pipe are primed, coated, and wrapped. A holiday detector is used to check the coating. The line is then properly supported, and backfilled if necessary. Always follow Pipeline and Hazardous Materials Safety Administration (PHMSA) regulations and company-specific procedures for replacing pipe.

Several types of equipment are used to move, lift, and position pipe in order to perform welding procedures. This equipment may include the following:

- External line-up clamps
- **Lowering-in belt**
- **Roller cradle**

3.3.1 Line-Up Clamps

After the pipe is cut and is ready for replacement, external line-up clamps can be used for positioning. Prefabricated pipe-welding clamps are available for both external and internal use to position pipe ends to be joined by welding.

Internal alignment clamps, as shown in *Figure 7 (A)*, are fitted with various types of wheels, depending on whether the pipe is coated internally, to allow the unit to easily be rotated and positioned inside the pipe. External clamps, as shown in *Figure 7 (B)*, are constructed to overlap the two

(A) INTERNAL CLAMP

(B) EXTERNAL CLAMP

62312-15_F07.EPS

Figure 7 Line-up clamps.

pipe faces to be joined, by fitting around the exterior surface. Both types of clamps hold the pipe faces firmly in place to permit welding around the complete circumference of the pipe without having to adjust the clamp position.

Different types of external clamps match up and secure the two pipe ends using a hand-actuated levering device or, in the case of some larger models, with hydraulic-jack assistance. Some internal clamp units have on-board compressed-air tanks to retract or extend wheels and to move the unit through the pipe.

3.3.2 Lowering-In Belts

Lowering-in belts (*Figure 8*) consist of a steel lifting iron and a fabric sling formed of several bands, or layers, of high-grade rip-resistant nylon. Each band should be manufactured to the same length, width, and stretching capacity as other bands in the sling.

The slings are flexible, easy to handle and conform to the pipe being lifted, even at very low temperatures. Lowering-in belts are available to handle a range of pipe sizes from 12 to 66 inches (30 to 168 cm) in diameter. Larger sizes may be special-ordered. They are designed to permit easy hooking and unhooking by one person. Each model can be disassembled for storage and transport to and from the job site.

Figure 8 Lowering-in belt.

3.3.3 Roller Cradles

Roller cradles (*Figure 9*), also called *wheel cradles*, are designed to lower in and lift out pipe during pipeline construction and maintenance activity. These cradles easily meet a broad range of lifting and other operational requirements. Cradle design is such that a range of pipe sizes can be handled by the same cradle. The pipe is protected from damage if loaded properly using the required number of cradle units to evenly distribute the load.

Roller cradles are available for lifting and handling pipe from 4 to 60 inches (10 to 153 cm) in diameter. Some models of cradles are fitted with polyurethane-coated rollers that have high load capacity, good wear resistance, and elasticity under load, to minimize breakage during use. Some models of cradles can also be used with a matching strongback (I-beam) to couple two cradles together.

3.4.0 Line Filling

The following steps outline the procedure for filling a line after work has been completed:

NOTE

The steps presented are general in nature and provided as a guideline. Always follow company-specific policies and procedures for performing this task.

Step 1 Ensure the sealing integrity of the valves:
- Close the drain valves and any vent valves.
- Tighten flanges as required.

Step 2 Unlock the main-line isolation valves.

Step 3 Open the main-line valves slowly to refill the pipe.

WARNING!

Avoid sudden pressure buildup and ensure a fire watch is standing by during a line fill.

Step 4 Ensure that there are no leaks as the section is flooded.

Step 5 Connect high-point vent valve(s) into a large container or tank truck.

Step 6 Connect bonding cable(s) between the pipe and the container or the tank truck.

Step 7 Open the vent valves and vent air into the container or tank truck.

CAUTION

Vent air as carefully and completely as possible. Trapped air can cause operational problems such as column separation.

NOTE

It may be necessary to stop and start a few times to ensure that all air is vented.

Step 8 When all air is vented, do the following:
- Close the vent valves.
- Disconnect the piping.
- Plug the vent valves.

Step 9 Return isolation valves to the original positions.

Step 10 Watch for leaks as the line is started and pressure builds up.

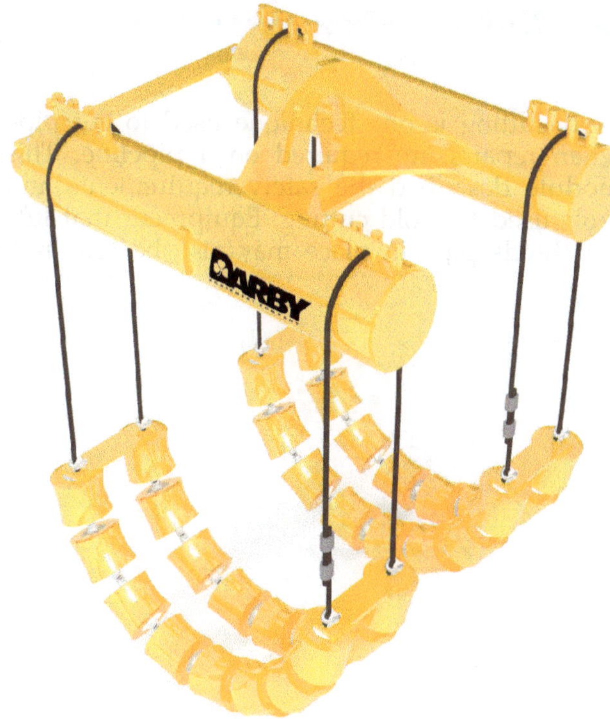

Figure 9 Roller cradle.

Additional Resources

It is assumed that you will learn to perform these procedures under the direct supervision of your crew leader, foreman, or instructor. For independent study, refer to the recommended Additional Resources at the back of this module.

3.0.0 Section Review

1. If mechanical damage is found when performing cold cutting, the proper response is to _____.
   a. request a shutdown
   b. determine the extent of the damage and notify appropriate company personnel
   c. call the 911 local emergency line
   d. evacuate the area

2. The final step of the cold cutting procedure is to _____.
   a. remove the bonding cables
   b. file the cut edges
   c. complete operator documentation
   d. perform any additional cuts necessary to free the pipe

3. The final welding pass is called a _____.
   a. tack weld
   b. cap weld
   c. plug weld
   d. fusion weld

4. The first step of a line filling procedure is to _____.
   a. open the vent valves
   b. connect the bonding cables to the tank truck
   c. unlock the main-line isolation valves
   d. ensure the sealing integrity of the valves

SUMMARY

Cold cutting is one technique used to cut pipe when repairs are required on a pipeline. This module described the safety requirements and tools used for cold cutting. Equipment that lifts and holds pipe in place may also be required.

This module outlined the general procedures for draining a pipe and performing cold cutting. It also explained how to replace a pipe and refill the line. Documentation is a critical final step in the cold cutting process.

Review Questions

1. Toxic-gas sensors in an atmospheric multigas monitor typically cover carbon monoxide and _____.
   a. carbon dioxide
   b. oxygen
   c. hydrogen sulfide
   d. lithium ions

2. Bonding cables should be made of stranded copper wire that is a minimum of _____.
   a. #4 AWG
   b. #6 AWG
   c. #8 AWG
   d. #10 AWG

3. The number of bonding cables used when cutting or separating a pipeline is _____.
   a. 2
   b. 3
   c. 4
   d. 5

4. Use a Wachs® wheel cutter in applications where it is important to avoid _____.
   a. electrical sparks
   b. metal chip contamination
   c. escape of volatile product
   d. corrosion

5. The life of a cutting blade can be increased by using _____.
   a. cutting liquid
   b. clamps
   c. a ratcheting action
   d. a blade sharpener

6. A pipeline must be shut down for draining unless there is a _____.
   a. vacuum
   b. very small drain volume
   c. firewatch
   d. bypass

7. When estimating the drain volume prior to draining, evacuate the line using _____.
   a. water
   b. air
   c. nitrogen
   d. helium

8. A threadolet fitting is installed when _____.
   a. evacuating the line
   b. isolating the pipe section
   c. configuring the drain connection
   d. mounting the cutter on the pipe

9. Large-diameter lines are usually cut with a _____.
   a. handheld cutter
   b. powered wheel cutter
   c. pneumatic cutter
   d. hydraulic cutter

10. Remove the pipeline coating after _____.
    a. installing bonding cables between the pipe sections
    b. mounting the cutter on the pipeline
    c. determining the locations of the cold cuts
    d. rolling or lifting the pipe out of position

11. When the welding has been completed, the weld is tested _____.
    a. with a pressure test
    b. using sound waves
    c. by radiography
    d. with a pig

12. The device used to position pipe prior to welding that is fitted with wheels to allow for easy rotation and positioning is the _____.
    a. roller cradle
    b. wheel cradle
    c. external line-up clamp
    d. internal alignment clamp

13. The slings in a lowering-in belt are made from _____.
    a. #4 AWG wire
    b. flexible steel strap
    c. polyurethane
    d. nylon

14. During line filling, large containers or tank trucks are used to _____.
    a. vent air
    b. plug valves
    c. disconnect piping
    d. check for leaks

Trade Terms Quiz

Fill in the blank with the correct term that you learned from your study of this module.

1. The liquid used to cool and lubricate cutting blades when cold cutting pipe is called a(n) ________.
2. The measurement standard used in North America to indicate the diameter of solid round electrically conducting wire is called ________.
3. To be protected against electrical sparks or discharges by the use of low-resistance conductors is to be ________.
4. The removal of product from a pipeline, typically performed when purging is neither preferable nor possible, is called ________.
5. The device consisting of a steel lifting iron and a multiband fabric sling that is used to lift and lower pipe segments into place is called a(n) ________.
6. Conditions within a pipeline system, identified by the operator, that may indicate (1) a malfunction of a component or (2) a deviation from normal operations that may indicate a condition exceeding design limits or result in a hazard to persons, property, or the environment are called ________.
7. A pipe cradle that uses rollers attached to the slings to allow the pipe to easily slide in and out of the cradle is called a(n) ________.
8. A technique for cutting pipe that does not create a spark that could ignite hazardous pipeline product is called ________.
9. A temporary weld used to hold two pieces of metal in position is called a(n) ________.
10. A fitting that provides a threaded connection for an outlet branch is called a(n) ________.

Trade Terms

Abnormal operating conditions (AOCs)
American Wire Gauge
Bonded
Cold cutting
Cutting liquid
Draining
Lowering-in belt
Roller cradle
Tack weld
Threadolet

Trade Terms Introduced in This Module

Abnormal operating conditions (AOCs): Conditions within a pipeline system, identified by the operator, that may indicate (1) a malfunction of a component or (2) a deviation from normal operations that may indicate a condition exceeding design limits or result in a hazard to persons, property, or the environment.

American Wire Gauge: A measurement standard used in North America to indicate the diameter of solid round electrically conducting wire. The larger the gauge number, the smaller the wire diameter.

Bonded: Protected against electrical sparks or discharges by the use of low-resistance conductors.

Cold cutting: A technique for cutting pipe that does not create a spark that could ignite hazardous pipeline product.

Cutting liquid: A liquid used to cool and lubricate cutting blades when cold cutting pipe.

Draining: The removal of product from a pipeline, typically performed when purging is neither preferable nor possible.

Lowering-in belt: A device consisting of a steel lifting iron and a multiband fabric sling that is used to lift and lower pipe segments into place.

Roller cradle: A pipe cradle that uses rollers attached to the slings to allow the pipe to easily slide in and out of the cradle; also called wheel cradle.

Tack weld: A temporary weld used to hold two pieces of metal in position.

Threadolet: A fitting that provides a threaded connection for an outlet branch.

Additional Resources

The following recommended resources can provide additional helpful information related to the requirements for this covered task:

API Recommended Practice 1161, Third Edition, © 2014 American Petroleum Institute. API Publishing Services, 1220 L Street NW, Washington, DC 20005. Distributed by Thompson Reuters (Scientific) LLC, **www.techstreet.com**

49 *CFR* 192, **www.ecfr.gov**.

49 *CFR* 195, **www.ecfr.gov**.

NCCER Module AOCCC-17, *Abnormal Operating Conditions Control Center*

NCCER Module AOCFG-17, *Abnormal Operating Conditions Field and Gas*

NCCER Module CT9_1-17, *Install Bonds*

NCCER Module CT14_1-17, *Locate Line*

NCCER Module CT32_0-17, *Observation of Excavation Activities*

NCCER Module CT38_3-17, *Visually Inspect that Welds Meet DOT Requirements*

NCCER Module CT39_0-17, *Backfilling a Trench Following Maintenance*

NCCER Module CT42_7-17, *Welding*

NCCER Module CT63_4-17, *Locally Operate Valves on a Liquid Pipeline System*

The following websites provide resources for products and training:

Pipeline and Hazardous Materials Safety Administration (PHMSA). **www.phmsa.dot.gov**

Figure Credits

Darby Equipment Company, Module Opener, Figures 7 and 9

Reprinted with permission, BASCO, Inc. bascousa.com, Figure 2

Bill Emery Praxair Services Inc. (tool shown in photo supplied by Ridgid Tool Company), Figure 4A

Courtesy of Reed Manufacturing Company, Figure 4B

Section Review Answer Key

Answer	Section Reference	Objective
Section One		
1. a	1.1.2	1a
2. b	1.2.1	1b
Section Two		
1. c	2.1.0	2a
2. b	2.2.0	2b
Section Three		
1. b	3.1.2	3a
2. c	3.2.0	3b
3. b	3.3.0	3c
4. d	3.4.0	3d

NCCER CURRICULA — USER UPDATE

NCCER makes every effort to keep its textbooks up-to-date and free of technical errors. We appreciate your help in this process. If you find an error, a typographical mistake, or an inaccuracy in NCCER's curricula, please fill out this form (or a photocopy), or complete the on-line form at **www.nccer.org/olf**. Be sure to include the exact module ID number, page number, a detailed description, and your recommended correction. Your input will be brought to the attention of the Authoring Team. Thank you for your assistance.

Instructors – If you have an idea for improving this textbook, or have found that additional materials were necessary to teach this module effectively, please let us know so that we may present your suggestions to the Authoring Team.

NCCER Product Development and Revision

13614 Progress Blvd., Alachua, FL 32615

Email: curriculum@nccer.org
Online: www.nccer.org/olf

❑ Trainee Guide ❑ Lesson Plans ❑ Exam ❑ PowerPoints Other ____________

Craft / Level: Copyright Date:

Module ID Number / Title:

Section Number(s):

Description:

Recommended Correction:

Your Name:

Address:

Email: Phone:

CTFB-17
Flange Bolting

Overview

The most common method of joining pipe is with welded connections, but systems that may need to be dismantled occasionally for maintenance or other reasons are joined with flanges. A flanged joint consists of two flanges, a gasket or other device to seal the joint, and the necessary fasteners (bolts or stud bolts) to hold the joint together. This module describes the components used in flange bolting/assembly, including how to properly store and protect these components, as well as the procedures used to install pipe flanges.

Trainees with successful module completions may be eligible for credentialing through the NCCER Registry. To learn more, go to **www.nccer.org** or contact us at 1.888.622.3720. Our website has information on the latest product releases and training, as well as online versions of our *Cornerstone* magazine and Pearson's product catalog.

Your feedback is welcome. You may email your comments to **curriculum@nccer.org**, send general comments and inquiries to **info@nccer.org**, or fill in the User Update form at the back of this module.

This information is general in nature and intended for training purposes only. Actual performance of activities described in this manual requires compliance with all applicable operating, service, maintenance, and safety procedures under the direction of qualified personnel. References in this manual to patented or proprietary devices do not constitute a recommendation of their use.

CTFB-17
Flange Bolting

Objectives

When you have completed this module, you will be able to do the following:

1. Describe the proper way to store and protect pipe and fittings.
   a. Describe the proper way to store pipe and materials.
   b. Describe the proper way to protect pipe and fittings.
2. Identify and describe the flanges used in pipeline applications.
   a. Identify and describe types of flanges.
   b. Identify and describe types of flange facings.
3. Identify and describe types of flange gaskets and gasket materials.
   a. Identify and describe types of flange gaskets.
   b. Identify and describe gasket materials.
4. Explain the procedures for installing pipe flanges.
   a. Explain the basic techniques involved in torquing and tensioning.
   b. Explain the preparation and procedures involved in installing gaskets.
   c. Explain the preparation and procedures involved in tightening flanges.
5. Describe the procedures used to perform flange bolting/assembly (CTFB-17).
   a. Identify the prerequisites, competencies, and abnormal operating conditions (AOCs) associated with flange bolting/assembly.
   b. Describe the procedures used to perform flange bolting/assembly.

Performance Task

Under the supervision of your instructor, you should be able to do the following:

1. Perform flange bolting/assembly (CTFB-17).
   - Identify potential abnormal operating conditions that may occur during performance of this CT, and know the appropriate actions to take in response to them.
   - Utilize the appropriate personal protective equipment according to relevant company procedures.
   - Ensure that the proper tools, materials, and equipment are present and that tools are in good working order.
   - Verify ANSI ratings of materials, as specified.
   - Inspect and prepare facing of the flange.
   - Inspect and clean threads on nuts and bolts.
   - Insert lineup pins to align flanges.
   - Place two bolts in bottom part of flange and drop in gasket.
   - Install remaining bolts and hand tighten nuts.
   - Verify proper flange tightening sequence and tightening procedure to attain specified torque and tighten accordingly.
   - Bolt up the flange to attain specified torque and tighten accordingly.
   - Complete appropriate documentation as required by operator's procedures.

Trade Terms

Abnormal operating conditions (AOCs)
American National Standards Institute (ANSI)
American Society of Mechanical Engineers (ASME)
ASTM International (ASTM)
Dunnage
Flange face
Gasket
Pressure vessel
Torque

Industry-Recognized Credentials

If you're training through an NCCER-accredited sponsor, you may be eligible for credentials from NCCER's Registry. The ID number for this module is CTCC-17. Note that this module may have been used in other NCCER curricula and may apply to other level completions. Contact NCCER's Registry at 888.622.3720 or go to **www.nccer.org** for more information.

Contents

Figures and Tables

Section One

1.0.0 Storing and Protecting Pipe and Materials

Objective

Describe the proper way to store and protect pipe and fittings.

a. Describe the proper way to store pipe and materials.
b. Describe the proper way to protect pipe and fittings.

Trade Terms

Dunnage: Hardwood blocking and pallets placed underneath materials to keep them off the ground and to allow access for forklifts or for placement of chokers and slings.

Flange face: The sealing surface machined onto flanges. Flange facings are governed by ANSI standards to ensure that all flanges of each type and rating have the same facing.

Gasket: A device that is used to make a pressure-tight connection and that is usually in the form of a sheet or a ring.

After piping materials are received from the supplier and before they are used on the job, they must be stored on the job site. Proper storage is critical. When planned carefully, it can help to maintain the quality of the pipe and materials, and will also save time by enabling workers to find materials easily. This section provides general guidelines for storing and protecting the piping materials used in flange bolting/assembly.

1.1.0 Storing Pipe and Materials

Pipe and materials may be stored inside or outside. Storage considerations vary depending on the location.

1.1.1 Inside Storage

If pipe and materials are stored inside a warehouse, a floor plan of the warehouse is needed to designate where different items are located and where they need to be placed when received. The floor space must be organized with several key factors in mind, such as safety, access to the material, distance from where the material is to be used, and any other special considerations dictated by the particular demands of the job site. Small materials, such as fittings, are usually ordered in large quantities and stored on shelves or in bins. They should be carefully sorted and clearly marked so that accurate inventory records can be kept. It is extremely important that materials be checked out of the warehouse only as they are needed. Do not take more supplies or materials than the job calls for.

1.1.2 Outside Storage

Most job sites store materials outside. The storage area should be well-planned and laid out in the same manner as a warehouse would be laid out, with adequate space between materials to allow access for forklift trucks and other vehicles. Consideration the flow of operations; if possible, the materials for a particular operation should be stored where they are easily accessible for that particular activity. It may be feasible to make several staging areas where the sequence of operations is mirrored in supply storage points.

Follow these general guidelines when storing pipe and materials outdoors:

- Place new materials in specially designated areas only.
- Label all materials clearly with identification tags or stenciled placards.
- Store all pipe and materials on appropriate hardwood dunnage to facilitate transportation of the materials and to keep them off the ground.
- Store valves in an upright position so that water does not collect in them and damage the seats.
- Store all materials on the basis of their compatibility with other materials. For example, do not store stainless steel pipe with carbon steel pipe because the rust from carbon steel can cause the stainless steel to corrode.

1.2.0 Protecting Pipe and Fittings

It is important to consider the effect that the weather will have on piping materials. Materials must be properly protected from rain, snow, heat, and humidity. The piping specifications often provide instructions on how the ends of pipe should be protected. If instructions are not included in the specifications, follow the general methods described here for protecting the flange face, male threaded ends, and female threaded ends.

1.2.1 Protecting Flange Faces

When storing flanges, the flange faces must always be protected. The flange face is the sealing surface machined onto the flange. Cover the flange faces with a wood, plastic, plywood, or hardboard disc. The disc can be strapped, pressed, bolted, or clipped into place. The protective disc needs to cover only the gasket surface of the flange face. If the specifications require that the flange cover be sealed weather-tight, use a rubber or vinyl disc between the flange and the cover, and bolt the cover in place with no less than four fasteners. *Figure 1* shows protected flange faces.

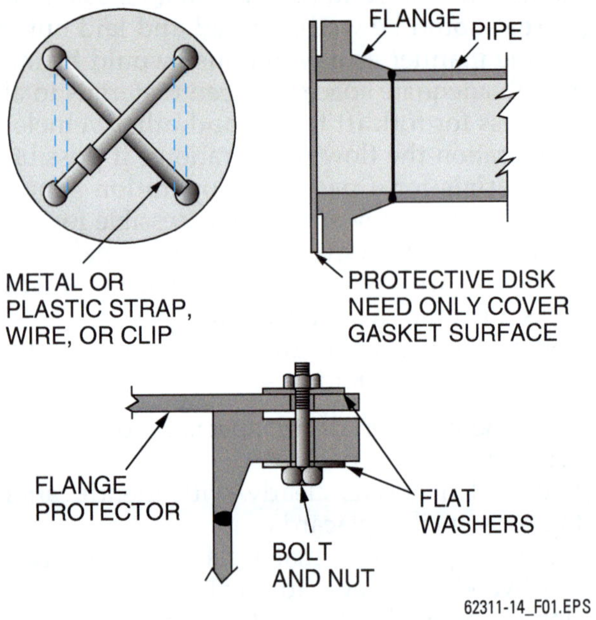

Figure 1 Protected flange faces.

1.2.2 Protecting Male and Female Threaded Ends

Male threaded ends should be protected by press-on plastic or light metal caps. *Figure 2* shows protected male threaded ends. Female threaded ends should be protected using a press-in plastic or light metal insert. If weatherproofing is required, seal the edge of the insert with at least two layers of waterproof tape. *Figure 3* shows protected female threaded ends.

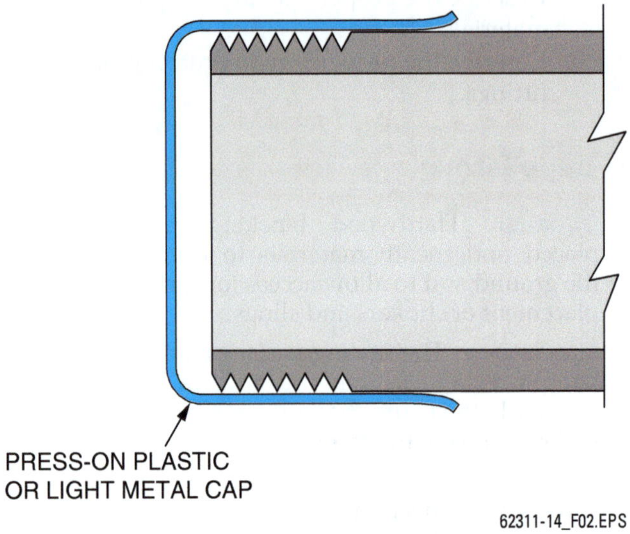

Figure 2 Protected male threaded ends.

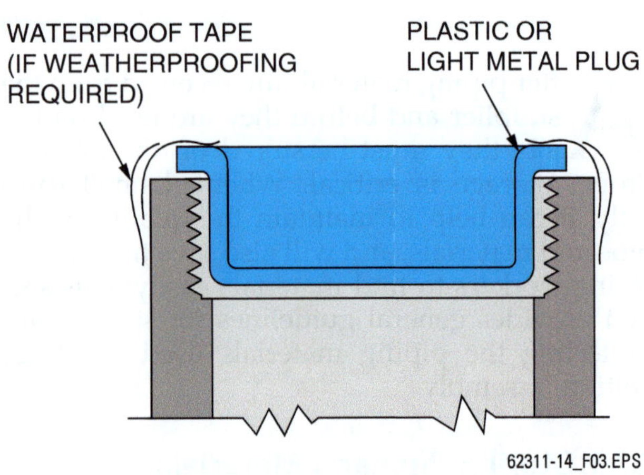

Figure 3 Protected female threaded ends.

Additional Resources

It is assumed that you will learn to perform these procedures under the direct supervision of your crew leader, foreman, or instructor. For independent study, refer to the recommended Additional Resources at the back of this module.

1.0.0 Section Review

1. To facilitate transportation and to keep materials off the ground, pipe and materials should be stored on _____.
   a. cement block
   b. plastic sheeting
   c. air bags
   d. dunnage

2. Special instructions on how the ends of pipe should be protected are often found in the _____.
   a. manufacturer's specifications
   b. piping specifications
   c. construction drawings
   d. job safety analysis

SECTION TWO

2.0.0 FLANGES

Objective

Identify and describe the flanges used in pipeline applications.

a. Identify and describe types of flanges.
b. Identify and describe types of flange facings.

Trade Terms

American National Standards Institute (ANSI): An organization that adopts standards for a variety of products, services, and processes in the United States to ensure consistency.

American Society of Mechanical Engineers (ASME): An organization that provides codes and standards, education and training, and knowledge sharing across the engineering disciplines, including the pipeline industry.

Pressure vessel: A metal container that can withstand high pressures.

Like most other piping components, flanges are made from several different types of materials, but the most commonly used is forged steel. The flanges are available in several different basic primary ratings. The primary rating is a pressure rating based on a pressure/temperature relationship, and is established by the American National Standards Institute (ANSI). *Table 1* shows the basic primary ANSI pressure ratings for forged-steel flanges, according to the American Society of Mechanical Engineers (ASME) standard B16.5. ASME is an organization that provides codes and standards, education and training, and knowledge sharing across the engineering disciplines, including the pipeline industry.

2.1.0 Types of Flanges

The flanges used most commonly in piping systems include blind flanges and threaded flanges.

2.1.1 Blind Flanges

Blind flanges are used to close off the ends of pipes, valves, and pressure vessel openings. From the standpoint of internal pressure and bolt loading, blind flanges are the most highly stressed of all flange types since the maximum stresses in a blind flange are bending stresses at the center. These flanges are available in the seven primary ratings, and their dimensions can be found in manufacturer's catalogs. *Figure 4* shows a blind flange.

Table 1. ANSI Flange Pressure Ratings

ANSI Pressure Rating (Forged Steel, based on −20°F to 120°F)	Maximum Operating Pressure (psi)
150	285
300	740
400	990
600	1,480
900	2,220
1,500	3,705
2,500	6,170

2.1.2 Threaded Flanges

Threaded flanges are available in all sizes and pressure ratings up to and including 12 inches and Class 2,500. However, the use of high-pressure threaded flanges should be limited to special applications. Their major advantage is that they can be assembled without welding, but the threaded joint will not withstand high temperatures or bending stresses. *Figure 5* shows a threaded flange.

2.2.0 Flange Facings

Facing refers to the sealing surface machined onto flanges. Flange facings are governed by ANSI standards to ensure that all flanges of each type and rating have the same facing. Typically, personnel should never use two different face styles at one joint, but this may be necessary when connecting pipe to manufactured equipment.

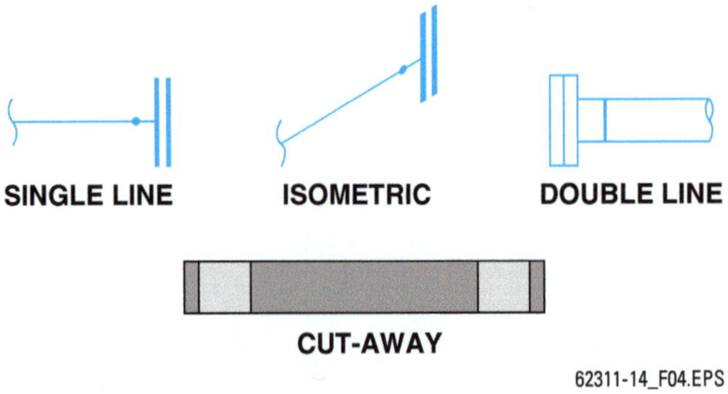

Figure 4 Blind flange.

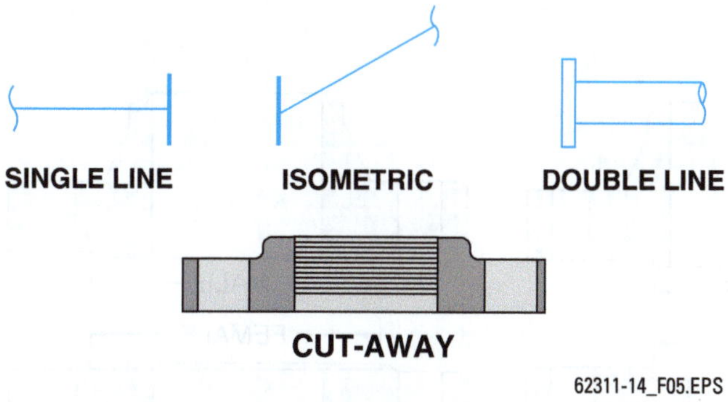

Figure 5 Threaded flange.

Two types of flange-facing finishes are the phonographic finish and the concentric serrated finish. The phonographic finish has a continuous narrow, shallow spiral groove in the face. The concentric serrated finish has a series of shallow circular grooves sharing a common center. The normal range of the concentric serrated finish is between 30 and 55 grooves per lateral inch of the diameter of the face. Both finishes allow soft gasket material to fill the grooves and keep liquids from leaking. Some special industrial applications may use a smooth finish, sometimes called a cold-water finish, to be used without a gasket.

Common flange facings include the following:

- Raised face
- Flat face
- Ring-joint-type
- Tongue-and-groove
- Male and female

Figure 6A and *Figure 6B* show flange-facing finishes and flanges.

2.2.1 Raised-Face Facing

The raised-face facing has a wide, raised rim around the center of the flange and is the most common facing used with steel flanges. The raised face on Class 150 and Class 300 flanges is $^{1}/_{16}$ inch high, and the face on flanges of all other pressure ratings is $^{1}/_{4}$ inch high. This facing uses a wide, flat ring gasket that is squeezed tightly between the flanges. The facing is machine-finished with tiny, spiral grooves that bite into and help hold the gasket.

2.2.2 Flat-Face Facing

The flat-face facing is flat across the entire face. This facing uses a gasket that has fastener holes cut in it and that covers the entire flange face. Flat-face facings are mainly used to join Class 150 and Class 300 forged-steel flanges. It is extremely important that the metals and pressure ratings of flange faces are compatible; if they are not, the flanges may crack or break. Refer to engineering specifications and follow company policies and procedures when selecting flanges.

2.2.3 Ring-Joint-Type Facing

The ring-joint-type (RJT) flange has a different type of facing because the contact surface of the seal is below the actual flange facing. This type of flange uses an oval, rectangular, or octagonal steel ring as a seal (*Figure 7*) that fits into a ring groove that is machined into the faces of the flanges. Only one steel ring is required for each pair of mating flanges.

The RJT flange facing is the most expensive standard facing, but for use in oil and refinery services, it is usually regarded as the most efficient because internal pressure acts on the ring to increase the sealing force. One disadvantage of using this type of facing and flange is that in order to remove a valve or a section of pipe from the system, the flanges have to be spread far enough apart to remove the ring. In tight hookups, this can cause problems.

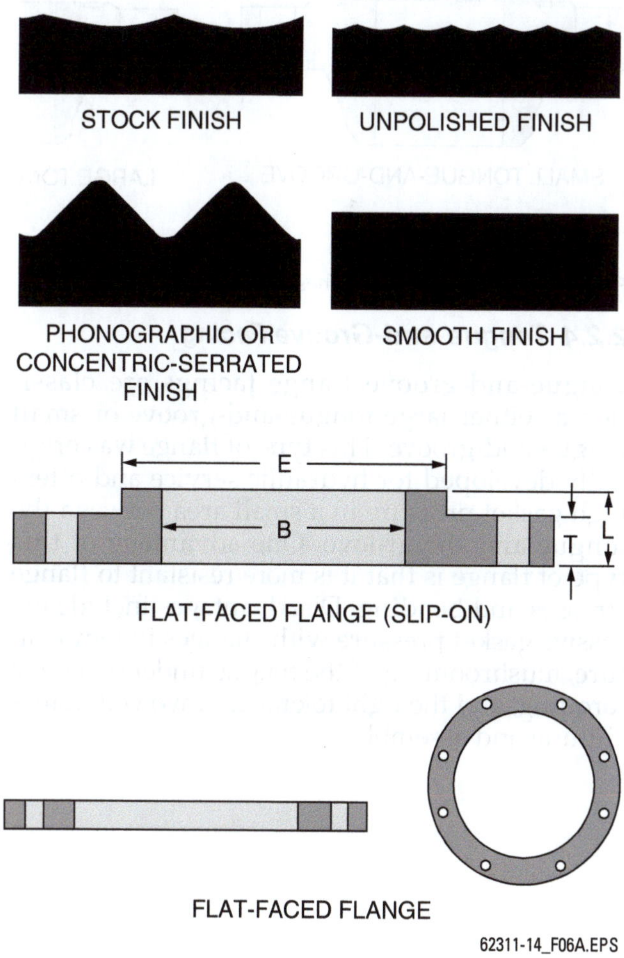

Figure 6A Flange-facing finishes and flanges. (1 of 2)

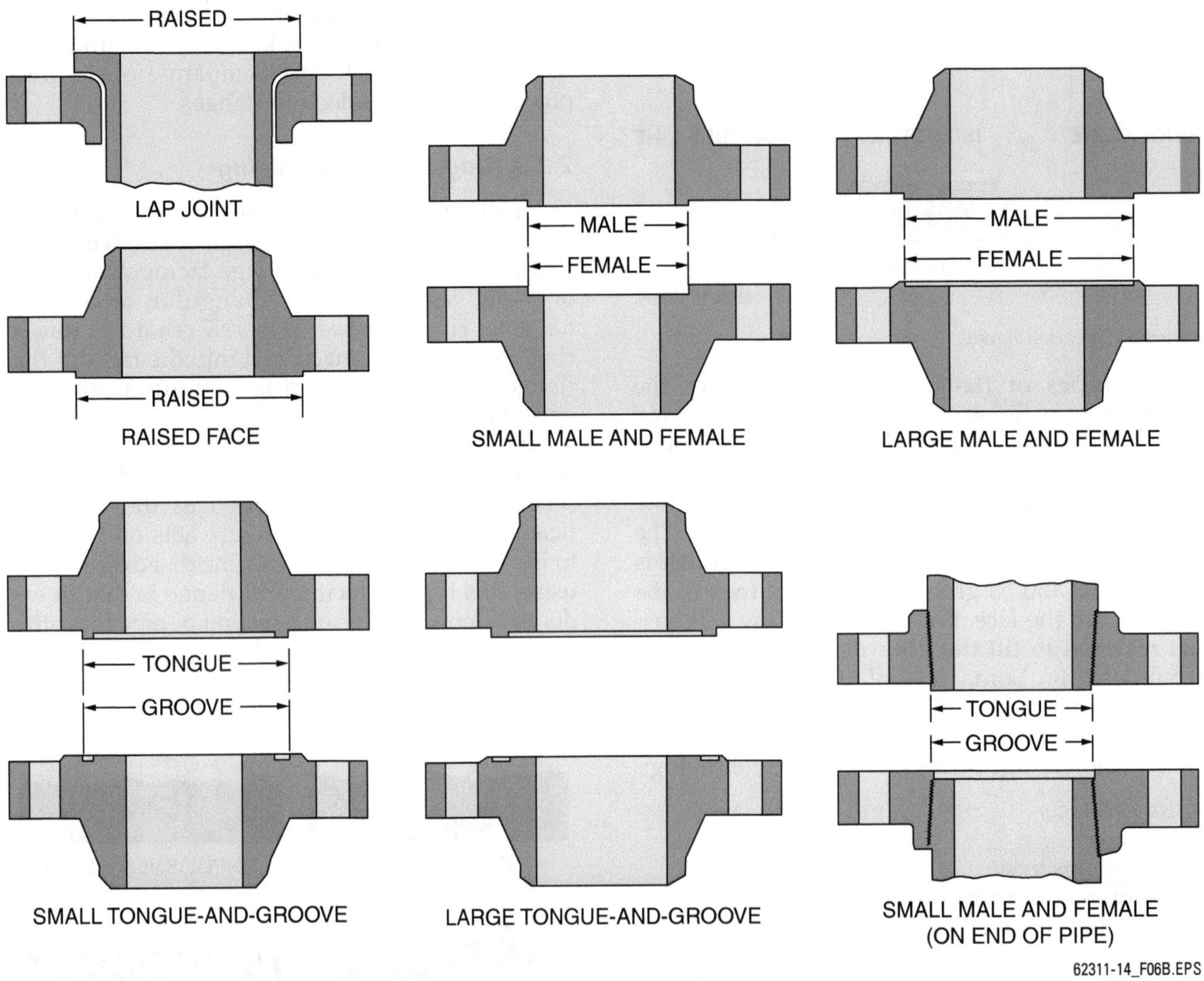

Figure 6B Flange-facing finishes and flanges. (2 of 2)

2.2.4 Tongue-and-Groove Facing

Tongue-and-groove flange facings are classified as either large tongue-and-groove or small tongue-and-groove. This type of flange was originally developed for hydraulic service and offers high gasket pressure in a small area between the tongue and the groove. One advantage of this type of flange is that it is more resistant to flange stresses and bending. Disadvantages include excessive gasket pressure with changes in temperature, mushrooming of the tongue under excessive torquing, and the tight tolerances involved in machining and assembly.

2.2.5 Male and Female Facing

Male and female flange facings fit together much like the tongue-and-groove flange facings. Male and female flanges are also classified as either large or small. A rim on the male face fits into a machined groove in the female flange. A flat ring gasket is used between the rim and the groove in the female flange. The disadvantages of tongue-and-groove flange facings also apply to male and female flange facings.

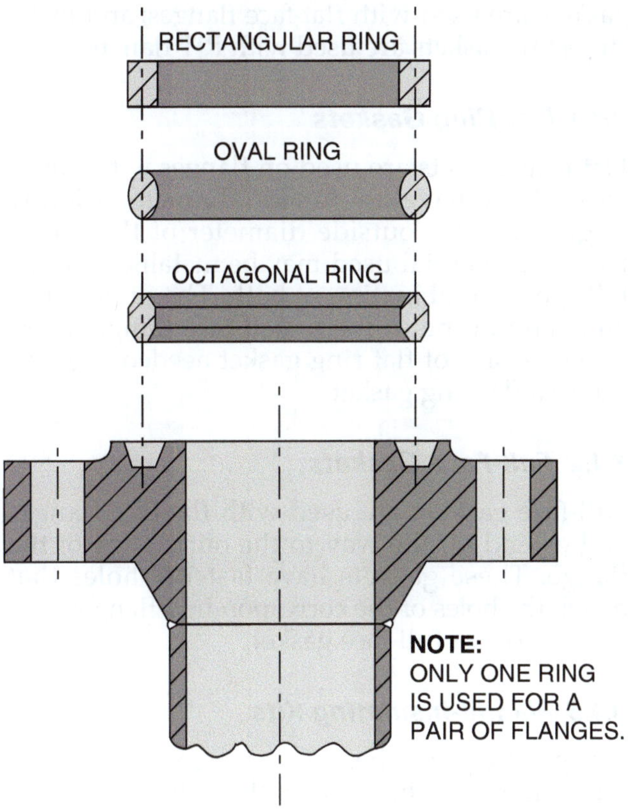

Figure 7 Metal ring-joint gaskets.

Additional Resources

It is assumed that you will learn to perform these procedures under the direct supervision of your crew leader, foreman, or instructor. For independent study, refer to the recommended Additional Resources at the back of this module.

2.0.0 Section Review

1. From the standpoint of internal pressure and bolt loading, the most highly stressed of all flange types is the _____.
   a. blind flange
   b. threaded flange
   c. hubbed flange
   d. lap-joint flange

2. The most common facing used with steel flanges is the _____.
   a. male
   b. tongue-and-groove
   c. raised-face
   d. ring-joint-type

Section Three

3.0.0 Flange Gaskets

Objective

Identify and describe types of flange gaskets and gasket materials.

a. Identify and describe types of flange gaskets.
b. Identify and describe gasket materials.

Trade Terms

ASTM International (ASTM): An international standards organization, formerly known as the American Society for Testing and Materials (ASTM), that develops and publishes technical standards to improve quality and safety across a variety of materials, products, systems, and services.

Gaskets are available in a variety of types and materials, as shown in *Figure 8*. The type of gasket used must be matched to the process characteristics and operating conditions to which it will be exposed. For example, different gasket materials are capable of withstanding different temperature and pressure ranges, and certain gasket materials are compatible with different process fluids. In piping systems, gaskets are placed between two flanges to make the joint leakproof. Gaskets are normally sized by the thickness of the gasket material in fractions of an inch.

To ensure that the composition of various types of gasket, gasket material, and some packing meets the required specifications, **ASTM International (ASTM)** devised a method of identifying different types of gasket material. ASTM, formerly known as the American Society for Testing and Materials, is an international standards organization that develops and publishes technical standards to improve quality and safety across a variety of materials, products, systems, and services, including those used in the pipeline industry.

3.1.0 Types of Flange Gaskets

The appropriate types of gaskets must be matched to different types of flanges. For example, flat ring gaskets are used with raised-face flanges; full-face gaskets are used with flat-face flanges; and metal ring-joint gaskets are used with RJT flanges.

3.1.1 Flat Ring Gaskets

Flat ring gaskets are used on flanges with raised faces. They have an outside diameter slightly larger than the outside diameter of the raised face. The material used may be metallic, nonmetallic, or a combination of both. The dimensions and class rating of the raised-face flange determine the size of flat ring gasket needed. *Figure 9* shows a flat ring gasket.

3.1.2 Full-Face Gaskets

Full-face gaskets are used with flat-face flanges and extend all the way to the outer edge of the flange. These gaskets have fastener holes that match the holes of the corresponding flanges. *Figure 10* shows a full-face gasket.

3.1.3 Flange-Insulating Kits

Many different types of flange-insulating kits are available to be used with gaskets. Fastener holes on standard flanges are 0.125" larger than the standard fasteners to allow full-length insulating sleeves to be used. As long as the sleeves are properly aligned, they will provide adequate insulation. *Figure 11* shows a typical insulated flange assembly.

Existing flanges can be insulated without taking the pipeline out of service if the existing gasket is made of a good insulating material. It is impossible to insulate an in-service flange if the existing gasket material is metal or graphite-impregnated. To insulate a fitting that has the appropriate gasket material, remove one fastener at a time from the flange, insulate it, and then reinstall it before removing the next fastener.

Sometimes existing flanges are not aligned closely enough for insulating standard fasteners. When this occurs, personnel use insulated steel washers and a half-length insulating sleeve on one half of the flange. A half-length sleeve goes through one side of the flange only, so it does not have to be aligned with the other side of the flange. Full fastener insulation can be used if a smaller fastener with a higher strength is used.

For new construction, insulated fittings are sometimes assembled and then tested before they are welded into the line. Prefabricated insulated flanges can be tested using an ohmmeter. The resistance values should be measured in the megohm range to indicate proper insulation.

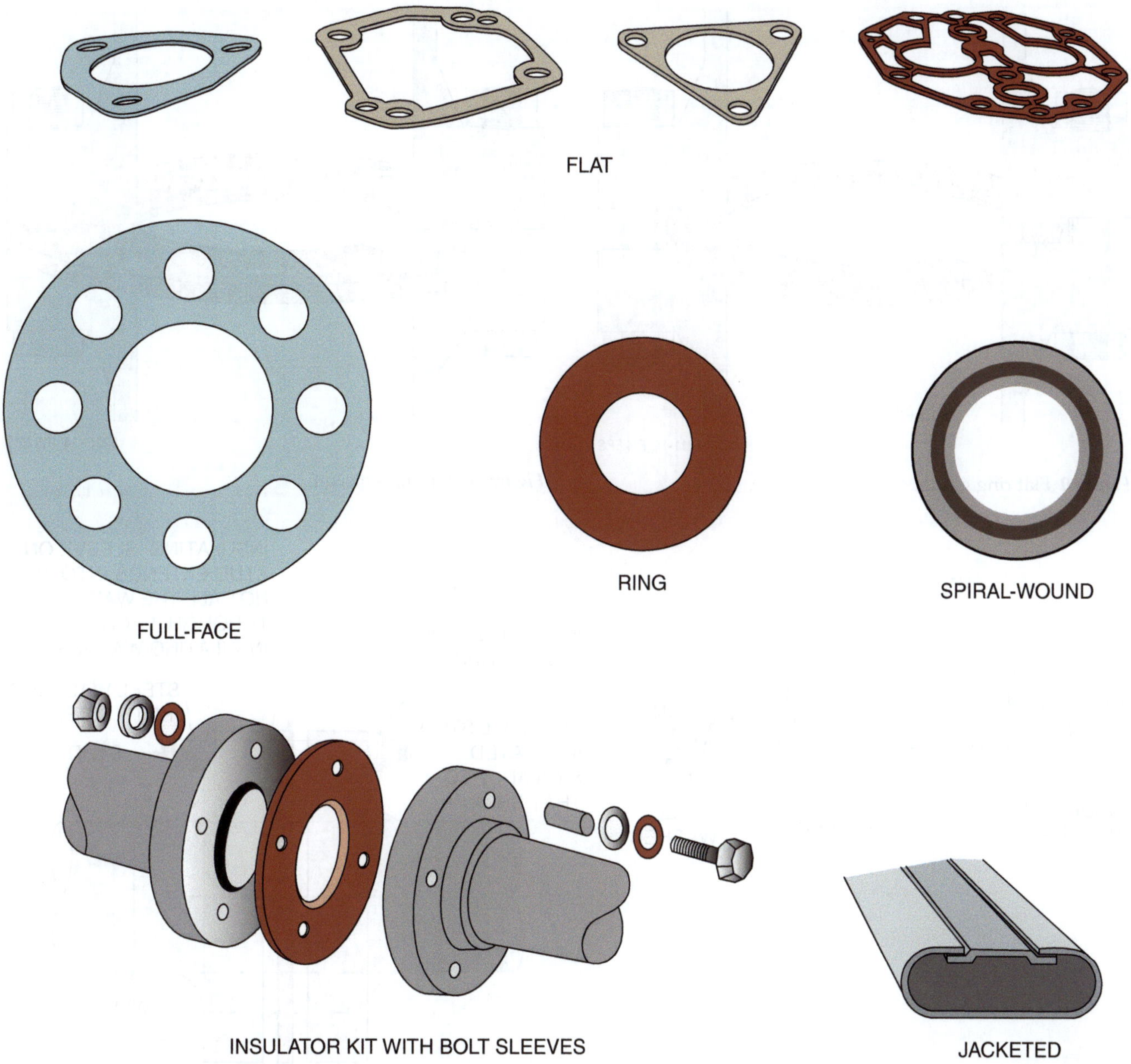

Figure 8 Different types of gaskets.

Pre-assembled flanges also allow for perfect alignment of the flange. This reduces the stresses placed on the insulating material when it is installed in the pipeline.

An insulated fitting has a substantial length of pipe on each side of the fitting. This reduces the resistance of several million ohms measured before installation to a resistance of only a few ohms after installation. The longer the length of the pipe on each side of the fitting, the lower the resistance will be. The resistance measured is also affected by the quality of the coating and the soil resistivity in the area. Measuring the resistance of an installed fitting means measuring the fitting resistance as well as the resistance to earth on each side of the fitting.

NOTE

Measurements of pipe-to-soil potential (CT1_1-17) on either side of the fitting may be used to identify possible shorts. Always use an insulator checker to identify whether the fasteners or the flanges are shorted.

If a defective insulated flange is found during testing, it may be possible to repair it without taking the pipeline out of service. To determine whether this is possible, personnel identify the

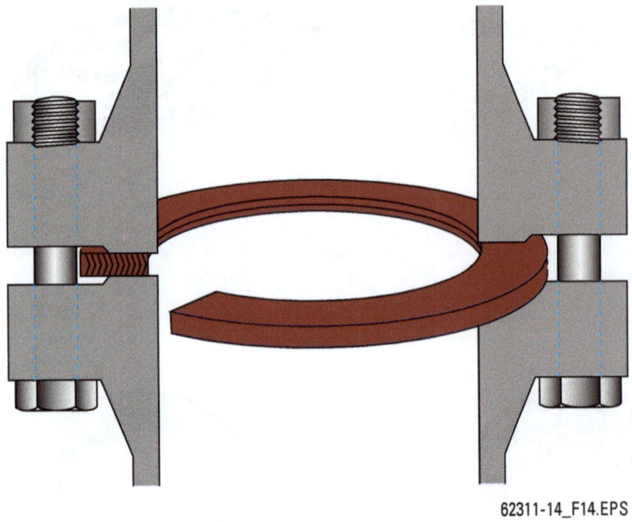

Figure 9 Flat ring gasket.

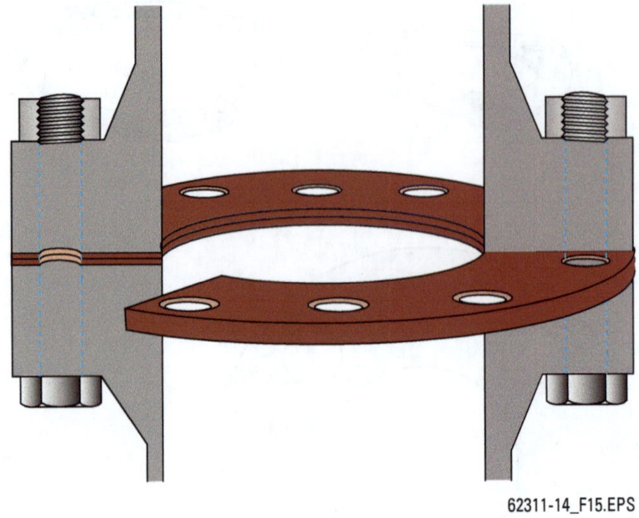

Figure 10 Full-face gasket.

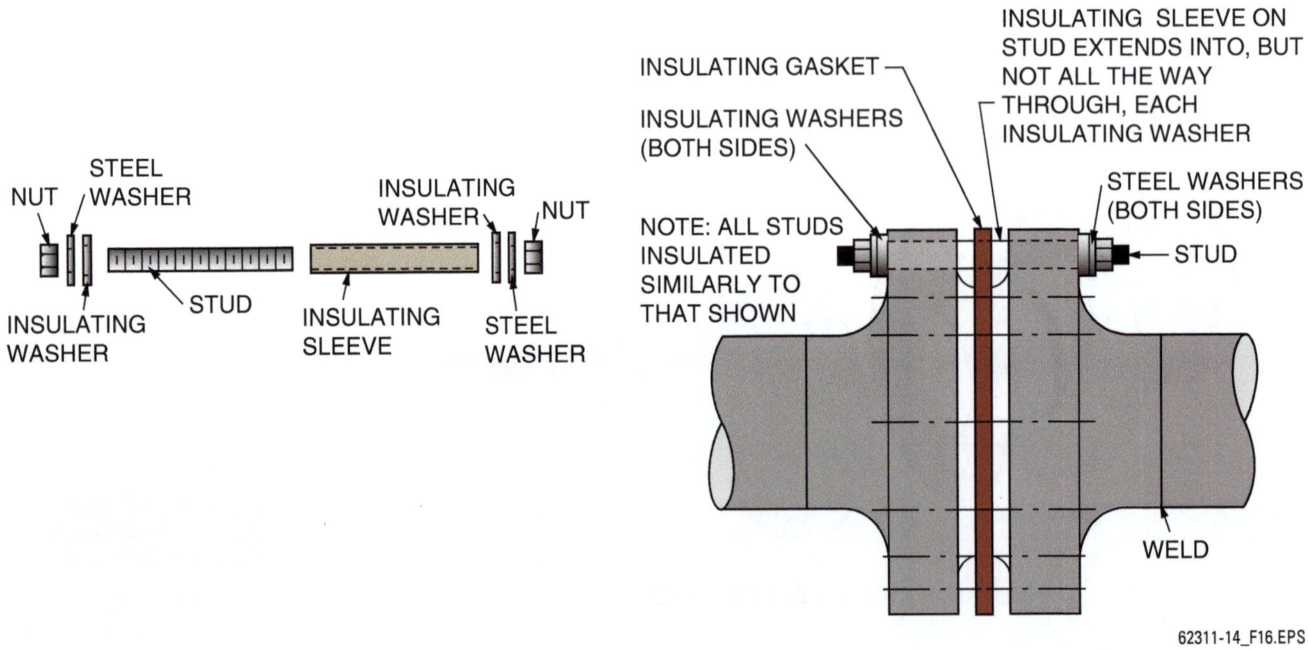

Figure 11 Insulated flange assembly.

cause of the short by checking the resistance between the flanges and each fastener. Fasteners that are shorted have a low or zero resistance. The insulation can then be replaced on the shorted fasteners individually. If all of the fasteners have a high resistance and any possible bypassing routes have been isolated, the cause is probably a shorted gasket.

If the short is a defective insulated fastener, the short may be fixed without taking the pipeline out of service. If the flange is shorted across the gasket, the pipeline must be taken out of service and the gasket replaced.

3.2.0 Gasket Materials

Gaskets are made of many different types of materials to meet the demands of the particular process system in which they are installed. Generally, one of the following four conditions will exist in a process system. These conditions affect the types of gaskets that can be used in the system.

- *High temperature/low pressure* – Generally includes temperatures from 500°F to 1,200°F and pressures up to 600 pounds per square inch (psi).

- *High temperature/high pressure* – Generally includes temperatures from 500°F to 1,200°F and pressures from 600 to 2,500 psi.
- *Low temperature/low pressure* – Generally includes temperatures up to 500°F and pressures up to 600 psi.
- *Low temperature/high pressure* – Generally includes temperatures up to 500°F and pressures from 600 to 2,500 psi.

Usually, the types of gaskets and gasket materials to be used on a project are listed in a special section of the piping specifications. Gasket materials are rated by ANSI specifications for applicable pressures.

Manufacturers use color codes to identify gaskets made of different materials. These color codes vary from one manufacturer to another. Some manufacturers use an additional color-coding scheme on the outside diameter (OD) of some gaskets so the material can be identified without breaking the seal. In such cases, the gasket body may be the same color as other types of gaskets, but the OD color will be different. Always check the color-coding scheme used by a particular manufacturer before replacing a gasket.

NOTE

The use of a commercial liquid gasket remover can make removing old gasket material easier. It reduces the amount of scraping and sanding needed, thus preventing possible damage to flange surfaces. Use a liquid gasket remover only if doing so is approved by company policies and procedures.

Since most process systems are subject to a combination of temperature and pressure, it is extremely important to match the gasket material to the operating conditions of the process system. The type of gasket material used is based on several factors. Generally, the gasket is chosen to withstand the temperature, pressure, and chemical properties of the process medium. Different gasket materials have different pressure and temperature ratings. These ratings are a reflection of the ability of a gasket material to withstand forces in a process system.

Common types of gaskets include the following:

- Polytetrafluoroethylene (PTFE)
- Fiberglass
- Acrylic fiber
- Metal
- Rubber
- Spiral-wound
- Vinyl
- Ceramic
- Asbestos
- Ethylene-propylene-diene terpolymer (EPDM)
- Neoprene
- Nitrile
- Silicone
- Viton®
- Gylon®

3.2.1 PTFE Gaskets

Polytetrafluoroethylene (PTFE) flange gaskets are nonporous, nongalvanic gaskets. They are usually white and are available in several pressure/temperature limits. PTFE is available in different forms and products. The biggest advantage of PTFE is that it is chemically inert. This means that it does not react chemically with the process medium.

Standard amorphous PTFE, the original form of PTFE, is available in sheet form, and can be cut into the various sizes necessary to fit the applications. However, amorphous PTFE is frequently associated with a phenomenon referred to as gasket creep and cold flow, which results in a loss of bolt load. When under continuous load, PTFE tends to continuously flow due to a lack of strength. This often creates a necessity for personnel to retighten fasteners in order to compensate for this material flow. The flow is slowed somewhat by using fillers, but other advantages are lost. Fillers may introduce characteristics that differ from the nonstick and chemically stable PTFE.

There is a process used to expand PTFE in a manner that dramatically reduces and nearly eliminates PTFE gasket creep and cold flow. These products are made from virgin PTFE resin, have no fillers, have the same chemical compatibility as the original PTFE, operate in the same temperature range (–450°F to 500°F), and have no colorizers or other additives. Expanded PTFE, due to its very high tensile-strength properties, can be used in pressure environments from full vacuum to 3,000 psi. Cord gasket material, made with expanded PTFE, is also available (*Figure 12*). It is made in rope form and sold in rolls to be applied to the joint by hand.

3.2.2 Fiberglass Gaskets

Fiberglass gaskets are treated so that they will remain soft and pliable. They sometimes include wire inserts for service for boiler handhold and manhole covers, as well as for tank heads and other high-pressure applications. Fiberglass gas-

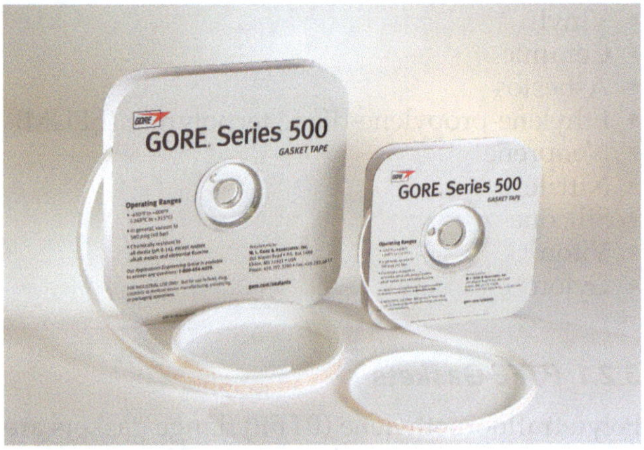

Figure 12 Cord gasket material.

kets are usually white in color. They can typically be used in applications up to 380°F and 180 psi.

3.2.3 Acrylic-Fiber Gaskets

Acrylic-fiber gaskets are used with oil, steam, weak acids, and alkalines. They are usually off-white in color. Acrylic-fiber gaskets can typically be used in applications up to 700°F and 1,200 psi. Many gaskets combine fibers of various materials filled with nitrile, EPDM, or neoprene. *Figure 13* shows a combination gasket made of fibers and other products.

3.2.4 Metal Gaskets

Metal gaskets are used for machine flanges and where metal-to-metal fits are required. They are

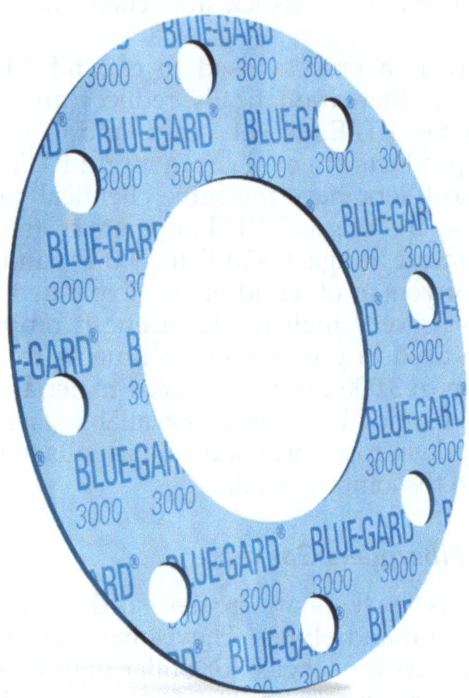

Figure 13 Combination gasket.

good for applications in which higher temperature fluctuations and pressures are normal. When using metal gaskets, the material must be suitable for use within the process system. Some of the types of metal gaskets available are described as follows:

- *Solid-metal gaskets* – Made from solid metal, these gaskets are used in processes suitable for the metal or alloy used to make the gasket. Common materials are copper, Monel®, steel, and iron. Copper gaskets are not used in high-pressure, high-temperature systems. Monel®, steel, and iron gaskets can be used in a variety of pressure and temperature applications.
- *Metal ring-joint gaskets* – The four types of standard ring-joint gaskets are shown in *Figure 14*. They are defined by the ring section: R octagonal; R oval; RX (eccentric octagonal); and BX (rectangular). The octagonal is considered superior. The rings are made of the softest carbon steel or iron available. For lower temperatures, the rings are made of plastic to resist corrosion or to insulate the joint from electric currents. The ring-joint gasket is effective at very high pressures.
- *Serrated metal gaskets* – These are solid metal gaskets with concentric ribs machined into their surfaces. With the contact area reduced to a few concentric lines, the required bolt load is reduced considerably. This design forms an efficient joint. Serrated gaskets are used with smooth-finished flange faces.
- *Corrugated metal gaskets* – Generally used on low-pressure systems where the flanges are smooth and fastener pressure is low. The ridges of the corrugations tend to concentrate the gasket loading along the concentric ridges.
- *Laminated metal gaskets* – These gaskets are made of metal with a soft filler. The laminate can be parallel to the flange face or spiral wound. They can be used in high-pressure, high-temperature applications.
- *Corrugated metal gaskets with asbestos inserted* – These gaskets are used for a variety of pressure and temperature applications in steam, water, gas, air, oil, oil vapor, and refrigerant systems.

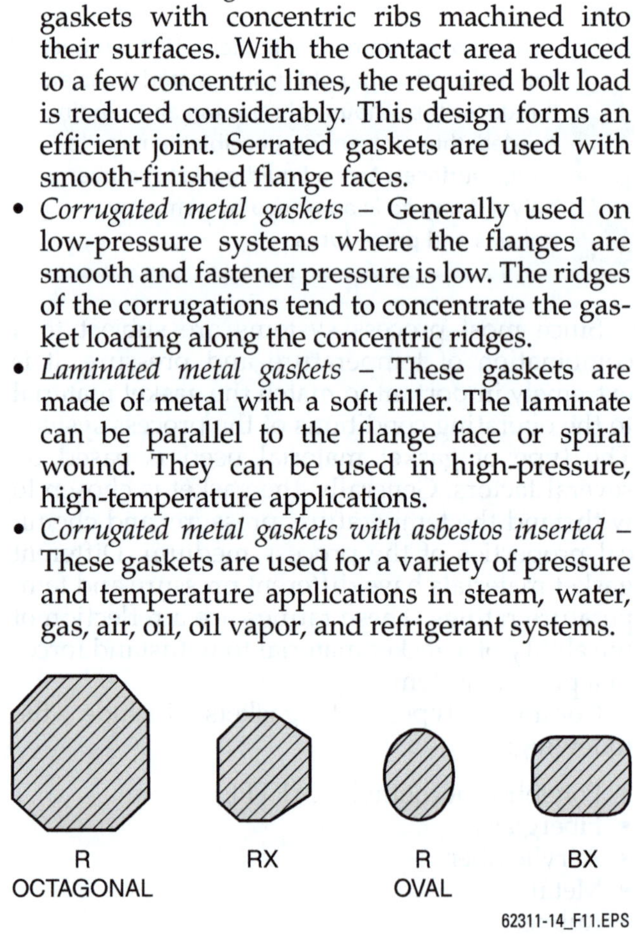

Figure 14 Standard ring-joint gasket profiles.

- *Flat metal-jacket gaskets with asbestos gasket material* – These gaskets are used in the same applications as corrugated metal gaskets, with asbestos inserted.
- *Corrugated metal-jacket gaskets with heat-resistant synthetic filler* – These gaskets are used for the same applications as corrugated metal gaskets with asbestos inserted. Synthetic replacements for asbestos are often referred to as *nonasbestos.*
- *Spiral-wound metal gaskets with heat-resistant synthetic filler* – Commonly made of stainless steel with a graphite insert, and used for a variety of pressure and temperature applications.

3.2.5 Natural-Rubber Gaskets

Natural rubber is seldom used in installations today. It has low solvent- and oil-resistance, but is excellent when used with water. Natural rubber has high resilience and may be used at temperatures up to 175°F. Rubber gaskets are available in a variety of pressure and temperature ratings, and generally come in thicknesses ranging from $\frac{1}{16}$ to 1 inch. They are used in low-pressure/low-temperature water, gas, air, and refrigerant systems. Common types of rubber gaskets include the following:

- *Standard black or red rubber gaskets* – Used for saturated steam up to 100 psi, with an approximate temperature range of –20°F to 170°F.
- *Reinforced rubber gaskets* – These gaskets are strengthened by polyester fabric plies and are commonly used for saturated steam and low-pressure steam. Reinforced rubber gaskets have an approximate temperature range of –40°F to 200°F. They are typically black.
- *High-test, reinforced rubber gaskets* – These are black gaskets used for pressures up to 500 psi. They have an approximate temperature range of –40°F to 200°F.

3.2.6 Spiral-Wound Gaskets

Spiral-wound gaskets are flat-face gaskets, commonly made of stainless steel with a graphite insert. They are used for a variety of pressure and temperature applications in the same systems as corrugated metal-jacket gaskets with heat-resistant synthetic filler. They are crushable gaskets that return to their original shape after being compressed, which allows the gaskets to adjust automatically to changes in line pressure, thermal shocks, vibration, and minor flange separation.

All spiral-wound metal gaskets are color-coded to identify the type of metal from which they are made. They usually have an inner ring, an outer ring, and a section of preformed windings in the middle that serve as the sealing element (*Figure 15*). The inner and outer rings are usually spot-welded, and serve as blowout preventers for the sealing element. The preformed shapes of the sealing element are not welded, and are usually slightly thicker through the gasket.

Table 2 shows the industry-standard color code for spiral-wound gaskets.

3.2.7 Vinyl Gaskets

Vinyl gaskets are specially fabricated for use as oil-resistant gasket, and also have good resistance to water, chemicals, oxidizing agents, ozone, and abrasion. Vinyl gaskets are black in color and have an approximate temperature range of 20°F to 160°F.

3.2.8 Ceramic Gaskets

Ceramic gaskets are used for high-temperature air applications, such as boiler systems, where other types of materials would fail due to the high temperature. Ceramic gaskets can typically be used for temperatures exceeding 1,500°F.

3.2.9 Asbestos Gaskets

Gaskets that contain encapsulated asbestos are still sold and used. They consist of a metal form with asbestos fibers inside, to be used in high-temperature contexts. If the metallic exterior is

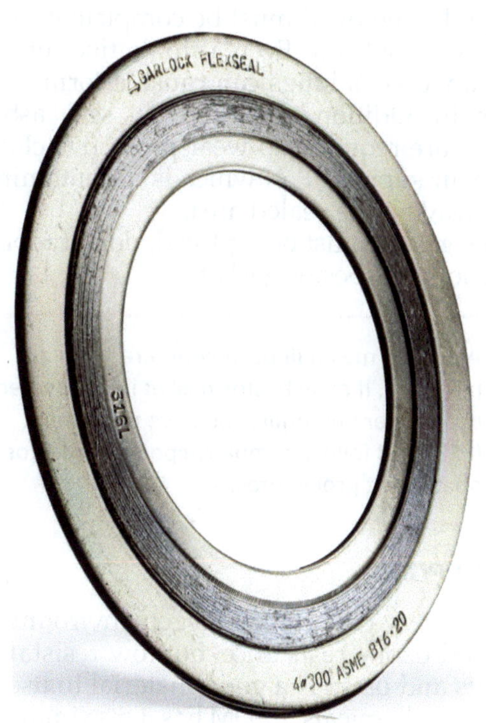

Figure 15 Spiral-wound gasket.

Table 2. Industry-Standard Color Code for Spiral-Wound Gaskets

Metal	Color
Type 304 stainless steel	Yellow
Type 316 stainless steel	Green
Type 317L stainless steel	Maroon
Type 321 stainless steel	Turquoise
Type 347 stainless steel	Blue
Monel®	Orange
Nickel	Red
Titanium	Purple
Alloy 20	Black
Carbon steel	Silver
Hastelloy® B	Brown
Hastelloy® C	Beige
Inconel® 600	Gold
Incoloy® 800	White
Filler Material	**Color of Stripe**
Polytetrafluoroethylene (PTFE)	White
Ceramic	Light green
Flexible graphite	Gray
Mica/graphite	Pink

broken and the white asbestos fibers are exposed, contact a supervisor.

In most locations, there are legal requirements for any action that may cause any possible exposure to asbestos. The area from which the asbestos is to be removed must be completely isolated from the outside air. Even a single fiber of asbestos in a worker's lungs can cause a form of lung cancer. In addition to an airtight seal, asbestos workers are required to wear protective clothing and an air supply, all of which is decontaminated before leaving the sealed area.

If the gasket must be replaced, do not replace it with another asbestos gasket.

WARNING!

Asbestos may still be encountered in older pipelines. If an asbestos gasket is discovered, use proper personal protective equipment (PPE) and follow company-specific asbestos abatement procedures.

3.2.10 EPDM

EPDM is very good in an oxygen environment. It has good ozone resistance, but low resistance to solvents and oil. It is a good material to use with water and chemicals. EPDM has a maximum temperature rating of 350°F, and has excellent flame resistance.

3.2.11 Neoprene

Neoprene, which has good resilience, is only used in noncritical conditions. Neoprene has fair oil and solvent resistance but poor chemical resistance, and it is not recommended for water service. It has a maximum temperature rating of 250°F.

3.2.12 Nitrile

Nitrile is used with medium-pressure oil and solvent services. It cannot be used with acetone or methyl ethyl ketone (MEK), but is an excellent material to use with water or alcohol. It has good chemical resistance but poor flame and ozone resistance. Nitrile has a maximum temperature rating of 250°F.

3.2.13 Silicone

Silicone is a widely used rubberlike material. There are more grades of silicone material than any other rubberlike material. It is not recommended for use with oil or solvents, or in water systems. Silicone has fair-to-good chemical and flame resistance, excellent ozone resistance, and poor-to-excellent resilience, depending on the grade used. It has a maximum temperature rating of 550°F.

3.2.14 Viton®

Viton® is a relatively inexpensive fluoro-elastomer gasket material, and the most commonly used material for chemical resistance. In addition, Viton® also has excellent resistance to oil, solvent, and ozone. It has a maximum temperature rating of 400°F.

3.2.15 Gylon® or Amerilon®

Under proprietary names like Gylon® and Amerilon®, there are materials that combine Kevlar® and PTFE to produce a very tough and resistant low-temperature gasket. This material is chemically inert, has a high resilience (45 percent recovery), and a maximum temperature rating of 500°F. Gylon® 3510 has one of the best sealability factors of any gasket material in the industry, and is one of the most commonly used. It is used heavily in the chemical and paper industries. *Figure 16* shows a gasket made of Gylon®.

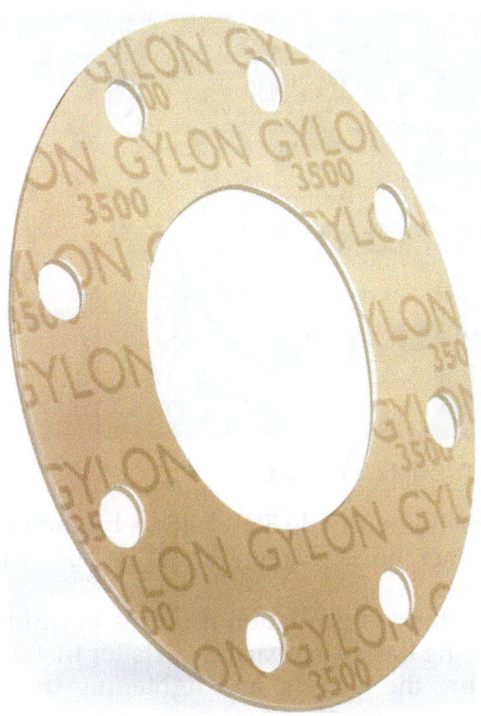

Figure 16 Gylon ® gasket.

3.2.16 Graphite-Impregnated Gaskets

Made of fiberglass and impregnated with graphite particles, these gaskets are used to provide a tight seal in some high-temperature applications. However, they are reactive to oxidation.

3.2.17 Soft-Metal Gaskets

Soft-metal gaskets are designed to compress into the concentric rings machined into the surface of carbon steel and stainless steel flanges. They are used for oil and gas pipelines, where their non-reactive materials and high wear-resistance allow for long periods of service without need of replacement.

Additional Resources

It is assumed that you will learn to perform these procedures under the direct supervision of your crew leader, foreman, or instructor. For independent study, refer to the recommended Additional Resources at the back of this module.

3.0.0 Section Review

1. The type of gasket used with flat-face flanges is the _____.
   a. metal-jacketed gasket
   b. flat ring gasket
   c. full-face gasket
   d. metal ring-joint gasket

2. Spiral-wound gaskets are _____.
   a. color coded
   b. made of PTFE
   c. suitable only for low temperature/low pressure applications
   d. octagonal in shape

SECTION FOUR

4.0.0 INSTALLING PIPE FLANGES

Objective

Explain the procedures for installing pipe flanges.

a. Explain the basic techniques involved in torquing and tensioning.
b. Explain the preparation and procedures involved in installing gaskets.
c. Explain the preparation and procedures involved in tightening flanges.

Trade Term

Torque: The amount of twisting force a fastener can withstand when threaded into a nut or a threaded blind hole drilled into a solid object.

ANSI has developed a standard that all manufacturers use as a guide for fastener hole locations on all flanges up to 24 inches. Flanges larger than 24 inches can have different-diameter fastener holes and a different number of fastener holes, depending on the type of flange. Check the pressure class rating of the flange and refer to flange dimensioning charts to determine the number and size of the fastener holes. The number of fastener holes required on flanges up to 24 inches also varies with the size and rating of the flange, although there is always an even number of holes in graduations of four. The fastener pattern for each size and rating is consistent in that each type, size, and rating has the same fastener-tightening sequence.

ANSI also has a standard to use as a guide for fabrication. The standard requires that flange fasteners always straddle the horizontal and vertical center lines unless otherwise indicated. This is known as the *two-hole method* (*Figure 17*). It is possible for a job specification to require that the fastener holes line up with the vertical and horizontal lines. This is known as the *one-hole method*. The general rule is to reference one center line only, depending on the type of drawing used. In elevation drawings, the usual reference is the vertical center line, and in plan drawings, the usual reference is the north-south center line. This rule ensures that all fabricators refer to the same center line. To determine the angular location of the fastener holes, divide 360 degrees by the number of fastener holes. *Figure 18* shows fastener hole locations.

Figure 17 Two-hole method.

Pipe flange installation requires torquing and tensioning basics, preparing parts for installation, installing the gaskets, and tightening the flange fasteners.

4.1.0 Torquing and Tensioning Basics

Torque is used to describe a twisting force. With fasteners, torque is associated with the amount of twisting force a fastener can withstand when threaded into a nut or a threaded blind hole drilled into a solid object.

4.1.1 Factors Affecting Torque Calculations

If a fastener is hand-threaded into a nut or blind hole, the amount of turning force is limited to the strength of the fingers doing the turning. When a wrench is used on the fastener (and the nut, if a nut is used), the amount of turning force can be increased. When a longer wrench is used, the turning motion can be increased. In all these situations, the amount of turning force is still in the hands of the individual performing the work. The amount of turning force or torque is limited to the manual strength of the individual, and the level of torque differs from one person to another. It may also vary from one fastener to another.

When a fastener is tightened to a specific amount of torque, it is considered properly tensioned. The term *tensioning* is often used in association with fastener-torquing activities, but tensioning applies more to stretching the fastener. The turning effect or torque (T) is expressed as a product of the force (F) and the radius (R) through which the force acts. The amount of torque (T) is expressed in pound-feet or foot-pounds. The following formula represents this expression:

$$T = F \times R$$

If 5 pounds of force is applied to the outermost end of a 6-inch wrench, the amount of torque is considered to be 30 inch-pounds ($2\text{-}\frac{1}{2}$ foot-pounds). If a 3-inch wrench is used, the torque would be only 15 inch-pounds. The torque on small fasteners may be expressed in inch-pounds. In the metric system, torque is expressed in Newton-meters. One Newton-meter equals 1.3558 foot-pounds.

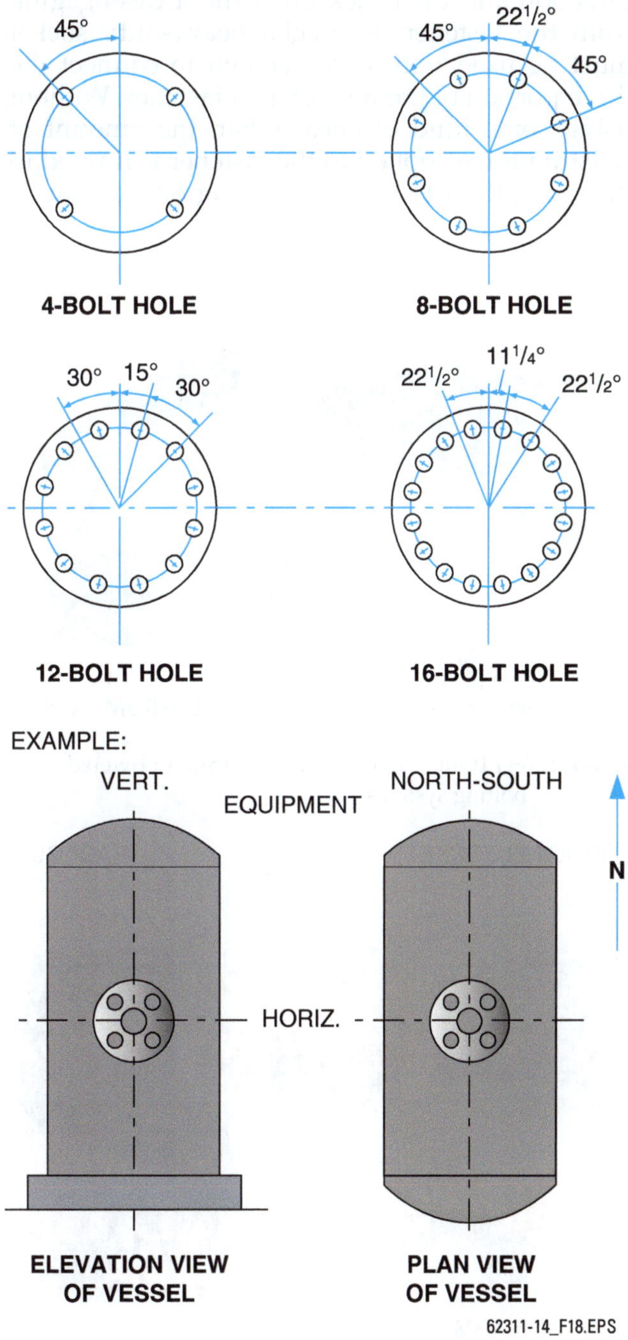

Figure 18 Fastener hole locations.

4.1.2 Bolt Charts

Most bolt charts provide torque values for fasteners up to one inch in diameter. Along with a fastener's torque value, the charts also may give the number of threads per inch (TPI) or pitch of a fastener, its clamping power, and even its minimum tensile strength. Such factors must be considered before torquing activities are performed. Having moisture or a lubricant between the fastener head or nut and the surface of the members being joined alters the torque value. Recommended torque values can vary by engineer and/or owner.

4.1.3 Alignment of Members

When two members, such as the flanged ends of two pipes, are bolted together, there should be no stress on either member when they are properly aligned and ready to be connected. Refer to any customer specifications for specific alignment requirements.

4.1.4 New or Used Fasteners and Nuts

Bolt charts and engineering drawings assume that the fasteners being installed are new fasteners. Any time that fasteners and nuts are mated and torqued, their threads stretch or deform a little from the condition they were in when new. When fasteners and nuts are re-used, that stretching increases, which means that the fastener/nut combination is not as strong as it was when the components were new. Ideally, a nut mated to a particular fastener should never be used on any other fastener.

Since human strength is not enough to properly torque such fasteners and nuts, special torquing and tensioning tools are used.

4.1.5 Torquing and Tensioning Hardware

The basic drive ratchet is a tool found in most maintenance technicians' toolboxes. Some maintenance personnel have their own torque wrenches (*Figure 19*), but they are rarely cali-

Figure 19 Typical torque wrench with socket.

brated. Companies that are serious about the calibration of their torque wrenches treat them as specialty tools. They keep them in protective boxes, carefully stored in the maintenance area. They are only taken out when manual torquing is necessary, and they are calibrated on a regular basis.

To handle the larger fasteners and nuts, most companies have one or more multifunctional torquing and tensioning tools that use hydraulic, pneumatic, or battery power. HYTORC® is one of several companies specializing in these devices. *Figure 20* shows two examples of HYTORC® hydraulic bolting systems.

WARNING!

To prevent personal injury, and to prevent damage to the tools or the equipment being tightened, only trained and qualified personnel are allowed to use hydraulic torquing and tensioning tools.

Figure 21 (A) shows a HYTORC® pneumatic 3-in-1 bolting system. A battery-powered lithium gun used for heavy industrial bolting is shown in *Figure 21 (B)*.

Another company, New World Technologies, builds and sells the RAD® torquing system. A RAD® gun being used with a reaction arm is shown in *Figure 22*. The reaction arm keeps the gun from turning in the opposite direction as the nut is tightened or loosened.

WARNING!

Always review the safety guidelines in the operator's manual for any torquing tool.

Torque wrenches have ratchet devices that allow them to tighten or loosen fasteners in one direction and then back up without disengaging from the fastener. A special heavy-duty socket must be used with each wrench to connect the drive portion of the wrench to a fastener. Without using some kind of cheater bar, the amount of twisting force applied to the fastener is limited to the strength of the person pulling on the handle of the wrench.

Figure 20 HYTORC ® hydraulic bolting systems.

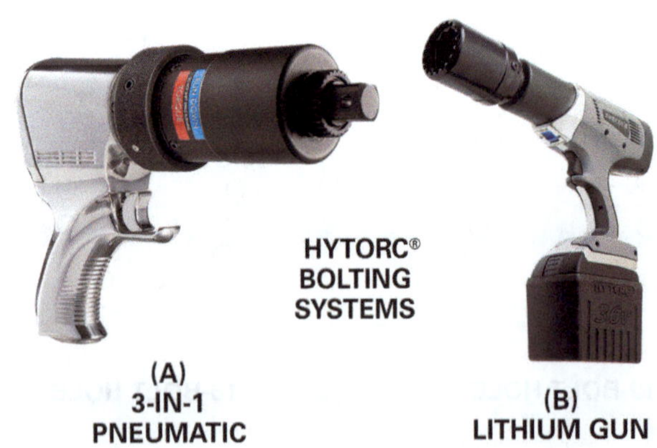

Figure 21 HYTORC ® pneumatic and lithium-powered bolting systems.

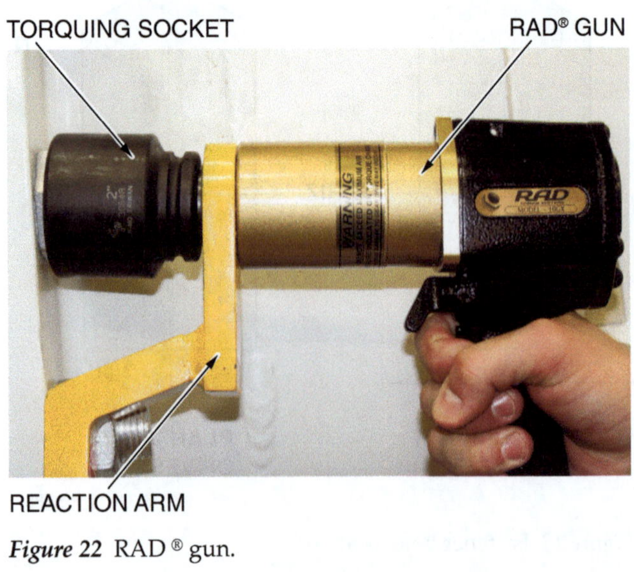

Figure 22 RAD ® gun.

A ratcheting box-end wrench allows the user to place the box end of the wrench onto a fastener or nut and tighten or loosen in one direction, and then back up without disengaging from the fastener or nut. Ratcheting wrenches are manual versions of hydraulic torquing and tensioning tools. The difference is that hydraulic fluid drives the gearing inside the drive portion of hydraulic tools. The head area of a basic hydraulic torque wrench is like the head area of ratcheting box-end wrenches.

A drive attachment made to fit into the drive head of the hydraulic torque wrench has a square drive end made to mate with the drive end of a socket. With the drive installed, the wrench is turned onto its side, and the square drive end is pushed into a socket selected to fit the fastener to be torqued. To tension a nut-and-fastener combination, a special washer or nut is used between the standard nut and the member being secured. Think of these devices as being special lock washers.

Some vessels may have a hundred fasteners and nuts securing only one cover. If each of those fasteners and nuts does not properly do its job, there may be a leak, and a leak can lead to a fire, an explosion, and possible injury to nearby workers. Torquing and tensioning devices are used to securely tighten all those fasteners and nuts to their proper levels of tightness.

Proper tightening techniques are critical. The number of torquing and tensioning tools available is normally limited by the maintenance budget. If only two tools are available and the job is to tighten the fasteners in a set of flanges with eight fasteners, the two fasteners positioned 180 degrees from each other are tightened before the tools are moved on to the next set of fasteners. The fasteners are only tightened small amounts during each cycle of tightening so that the flanges are not warped. *Figure 23* shows the recommended tightening patterns found on most vessels.

When multiple torquing and tensioning tools are used, hydraulic hoses are connected to form a daisy chain from one tool to the next with a hose going from the last tool back to the pump. All hydraulic hoses and tools are fitted with quick-disconnect connectors. Each tool has an input and an output connector. Some tools have multiple connectors.

Fasteners are usually only torqued part way during the first round of torquing. Consult the torquing specifications for whatever equipment

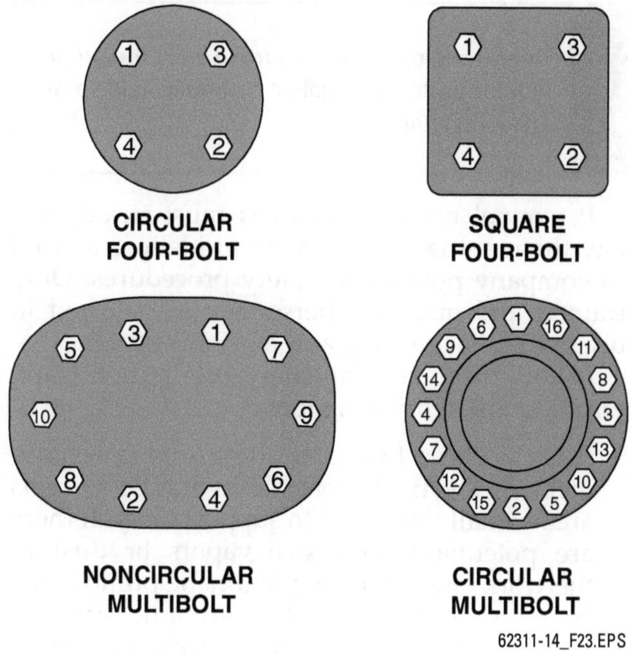

Figure 23 Recommended tightening patterns.

items are being worked on. When the tool is first placed onto the nut and activated, it clicks. The clicking sound indicates that the hydraulic cylinder inside the tool has tightened the nut as far as it could with that stroke. Releasing the applied pressure retracts the cylinder and resets the tool for another try.

Not all hydraulic tension wrenches can be set to stop at a particular pressure. With models that do not automatically stop, watch the pressure gauge to stop the increase. Extending and then retracting the cylinder one time is considered to be one cycle. During the final round of torquing, the cycles should be repeated until the tool stalls at the final preset torque/psi setting of the hydraulic system. To ensure that the tool has stalled, try one more cycle before removing the tool from the nut.

4.1.6 Torquing and Tensioning Safety

Torquing and tensioning devices, especially high-pressure hydraulic tools, can be dangerous. During torquing and tensioning activities, parts can break and fly anywhere. Torquing and tensioning tools can become disengaged, which can turn them into hazardous flying objects. Members being disconnected may be under stress and may move quickly and dangerously when the fasteners are released. Care must be taken not to drop any tool or member on personnel during these activities.

WARNING!

To prevent injury, make sure to wear the proper PPE for protection against hydraulic fluid spray and flying parts.

Personnel using these tools must strictly follow the manufacturer's safety guidelines as well as company policy and safety procedures. Only trained and qualified personnel are allowed to use hydraulic torquing and tensioning tools.

Follow these general safety rules when using torquing and tensioning tools:

- Because these tools are often used on equipment in the field, portable hydraulic systems are normally required to support them. If there are potentially explosive vapors or dust in the work environment, pneumatic pumps may be required to drive the hydraulic systems. Review and follow all manufacturer recommendations for the safe use of their tools.
- To prevent electrical shock, ensure that all equipment is properly grounded.
- Ensure that all hydraulic lines and connectors are in good condition and properly connected.
- Do not kink or pinch hoses.
- Inspect each tool and make sure that it is in good operating condition with all guards in place.
- Verify that each tool has been calibrated within the past year, at a minimum.
- Ensure that the pump and tool(s) are operated only by a designated operator.
- Call "All Clear" prior to starting the pump.
- Before placing a tool onto a nut, test it to ensure that it is functioning properly.
- Ensure that the right socket is selected for each nut being tightened.
- When positioning a torquing-and-tensioning tool for torquing, make sure that the tool is properly positioned for tightening (remember that they must be turned over when loosening nuts) and that the hydraulic hoses to it are not in a position to be pinched.
- Attach the tool and socket to the nut being tightened.
- Pressurize the hydraulic system for just a moment to see how the tool reacts. If it rides up or creeps, stop and re-adjust the reaction arm to a more secure location. If everything appears to be correct, begin the process.

Before starting any torquing or tensioning job, review the engineering specifications for the process equipment on which the work will be performed. Review and follow site safety procedures, and make sure that the following requirements are met:

- All torquing and tensioning levels are clearly understood.
- All work-area hazards are understood, and appropriate precautions are taken.
- The hydraulic pumping system being used has sufficient power to support the tools performing the work.
- All fasteners and nuts are new, or approved by the engineer responsible.

4.1.7 Torquing Conventional Nuts

Conventional nuts and fasteners may or may not have washers placed between the nut and the surface of the components being connected. The nuts should be threaded down their fasteners far enough to take up any slack before the torquing tool is installed. Preloading the fasteners and nuts ensures that the pieces are properly mated before the torquing process begins. If the fasteners are loose on the surface of the opposite flange member, use a wrench strong enough to secure the fastener head during the torquing activities. Also make sure that the wrench is securely installed. The hydraulic hoses should have already been connected between the hydraulic system and the tools.

Before any kind of lubricants are applied to the fastener threads or the nuts, check assembly specifications for the members being assembled.

Perform the following general procedure to torque conventional nuts:

Step 1 Obtain a socket to fit the nut being tightened. If needed, also make sure a wrench is properly installed and secured on the fastener head at the opposite side of the member being attached.

Step 2 Insert the square drive attachment into the socket and place the socket onto the nut.

Step 3 Set the pressure of the hydraulic system to the first desired torquing level and lock the pump setting.

Step 4 Apply pressure to the tool in two or three quick shots, and cycle the tool to ensure that it is operating properly.

Step 5 Position the torquing tool into the torquing position and lower it onto the square drive attachment.

Step 6 Ensure that the tool is properly coupled to the square drive attachment.

Step 7 Ensure that the tool is positioned with its reaction pad securely mounted against a firm object.

Step 8 Ensure that all hoses are free of any obstacles.

Step 9 Apply pressure to the tool and ensure that the reaction point is stable.

Step 10 Continue applying pressure until a click is heard.

Step 11 Repeat the cycles until the first specified level of torque is applied to the nut and fastener.

Step 12 Release pressure to the tool and remove the tool from the square drive attachment.

Step 13 Remove the socket and drive attachment from the nut and verify that neither the nut or tool is damaged.

Step 14 Move the socket and drive attachment to the next nut and repeat the process.

Step 15 Continue torquing all the nuts until they all have been torqued to the initial point.

Step 16 Reset the hydraulic system for the next level of torque, and then repeat the process on all the fasteners.

Step 17 Remove all the tools and associated hardware and inspect them for damage.

Step 18 Store tools and associated hardware as directed.

4.1.8 Detensioning Nuts and Fasteners

The detensioning process is similar to the loosening process. Perform the following general procedure to detension nuts and fasteners:

Step 1 Install the holding socket that fits the nut and position its L-T switch to the L position.

Step 2 Hand-twist the socket to seat it onto the nut and washer.

Step 3 Flip the torquing-and-tensioning tool over so that it is positioned to loosen, and then engage it to the square drive assembly.

Step 4 Make sure that the reaction pad of the tool is securely positioned.

Step 5 Pressurize the tool momentarily to see if the socket engages. If the socket does not fully engage, remove the tool and reseat the socket. Also make sure that there is no grease or oil on the surfaces.

Step 6 Press and hold the control device, applying pressure to the tool, and listen for a click indicating that the hydraulic cylinder in the tool has fully extended.

Step 7 Release the applied pressure and listen for the cylinder to click again when it fully retracts.

Step 8 Repeat this process until the nut and washer can be turned by hand.

Step 9 Remove all the tools and associated hardware when finished, and inspect them for damage.

Step 10 Store tools and associated hardware as directed.

4.1.9 Maintaining Torquing and Tensioning Tools

Many torquing and tensioning tools have seals that may deteriorate or become damaged over time. Those seals must be replaced. Mechanical parts inside the tools wear and eventually break. When a tool fails, refer to the troubleshooting charts in the operator's manual for guidance.

Even new hydraulic devices tend to leak, which means that occasionally the tools may be slick. Dirt and dust collect on any oily surface, and the environments in which these tools are used are not always the cleanest. However, these tools must be kept clean. Along with the portable hydraulic units, they should be tested in the shop before they are carried out to a work site.

Records must be maintained as to when and where these tools were used. Regulatory guidelines often mandate that these tools be calibrated on a specific basis. Do not use uncalibrated tools. If a tool is found to be uncalibrated or unusable, tag it and turn it in for repair or calibration.

4.2.0 Preparing Flanged Joints and Installing Gaskets

Before installing gaskets, personnel must perform a number of steps to ensure proper installation. They must first clean and inspect the parts, lubricate the parts as appropriate, and align the parts.

4.2.1 Cleaning and Inspecting Parts

As with all pipe joints, flanged joints must be cleaned before installation to ensure the best possible fit. Flanged joints are designed so that the face of each flange matches up squarely with a gasket to form a tight seal. All rust, debris, or other contamination must be removed from all flange and gasket surfaces; these obstructions could cause the flanges to connect improperly or cause the seal to break. Flanges from the factory and on other equipment often have a gum, paint, or residue that must be scraped off the face. Flanges with grooved faces must be clean and free of dirt so that the gasket will fit properly. It is sometimes necessary to use solvents to remove the protective coating put on flanges at the factory.

Always follow company policies and procedures for the proper cleaning process and tools, and always check the engineering specifications before applying any type of solvent to the flanges. In addition, be sure to visually inspect the joints to ensure there is no mechanical damage or corrosion, and to ensure that they have the appropriate ANSI rating for the intended service.

4.2.2 Lubricating Parts

When preparing joints for gasket installation, lubrication is a critical step. Lubrication reduces friction, and less torque is then required to achieve a given tension in the joint. It also allows for a consistent load across the fasteners within the joint, and makes disassembly of the fasteners easier. All internal and external threaded surfaces and nut/washer faces should be coated generously with a lubricant.

Personnel use different types of lubricants depending on the specific application of the fasteners. There are different torque ratings for each type of lubricant. Always follow company policy and procedures regarding lubrication.

4.2.3 Aligning Parts

After it is cleaned, the pipe is put into place and properly supported. The jacks and stands must be adjusted until the flanges are properly aligned. *Figure 24* shows good and bad alignment. Drift pins can be used to align the fastener holes; personnel should never try to align the fastener holes with their fingers. Specifications and standards will give tolerances for misalignment. Use a Hi-Lo® gauge to measure pipe and fitting misalignment during installation, as shown in *Figure 25.*

4.2.4 Installing Gaskets

After the parts are aligned, the gaskets should be selected and put in place. As explained earlier, each flange has its own gasket type. Follow this general procedure to install a full-face gasket between flanges:

Step 1 Check the specification to determine the type of gasket needed.

Step 2 Find the proper gasket, based on the specification. Lubricate if necessary, per company policy.

Step 3 Place the gasket between the flange faces.

Step 4 Hold the gasket in place, and insert the bottom bolts into the flanges.

WARNING!

Personnel should never use their fingers to hold the gasket in place; use a drift pin instead.

Step 5 Insert all the fasteners through the flange fastener holes.

Step 6 Place the nuts on the fasteners, and tighten the nuts finger-tight.

Follow this general procedure to install a gasket between two raised-face flanges:

Step 1 Check the specifications to determine the type of gasket needed.

Step 2 Find the proper gasket, based on the specification. Lubricate if necessary, per company policy.

Step 3 Insert the bottom fasteners into the flanges.

Step 4 Place the gasket between the flange faces.

Step 5 Insert all the fasteners through the flange fastener holes.

Step 6 Place the nuts on the fasteners, and tighten them finger-tight.

4.3.0 Tightening Flanges

Flanges must be tightened correctly to provide an evenly distributed clamping force that will prevent leakage around bolted connections. Proper stud tension, correct tightening pattern, and clean, rust-free studs are necessary for properly tightened bolted connections.

A hydraulic fastener tensioning system provides the most accurate method of applying proper fastener stress, and is highly recommended for tightening pump flanges and other critical flanged connections. This system must be used according to the manufacturer's operating

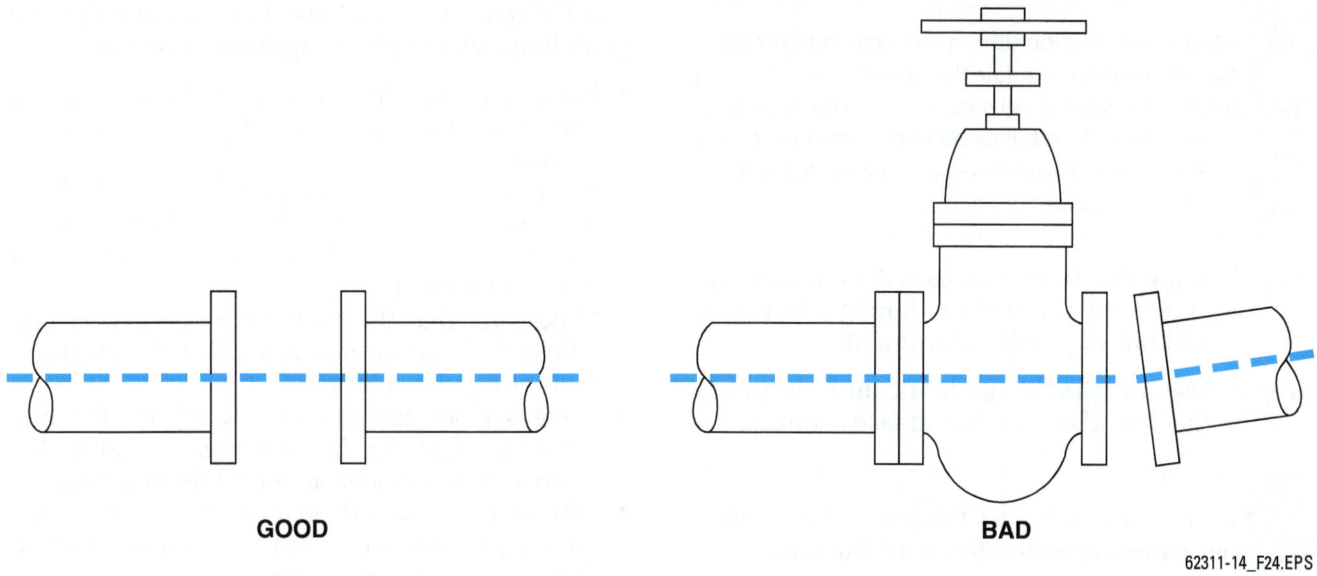

Figure 24 Good and bad alignment.

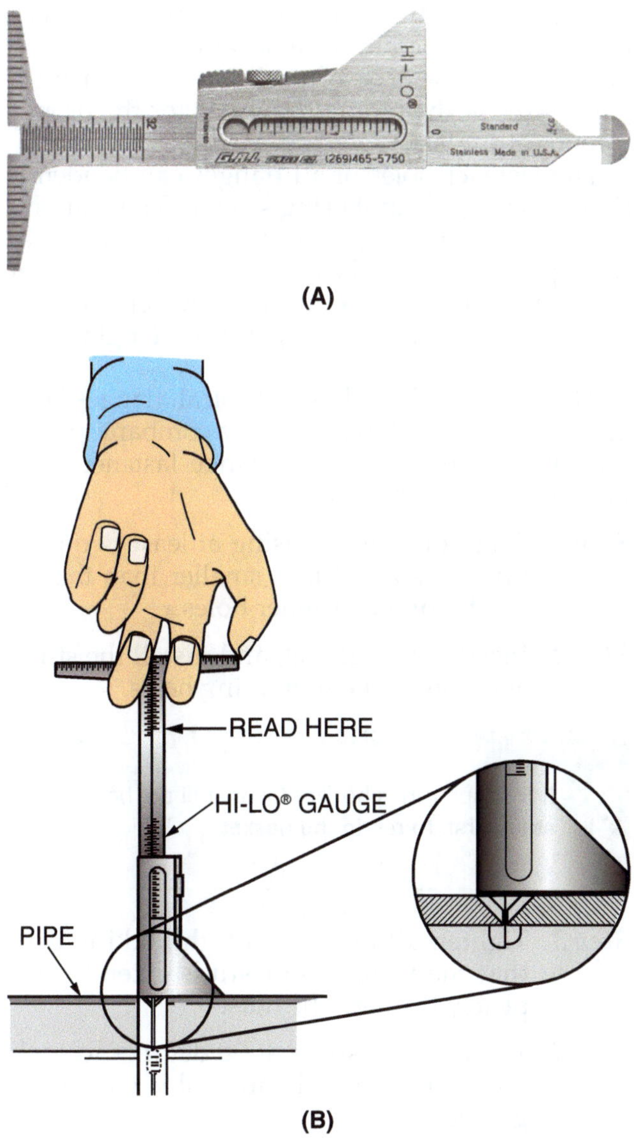

Figure 25 Using a Hi-Lo ® gauge.

procedures. This section addresses the preparation and procedures for tightening flanges.

4.3.1 Flange-Torquing Preparation

This section describes typical preparations for flange torquing. Always follow company policy and procedures. To prepare for flange torquing, follow this general procedure:

Step 1 Ensure that the piping is free of foreign materials and construction debris.

Step 2 Examine the studs for correct size and material specification.

Step 3 Examine the studs for mechanical damage, such as burrs or nicks, dirt, or rust. Replace if necessary.

Step 4 Check that the studs have sufficient length for the tensioning tool.

Step 5 Ensure that the gaskets are the correct size, type, and material for the application.

> **NOTE**
>
> Used studs that are in good condition may be re-used. Never install a used gasket.

Step 6 Clean the flange faces using a steel scraper or wire brush. The appropriate tool needed may vary depending on the pipe material.

WARNING!

Always use caution during the cleaning process. Never mix the materials that are cleaned off the flange faces, as this can create a hazardous atmosphere. Also be aware that cleaning with a steel scraper or wire brush may cause a spark, creating an ignition source.

Step 7 Wipe the flange faces with a clean rag, and check for surface damage, trueness, and out-of-parallel alignment.

Step 8 Use an appropriate lubricant to lubricate the threads of studs and faces of nuts.

NOTE

All internal and external threaded surfaces and nut/washer faces should be coated generously with a lubricant.

4.3.2 Tightening Flange Fasteners

Flange fasteners must be tightened carefully to avoid warping the flange. If the fasteners are not tightened correctly, the joint may leak or crack. Pressure must be applied evenly, in the correct amount and in the correct torquing sequence, to make a good seal.

CAUTION

Flange fasteners must be tightened to an even torque. Overtightening can deform the flange, the gasket, or both. Overtightening can also damage the fastener itself. Do not exceed the torque recommendations given for the flange and gasket.

Many tools are used for flange tightening, including:

- Torque wrenches
- Hammer wrenches
- Hydraulic torque wrenches
- Open-end box wrenches
- Air impact wrenches
- Combination wrenches
- Drift pins (*Figure 26*)
- Line-up pins
- Sockets

A torque wrench is used to apply equal pressure at all points around the flange. With the aid of a torque wrench, personnel can properly adjust and tighten a flanged joint. Follow these general guidelines when tightening flange fasteners:

- Make sure that the threads of the flange, nut, and fastener are clean and lightly oiled before torquing.
- If the fasteners of a flange have already been tightened, loosen and retighten them. Loosen each fastener one full turn, using an open-end or socket wrench.
- Make sure that the torque wrench is correctly adjusted. Most companies calibrate precision tools at regular intervals.
- Always place the torque wrench on the nut, not on the fastener. Use an open-end or socket wrench to hold the fastener while torquing.
- Always use a smooth, even motion when using the torque wrench. A hurried, uneven motion does not give a correct measurement.

Each piping system may use different torque settings, depending on the sizes of the flanges and fasteners, the pressure or temperature of the system, and the type of gasket used. Consult a supervisor or the piping drawings for the proper torque setting.

The fastener holes of all flanges can be identified by a simple numbering system. There are no numbers on the flanges, but each hole, starting at the top and moving clockwise, is referred to by a number. To properly torque a flange joint, use the crossover method, or star pattern, of tightening the fasteners.

This section describes a typical flange-tightening procedure. Always follow company policy and procedures. To tighten flange fasteners, follow this general procedure:

Step 1 Align the flanges using at least two line-up pins sized 2 mm smaller than the diameter of the fastener holes.

Step 2 Insert a new gasket, and install the studs and nuts in the remaining holes.

NOTE

If the pipe run is horizontal, install the bottom studs first, to retain the gasket.

Step 3 Tighten all nuts hand-tight, taking care that the fasteners or studs extend completely through the nuts.

Step 4 Tighten four studs at 90 degrees to each other until there is no visible gap at the gasket.

Figure 26 One type of drift pin (common barrel pin).

> **NOTE**
> It may be necessary to tighten more than four studs to close the gap.

Step 5 Measure the distance from face to face at the flange outside diameter, moving around the flange circumference to ensure that they come together evenly.

> **NOTE**
> The flange is now faced.

Step 6 Number the studs according to the numbering-and-torque diagram, which provides the proper torquing sequence. (*Figure 27*).

Step 7 Tighten the studs in numbered sequence (diametrically opposed) to approximately 30 percent of the torque value for the stud size. Skip the line-up pins; leave the pins in place to ensure that the flange is aligned. Refer to company-specific engineering specifications for the appropriate torque values.

> **NOTE**
> If tightening pump flanges, some variation in sequence may be necessary to avoid excessive pump-shaft deflection.

Step 8 Remove the line-up pins, and install the studs and nuts.

Step 9 Tighten all the studs in numbered sequence to approximately 60 percent of the torque value provided in the specifications and/or as defined by company policy.

Step 10 Tighten the studs in numbered sequence to 100 percent of the torque value.

Step 11 Check the torque of each nut, moving to each adjacent nut around the flange in clockwise rotational sequence. Use one or more passes to ensure that the studs are evenly tightened.

Step 12 Slowly pressurize the facility to working pressure, and visually check for leaks.

Step 13 If there is a leak, check the stud tension. If the leak cannot be stopped by further tightening:

- Relieve the pressure
- Remove the studs and gaskets
- Check for flange-face defects and gasket defects
- Make the necessary repairs or replacements
- Repeat the installation

> **NOTE**
> If the flange assembly is in a high-pressure or high-temperature application, retorquing may be necessary after startup to compensate for any relaxation or creep in the assembly.

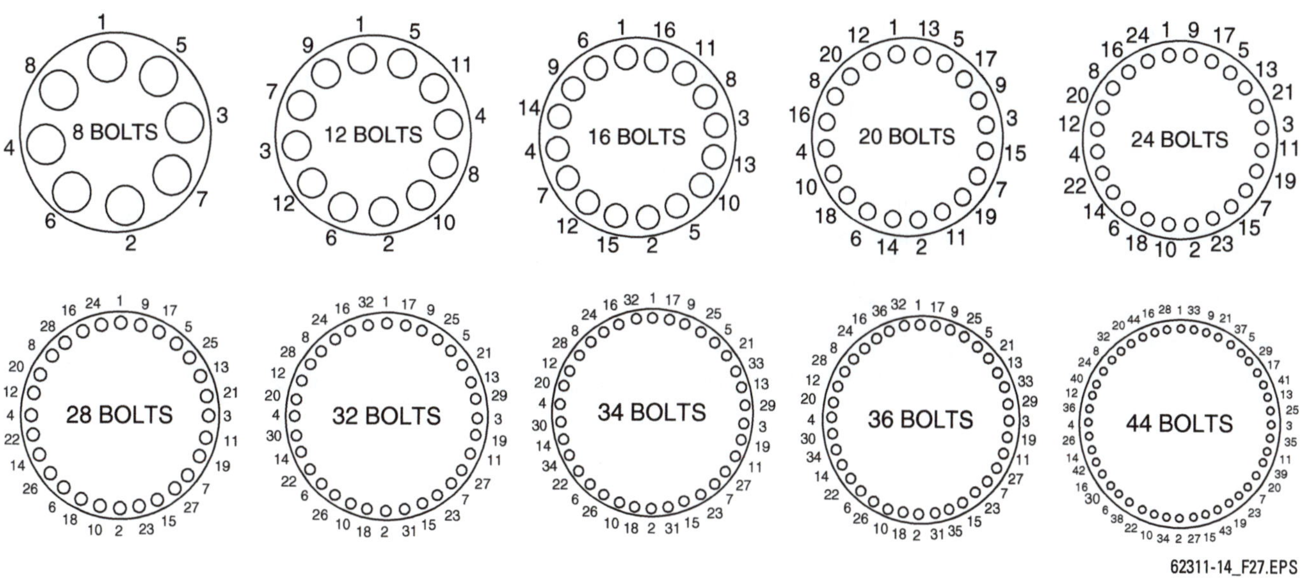

Figure 27 Numbering-and-torque diagram.

Additional Resources

It is assumed that you will learn to perform these procedures under the direct supervision of your crew leader, foreman, or instructor. For independent study, refer to the recommended Additional Resources at the back of this module.

4.0.0 Section Review

1. When fasteners and nuts are re-used, the fastener/nut combination is _____.
   a. not as strong as it was when the components were new
   b. the same as it was when the components were new
   c. stronger than it was when the components were new
   d. more reliable than it was when the components were new

2. The first step in installing gaskets is to _____.
   a. align and lubricate the parts
   b. lubricate and install the parts
   c. align and install the gasket
   d. clean and inspect the parts

3. To properly torque a flange joint, personnel should tighten the fasteners _____.
   a. clockwise
   b. using the crossover method
   c. counterclockwise
   d. in no particular order

SECTION FIVE

5.0.0 FLANGE BOLTING AND ASSEMBLY

Objective

Describe the procedures used to perform flange bolting/assembly (CTFB-17).

a. Identify the prerequisites, competencies, and abnormal operating conditions (AOCs) associated with flange bolting/assembly.
b. Describe the procedures used to perform flange bolting/assembly.

Performance Task

Perform flange bolting/assembly (CTFB-17).

- Identify potential abnormal operating conditions that may occur during performance of this CT, and know the appropriate actions to take in response to them.
- Utilize the appropriate personal protective equipment according to relevant company procedures.
- Ensure that the proper tools, materials, and equipment are present and that tools are in good working order.
- Verify ANSI ratings of materials, as specified.
- Inspect and prepare facing of the flange.
- Inspect and clean threads on nuts and bolts.
- Insert lineup pins to align flanges.
- Place two bolts in bottom part of flange and drop in gasket.
- Install remaining bolts and hand tighten nuts.
- Verify proper flange tightening sequence and tightening procedure to attain specified torque and tighten accordingly.
- Bolt up the flange to attain specified torque and tighten accordingly.
- Complete appropriate documentation as required by operator's procedures.

Trade Term

Abnormal operating conditions (AOCs): Conditions within a pipeline system, identified by the operator, that may indicate (1) a malfunction of a component or (2) a deviation from normal operations that may indicate a condition exceeding design limits or result in a hazard to persons, property, or the environment.

This section describes the steps required to prepare pipe and flanges for bolting and the procedures for assembling and bolting flanges. It also describes the abnormal operating conditions (AOCs) that can occur during these procedures and the proper reaction for each. Always follow company safety policy and procedures when performing flange bolting/assembly.

5.1.0 Prerequisites, Competencies, and Abnormal Operating Conditions

Personnel must possess a variety of skills in order to successfully perform flange bolting/assembly. The various operations require a series of steps, which technicians must be competent to perform. Personnel must also be aware of the particular abnormal operating conditions (AOCs) that may occur during the procedure and know the proper responses to them.

5.1.1 Prerequisites and Competencies

Personnel must possess the following basic knowledge and skills prior to performing flange bolting/assembly:

- Be competent with standard hand tools and any specialized tools or equipment recommended by the manufacturer.
- Be able to follow instructions and interpret engineering drawings.
- Be familiar with industry-standard color codes for gaskets.
- Be able to recognize and follow operator procedures when working with asbestos gaskets.
- Be familiar with entry permits and procedures for a confined space.
- Be able to perform an atmospheric test before, during, and after the work has been completed.
- Know how to properly use personal protective equipment (PPE).

5.1.2 Abnormal Operating Conditions

It is critical that personnel know how to recognize AOCs that may occur during these procedures, and the proper responses to each.

- *Asbestos is encountered* – If an asbestos gasket is discovered, use proper personal protective equipment (PPE) and follow company-specific asbestos abatement procedures.
- *Corrosion or mechanical damage is present* – Document the condition as required and notify the appropriate personnel to take the required corrective action.

- *Electrical shock* – Avoid electrical shocks by following company procedures and by wearing the appropriate personal protective equipment (PPE). If a severe shock does occur, notify 911 and/or the appropriate personnel.
- *A leak of hazardous gases or liquids occurs* – Evidence of a release may include and odor of hazardous gas or liquids, dead vegetation, audible cues (e.g., roaring, hissing), and/or a sheen on water. Take appropriate action using the proper PPE to protect people and the environment, and limit access to the area as much as possible. Notify the proper authorities and company personnel. Complete any required documentation and/or reporting according to company procedures.
- *A fire and/or explosion or ignition of hazardous, flammable liquids or gases occurs* – Evacuate the area and take only defensive action using the proper PPE. Contact the control center to isolate or shut down the line as quickly as possible. Engage the proper emergency response services.

5.2.0 Procedure

This procedure describes the steps typically involved in flange bolting/assembly. Always follow company policy and procedures. If you have any questions, contact your supervisor.

WARNING! Perform any preliminary tasks required to ensure safe working conditions. Always follow company policy and procedures for preparation and installation.

5.2.1 Flange Bolting Procedure

Follow this general procedure for performing a flange bolting/assembly:

Step 1 Ensure that the proper tools, materials, and equipment are present and that tools are in good working order.

Step 2 Verify ANSI ratings of materials, as specified. This ensures that the materials conform to the temperature and pressure ratings of the pipeline.

Step 3 Inspect and prepare the facing of the flange. Rust, debris, or other contaminants must be removed from all flanges and gaskets because these obstructions could cause the flanges to connect improperly or cause the seal to break. Always follow company policies and procedures for the proper cleaning process and tools, and always check the engineering specifications before applying any type of solvent to the flanges.

Step 4 Inspect and clean threads on nuts and bolts. All internal and external threaded surfaces and nut/washer faces should be coated generously with a lubricant.

Step 5 Insert lineup pins to align flanges.

Step 6 Place two bolts in the bottom part of the flange and drop in the gasket.

Step 7 Install remaining bolts and hand tighten the nuts.

Step 8 Verify the proper flange tightening sequence and tightening procedure to attain specified torque and tighten accordingly. Proper stud tension, correct tightening pattern, and clean, rust-free studs are necessary for properly tightened bolted connections.

Step 9 Bolt up the flange to attain specified torque and tighten accordingly. To properly torque a flange joint, use the crossover method, or star pattern, of tightening the fasteners.

Step 10 Complete appropriate documentation as required by operator's procedures.

Additional Resources

It is assumed that you will learn to perform these procedures under the direct supervision of your crew leader, foreman, or instructor. For independent study, refer to the recommended Additional Resources at the back of this module.

5.0.0 Section Review

1. If a fire or explosion occurs during inspection, installation, or testing, defensive action consists of _____.
   a. using a fire extinguisher
   b. consulting the appropriate emergency checklist
   c. using the proper PPE
   d. using a hazardous vapor detector

2. The final step of performing a flange bolting/assembly is to _____.
   a. complete appropriate documentation
   b. inspect the nut and bolts
   c. verify the ANSI ratings of materials
   d. check the engineering specifications

SUMMARY

A flanged joint consists of two flanges, a gasket or other device to seal the joint, and the necessary fasteners to hold the joint together. This module described the flanges and gaskets used in flange bolting/assembly. It emphasized the importance of properly storing and protecting assembly components and explained how to do so. A wide variety of flanges are used in the pipeline industry, such as raised-faced flanges and ring-joint-type flanges. Many different gasket materials are used as well, from PTFE gaskets to combination gaskets.

This module also described the procedures used to install pipe flanges. This process involves first installing gaskets and then tightening flange fasteners. The components must be cleaned, inspected, lubricated, and aligned properly before the gasket is installed. Tightening the flange fasteners must be done according to a specific torquing sequence. Always follow company policy and procedures for all aspects of the flange bolting/assembly process.

Review Questions

1. Proper storage procedures must be implemented to maintain the quality of the pipe and materials and _____.
   a. keep the workers busy
   b. make more work
   c. save time
   d. keep theft to a minimum

2. Store valves in a(n) _____.
   a. open basket
   b. shipping crate
   c. horizontal position
   d. upright position

3. Flange protective discs must cover the _____.
   a. entire flange
   b. fastener holes
   c. gasket surface
   d. bore

4. Male threaded ends should be protected by _____.
   a. press-on caps
   b. press-in plugs
   c. wooden plugs
   d. duct tape

5. Using different flange-face styles at one joint is _____.
   a. sometimes necessary
   b. never necessary
   c. always required
   d. never acceptable

6. If the metals and pressure ratings of flange faces are not compatible, the flange faces may _____.
   a. become too hot
   b. crack
   c. become loose
   d. explode

7. A disadvantage of the RJT flange is that in order to remove a fitting, the flanges must be _____.
   a. pushed together
   b. welded together
   c. spread far enough apart to remove the ring
   d. sawed apart to remove the ring

8. Temperatures below 500°F and pressures below 600 psi are considered _____.
   a. high temperature/low pressure
   b. high temperature/high pressure
   c. low temperature/low pressure
   d. low temperature/high pressure

9. PTFE gaskets are chemically _____.
   a. sensitive
   b. stable
   c. explosive
   d. reactive

10. Corrugated metal gaskets are generally used on _____.
    a. concentric serrated flanges and high-pressure applications
    b. concentric serrated flanges and low-pressure applications
    c. smooth flanges and low-pressure applications
    d. phonographic serrated flanges and high-pressure applications

11. Spiral-wound metal gaskets are color-coded to identify the _____.
    a. size
    b. type of seal
    c. number of rings
    d. type of metal

12. Nitrile gaskets could be used with _____.
    a. acetone
    b. MEK
    c. severe service applications
    d. oil

13. Graphite-impregnated gaskets are used for _____.
    a. pure oxygen atmospheres
    b. high temperatures
    c. extremely high pressures
    d. fire protection systems

14. During the second pass of tightening the fasteners of a flange, the torque wrench must be set at _____.
    a. 20 percent
    b. 40 percent
    c. 60 percent
    d. 75 percent

15. The purpose of using line-up pins is to _____.
    a. install gaskets
    b. lubricate parts
    c. add pressure for torquing
    d. ensure that the flange is aligned

16. When tightening flanges, always place the torque wrench on the _____.
    a. fastener
    b. nut
    c. flange
    d. gasket

17. The numbering-and-torque diagram illustrates _____.
    a. proper torquing technique
    b. proper torquing sequence
    c. correct fastener size for flanges
    d. gasket and flange matches

18. When checking the torque of each nut, move to each nut around the flange in a _____.
    a. diametrically opposed sequence
    b. counterclockwise rotational sequence
    c. clockwise rotational sequence
    d. horizontal/vertical sequence

19. If a fire or explosion occurs during installation, the first step is to _____.
    a. contact the control center
    b. use the proper PPE
    c. evacuate the area
    d. shut down the pipeline

20. Before installing the gasket, two bolts should be inserted on what part of the flange?
    a. The bottom
    b. The right side
    c. The top
    d. The left side

Trade Terms Quiz

Fill in the blank with the correct term that you learned from your study of this module.

1. A device that is used to make a pressure-tight connection and is usually in the form of a sheet or a ring is a(n) ________.
2. Hardwood blocking and pallets placed underneath materials to keep them off the ground and to allow access for forklifts or for placement of chokers and slings is referred to as ________.
3. A metal container that can withstand high pressures is a(n) ________.
4. An organization that adopts standards for a variety of products, services, and processes in the United States to ensure consistency is the ________.
5. The sealing surface machined onto flanges which is governed by ANSI standards is the ________.
6. An organization that provides codes and standards, education and training, and knowledge sharing across the engineering disciplines, including the pipeline industry is the ________.
7. The amount of twisting force a fastener can withstand when threaded into a nut or a threaded blind hole drilled into a solid object is referred to as ________.
8. Conditions within a pipeline system, identified by the operator, that may indicate (1) a malfunction of a component or (2) a deviation from normal operations that may indicate a condition exceeding design limits or result in a hazard to persons, property, or the environment are called ________.
9. An international standards organization that develops and publishes technical standards to improve quality and safety across a variety of materials, products, systems, and services, is the ________.

Trade Terms

Abnormal operating conditions (AOCs)
American National Standards Institute (ANSI)
American Society of Mechanical Engineers (ASME)
ASTM International (ASTM)
Dunnage
Flange face
Gasket
Pressure vessel
Torque

Trade Terms Introduced in This Module

Abnormal operating conditions (AOCs): Conditions within a pipeline system, identified by the operator, that may indicate (1) a malfunction of a component or (2) a deviation from normal operations that may indicate a condition exceeding design limits or result in a hazard to persons, property, or the environment.

American National Standards Institute (ANSI): An organization that adopts standards for a variety of products, services, and processes in the United States to ensure consistency.

American Society of Mechanical Engineers (ASME): An organization that provides codes and standards, education and training, and knowledge sharing across the engineering disciplines, including the pipeline industry.

ASTM International (ASTM): An international standards organization, formerly known as the American Society for Testing and Materials (ASTM), that develops and publishes technical standards to improve quality and safety across a variety of materials, products, systems, and services.

Dunnage: Hardwood blocking and pallets placed underneath materials to keep them off the ground and to allow access for forklifts or for placement of chokers and slings.

Flange face: The sealing surface machined onto flanges. Flange facings are governed by ANSI standards to ensure that all flanges of each type and rating have the same facing.

Gasket: A device that is used to make a pressure-tight connection and that is usually in the form of a sheet or a ring.

Pressure vessel: A metal container that can withstand high pressures.

Torque: The amount of twisting force a fastener can withstand when threaded into a nut or a threaded blind hole drilled into a solid object.

Additional Resources

The following recommended resources can provide additional helpful information related to the requirements for this covered task:

API Recommended Practice 1161, Third Edition, © 2014 American Petroleum Institute. API Publishing Services, 1220 L Street NW, Washington, DC 20005. Distributed by Thompson Reuters (Scientific) LLC, **www.techstreet.com**

NCCER module AOCCC-17, *Abnormal Operating Conditions Control Center*

NCCER module AOCFG-17, *Abnormal Operating Conditions Field and Gas*

The following websites offer resources for products and training:

Alyeska Pipeline Service Company maintains a website on the Trans Alaska Pipeline System, the most famous aboveground pipeline, **http://www.alyeska-pipe.com**

American National Standards Institute (ANSI), **www.ansi.org**

American Society of Mechanical Engineers (ASME), **www.asme.org**

ASTM International (ASTM), **www.astm.org**

Figure Credits

© W.L. Gore & Associates, Figure 12

Photo courtesy of the Garlock family of companies, Figure 13 and Figures 15-16

STARCON a CIANBRO company, Figure 19

HYTORC Industrial Bolting System, Figures 20-21

G.A.L. Gage Co., Figure 25A

Courtesy of Klein Tools, Inc., Figure 26

Reprinted from ASME B16.5-2013, by permission of The American Society of Mechanical Engineers. All rights reserved, Table 1

Section Review Answer Key

Answer	Section Reference	Objective
Section One		
1. d	1.1.2	1a
2. b	1.2.0	1b
Section Two		
1. a	2.1.1	2a
2. c	2.2.1	2b
Section Three		
1. c	3.1.2	3a
2. a	3.2.6	3b
Section Four		
1. a	4.1.4	4a
2. d	4.2.0	4b
3. b	4.3.2	4c
Section Five		
1. c	5.1.2	5a
2. a	5.2.1	5b

NCCER CURRICULA — USER UPDATE

NCCER makes every effort to keep its textbooks up-to-date and free of technical errors. We appreciate your help in this process. If you find an error, a typographical mistake, or an inaccuracy in NCCER's curricula, please fill out this form (or a photocopy), or complete the online form at **www.nccer.org/olf**. Be sure to include the exact module ID number, page number, a detailed description, and your recommended correction. Your input will be brought to the attention of the Authoring Team. Thank you for your assistance.

Instructors – If you have an idea for improving this textbook, or have found that additional materials were necessary to teach this module effectively, please let us know so that we may present your suggestions to the Authoring Team.

NCCER Product Development and Revision

13614 Progress Blvd., Alachua, FL 32615

Email: curriculum@nccer.org

Online: www.nccer.org/olf

❑ Trainee Guide ❑ Lesson Plans ❑ Exam ❑ PowerPoints Other ____________

Craft / Level: Copyright Date:

Module ID Number / Title:

Section Number(s):

Description:

Recommended Correction:

Your Name:

Address:

Email: Phone:

CTMP-17

Mud Plugging

Overview

When a pipe section needs to be repaired or replaced, vapors in the pipeline must be isolated so that work can be completed safely. Mud plugging is a vapor-isolation method in which a wall of mud is installed inside the pipe to seal an open section and act as a vapor barrier. Mud plugs cannot be used as the sole vapor-isolation method for piping that cannot be vented, for vertical piping, or for piping with an angle greater than 30 degrees. This module describes how to prepare a mud mixture and the procedures for installing a mud plug.

CTMP-17

Mud Plugging

Objectives

When you have completed this module, you will be able to do the following:

1. Explain the proper procedures for preparing the mud mixture.
   a. Explain how to determine the Bentonite-to-water ratio.
   b. Explain how to mix the mud and form the mud balls.
   c. Explain how to store the mud balls properly.
2. Describe the procedures used to form and install a mud plug (CTMP-17).
   a. Identify the prerequisites, competencies, and abnormal operating conditions (AOCs) associated with installing a mud plug.
   b. Describe the procedures used to install a mud plug.

Performance Task

Under the supervision of your instructor, you should be able to do the following:

1. Form and install a mud plug (CTMP-17).
   - Identify potential abnormal operating conditions that may occur during performance of this CT, and know the appropriate actions to take in response to them.
   - Utilize the appropriate personal protective equipment according to relevant company procedures.
   - Prepare the mud balls.
   - Monitor the inside of the pipe and surrounding area for hazardous vapors.
   - Clean the interior of the pipe.
   - Determine the starting point of the mud plug inside the pipe.
   - Mark the beginning and ending point of the mud plug.
   - Create an initial ring of mud balls around the inside perimeter of the pipe.
   - Working toward the pipe opening, continue packing mud balls into the pipe in a series of rings until the desired length of the mud plug is reached.
   - Perform atmospheric testing at the pipe opening with a multi-gas monitor.
   - Check upstream vents to make sure they are open and free of product.
   - Complete appropriate documentation as required by operator's procedures.

Trade Terms

Abnormal operating conditions (AOCs)
API Grade Bentonite
Base disk
Mud balls
Mud plug

Industry-Recognized Credentials

If you're training through an NCCER-accredited sponsor, you may be eligible for credentials from NCCER's Registry. The ID number for this module is CTMP-17. Note that this module may have been used in other NCCER curricula and may apply to other level completions. Contact NCCER's Registry at 888.622.3720 or go to **www.nccer.org** for more information.

Contents

Figures

SECTION ONE

1.0.0 PREPARING THE MUD MIXTURE

Objective

Explain the proper procedures for preparing the mud mixture.

a. Explain how to determine the Bentonite-to-water ratio.
b. Explain how to mix the mud and form the mud balls.
c. Explain how to store the mud balls properly.

Trade Terms

API Grade Bentonite: An absorbent clay, available in powder form, which is mixed with potable water to create a mud mixture that is used in the pipeline industry to create a mud plug. API Grade Bentonite is manufactured to exceed API 13A, Section 9 specifications.

Mud balls: Small balls (approximately 3 inches in diameter) formed from a Bentonite-and-water mud mixture and packed into a pipe to create a mud plug.

Mud plug: A wall of mud installed inside a pipe to seal an open pipe section and act as a vapor barrier so that work can be safely completed.

To create a mud plug, a mud mixture is packed into an open pipe section that has been isolated, drained, and vented, in order to seal the pipe and prevent vapors in the pipeline from escaping. Before a mud plug can be installed, the mud mixture must be prepared. To do so, personnel must first ensure they have the proper personal protective equipment (PPE) and proper materials including, but not limited to:

- Gloves (usually latex)
- Dust mask
- Safety glasses
- Container for mixing
- API grade Bentonite powder
- Potable water (water that is suitable for drinking)

NOTE

The Bentonite used must be API Grade Bentonite, (manufactured to exceed *API 13A, Section 9* specifications), typically denoted as Premium Gel, since it is readily available. The Bentonite is a fine powder (>70% passing through a #200 sieve [75 μm]), has a pH of 9 ± 1, and has a powder density between 50 to 70 lb/ft^3 (0.8 to 1.1 g/cm^3), depending on the degree of compaction of the powder.

NOTE

The unit of measure represented by μm is known as a *micrometer*, and is equal to one millionth of a meter.

Always follow company policy and procedures for required PPE.

Mud mixtures may be made prior to use, formed into balls, and stored in sealed containers to prevent evaporation. Preparing the mud mixture involves determining the correct Bentonite-to-water ratio, mixing the mud, forming mud balls, and storing the mud balls properly.

1.1.0 Determining the Bentonite-to-Water Ratio

The correct ratio for the Bentonite-and-water mixture used for mud balls is 50-percent Bentonite to 50-percent water, by weight. This ratio results in a mixture that exhibits good processing characteristics, such as strength, moldability, and adhesion. Variations of up to 10-percent are acceptable (for example, a ratio of 40-percent Bentonite to 60-percent water). Keep in mind that the drier the mixture, the more difficult it may be to form mud balls.

To determine the total amount of mud mixture needed to create a specific mud plug, use the following formula:

$$\text{total lbs of mud} = \frac{(\text{pipe diameter in inches})^3}{16}$$

As an example, use this formula to determine the amount of Bentonite and water needed to create a mud plug for a 10-inch pipe. First, calculate the total amount of mud mixture needed for this size pipe:

$$\text{total lbs of mud} = \frac{(\text{pipe diameter in inches})^3}{16}$$

$$\text{total lbs of mud} = \frac{(10\text{ in})^3}{16}$$

$$\text{total lbs of mud} = \frac{10 \times 10 \times 10}{16}$$

$$\text{total lbs of mud} = \frac{1000}{16} = 62.5\text{ lbs}$$

Using this information along with the 50-percent Bentonite-to-water ratio, you can determine the amount of Bentonite and water needed. Because 50 percent of 62.5 lbs is 31.25 lbs (62.5 ÷ 2 = 31.25), 31.25 lbs of Bentonite and 31.25 lbs of water will be needed to make enough mud mixture for a 10-inch pipe.

Now that you know the amount of water needed in pounds, you must convert this amount to gallons (1 gallon of water weighs 8.33 pounds):

$$31.25\text{ lbs} \div 8.33\text{ lbs} = 3.75\text{ gal}$$

Therefore, to install a mud plug for a 10-inch pipe, 31.25 lbs of Bentonite and 3.75 gallons of water are needed. The amount of the resulting mud mixture should be enough to form a mud plug that has a length roughly equal to 1.5 times the diameter of the pipe (15-inch-long mud plug for a 10-inch pipe).

NOTE

Do not rely on packaging labels to determine the Bentonite powder weight, because powder density can vary. Always weigh the Bentonite powder before mixing it with the water.

1.2.0 Mixing the Mud and Forming Mud Balls

Containers such as buckets, washtubs, or wheelbarrows are commonly used to mix the mud, depending on the amount of mud needed. Ensure that the mixing container is free of debris such as rocks, sticks, grass, or trash.

When mixing the Bentonite and water, use a one-to-one ratio mixture. For best results, add the Bentonite to the water (rather than adding the water to the Bentonite). *Figure 1* shows how the mixture should appear initially. Mix until a mud is created that is cohesive and easily moldable, as shown in *Figure 2*. The mixture may appear too dry at first; mix thoroughly before adding water.

Figure 1 Bentonite powder added to water.

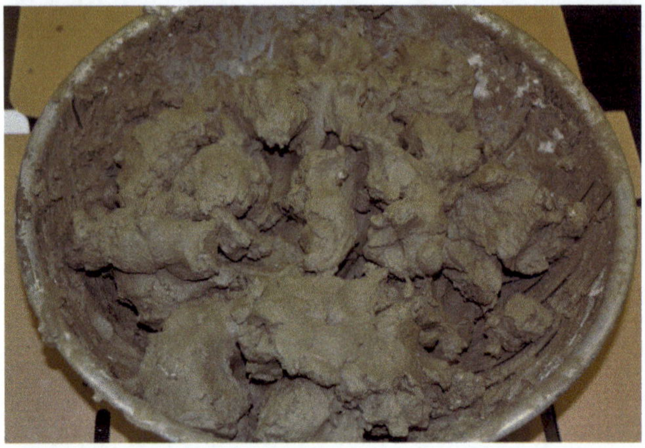

Figure 2 Mud mixture that is properly mixed.

Inspect the mixture for any dry or powdery areas, making sure the powder is fully mixed into the water. After mixing, keep the mud in a covered or shaded area to minimize drying.

WARNING!

Never add antifreeze or sand to the Bentonite mixture. Doing so can affect the ability of the mud plug to adhere to the pipe, creating an ineffective seal.

For ease of installation, shape the mud into balls. The balls can be more easily stored and then transported to the installation location. To form mud balls properly, mold them into spheres that are approximately 3 inches in diameter (*Figure 3*), roughly the size of a baseball. If the mud balls will not be installed in the pipe for several hours or more, inspect the balls for dryness before installation.

62211-14_F03.EPS

Figure 3 A properly formed mud ball.

Figure 4 Storing mud balls properly.

1.3.0 Storing Mud Balls

Mud mixtures may be made prior to use, formed into balls, and stored in sealed containers to prevent evaporation (*Figure 4*). They must also be stored at room temperature until they are installed. Proper storage ensures that the balls remain sufficiently moist and the water they contain does not evaporate. Balls that become cold, frozen, or dried out should not be used, as they may not form an effective vapor barrier when installed.

Additional Resources

It is assumed that you will learn to perform these procedures under the direct supervision of your crew leader, foreman, or instructor. For independent study, refer to the recommended Additional Resources at the back of this module.

1.0.0 Section Review

1. The mixture used for mud plugging can be stored _____.
   a. in a freezer
   b. at room temperature
   c. in an open container
   d. in a hot storage area

2. Antifreeze may be added to the mud mixture to help it adhere better. True or false?
   a. True
   b. False

3. If mud balls become frozen, they _____.
   a. can be stored indefinitely
   b. can be used "as is" after thawing
   c. must be slowly thawed, then re-formed
   d. should not be used

SECTION TWO

2.0.0 INSTALLING A MUD PLUG

Objective

Describe the procedures used to form and install a mud plug (CTMP-17).

a. Identify the prerequisites, competencies, and abnormal operating conditions (AOCs) associated with installing a mud plug.
b. Describe the procedures used to install a mud plug.

Performance Task

1. Form and install a mud plug (CTMP-17).
   - Identify potential abnormal operating conditions that may occur during performance of this CT, and know the appropriate actions to take in response to them.
   - Utilize the appropriate personal protective equipment according to relevant company procedures.
   - Prepare the mud balls.
   - Monitor the inside of the pipe and surrounding area for hazardous vapors.
   - Clean the interior of the pipe.
   - Determine the starting point of the mud plug inside the pipe.
   - Mark the beginning and ending point of the mud plug.
   - Create an initial ring of mud balls around the inside perimeter of the pipe.
   - Working toward the pipe opening, continue packing mud balls into the pipe in a series of rings until the desired length of the mud plug is reached.
   - Perform atmospheric testing at the pipe opening with a multigas monitor.
   - Check upstream vents to make sure they are open and free of product.
   - Complete appropriate documentation as required by operator's procedures.

Trade Terms

Abnormal operating conditions (AOCs): Conditions within a pipeline system, identified by the operator, that may indicate (1) a malfunction of a component or (2) a deviation from normal operations that may indicate a condition exceeding design limits or result in a hazard to persons, property, or the environment.

Base disk: The initial ring of mud balls that is installed inside a pipe section to form a mud plug. The base disk is installed at the farthest point of the mud plug, away from the pipe opening.

To install a mud plug, personnel pack mud balls into the open pipe section by hand. The term *mud plug* refers to the wall of mud that is constructed inside the pipe. A properly installed mud plug seals the pipe, preventing vapors in the pipeline from escaping. Mud plugging can be used for pipe that has any wall thickness, but many operators limit the use of mud plugging to pipe that has a diameter of 6 inches or larger. Always follow company-specific policies and procedures. Mud plugs cannot be used as the only vapor-isolation method for piping that cannot be vented, vertical piping, or piping with an angle greater than 30 degrees.

After the decision to perform mud plugging is made, designated personnel must first isolate the repair area and drain the pipe to ensure that the mud plug does not come into contact with petroleum liquids. To isolate the repair area, designated personnel close, lock, and tag out upstream and downstream valves. To relieve vapor pressure, vents are installed on the upstream and downstream side of the mud plug location to ensure pressure does not build up. Pressure buildup could cause the mud plug to be ejected from the pipe. The vents should be directed to a safe location downwind of the work site. Designated personnel must monitor the vents continuously during the work activity to verify that pressure is not increasing. *Figure 5* illustrates the location of closed, locked valves, vents, and mud plugs.

The following steps indicate where mud plugging fits into the overall pipe repair process:

Step 1 Shut down the pipeline.

Step 2 Isolate the pipe.

Step 3 Drain the pipe.

Step 4 Install vents.

Step 5 Install the mud plug.

Step 6 Complete the required work.

Step 7 Close all vents and put the pipe back into service, which will displace the mud plug.

Note that the mud plug installation is one of the last steps completed prior to performing the work.

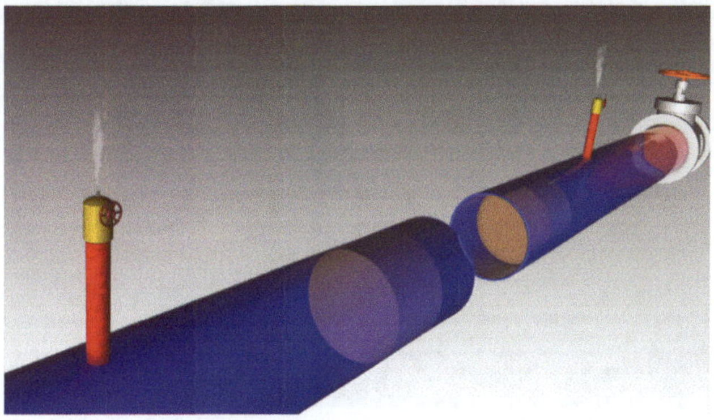

62211-14_F05.EPS

Figure 5 Location of closed, locked valves, vents, and mud plugs.

This section describes the steps required to prepare the pipe for mud plug installation, and the procedures for packing the mud in the pipe to create the mud plug. It also describes the abnormal operating conditions (AOCs) that can occur during mud plug installation, and the proper reaction for each. Always follow company-specific safety policies and procedures when installing mud plugs.

2.1.0 Prerequisites, Competencies, and Abnormal Operating Conditions

Personnel must possess a variety of skills in order to successfully prepare and install mud plugs. The various operations require a series of steps, which technicians must be competent to perform. Personnel must also be aware of the particular abnormal operating conditions (AOCs) that may occur during the procedure and know the proper responses to them.

2.1.1 Prerequisites and Competencies

Personnel must possess the following basic knowledge and skills prior to preparing and installing mud plugs:

- Be competent with standard hand tools and any specialized tools or equipment.
- Be able to follow instructions and interpret drawings.
- Be familiar with how to shut down, isolate, and drain the pipeline.
- Be familiar with how to install vents in the pipeline.
- Be familiar with entry permits and procedures for a confined space.
- Be able to perform an atmospheric test before, during, and after the work has been completed.
- Know how to properly use personal protective equipment (PPE).

2.1.2 Abnormal Operating Conditions

If the isolation process fails at any time during mud plug installation, an AOC can occur. It is critical that personnel know how to recognize and react properly should an AOC occur.

- *A leak of hazardous gases or liquids* – Evidence of a release may include an odor of hazardous gas or liquids, dead vegetation, audible cues (*e.g.*, roaring, hissing), and/or a sheen on water. Take appropriate action using the proper PPE to protect people and the environment, and limit access to the area as much as possible. Notify the proper authorities and company personnel. Complete any required documentation and/or reporting according to company procedures.
- *A fire and/or explosion or ignition of hazardous, flammable liquids or gases* – Evacuate the area and take only defensive action using the proper PPE. Contact the control center to isolate or shut down the line as quickly as possible. Engage the proper emergency response services.

2.2.0 Procedure

The following procedure is a basic guideline for installing a mud plug. Always follow company policy and procedures. If you have any questions, contact your supervisor.

WARNING!

Perform any preliminary tasks required to ensure safe working conditions. Always follow company policy and procedures during preparation, installation, and testing.

Preparing the pipe for installation includes taking the appropriate safety precautions and ensuring that the inside of the pipe is clean and dry.

2.2.1 Safety Precautions

Before beginning the installation process, personnel must make sure that they have the appropriate personal protective equipment (PPE), which generally includes, but is not limited to the following:

- Gloves (usually latex)
- Safety glasses
- Fire-retardant clothing
- Respiratory protection
- Multi-gas monitor

Always follow company policy and procedures for required PPE.

Before beginning any work near the pipe opening, personnel must use a multi-gas monitor to check the atmosphere inside the pipe and surrounding area for hazardous vapors (*Figure 6*). The atmosphere must be monitored during all work activity to ensure that it is safe to continue working.

WARNING!

Do not work on the upstream side of the mud plug.

WARNING!

Mud plugs are designed to provide a vapor seal only. They do not protect against pressure or free liquid.

2.2.2 Cleaning the Pipe

Before the mud plug is installed, the interior of the pipe must be clean and dry. Remove anything that would affect the ability of the mud to adhere to the pipe, such as residual hydrocarbons, particles, or debris.

When cleaning large-diameter pipe, personnel may need to climb into the pipe, creating confined-space safety issues. Respirators may be required. Always follow company policy and procedures for confined-space entry.

2.2.3 Packing the Mud

To pack the mud into the pipe to create the mud plug, use the following general procedure:

Step 1 Determine the starting point of the mud plug inside the pipe. The length of the mud plug should equal to one and a half times the pipe diameter by measuring from the open end of the pipe. Factor in a gap at the opening of the pipe to allow appropriate space for the cutting and/or welding work. The gap may vary depending on the nature of the work to be completed.

Figure 6 Checking the atmosphere inside and around an open pipe section.

Step 2 Mark the beginning and ending point of the mud plug. Mud balls can be pressed in place in both locations to serve as markers.

WARNING!

When working with large-diameter pipe, it may be necessary to enter the pipe, creating confined-space safety issues. Follow company policy and procedures for confined-space entry.

Step 3 Create an initial ring of mud balls around the inside perimeter of the pipe. To ensure proper adhesion of the mud plug to the pipe, apply a thin layer of mud to the interior surface of the pipe by rubbing a mud ball over it. Continue by adhering a single ring of mud balls around the inside perimeter of the pipe, pressing each one into the previous one and smoothing the mud into an even surface. This initial ring of mud balls is referred to as the **base disk** (*Figure 7*).

Step 4 Working toward the pipe opening, continue packing mud balls into the pipe in a series of rings. Each mud ball that is added should be molded into the existing mud in the pipe to form a solid disk, ensuring that no gaps exist.

CAUTION

Always place the mud balls inside the pipe with care. Using excessive pressure could compromise the integrity of the mud plug. Never throw mud balls into the pipe.

Figure 7 A properly installed base disk.

Step 5 When the desired length of the mud plug is reached, the completed mud plug should have no cracks or gaps. There should be some space between the mud plug and the opening of the pipe to allow room for necessary work to take place. *Figure 8* shows a properly installed, completed mud plug.

Step 6 Perform atmospheric testing at the pipe opening with a multi-gas monitor to ensure that the mud plug has created an effective vapor seal. If vapors are detected, stop all work and visually inspect the mud plug for cracks or shrinkage that may be compromising the seal. If problems with the plug are discovered, rework the mud plug and continue to check for vapors until none are detected. Note that it is acceptable for the visible end of the mud plug to show some cracking and shrinking after several hours. The interior of the plug should still remain intact.

Figure 8 A properly installed, completed mud plug.

WARNING!

In most cases, elevated or below-freezing temperatures will not reduce the effectiveness of a properly constructed mud plug. Always perform atmospheric testing periodically during any work activity to ensure the mud plug is providing an effective vapor seal.

Step 7 Check upstream vents to make sure they are open and free of product.

Step 8 Complete appropriate documentation as required by operator's procedures.

NOTE

Documenting data for compliance, historical review, and trending is an important element of maintenance.

Additional Resources

It is assumed that you will learn to perform these procedures under the direct supervision of your crew leader, foreman, or instructor. For independent study, refer to the recommended Additional Resources at the back of this module.

2.0.0 Section Review

1. If a fire or explosion occurs during inspection, installation, or testing, defensive action consists of _____.
   a. using a fire extinguisher
   b. consulting the appropriate emergency checklist
   c. using the proper PPE
   d. using a hazardous vapor detector

2. A mud plug is installed to provide a temporary seal against _____.
   a. vapor
   b. pressure changes
   c. liquid infiltration
   d. atmospheric contamination

3. When installing a mud plug in a large-diameter pipe, personnel may be required to wear respirators. True or false?
   a. True
   b. False

4. The mud plug is built _____.
   a. out from the sides of the pipe
   b. as a series of complete rings
   c. with two interlocking layers
   d. upward from the bottom of the pipe

Summary

Mud plugging is a vapor-isolation method that is used when pipe must be repaired or replaced. This module described how to prepare a mud mixture and how to install a mud plug. The correct ratio for a mud mixture is 50-percent API Grade Bentonite to 50-percent water, by weight. This ratio yields a mixture with the proper strength, moldability, and adhesion. Mud balls are formed, roughly the size of a baseball, for ease of installation. Mud mixtures may be made prior to use, formed into balls, and stored in sealed containers to prevent evaporation.

Before mud plugs are installed, the pipeline must be shut down, isolated, drained, and vented. Personnel must use a multi-gas monitor to check the atmosphere in and around the pipe opening for hazardous vapors before beginning any work. The pipe must also be cleaned and dried before mud plugging begins. Personnel must follow specific procedures when packing mud balls into the pipe to form an effective mud plug. This module also described AOCs that may occur during mud plugging, and the proper response. Always follow company policy and procedures for mud plug installation.

Review Questions

1. Gloves worn to protect your hands when preparing a Bentonite-water mixture are usually made from _____.
   a. canvas
   b. latex
   c. rubber
   d. leather

2. The preferred ratio when preparing a mud mixture is _____.
   a. 75% Bentonite to 25% water
   b. 65% Bentonite to 35% water
   c. 50% Bentonite to 50% water
   d. 25% Bentonite to 75% water

3. A mud plug for a 10-inch pipe, when mixed in the recommended ratio, requires 31.25 pounds of Bentonite and _____.
   a. 6.25 gallons of water
   b. 5.50 gallons of water
   c. 4.85 gallons of water
   d. 3.75 gallons of water

4. When plugging a 12-inch diameter pipe, enough mud should be mixed to make a plug with a length of ____.
   a. 15 inches
   b. 18 inches
   c. 21 inches
   d. 24 inches

5. If mud balls have been stored for several hours, they should be inspected before use to ensure there is no _____.
   a. shrinkage
   b. separation
   c. dryness
   d. foreign matter

6. Prepared mud balls should be stored _____.
   a. at room temperature
   b. in direct sunlight
   c. in very low humidity
   d. in temperatures of 85°F or higher

7. Mud plugs cannot be used in piping inclined at an angle of greater than _____.
   a. 15 degrees
   b. 30 degrees
   c. 45 degrees
   d. 60 degrees

8. Vents are installed to relieve vapor pressure _____.
   a. in a confined-entry situation
   b. when a respirator is not available
   c. only rarely, as needed
   d. upstream and downstream of the mud plug

9. If a leak of hazardous gases or liquids is encountered, the final step of the proper response is to _____.
   a. limit access to the area
   b. call 911
   c. use the proper PPE
   d. complete any required documentation

10. To ensure that the mud plug will adhere properly, the inside pipe surface must be _____.
    a. dampened
    b. slightly roughened
    c. clean and dry
    d. polished

11. When the mud plug has been installed, a multi-gas monitor is used to conduct a(n) _____.
    a. pressure inspection
    b. atmospheric test
    c. seal integrity check
    d. toxicity exam

Trade Terms Quiz

Fill in the blank with the correct term that you learned from your study of this module.

1. A wall of mud installed inside a pipe to seal an open pipe section and act as a vapor barrier is a ________.
2. An absorbent clay used to create a mixture for mud plugging is ________.
3. Conditions within a pipeline system, identified by the operator, that may indicate (1) a malfunction of a component or (2) a deviation from normal operations that may indicate a condition exceeding design limits or result in a hazard to persons, property, or the environment are called ________.
4. The initial ring of mud balls that is installed inside a pipe section to form a mud plug is the ________.
5. Small spheres formed from a mud mixture and packed into a pipe to create a mud plug are called ________.

Trade Terms

Abnormal operating conditions (AOCs)
Base disk
API Grade Bentonite
Mud balls
Mud plug

Trade Terms Introduced in This Module

Abnormal operating conditions (AOCs): Conditions within a pipeline system, identified by the operator, that may indicate (1) a malfunction of a component or (2) a deviation from normal operations that may indicate a condition exceeding design limits or result in a hazard to persons, property, or the environment.

API Grade Bentonite: An absorbent clay, available in powder form, which is mixed with potable water to create a mud mixture that is used in the pipeline industry to create a mud plug. API Grade Bentonite is manufactured to exceed API 13A, Section 9 specifications.

Base disk: The initial ring of mud balls that is installed inside a pipe section to form a mud plug. The base disk is installed at the farthest point of the mud plug, away from the pipe opening.

Mud balls: Small balls (approximately 3 inches in diameter) formed from a Bentonite-and-water mud mixture and packed into a pipe to create a mud plug.

Mud plug: A wall of mud installed inside a pipe to seal an open pipe section and act as a vapor barrier so that repair work can be safely completed.

Additional Resources

The following recommended resources can provide additional helpful information related to the requirements for this covered task:

API Recommended Practice 1161, Third Edition, © 2014 American Petroleum Institute. API Publishing Services, 1220 L Street, NW, Washington, DC 20005. Distributed by Thompson Reuters (Scientific) LLC, **www.techstreet.com**.

NCCER module AOCCC-17, *Abnormal Operating Conditions Control Center*

NCCER module AOCFG-17, *Abnormal Operating Conditions Field and Gas*

Marathon Pipeline Mud Plugging Video, **www.ttoolboxes.com/Documents/videos.cfm**

Figure Credits

Marathon Pipeline LLC, Module Opener, Figures 1–8

Section Review Answer Key

Answer	Section Reference	Objective
Section One		
1. b	1.0.0	1
2. b	1.2.0	1b
3. d	1.3.0	1c
Section Two		
1. c	2.1.2	2a
2. a	2.2.1	2b
3. a	2.2.2	2b
4. b	2.2.3	2b

NCCER CURRICULA — USER UPDATE

NCCER makes every effort to keep its textbooks up-to-date and free of technical errors. We appreciate your help in this process. If you find an error, a typographical mistake, or an inaccuracy in NCCER's curricula, please fill out this form (or a photocopy), or complete the online form at **www.nccer.org/olf**. Be sure to include the exact module ID number, page number, a detailed description, and your recommended correction. Your input will be brought to the attention of the Authoring Team. Thank you for your assistance.

Instructors – If you have an idea for improving this textbook, or have found that additional materials were necessary to teach this module effectively, please let us know so that we may present your suggestions to the Authoring Team.

NCCER Product Development and Revision

13614 Progress Blvd., Alachua, FL 32615

Email: curriculum@nccer.org
Online: www.nccer.org/olf

❏ Trainee Guide ❏ Lesson Plans ❏ Exam ❏ PowerPoints Other ______________

Craft / Level: Copyright Date:

Module ID Number / Title:

Section Number(s):

Description:

Recommended Correction:

Your Name:

Address:

Email: Phone:

CTTB-17
Tubing

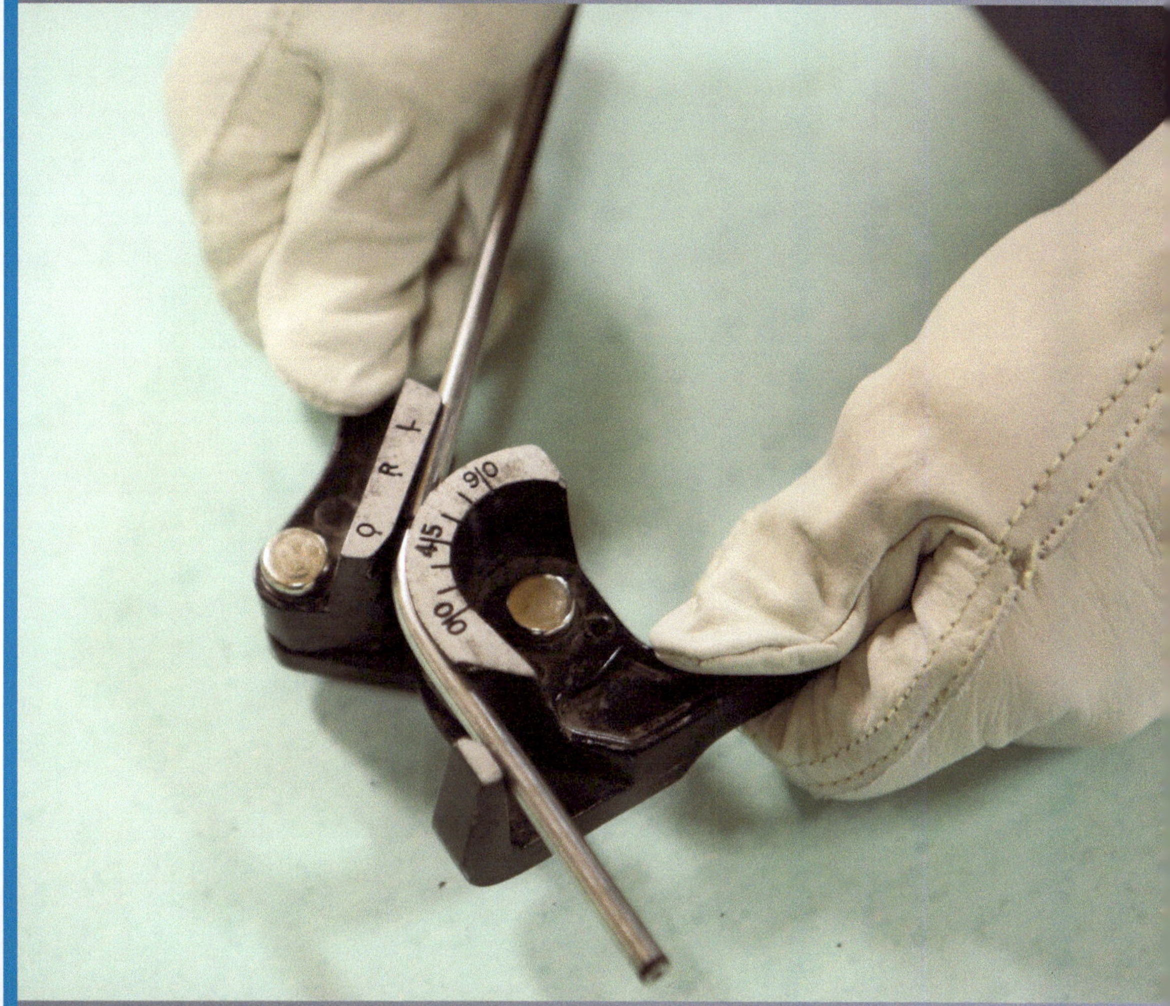

Overview

The terms *piping* and *tubing* are sometimes used interchangeably. However, tubing has thinner walls and is typically more flexible than piping. Its flexibility allows tubing to be bent to fit into tight spaces. In certain cases, soft annealed tubing is the most effective choice for pipeline applications. This module introduces the sizes and types of tubing used in pipeline maintenance, methods for storing and handling tubing properly, tools and methods for cutting, bending, and joining tubing, and methods for installing tubing.

Trainees with successful module completions may be eligible for credentialing through the NCCER Registry. To learn more, go to **www.nccer.org** or contact us at 1.888.622.3720. Our website has information on the latest product releases and training, as well as online versions of our Cornerstone magazine and Pearson's product catalog.

Your feedback is welcome. You may email your comments to curriculum@nccer.org, send general comments and inquiries to **info@nccer.org**, or fill in the User Update form at the back of this module.

This information is general in nature and intended for training purposes only. Actual performance of activities described in this manual requires compliance with all applicable operating, service, maintenance, and safety procedures under the direction of qualified personnel. References in this manual to patented or proprietary devices do not constitute a recommendation of their use.

CTTB-17
Tubing

Objectives

When you have completed this module, you will be able to do the following:

1. Describe the general properties of tubing and explain the proper and safe methods of storing and handling tubing.
   a. Describe the properties and advantages of tubing.
   b. Describe sizing measurements.
   c. Describe tubing materials used in the pipeline industry.
   d. Describe tubing standards and specifications.
   e. Explain the proper and safe methods for storing tubing.
   f. Explain the proper and safe methods for handling tubing.
2. Describe the procedures and components used in tubing fabrication.
   a. Describe the tools and methods used to cut tubing.
   b. Describe the tools and methods used to bend tubing.
   c. Describe the tools and methods used to connect tubing.
3. Describe the procedures used to inspect, install, and test tubing (CTTB-17).
   a. Identify the prerequisites, competencies, and abnormal operating conditions (AOCs) associated with installing tubing.
   b. Describe the procedures used to inspect, install, and test tubing.

Performance Task

Under the supervision of your instructor, you should be able to do the following:

1. Install tubing (CTTB-17).
   - Identify potential abnormal operating conditions that may occur during performance of this CT, and know the appropriate actions to take in response to them.
   - Utilize the appropriate personal protective equipment according to relevant company procedures.
   - Inspect the tubing and components prior to installation.
   - Install the tubing, providing support as needed.
   - Perform a gap inspection test.
   - Perform a leak and pressure test.
   - Complete appropriate documentation as required by operator's procedures.

Trade Terms

Abnormal operating conditions (AOCs)
Cap
Compression
Drawn
Ferrule
Fitting
Flare
Plug
Reamer
Tubing bender
Tubing cutter

Industry-Recognized Credentials

If you're training through an NCCER-accredited sponsor, you may be eligible for credentials from NCCER's Registry. The ID number for this module is CTTB-17. Note that this module may have been used in other NCCER curricula and may apply to other level completions. Contact NCCER's Registry at 888.622.3720 or go to **www.nccer.org** for more information.

Contents

Figures and Tables

[illegible]

Section One

1.0.0 General Properties of Tubing

Objective

Describe the general properties of tubing and explain the proper and safe methods of storing and handling tubing.

a. Describe the properties and advantages of tubing.
b. Describe sizing measurements.
c. Describe tubing materials used in the pipeline industry.
d. Describe tubing standards and specifications.
e. Explain the proper and safe methods for storing tubing.
f. Explain the proper and safe methods for handling tubing.

Trade Terms

Cap: A fitting used to close off the end of tubing.

Compression: The process of mechanically connecting a length of tubing by compressing or squeezing a ferrule between a fitting and nut by tightening the nut.

Drawn: Soft tubing that is pulled or stretched, by hand or by mechanically feeding it through a die, to reduce diameter and wall thickness.

Fittings: Devices (such as elbows, tees, or unions) used with tubing to make connections to instruments, piping, or other tubing. Fittings are also used to close off tubing and/or to branch tubing.

Flare: The angle made at the enlarged end of a piece of tubing to match a fitting.

Although the terms *tubing* and *piping* can sometimes be used interchangeably, tubing is technically more flexible than piping, and can be bent into shape for certain applications. Unlike piping, it is not rigid and does not require fittings to change direction.

For example, you may be familiar with the flexible copper tubing used to connect a refrigerator's icemaker to the water supply. It comes in a long coil, and can be cut to length and bent into shape. It is then connected at each end with suitable fittings. In contrast, copper piping comes in rigid sections of a certain length. These must be cut to length and soldered to suitable fittings at each change in direction or to make connections.

It is important to understand when and why tubing should be used in certain pipeline operations. When working with tubing, it is also critical that personnel understand the proper size and type of tubing be used for specific applications. The engineering specifications, piping specifications, installation detail drawings, and other information for a project usually indicate what size and type of tubing to use. However, personnel must be able to identify and verify the size and type of the tubing once they know the requirements.

Understanding proper and safe methods for storing and handling tubing is extremely important. Taking care of tubing materials prevents damage to the tubing, reduces personal injuries, and saves time and money on the job.

1.1.0 Advantages of Tubing

The need to use soft annealed tubing is based on a number of different factors, such as pressure and temperature conditions, corrosiveness of the environment, vibration in the system, system flow requirements, compatibility with the process fluids, and cost. Soft annealed tubing has the following advantages over piping:

- Because soft annealed tubing has considerable give, it can absorb shock and vibration that occur in some systems, which can damage components or cause leaks.
- Since tubing may be bent, fewer fittings are needed. Using tubing reduces installation time, decreases the number of potential places for leaks, and generally results in a more efficient installation. *Figure 1* shows how bends in typical tubing run can eliminate several fittings that would be necessary if piping were used for the same installation.
- Since hard tubing connections are usually not threaded like pipe, wall thickness is not reduced and installation time is often decreased.
- Soft tubing can be joined using flare or compression fittings, which are strong yet quick and easy to install.
- Since the inside of tubing is usually smoother than the inside of piping, flow through the tubing may be more even and less turbulent.

> **NOTE**
> The information given in this module is general in nature. When selecting tubing, it is important to make sure that the selection is compatible with the process or application where it is to be used. When in doubt, always consult the manufacturer.

Figure 2 shows a typical tubing installation.

1.2.0 General Sizing Measurements for Tubing

Tubing is generally sized according to its outside diameter (OD) and wall thickness. The OD is normally expressed in fractions of an inch, such as $\frac{1}{4}$ inch, $\frac{3}{8}$ inch, or $\frac{1}{2}$ inch. The OD of most types of tubing ranges from $\frac{1}{16}$ inch to 2 inches. The wall thickness of tubing is usually expressed in decimal units to the thousandth of an inch, such as 0.035 inch or 0.068 inch. The normal range of wall thickness for tubing is 0.010 inch to 0.188 inch. *Figure 3* shows the OD, wall thickness, and inside diameter (ID) dimensions for tubing.

Normally, the manufacturer will print or stamp the OD and wall thickness on the tubing.

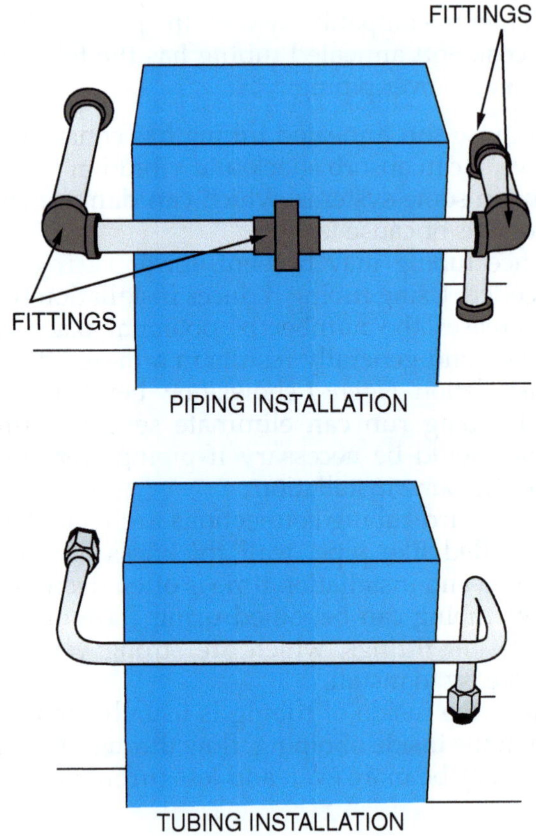

Figure 1 Tubing versus piping installation.

However, if personnel need to measure the OD of tubing, they can use a rule, outside caliper, or vernier caliper as shown in *Figure 4*.

> **NOTE**
> When using a rule to measure the OD of tubing, remember to measure the diameter across the center of the tubing.

While most tubing in the United States is measured in inches and fractions of an inch, tubing sizes may sometimes be marked in metric units, measured as millimeters. If personnel are working with metric-sized tubing, remember to use a metric rule or other metric measuring tools to obtain measurements.

Another common measurement for tubing is the inside diameter (ID) of the tubing. As shown in *Figure 3*, the ID is the distance across the middle of the hole in the tubing from an inside wall to the opposite inside wall. The inside diameter of any tube is equal to the outside diameter minus twice the wall thickness. For example, tubing with a $\frac{1}{2}$-inch OD and a wall thickness of 0.049 inch would have an ID of 0.402 inches, as calculated here:

$$\text{Tubing ID} = \text{OD} - (2 \times \text{wall thickness})$$
$$0.402 = 0.500 - (2 \times 0.049)$$

For any given OD of tubing, the ID decreases as the wall thickness increases. In other words, as the wall of the tube gets thicker, the hole in the middle gets smaller. *Table 1* shows typical OD sizes of tubing and their wall thicknesses and corresponding IDs.

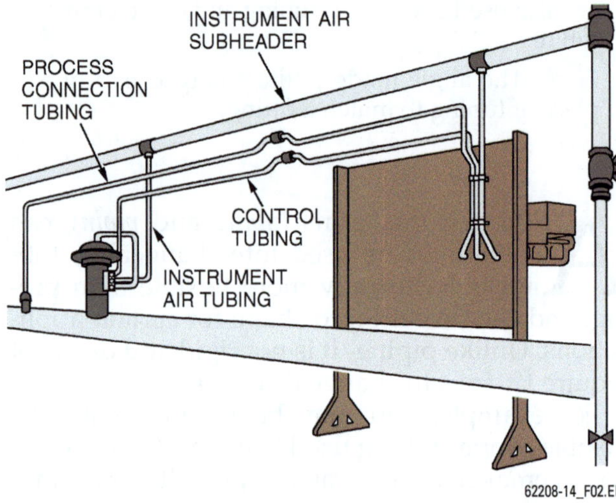

Figure 2 A typical tubing installation.

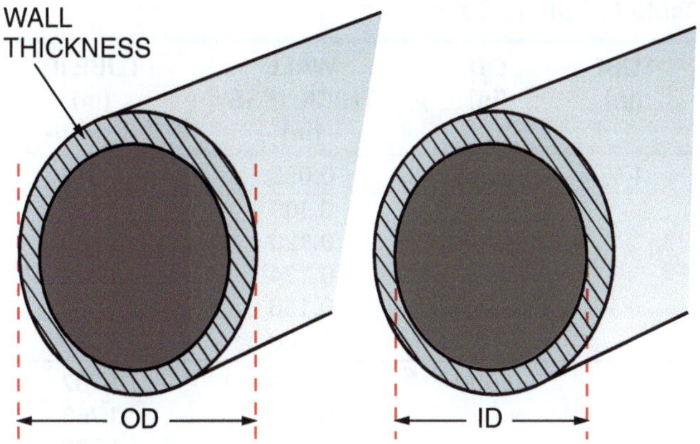

Figure 3 Tubing size measurements.

1.3.0 Tubing Materials

Tubing is made from a wide variety of materials including different types of metals and poly or plastics. Several organizations have published standards and specifications that guide the selection of different types of tubing. Each material has certain characteristics that make it desirable for particular applications. In the pipeline industry, the material that is primarily used is stainless steel.

1.3.1 Characteristics of Stainless Steel Tubing

Stainless steel is a steel alloy that contains at least 12 percent chromium by weight. Stainless steel tubing is stiffer and stronger than copper tubing. It is used to handle many types of corrosive fluids. It performs well in high-pressure and high-temperature applications.

Stainless steel tubing is designated by a number that refers to the alloy content of the metal. Two common types of stainless steel tubing used in pipeline applications are identified by the numbers 304 and 316. They each contain chromium and nickel but not manganese. The stainless steel tubing size, wall thickness, and compositional alloys determine the temperature and pressure ranges the tubing can withstand.

Even though stainless steel tubing is more expensive than some other tubing materials, it has several important qualities that make it practical for specific applications. Stainless steel is corrosion resistant and can easily be kept sterile to prevent contamination. Stainless steel tubing used in pipeline work can range in size from ⅛ inch to 1 inch. It is available in different wall thicknesses.

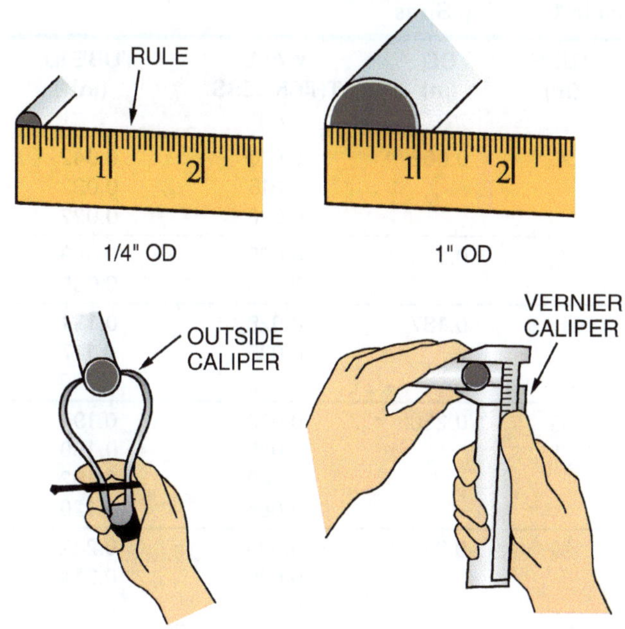

Figure 4 Measuring outside diameter of tubing.

1.3.2 Welded and Seamless Tubing

Metal tubing, including stainless steel tubing, is available in two general types, welded and seamless. Welded tubing is made by rolling flat strips of metal and then forming the metal into a tube. The seam is then welded. While the welded seam may not be visible on the outside of the tubing, it can usually be seen on the inside surface of the tubing.

Seamless metal tubing is made by passing a cylinder of metal through a die that cuts out the inside of the tube. This process is called extruding. Usually seamless tubing is **drawn** through additional die sets to get the proper wall thickness and to meet other specifications for the tubing.

Metal tubing comes in straight lengths (normally 20 feet), while other lengths can be obtained from 12 feet to 40 feet. Some types of tubing also come in rolls or coils, usually ranging from 50 feet to 100 feet.

1.4.0 Tubing Standards and Specifications

Several organizations and professional societies, including the American Society of Testing Materials (ASTM), the American National Standards Institute (ANSI), and the American Society of Mechanical Engineers (ASME), have published standards and specifications for tubing. These

Table 1 Tubing Sizes

TUBE (in)	OD (in)	WALL THICKNESS (in)	TUBE ID (in)
1/16	0.062	0.010	0.042
		0.015	0.032
		0.020	0.022
1/18	0.125	0.028	0.069
		0.035	0.055
3/6	0.187	0.028	0.131
		0.035	0.117
		0.049	0.089
1/4	0.250	0.028	0.194
		0.035	0.180
		0.049	0.152
		0.065	0.120
5/6	0.312	0.035	0.242
		0.049	0.214
		0.065	0.182
3/8	0.375	0.035	0.305
		0.049	0.277
		0.065	0.245
1/2	0.500	0.035	0.430
		0.049	0.402
		0.065	0.370
		0.083	0.334
5/8	0.625	0.049	0.527
		0.065	0.495
		0.083	0.459
		0.095	0.435
3/4	0.750	0.049	0.652
		0.065	0.620
		0.083	0.584
		0.095	0.560
		0.109	0.532
7/8	0.875	0.049	0.777
		0.065	0.745
		0.083	0.709
		0.095	0.685
		0.109	0.657
1	1.000	0.065	0.870
		0.083	0.834
		0.095	0.810
		0.109	0.782
		0.120	0.760
1-1/4	1.250	0.083	1.084
		0.095	1.060
		0.109	1.032
		0.120	1.010
		0.134	0.982
		0.056	0.938
1-1/2	1.500	0.095	1.310
		0.109	1.282
		0.120	1.260
		0.134	1.232
		0.156	1.188
		0.188	1.124
2	2.000	0.109	1.782
		0.120	1.760
		0.134	1.732
		0.156	1.688
		0.188	1.624

standards describe the various characteristics, applications, and capabilities of different types and sizes of tubing. Engineers and designers use these standards in selecting and identifying the type and size of tubing for particular applications.

ASTM has a series of standards commonly used to identify and specify tubing. Some of the standards provide general requirements about certain types of tubing. Others contain specific requirements for finish, hardness, alloy, working pressures and temperatures, method of manufacture, etc., for particular types of tubing. In the ASTM system, each standard has an identification number consisting of a letter and a series of numbers. If the metal contains iron (ferrous metal), the identification number begins with the letter A. If the metal does not contain iron (nonferrous metals), the identification number begins with the letter B. For example, A213 is a specification for seamless stainless steel tubing, and B68 is a specification for high-quality, soft, annealed copper tubing.

An example of a general ASTM standard is ASTM A450, which covers general requirements such as OD tolerances for steel and stainless steel tubing, both welded and seamless. The ASTM identification number is often printed or stamped on the tubing.

When the project specifications list the required ASTM, ASME, or ANSI tubing identification number, that type of tubing must be used. When the working temperature of the process increases, the allowable working pressure of the tubing decreases. Different pressure charts are used for different types of tubing. The ASME standard B31.3, *Process Piping*, is often used to determine the required wall thickness of tubing needed for different system working pressures. *Table 2* shows a typical allowable pressure chart used to determine the size of stainless steel tubing for various working pressures. *Table 3* is used for determining tubing pressure ratings at elevated temperatures.

1.5.0 Proper and Safe Methods for Storing Tubing

There are correct ways to store tubing. Proper storage helps avoid costly damage, saves time in identifying stored tubing, and reduces the chance of injury to personnel.

Tubing racks like the one shown in *Figure 5* are commonly used for storing hard-drawn tubing. The racks provide support to prevent bending, sagging, distortion, scratching, or marring of tubing surfaces. Most racks have separate levels or compartments where different types and sizes of tubing can be stored for ease of identification and selection. Tubing racks are usually elevated to help avoid damage that might occur at floor level. Tubing that is stored at floor level is easily damaged or contaminated by people, materials, moisture, or equipment in the area.

Always inspect tubing before storing it to be sure it is undamaged and clean inside and out. The ends of stored tubing should be sealed to prevent damage. Tubing ends can be temporarily capped, taped, or plugged using any method that prevents unwanted entry. To prevent contamination and corrosion, tubing should be covered while in storage.

> **WARNING!**
> Always ensure that stored tubing is securely held in place to prevent it from falling or rolling in any way that could cause injury.

1.6.0 Proper and Safe Methods for Handling Tubing

Tubing is made to rigid specifications, but it can be degraded by careless handling. From the time tubing is delivered to a job site until its installation is complete, proper and safe handling methods must be used to ensure quality tubing installations. Some basic guidelines for handling tubing are listed below:

Table 2. Stainless Steel Tubing Working Pressure Chart

Tube OD mm	Tube Wall Thickness, mm Working Pressure, bar Note: For gas service, select a tube wall thickness outside of the shaded area.													
3	670													
6	310	420	540	710										
8		310	390	520										
10		240	300	400	510	580								
12		200	250	330	410	470								
14		160	200	270	340	380	430							
15		150	190	250	310	360	400							
16			170	230	290	330	370	400						
18			150	200	260	290	320	370						
20			140	180	230	260	290	330	380					
22			140	160	200	230	260	300	340					
25					180	200	230	260	290	320				
28						180	200	230	260	280	330			
30						170	180	210	240	260	310			
32						160	170	200	220	240	290	330		
38							140	160	190	200	240	270	310	
50										150	180	210	240	270

Table 3. Working Pressure and Temperature Factors for Elevated Temperatures

To determine the allowable pressure for aluminum at elevated temperatures, multiply the allowable working pressure from Table 2 by the appropriate factor from the chart below.

Example: The allowable working pressure for 1⁄2" OD 3 0.035 wall tubing is 760 psi at 400°F.

		FACTORS	
°F	°C	304SS	316SS
200	93	1.00	1.00
400	204	0.93	0.96
600	316	0.83	0.85
800	427	0.76	0.79
1000	538	0.69	0.76
1200	649	0.30	0.37

- Never drag tubing off a delivery truck, storage rack, or on the ground. Dragging tubing across any rough surface can scratch or dent the outside surface, damage the ends, or contaminate the tubing. Scratches or nicks on tubing are potential paths for leaks.
- Keep tubing away from any material that might contaminate it during handling.
- Keep the end caps on during handling or transporting of tubing to avoid damage to ends and prevent contamination from entering the tubing.
- Wear gloves when handling tubing to avoid injury due to rough or sharp edges of the tubing.

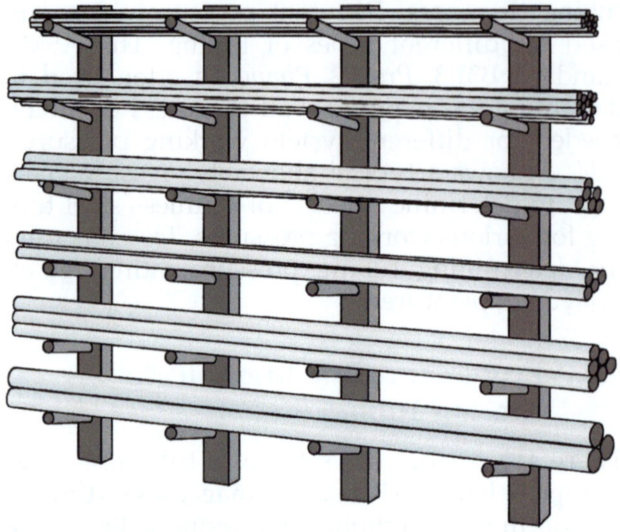

Figure 5 Typical tubing rack.

- Do not use a vise to hold tubing unless it is equipped with special jaws for holding and protecting the tubing.
- Put flags on the ends of long lengths of tubing during handling and transporting so that personnel can easily see the tubing ends and avoid damage and/or accidents.
- Be aware of people, materials, and equipment in the area when handling or transporting tubing so that the tubing does not hit anyone or anything.
- Place straight lengths of tubing inside polyvinyl chloride (PVC) piping or on an aluminum tray while transporting them to prevent bending or damage.

Additional Resources

It is assumed that you will learn to perform these procedures under the direct supervision of your crew leader, foreman, or instructor. For independent study, refer to the recommended Additional Resources at the back of this module.

1.0.0 Section Review

1. Because the inside of tubing is smoother than the inside of pipe, liquid or gas flows may be less _____.
   a. rapid
   b. agitated
   c. turbulent
   d. rotational

2. All of the following can be used to measure the OD of tubing except a(n) _____.
   a. rule
   b. inside caliper
   c. outside caliper
   d. vernier caliper

3. A stainless steel commonly used in pipelines is designated as Type _____.
   a. 302
   b. 304
   c. 306
   d. 308

4. An ASTM, ASME, or ANSI tubing identification number listed in the project specifications indicates that use of the identified tubing is _____.
   a. permitted
   b. acceptable
   c. recommended
   d. required

5. To prevent damage or injuries when transporting long lengths of tubing, _____.
   a. paint the ends a bright color
   b. attach flags to the ends
   c. cover the ends with padded caps
   d. hang flashing lights on the ends

6. A vise used to clamp tubing must be equipped with special jaws. True or false?
   a. True
   b. False

SECTION TWO

2.0.0 TUBING FABRICATION

Objectives

Describe the procedures and components used in tubing fabrication.

a. Describe the tools and methods used to cut tubing.
b. Describe the tools and methods used to bend tubing.
c. Describe the tools and methods used to connect tubing.

Trade Terms

Ferrule: A sleeve- or ring-type component of a tubing fitting, often referred to as a barrel, which forms an intricate part of the tubing connection and sealing process.

Plug: A tubing fitting used to close off a port of another fitting.

Reamer: A device used to remove burrs and ridges from the inside of tubing.

Tubing bender: A device used to bend tubing without kinking or crimping. Tubing benders may be hand operated, table mounted, or hydraulically powered.

Tubing cutter: A device that is rotated around tubing to separate it by using a cutting wheel or wheels and rollers.

To fabricate a tubing installation, personnel must know how to cut and bend tubing using a variety of tools and methods. They must also be very familiar with the fittings available to connect tubing and understand the purpose for each type of fitting.

2.1.0 Cutting Tubing

Before tubing can be installed, it must be properly cut to the correct length. Tubing in the pipeline industry is usually cut with a tool called a tubing cutter. Several types of cutters exist. Cutting tubing also involves the process of deburring.

2.1.1 Types and Sizes of Tubing Cutters

Tubing cutters are available in several different types and sizes. Generally, a tubing cutter consists of a cutting wheel, a handle, and one or more rollers. Some cutters have a sliding cutting wheel that is moved against the outside of the tubing. On sliding-roller types, the rollers are tightened to move the outside of the tubing against the cutting wheel. Internal tubing cutters cut the tubing from the inside. These internal tubing cutters are used where there is not enough room to use outside cutters. *Figure 6* shows all three types of cutters. Tubing cutters are commonly used for cutting stainless steel tubing.

2.1.2 Cutting Tubing with a Tube Cutter

It is always good practice for personnel to check the manufacturer's instructions before using any tubing cutter, to make sure that they know the proper and safe way to use the cutter. The manufacturer's instructions may also specify the type and size of cutter and wheel needed for the specific cutting job.

Before using a tubing cutter, the tubing should be marked at the exact place where the cut is to be made. A sharp pencil, colored felt-tip pen, or silver marking pencil are good tools for marking tubing because they make a thin line without scratching. A mark should be made at the starting point and ending point of the measurement (*Figure 7*). It is best to make the mark all the way around the tubing so that it is always visible while cutting. A sleeve or ferrule can be used as a guide for making the mark around the tubing. A tubing cutter is then rotated around the tubing to cut it, as shown in *Figure 8*.

NOTE

Do not use grease pencils or crayons to mark metal tubing because the mark will be too wide. Wide marks may affect the accuracy of the cut. Do not use sharp items like pens or scribes to mark metal tubing, since the scratch marks are points where corrosion or stress can dangerously weaken the tubing.

WARNING!

Tubing cutters and ends of cut tubing have exposed sharp edges. Take care to avoid cuts to hands and fingers. To protect the eyes when cutting tubing, always wear approved eye protection.

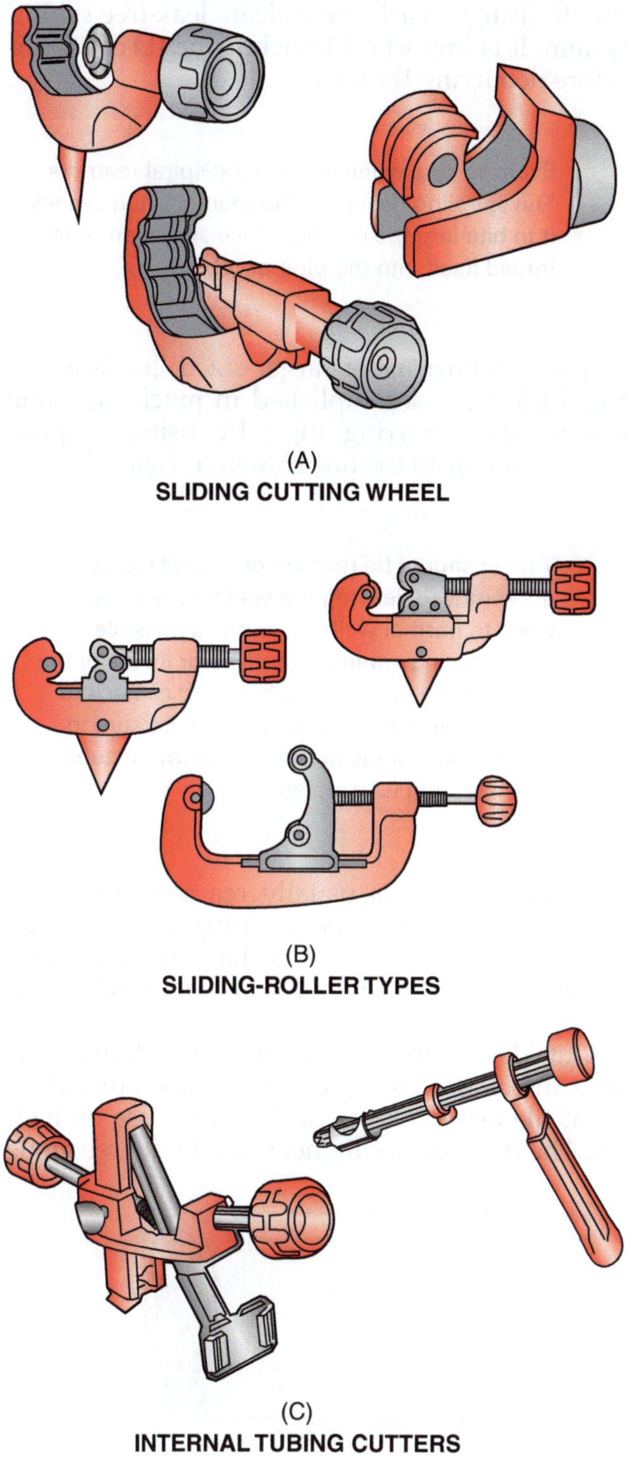

62208-14_F06.EPS

Figure 6 Tubing cutters.

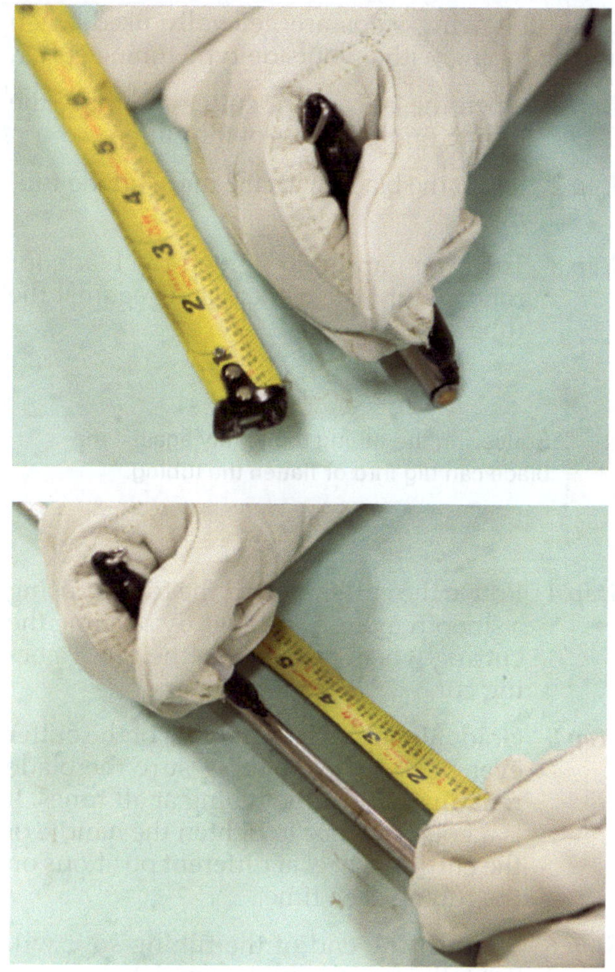

62208-14_F07.EPS

Figure 7 Marking tubing.

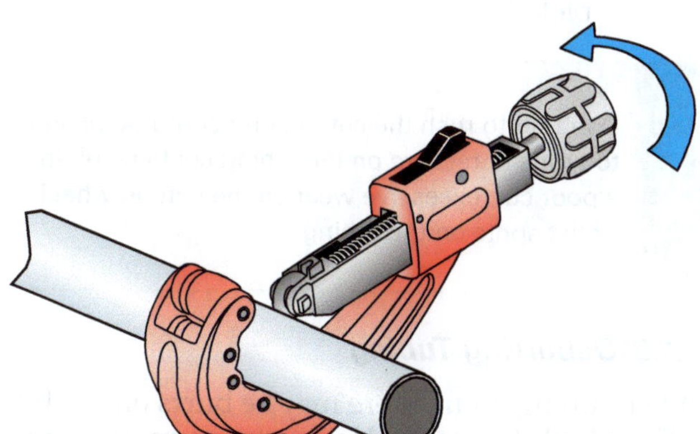

62208-14_F08.EPS

Figure 8 Using a tubing cutter.

The following steps are generally used when cutting tubing with an outside-type tubing cutter.

Step 1 Open the jaws of the cutter and place the tubing on the rollers of the cutter.

Step 2 Align the blade with the mark on the tubing.

Step 3 Turn the handle to tighten the blade (or rollers) until the blade is snug against the tubing.

CAUTION

Squeezing the tubing too tightly against the blade can dig into or flatten the tubing.

Step 4 Rotate the cutter around the tubing, using a smooth motion and slightly pushing the cutting wheel farther into the groove being cut.

Step 5 Gradually tighten the handle of the cutter every few turns, making sure the blade remains against the tubing at all times. It is a good practice to tighten the handle of the cutting wheel at different positions on the tubing each time.

Step 6 Support the end of the tubing so it will not fall when the cut is completed.

Step 7 Keep turning the cutter and tightening the handle until the tubing cut is completed.

NOTE

Don't try to rush the cutting procedure. Applying too much pressure on the tubing could result in a poor cut, excessive wear on the cutting wheel, or flat spots on the tubing.

2.1.3 Deburring Tubing

When tubing is cut, there may be burrs on the ID, OD, or both (*Figure 9*). Hacksaws normally leave burrs since the cutting action is not as smooth as that of other cutting tools. If not removed, burrs on the ID of tubing can break loose, causing chips and slivers to enter the tubing system. Chips and slivers can plug small pilot holes or vents, scratch valve seats and stem tips, or damage soft seals such as O-rings. The tubing ID can also be reduced because of the burrs. A smaller ID may affect flow through the tubing. Burrs on the OD of tubing make it difficult or impossible to properly install fittings to achieve a clean, leak-free tubing system. It is important to debur the ID of tubing before deburring the OD.

CAUTION

Exercise care when using hand spiral reamers. The spiraling design of the reamer often causes it to bite into the metal surface and attempt to thread itself into the pipe or tubing.

Inside deburring of larger-sized, hard-drawn metal tubing is accomplished in much the same manner as deburring pipe, by using a spiral **reamer** similar to the one shown in *Figure 10*.

CAUTION

Tubing should be reamed only until the burrs are removed, leaving the wall thickness as close to normal wall thickness as possible. Overreaming either the outside or inside of the tubing can weaken the tubing and cause it to split and/or leak when connected. Be sure to remove any filings or particles from the tubing after reaming is complete.

Smaller tubing is usually reamed (deburred) using specially designed reaming tools, like the one shown in *Figure 11*, that have the capability of reaming both the inside and the outside of the tubing.

Outside deburring of larger, hard-drawn tubing can be done using several methods including a flat file or sandpaper designed for use on metals. The file is normally held at a 45-degree angle

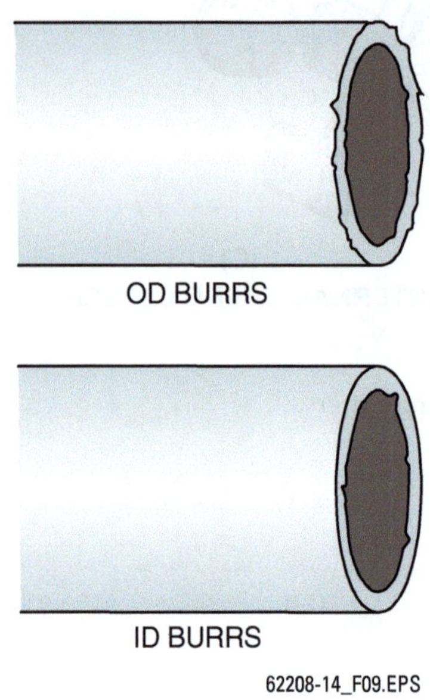

Figure 9 Tubing burrs.

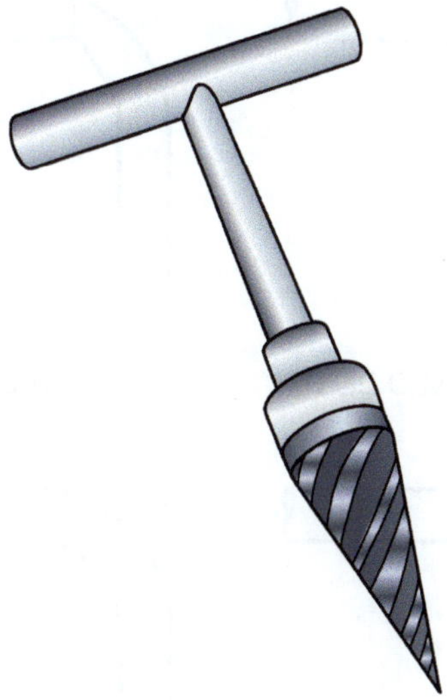

Figure 10 Tubing reamer.

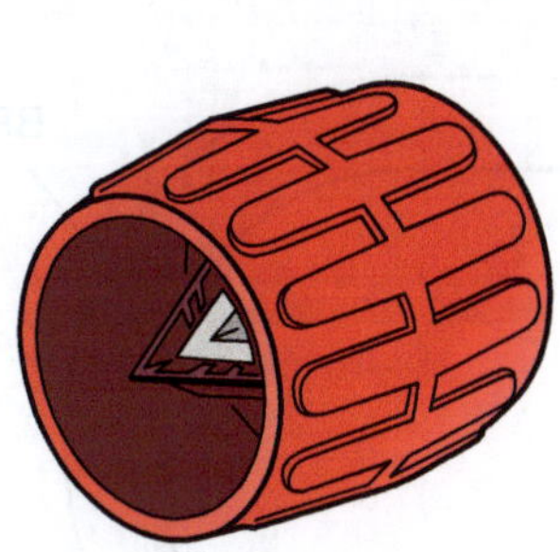

62208-14_F11.EPS

Figure 11 Tube inner-outer reamer tool.

while applying forward strokes with the file, making a uniform pattern completely around the end of the tubing. Sandpaper wrapped around the end of the tubing while manually turning the tubing will also remove the burrs.

WARNING!

Care must be exercised when using the sandpaper method to deburr tubing in order to prevent sharp metal burrs from penetrating the sandpaper and contacting the hand, possibly causing cuts.

2.2.0 Bending Tubing

One of the advantages of tubing is that it can be bent and shaped for various types of installations. Layout, cutting, bending, and attaching fittings to tubing requires practice and a number of skills to become an expert. This section describes common types of bends that are made in tubing installations and common types of tubing benders.

2.2.1 Standard Tubing Bends

Tubing installations require a number of different types of bends (*Figure 12*). The following standard bends are used for specific purposes in tubing installations:

- Angle bend – Used to change the direction of the tubing run
- Offset bend – Used to move the tubing run to one side to avoid an interference
- Crossover bend – Used to allow room for other tubes to cross (saddle)
- U-bend – Used to turn the tubing run around
- Expansion bend – Used to allow room for the tubing line to expand or absorb vibration

2.2.2 Tubing Bending Methods

Tubing must be bent carefully and correctly. A good tubing bend is made using the right equipment and proper methods, and is smooth all the way around the bend, providing even flow through the tubing. Whenever tubing is bent, the metal at the back of the bend is stretched and the metal in the throat is compressed. A properly made bend will be smooth, stay in plane, and lie flat on a smooth surface. If the bend is too sharp, one of two things will happen (*Figure 13*):

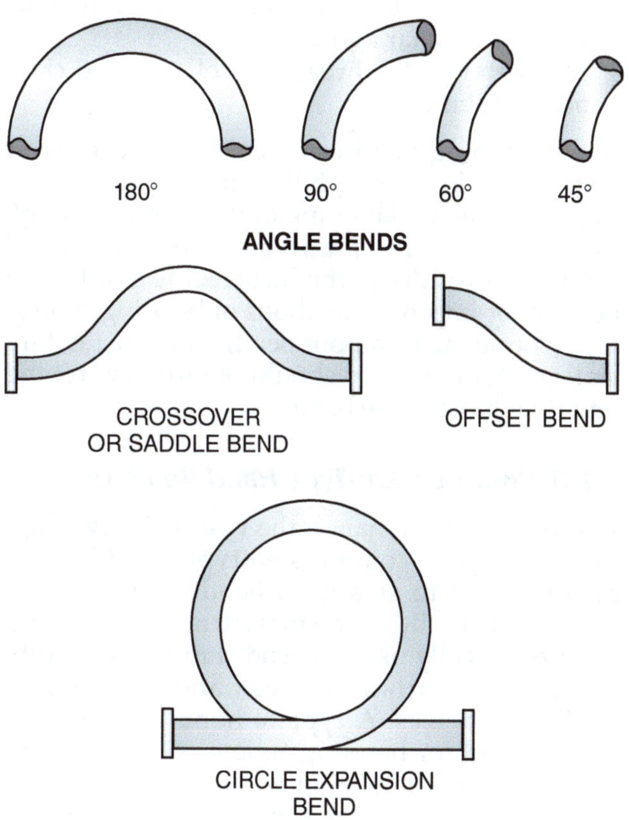

62208-14_F12.EPS

Figure 12 Standard tubing bends.

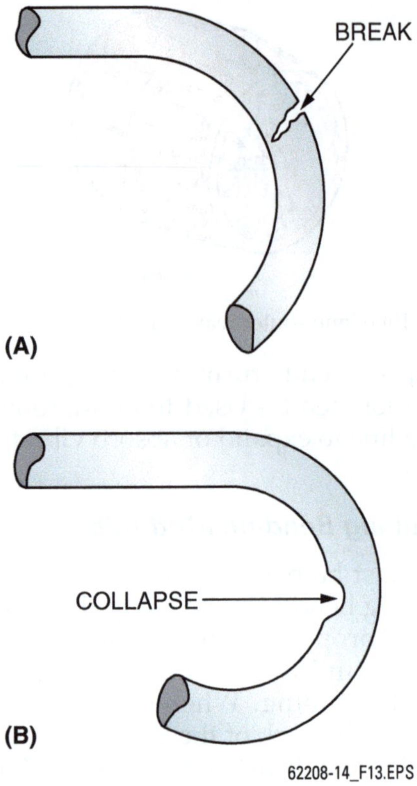

Figure 13 Damaging effects of overstretching and compressing in tubing.

- The metal at the back of the bend will be stretched too far and will break
- The metal in the throat will collapse and close off the tubing

A poor tubing bend can weaken the tubing or restrict the flow through the tubing. A good tubing bend has no flattening of the tubing through the bend. *Figure 14* shows a good bend contrasted with some results of the incorrect use of tubing benders or improper methods of bending tubing. If personnel make a poor bend, they must not install the tubing. They should obtain new tubing and then remake the bend.

2.2.3 Compression-Type Hand Benders

In pipeline maintenance, the vast majority of applications use a compression-type hand bender, also referred to as a lever bender. These hand benders are easily stored in the truck. Hand benders are typically used to bend stainless steel tubing up to 1⁄2 inch in diameter and are available in different sizes. A type of hand bender commonly used for bending tubing of 34 inches or less in length is shown in *Figure 15*. It is the most commonly used type of bender. Select the bender based on the size of tubing to be used.

The compression-type hand bender consists of a slide block, slide-block handle, forming wheel,

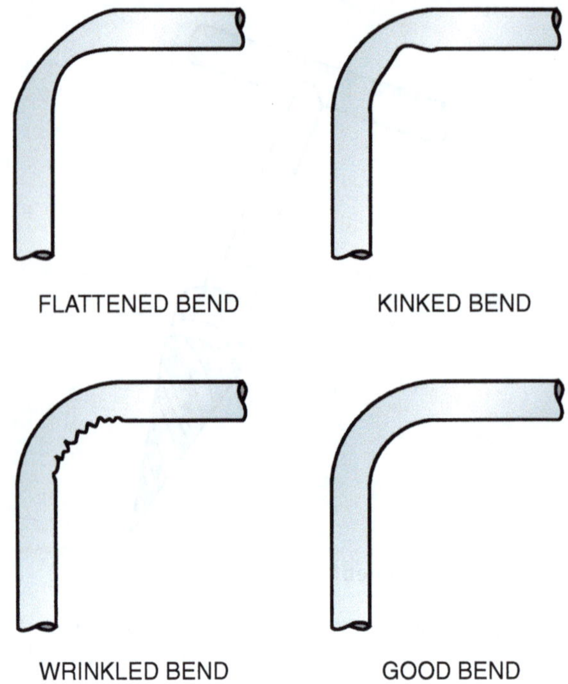

Figure 14 Improper tubing bends and a proper tubing bend.

forming-wheel handle, and tube clamp. Before using a tubing bender, it is important to check the manufacturer's manual to make sure the bender is the correct size for the tubing being bent. Using a bender that does not fit the size of the tubing will generally result in a flattened or kinked bend.

WARNING!

Personnel should always wear approved eye protection when bending tubing to protect eyes from tubing ends when in the bending sweep. On compression-type hand benders, personnel should place fingers so as to prevent pinching them between the slide block, forming wheel, and forming-wheel handle.

The tubing to be bent is positioned in the groove in the forming wheel. The tubing is held in place by the tube clamp. The markings on the slide block are used for properly aligning the tubing in the bender. A groove along the outside of the forming wheel supports the tubing as it is bent. The numbers around the forming wheel refer to the angle of the bend that will be formed, such as a 45-degree-bend or a 90-degree-bend. The bend is formed by moving the slide block around the outside of the tubing in the groove to the desired position. As the slide block moves, it applies compression to bend the tubing.

To form a bend with a compression-type hand bender, follow these steps:

Step 1 Measure and mark the tubing exactly (*Figure 16*). Use three marks to guide the bending process: a reference mark (made at the end of the tubing, which will be placed in the bender first); measurement marks (made at the vertex of the bend and the ending point of the tubing); and a directional mark (which indicates which direction the tubing is to be bent).

Step 2 Note that the slide block is marked with 0, R, and L. Insert the tubing in the bender, aligning the mark made on the tubing with the L or R on the slide block, based on the direction from which the tubing was measured. If the measurement was made from the left, place the tubing mark on the L. If the measurement was made from the right, place the tubing mark on the R. *Figure 17* shows the tubing mark aligned with the L mark on the slide block.

Step 3 Clamp the tubing securely to the bender by dropping the clamp over the tubing to hold it in place.

Step 4 Align the 0 on the slide block with the 0 on the forming wheel, as shown in *Figure 17*.

Step 5 Note that the forming wheel is marked with 00, 45, and 90, which represent degree marks. Move the slide-block handle around the tubing, forcing it to conform to the shape of the forming wheel until the 0 on the slide block aligns with the proper degree mark on the forming wheel. For example, to make a 45-degree bend, align the mark on the tubing with the 45. *Figure 18* shows the bending process to create a 45-degree bend. To make a 90-degree bend, align the mark on the tubing with the 90 (*Figure 19*). For bends less than 33 degrees, move the mark on the tubing to the appropriate mark closer to the 0 on the slide block.

> **NOTE**
>
> Bend the tubing in one direction only so as not to overstretch or kink it. Also, when bending harder metal tubing, bend it slightly past the mark to allow for springback.

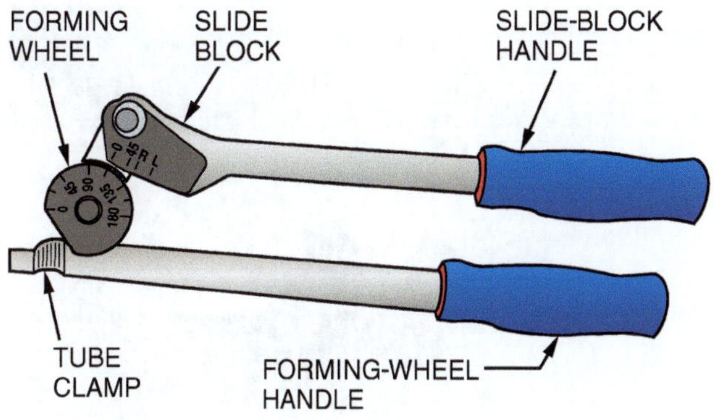

62208-14_F15.EPS

Figure 15 Compression-type hand bender.

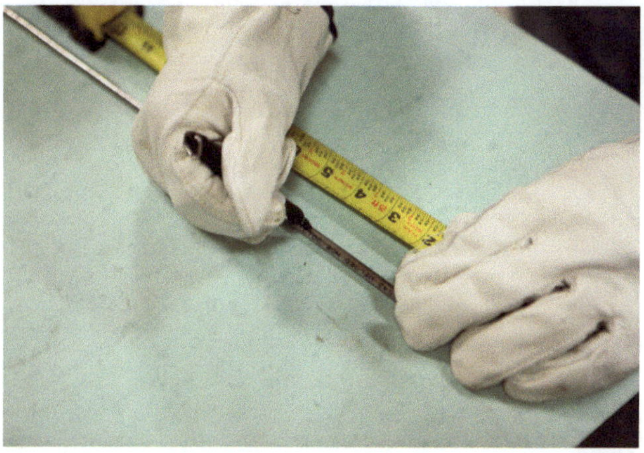

62208-14_F16.EPS

Figure 16 Measuring and marking tubing.

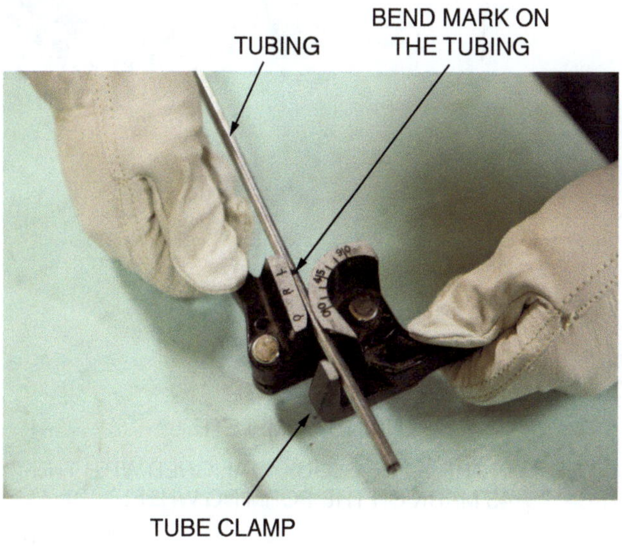

62208-14_F17.EPS

Figure 17 Aligning the tubing mark with the slide block.

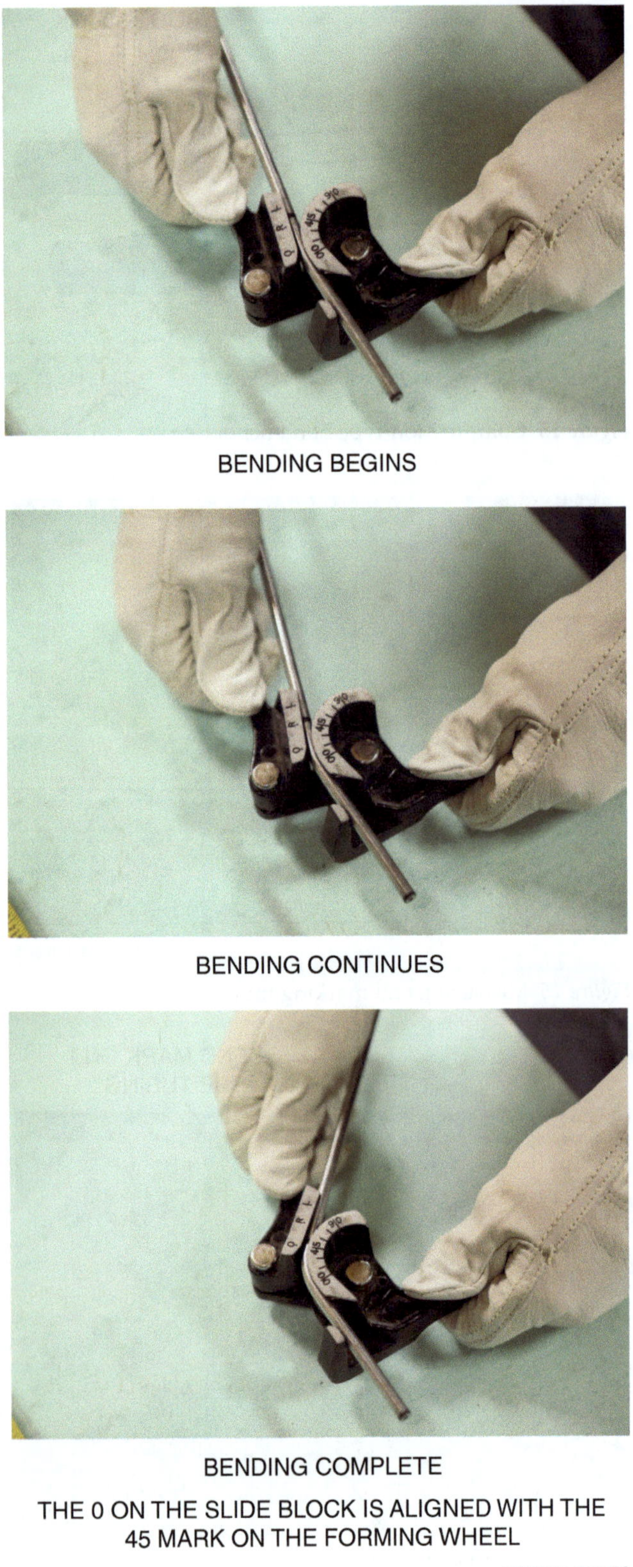

Figure 18 Making a 45-degree bend

Step 6 Lift the handle, raise the clamp, and remove the tubing.

Step 7 Check the tubing to be sure the correct angle for the bend was achieved. Also check that the length from the end of the tubing to the center of the bend is correct. When the bending is complete, the tubing mark should be at the center of the bend.

2.2.4 Table-Mounted Tubing Benders

There are several types and sizes of table-mounted tubing benders that are used to bend tubing of medium and heavy wall thicknesses. These benders are often used for tubing that is ¾ inch or larger. They are also used when a large number of bends are being made. Most table benders work similarly to a hand bender except that there is some type of gear assembly used to power the bender.

2.2.5 Hydraulic Tubing Benders

Most hydraulic tubing benders have a cylinder and ram assembly that bends the tubing with power supplied by a hydraulic pump. Hydraulic tubing benders are used to bend harder metals and larger types of tubing. Hydraulic benders are usually used for ¾-inch to 2-inch hard tubing, such as steel or stainless steel with heavy wall thicknesses.

> **NOTE**
> When using hydraulic methods to bend tubing, always operate the particular type of hydraulic bender in accordance with the manufacturer's instructions; most models operate somewhat differently.

2.3.0 Connecting Tubing

Tubing is connected to other components with tubing fittings to make a tubing run. For example, a tubing fitting might be used to make the connection between tubing and tubing, tubing and piping, tubing and a valve, or tubing and a bulkhead. The sizes and types of fittings to be used for a specific job are usually specified in

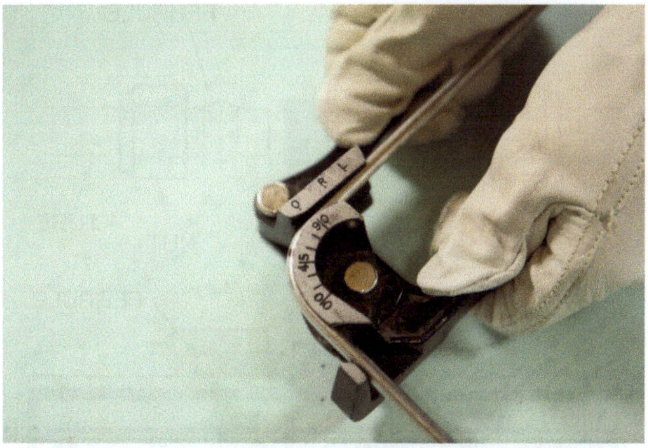

62208-14_F19.EPS

Figure 19 Making a 90-degree bend.

the engineering specifications, or on the installation drawings for a project. Most tubing fittings are sized by the outside diameter of the tubing to which they are to be connected. For example, ½-inch tubing would take a ½-inch tubing fitting. Fittings are specified to match the tubing material being used and the application. Factors such as pressure, temperature, and process characteristics are taken into consideration when selecting tubing fittings. Never connect tubing of one material to a fitting of a different material. Never mismatch tubing and tubing fittings. These should be from the same manufacturer to ensure a proper fit.

Fitting manufacturers use codes to identify the fitting material. The codes are sometimes stamped on the fitting. These codes may also indicate the heat (batch) number of the traceable lots. Some of these identification codes and other specifications have been developed by ASTM, ASME, and ANSI, like the codes for tubing materials. Other codes are manufacturer specific. Refer to the fitting manufacturer's catalog to determine the material codes. The four most common types of tubing fittings used in pipeline work are:

- Flare fittings
- Compression fittings
- Socket-welded fittings
- Butt-welded fittings

Additional fittings used to connect tubing are also described in this section.

2.3.1 Flare Fittings

A flare fitting (*Figure 20*) has its name because the end of the tubing is enlarged, or flared, slightly to create a rim against which the sleeve and body of the fitting can seal. Flare fittings can be assembled and disassembled repeatedly. They come in a variety of types, sizes, and materials for specific applications.

Most flare fittings consist of three main parts: a body, a flare nut, and a flare sleeve. However, not all flare fittings include a sleeve. Sleeves are used primarily in flare fittings used for high-pressure applications. The fitting body of the flare fitting shown in *Figure 20* is part of a piping-to-tubing connector. The tip of the fitting body has a tapered angle of 37 degrees. Another commonly used fitting has a 45-degree angle taper. The tubing used with a particular flare fitting must be flared at the same angle as the tip of the fitting body. The 37-degree flare fittings are typically used for high-pressure applications such as hydraulic oil systems. The 45-degree flare fittings are usually used for low-pressure water and air systems.

The sleeve in a flare fitting acts as a buffer between the nut and the flared end of the tubing. The nut holds the tubing flare against the fitting body. When the nut is tightened, it applies pressure against the sleeve. The sleeve, in turn, applies pressure against the tubing, securing and sealing the tubing against the body. *Figure 21* shows a cutaway view of an assembled flare fitting.

In high-vibration applications, flare fittings may fail. Because the tubing is flared away from the center line, the wall thickness of the pipe is

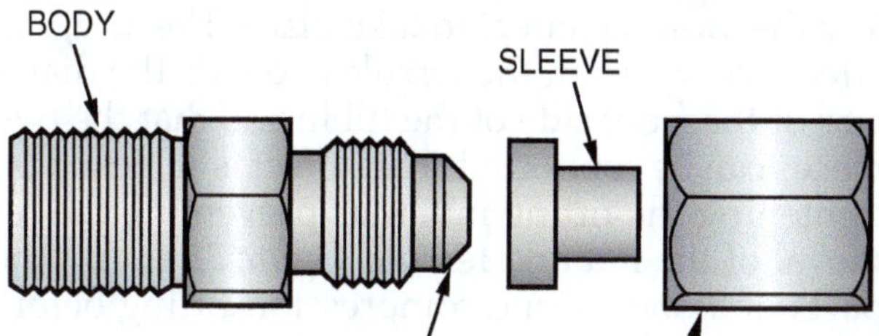

62208-14_F20.EPS

Figure 20 Parts of a flare fitting.

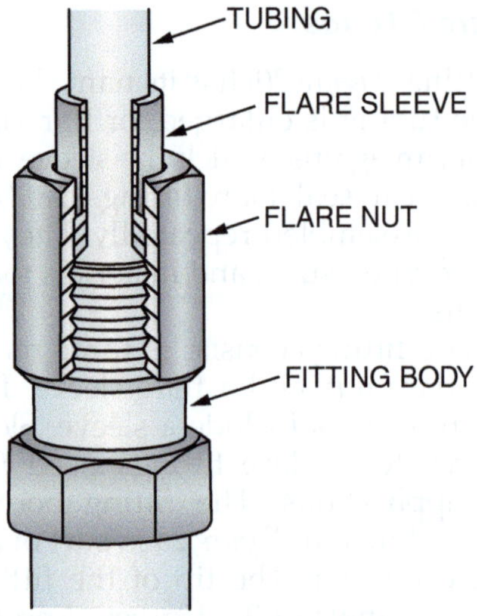

Figure 21 Assembled flare fitting.

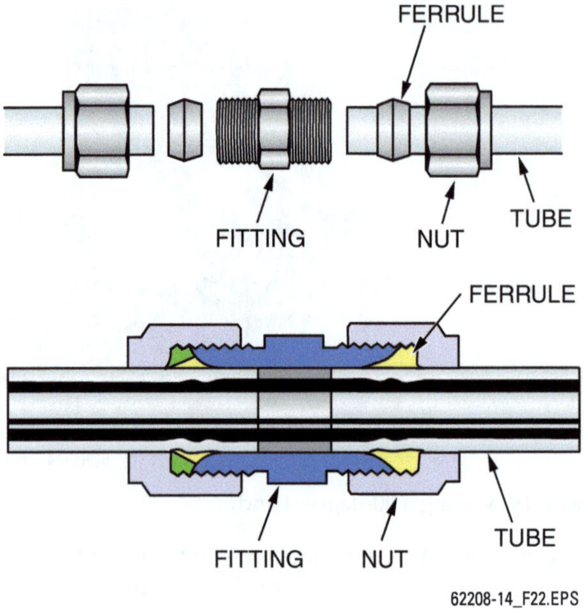

Figure 22 One-ferrule compression fitting.

decreased and cracks can form. If retightened, the cracks can become larger, causing the failure of the fitting. Because of this potential for failure, it is extremely important to brace the tubing.

2.3.2 Compression Fittings

Compression fittings are considered very reliable and relatively easy to install. They can usually be disconnected and reconnected without damaging the tubing or the fitting. They come in a variety of designs, sizes, and materials to fit different tubing applications. When using compression fittings, the end of the tubing is not flared. The tubing is fitted with a compression ring, or ferrule, that pinches the tubing as the nut is tightened on the body of the fitting. Compression fittings have either one or two ferrules for each joint. The ferrule of the fitting must be harder than the tubing to allow the swaging effect to take place. The swaging effect occurs when the ferrule expands the diameter of the front side of the tubing so that the two pieces can be connected. This effect is achieved by tightening the fitting nut, thus preventing the removal of the nut and ferrule. *Figure 22* shows the parts of a one-ferrule compression fitting before and after making a tubing connection.

CAUTION

To avoid possible rupture, always make sure that the flare angle matches the angle of the sleeve.

Two-ferrule compression fittings have a back ferrule and a front ferrule for each joint. *Figure 23* shows the parts of a two-ferrule compression fitting before and after making a tubing connection.

2.3.3 Socket-Welded Fittings

A socket-welded fitting has a hole, or socket, into which the tubing is placed. A fillet weld is made where the outside of the tubing wall protrudes from the end of the fitting. Socket-welded fittings can either be socket-to-socket or may connect the socket to some other form of tubing connector, as shown in *Figure 24*.

Socket-welded fittings are usually used with high-grade, heavy-walled tubing that may be difficult to flare or use with a compression-type fitting. They are often used when there is concern that vibration in the system might loosen the connection. Socket-welded fittings ensure there will be no leaks in the system. When welding a socket-welded fitting, it is important that the other end of the tubing be protected, since it can be damaged by the hot spatter from the welding process, particularly if the fitting is a socket-to-flare or socket-to-compression type.

NOTE

When inserting tubing into a socket-weld tubing fitting, do not let it bottom out. Allow $\frac{1}{16}$ inch to $\frac{1}{8}$ inch for expansion space.

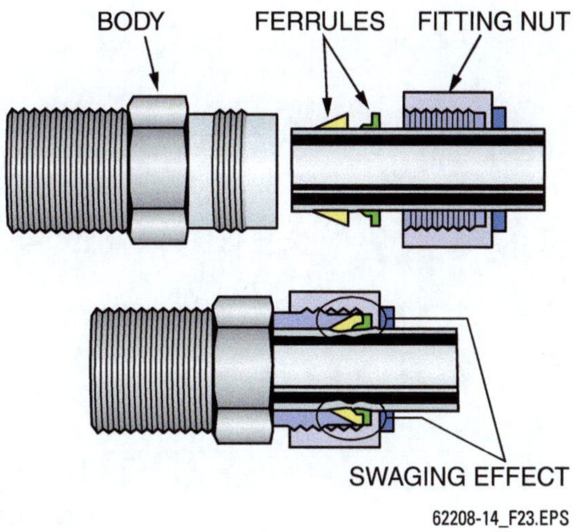

Figure 23 Two-ferrule compression fitting.

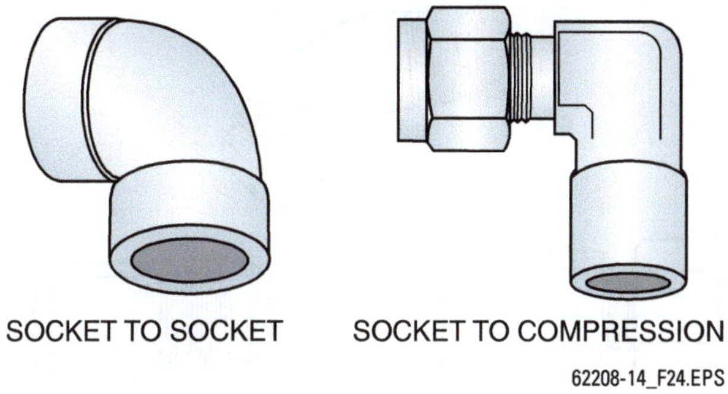

Figure 24 Socket-welded fittings.

2.3.4 Butt-Welded Fittings

Butt-joint or butt-welded fittings can be used for making tubing-to-tubing connections or to change a run from butt-welded piping to tubing. When butt-welding tubing or piping, the ends to be joined are beveled, lined up, and welded around the outside of the joint. A bevel is an angle cut on the end of the tubing.

Sometimes the terms *bevel* and *chamfer* are improperly used interchangeably. They are not the same. A chamfer is an angle cut only on the edge of the tubing. The flat edge of the tubing left after a chamfer is cut is called the land. A beveled edge is cut at the diagonal and has no land. *Figure 25* shows beveled and chamfered ends of tubing.

In the piping-to-tubing type of connection, the butt-joint is often welded on the piping side of the fitting. The tubing side is fitted with some other type of fitting, such as a flared or compression fitting.

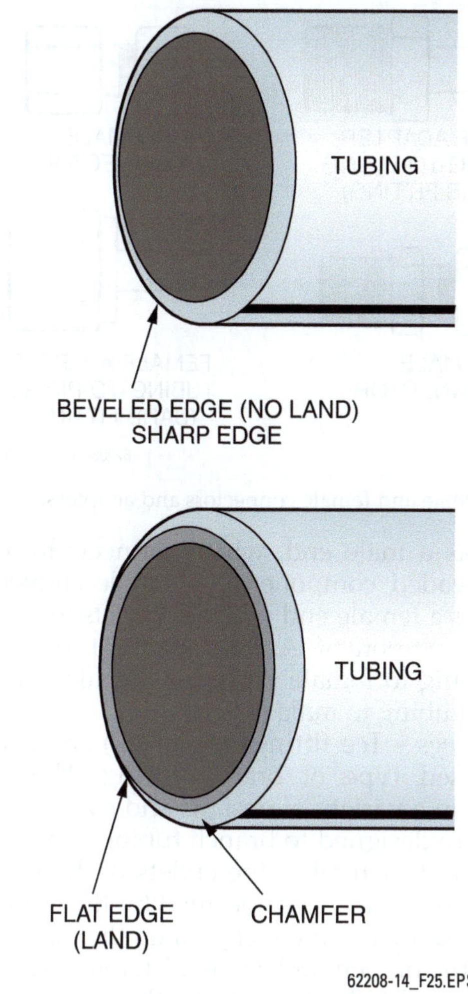

Figure 25 Beveled and chamfered ends of tubing.

2.3.5 Other Types of Tubing Fittings

There are numerous types of fittings used to change direction of tubing runs, provide branch connections, close ends of tubing, and connect tubing to tubing, tubing to piping, and tubing to instruments. The following sections describe several of the most commonly used types of tubing fittings.

Male and Female Connectors and Adapters – The term *male fitting* refers to fittings that have outside threads that connect tubing to a component with a female, or internal, thread. The term *female fitting* refers to fittings that have internal threads that connect tubing to a component with male, or outside, threads. While many types of fittings are referred to as male- or female-type fittings, two of the most common are male and female connectors and adapters (*Figure 26*). A male connector

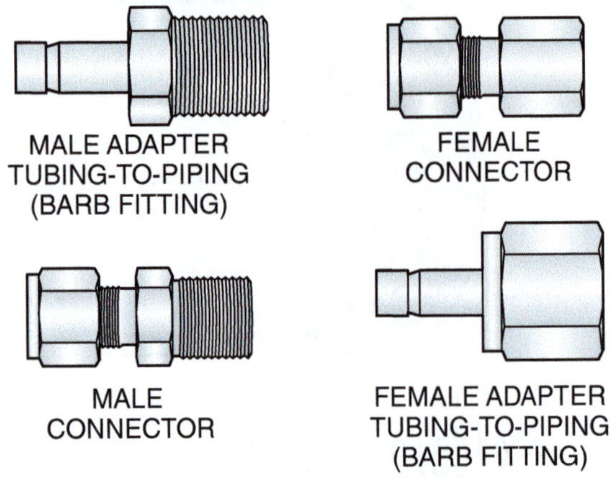

62208-14_F26.EPS

Figure 26 Male and female connectors and adapters.

fitting has a male end, which connects to a female threaded component. A female connector fitting has a female end which connects to a male threaded component. A male adapter fitting connects tubing to female piping. A female adapter connects tubing to male piping.

Tee fittings – Tee fittings (*Figure 27*) are a commonly used type of branch fitting. They are available in a variety of designs and sizes. All tee fittings are designed to branch tubing at a 90-degree angle. When all of the outlets of the tee are the same size, the tee is specified by the nominal size (for example, 1-inch tee). When the main outlets and the branch outlets are different sizes, the tee is referred to by first giving the main tubing run size, and then the branch run size (for example, 1-inch × 1⁄4-inch tee).

Elbow fittings – Elbow fittings (*Figure 28*) are curved fittings that connect tubing to tubing or piping to tubing at various angles. An elbow fitting is commonly referred to as an ell. The standard angles are 45 degrees and 90 degrees, although other angle sizes are available.

Coupling fittings – Coupling fittings (*Figure 29*) connect two pieces of tubing, or tubing and piping, without branching or stopping the flow through the run. Unlike union fittings, coupling fittings cannot be removed without disturbing the position of the tubing or piping. Coupling fittings may be connected by butt or socket welding or threaded ends.

Union fittings – Union fittings (*Figure 30*) connect two pieces of tubing, or tubing and piping, without branching or stopping the flow through the run. Union fittings can be connected and

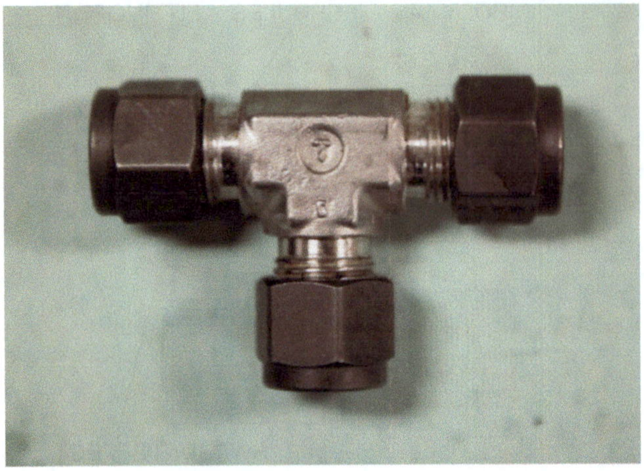

62208-14_F27.EPS

Figure 27 Tee fittings.

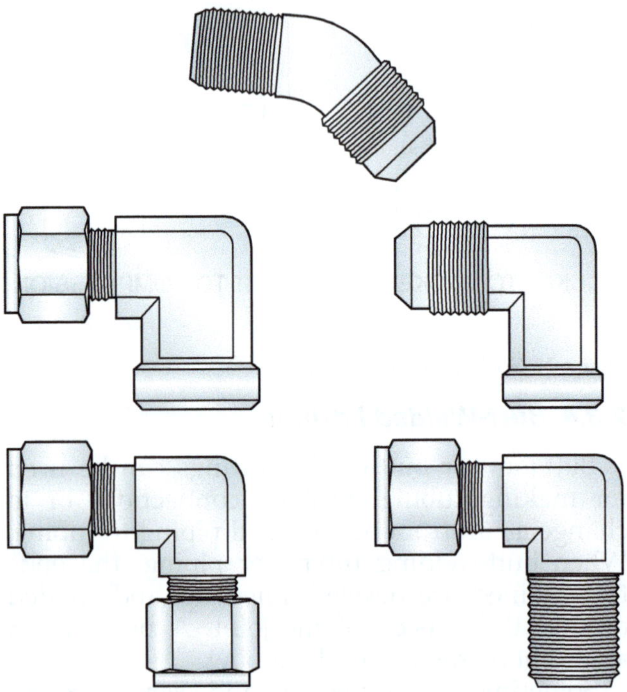

62208-14_F28.EPS

Figure 28 Elbow fittings.

disconnected without disturbing the position of the tubing or piping.

Cross fittings – Cross fittings (*Figure 31*) have two branch outlets that form the shape of a cross, providing four outlets spaced 90 degrees from each other. Cross fittings come in a variety of sizes and designs. All four outlets in a straight cross fitting are equal in dimension. In a reducing cross fitting, the two branch outlets are smaller than the run outlets. This type of cross fitting is used to obtain the desired flow and pressure through the system.

Tubing caps and plugs – Caps are used to close off the end of tubing. **Plugs** are used to close off an unused port of a tubing fitting. Caps and plugs come in various sizes and shapes. *Figure 32* shows one type of tubing cap and plug.

> **NOTE**
>
> The term *plugs* can sometimes be confusing, especially when talking about tubing. Plugs are usually thought of by most technicians as being a device that is installed inside of something. However, in the case of plugs used with tubing, the plugs are installed over the male portion of a fitting, not inside it.

Bulkhead fittings – Bulkhead fittings are used to connect two pieces of tubing through a plate, junction box, or control panel by use of a locknut. Bulkhead fittings come in several types, sizes, and materials. *Figure 33* shows some typical bulkhead fittings.

Thermocouple fittings – Thermocouples are widely used devices for measuring temperature in industrial processes. Most industrial thermocouples have a female connector at the end of the metal sheath that surrounds the thermocouple wires.

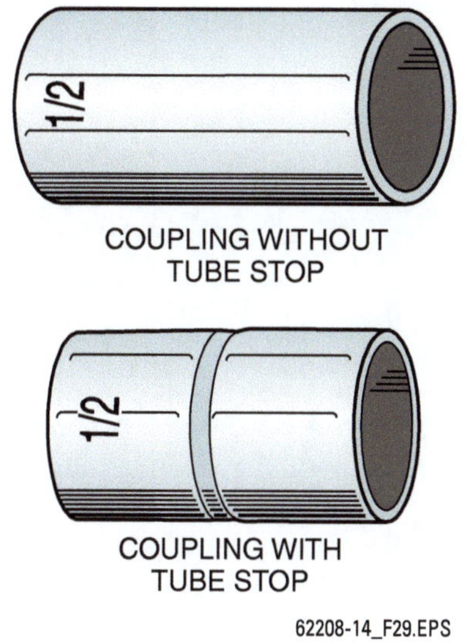

Figure 29 Coupling fittings.

Figure 30 Union fitting.

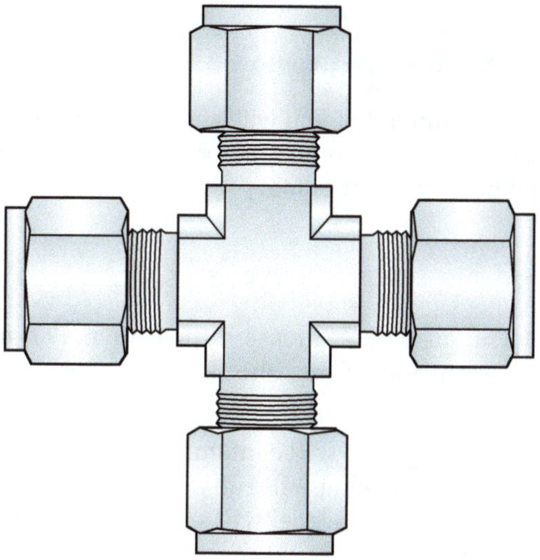

Figure 31 Cross fitting.

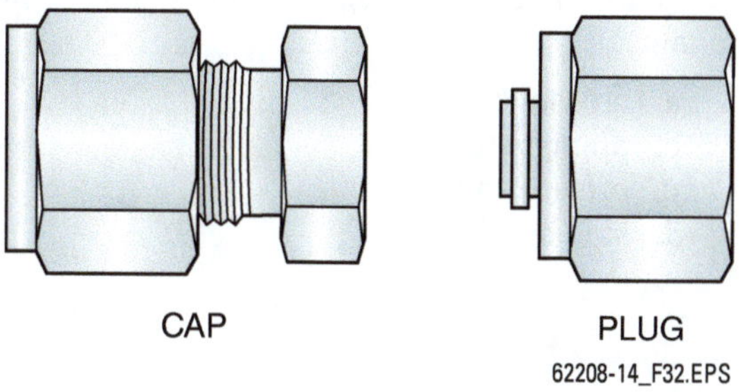

Figure 32 Cap and plug fittings.

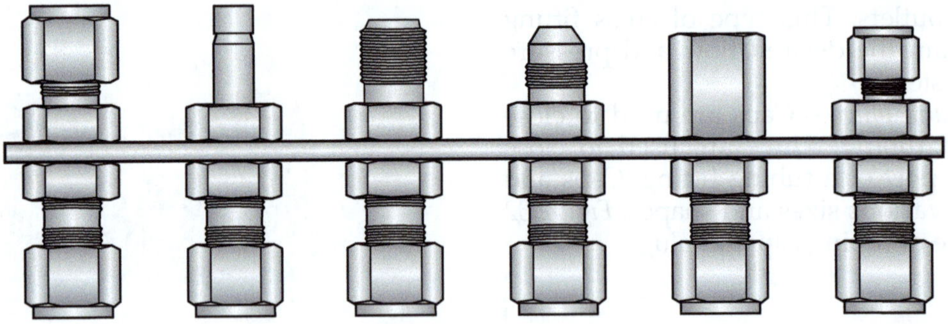

Figure 33 Bulkhead fittings.

Additional Resources

It is assumed that you will learn to perform these procedures under the direct supervision of your crew leader, foreman, or instructor. For independent study, refer to the recommended Additional Resources at the back of this module.

2.0.0 Section Review

1. The use of a grease pencil to make marks for cutting tubing should be avoided because _____.
   a. the grease will contaminate the cut
   b. the marking will rub off
   c. the cutter wheel will slip on the greasy surface
   d. the mark will be too wide for accuracy

2. The type of bend used to avoid an interference by moving the tubing run to one side is called a(n) _____.
   a. crossover bend
   b. angle bend
   c. offset bend
   d. side bend

3. When inserting tubing into a socket weld tubing fitting, allow $\frac{1}{16}$" to $\frac{1}{8}$" for _____
   a. rotation
   b. expansion
   c. contraction
   d. fluxing

SECTION THREE

3.0.0 TUBING INSTALLATION

Objectives

Describe the procedures used to inspect, install, and test tubing (CTTB-17).

a. Identify the prerequisites, competencies, and abnormal operating conditions (AOCs) associated with installing tubing.
b. Describe the procedures used to inspect, install, and test tubing.

Performance Task

Install tubing (CTTB-17).

- Identify potential abnormal operating conditions that may occur during performance of this CT, and know the appropriate actions to take in response to them.
- Utilize the appropriate personal protective equipment according to relevant company procedures.
- Inspect the tubing and components prior to installation.
- Install the tubing, providing support as needed.
- Perform a gap inspection test.
- Perform a leak and pressure test.
- Complete appropriate documentation as required by operator's procedures.

Trade Terms

Abnormal operating conditions (AOCs): Conditions within a pipeline system, identified by the operator, that may indicate (1) a malfunction of a component or (2) a deviation from normal operations that may indicate a condition exceeding design limits or result in a hazard to persons, property, or the environment.

Personnel must inspect tubing components before performing installation. They must carefully follow installation instructions for the fittings being used. They must also inspect the tubing installation after it is complete to ensure that it is safe for operation.

3.1.0 Prerequisites, Competencies, and Abnormal Operating Conditions

Personnel must possess a variety of skills in order to successfully inspect, install, and test tubing. The various operations require a series of steps, which technicians must be competent to perform. Personnel must also be aware of the particular **abnormal operating conditions (AOCs)** that may occur during the procedure and know the proper responses to them.

3.1.1 Prerequisites and Competencies

Personnel must possess the following basic knowledge and skills prior to installing tubing:

- Be competent with standard hand tools and any specialized tools recommended by the manufacturer.
- Be able to follow instructions and interpret drawings provided in the manufacturer's service manual.
- Be familiar with the types of tubing and associated components.
- Be familiar with entry permits and procedures for a confined space.
- Be able to perform a gap, leak, and pressure test once all work has been completed.
- Know how to properly use personal protective equipment (PPE)

3.1.2 Abnormal Operating Conditions

Personnel inspecting tanks must be able to identify the following potential AOCs and know the proper response to each, should they occur:

- *A leak of hazardous gases or liquids occurs – Take appropriate action using the proper PPE to protect people and the environment. Notify the proper authorities and company personnel. This situation can occur due to bends, kinks, or wrinkles in the tubing, using an incorrect fitting, and/or improperly secured fittings.*
- *A fire and/or explosion or ignition of hazardous, flammable liquids or gases occurs* – Evacuate the area and take only defensive action using the proper PPE. Contact the control center to isolate or shut down the line as quickly as possible. Engage the proper emergency response services.

3.2.0 Procedure

The following procedure is a basic guideline for inspecting, installing, and testing tubing. Always follow company-specific procedures. If you have any questions, contact your supervisor.

> **WARNING!**
>
> Perform any preliminary tasks required to ensure safe working conditions. Always follow company policy and procedures both before and during the inspection, installation, and testing.

3.2.1 Inspecting Tubing Components before Installation

Before installing tubing, personnel must inspect the components. They should complete the following steps:

Step 1 Make sure the fitting is the proper size for the tubing being used.

Step 2 Inspect the fittings to ensure that three lines are visible on the fitting. This ensures that the front and back ferrules are in place.

> **CAUTION**
>
> Do not disassemble the tube fittings before installation. The fittings should already be finger tight. Disassembling the fitting can allow dirt or other foreign particles into the fitting, which may affect sealing.

3.2.2 Installing Tubing

To install fittings from finger-tight for tubing that is 1 inch in diameter or smaller, do the following:

Step 1 Insert the tube into the fitting.

Step 2 Make sure that the nut is finger tight and that the tube sits tightly on the shoulder of the tube fitting body. Depending on length of tubing fabricated, supports may be needed to help deal with weight and vibration. The supports must allow easy access to tubing but not get in the way of daily operation.

Step 3 Scribe the nut at the six o'clock position.

Step 4 Hold the body of the fitting tightly and tighten the nut 1 ¼ turns to the nine o'clock position. For 1⁄16-, 1⁄8-, and 3⁄16-in. and 2-, 3-, and 4-mm tube fittings, tighten the nut ¾ turn to the three o'clock position. *Figure 34* shows a wrench being used to tighten a nut.

> **NOTE**
>
> It is very important to not over-tighten a fitting. Always follow the directions for tightening each size fitting as indicated.

To install fittings from snug for tubing that is 1 inch in diameter or smaller, do the following:

Step 1 Insert the tube into the fitting.

Step 2 Tighten the nut until the tube will not turn by hand or move inside the fitting.

Step 3 Scribe the nut at the six o'clock position.

Step 4 Hold the body of the fitting tightly and tighten the nut 1 ¼ turns to the nine o'clock position. For 1⁄16-, 1⁄8-, and 3⁄16-in. and 2-, 3-, and 4-mm tube fittings, tighten the nut ¾ turn to the three o'clock position.

3.2.3 Pre-swaging

Personnel often use a pre-swaging tool (*Figure 35*) in installation areas where work space is tight or the work must be done overhead. In these cases, the pre-swaging tool makes the installation process much easier. To use the pre-swaging tool, do the following:

62208-14_F34.EPS

Figure 34 Tightening a nut with a wrench.

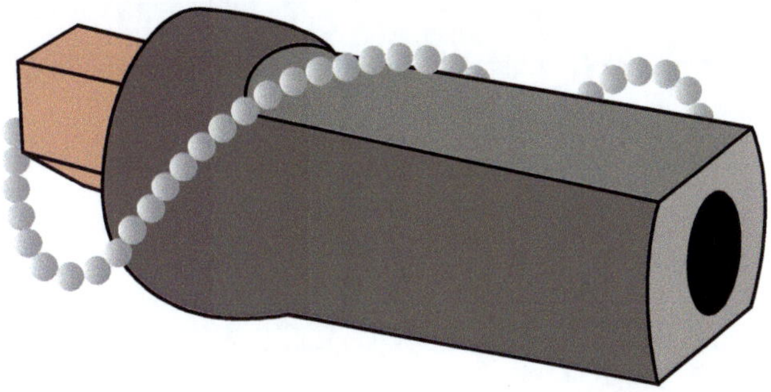

62208-14_F35.EPS

Figure 35 Pre-swaging tool.

Step 1 Install the nut and ferrules onto the pre-swaging tool.

Step 2 Insert the tubing into the pre-swaging tool.

Step 3 Make sure that the nut is finger tight and that the tube sits tightly on the shoulder of the pre-swaging tool.

Step 4 Scribe the nut at the six o'clock position.

Step 5 Hold the pre-swaging tool firmly and tighten the nut 1 $^1/_4$ turns to the nine o'clock position. For $^1/_{16}$-, $^1/_8$-, and $^3/_{16}$-in. and 2-, 3-, and 4-mm tube fittings, tighten the nut $^3/_4$ turn to the three o'clock position.

Step 6 Loosen the nut.

Step 7 Remove the tube, now with pre-swaged ferrules, from the pre-swaging tool.

Step 8 Insert the tube into the body of the fitting, ensuring that the front ferrule seats.

Step 9 Rotate the nut with a wrench to the pulled-up position.

Step 10 Tighten the nut slightly with a wrench.

3.2.4 Inspecting Tubing after Installation

An extremely important step in proper tubing installation is to ensure that a fitting is sufficiently tight after initial installation. To do so, personnel first perform a gap inspection test. After ensuring that the fittings are tight, they perform a leak test and a pressure test.

3.2.5 Gap Inspection Test

Personnel use a gap inspection gauge to test the fitting accordingly (*Figure 36*). Gauge sizes vary; always use the gauge that fits the fitting being inspected. Use the gap inspection gauge after properly pulling up on the nut. Place the gauge next to the gap between the nut and fitting body. If the gauge will not enter this gap, the fitting is properly tightened. If the gauge does enter the gap, additional tightening is needed. The gaugeability of a fitting is a measure of the amount of tightening required. Check the manufacturer's specifications for the proper gaugeability matrix based on the installation method and product size you are using.

NOTE

Each gap inspection gauge has a thick end and a thin end. Use the thick end to test fittings installed with the finger-tight method. Use the thin end to test fittings installed from snug.

CAUTION

Never use a gap inspection gauge on reassembled fittings or on fittings that were assembled with the pre-swaging tool.

3.2.6 Leak Test and Pressure Test

To perform a leak test on a connection within a tubing installation, personnel spray the fittings with a soapy substance and then apply pressure. If a leak exists, a bubble will form (*Figure 37*). If a leak does not exist, no bubble will form (*Figure 38*).

To pressure test tubing fittings, personnel test them in one of two ways: pneumatically (with nitrogen or air) or hydraulically.

62208-14_F36.EPS

Figure 36 Gap inspection gauge.

62208-14_F37.EPS

Figure 37 Failed leak test (bubble forms).

62208-14_F38.EPS

Figure 38 Passed leak test (no bubble forms).

Additional Resources

It is assumed that you will learn to perform these procedures under the direct supervision of your crew leader, foreman, or instructor. For independent study, refer to the recommended Additional Resources at the back of this module.

3.0.0 Section Review

1. If a fire or explosion occurs during inspection, installation, or testing, what must you use in order to take defensive action?
   a. A fire extinguisher
   b. The appropriate emergency checklist
   c. The proper PPE
   d. A hazardous vapor detector

2. During the pre-installation inspection, make sure that the number of lines visible on the fitting is _____.
   a. one
   b. two
   c. three
   d. four

3. From finger-tight, the nut on a $\frac{3}{16}$" tubing fitting should be tightened _____.
   a. one full turn
   b. $\frac{3}{4}$ turn
   c. $\frac{1}{2}$ turn
   d. $\frac{1}{4}$ turn

4. The thin end of a gap inspection gauge is used to check fittings installed with the finger-tight method. True or false?
   a. True
   b. False

SUMMARY

In some cases, tubing is the most effective material for pipeline applications. In this training module, you learned about the various types and sizes of tubing. You learned the importance of properly handling and storing tubing to protect people and property and in saving time and money on the job. You learned the procedures for cutting, deburring, bending, connecting, and installing tubing. You learned about the common types of fittings used with tubing. Tubing installation is an important part of pipeline maintenance. The knowledge and skills covered in this module provide basic and fundamental competencies required to effectively work with tubing installations.

Review Questions

1. The OD of tubing is generally expressed in _____
   a. decimal units
   b. millimeters
   c. fractions of an inch
   d. metric fractions

2. If the OD of a piece of tubing is ¾" and the ID is 0.584", the wall thickness is _____.
   a. 0.081"
   b. 0.083"
   c. 0.092"
   d. 0.098"

   Tubing ID = OD – (2 × wall thickness)

   0.584″ = ¾" – (2 × wall thickness)

   0.584″ = 0.75″ – (2 × wall thickness)

   (2 × wall thickness) = 0.75″ – 0.584″

   (2 × wall thickness) = 0.166″

   Wall thickness = 0.166″ ÷ 2

   Wall thickness = 0.083″

3. The chromium content of stainless steel, by weight, must be at least _____.
   a. 6 percent
   b. 8 percent
   c. 10 percent
   d. 12 percent

4. Types 304 and 316 stainless steel tubing are made from alloys that contain chromium and nickel, but not ____.
   a. magnesium
   b. molybdenum
   c. manganese
   d. mercury

5. Straight lengths of metal tubing are usually produced with a length of _____.
   a. 24 feet
   b. 20 feet
   c. 16 feet
   d. 8 feet

6. Organizations that have developed and published standards for tubing include ASTM, ASME, and _____.
   a. ANSI
   b. ANMS
   c. NSTA
   d. ASHO

7. The ASTM identification number designating a specific type of copper tubing would begin with the letter _____.
   a. C
   b. A
   c. B
   d. T

8. Improper handling of tubing can cause scratches or nicks, which are _____.
   a. not considered a problem
   b. possible contamination sites
   c. potential paths for leaks
   d. cause for rejecting the material

9. Stainless steel tubing is generally trimmed to the desired length by using a _____.
   a. hacksaw
   b. oxyacetylene torch
   c. cutoff saw
   d. tubing cutter

10. As a guide for marking the cutting line all the way around the tubing, you can use a(n) _____.
    a. scriber
    b. sleeve or ferrule
    c. paper collar
    d. o-ring

11. When using a tubing cutter, the cutting wheel handle should be tightened _____.
    a. every other turn
    b. at the same point on each turn
    c. at a different spot on each turn
    d. every third turn

12. To remove internal burrs on larger, hard-drawn tubing, the usual tool is a _____.
    a. spiral hand reamer
    b. rat-tail file
    c. power reamer bit
    d. wire wheel

13. The specialized small tool shown in Review Question *Figure RQ01* is a(n) _____.

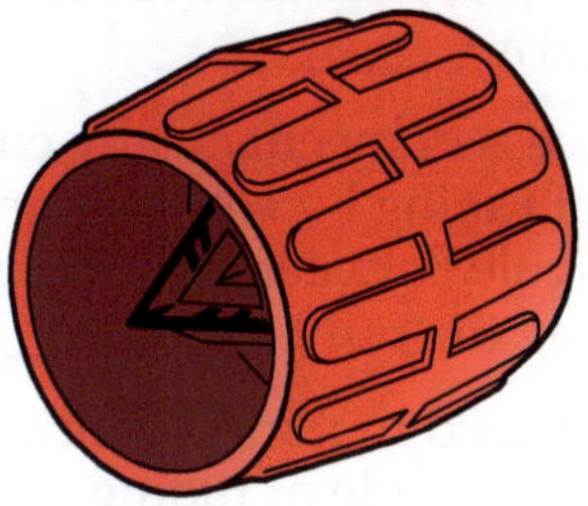

Figure RQ01

    a. internal deburring tool
    b. centering guide
    c. tubing scriber
    d. inside-outside reamer

14. A crossover bend in tubing is also known as a(n) _____.
    a. U-bend
    b. saddle bend
    c. expansion bend
    d. double return bend

15. A good tubing bend will show no _____.
    a. stretch marks
    b. discoloration
    c. flattening
    d. tool marks

16. Compression-type hand benders are used to bend stainless steel tubing with a diameter of up to _____.
    a. ¼"
    b. ½"
    c. ¾"
    d. 1"

17. Flare fittings used in high-pressure applications have a fitting body with a tip that is tapered _____.
    a. 16 degrees
    b. 28 degrees
    c. 37 degrees
    d. 45 degrees

18. When the nut of a compression fitting is tightened, it causes the tubing to be pinched by a ring known as a(n) _____.
    a. compressor
    b. dilation band
    c. expander
    d. ferrule

19. To prepare for tubing installation "from snug," the fitting nut is tightened _____.
    a. three full turns
    b. until the tubing cannot be pulled out of the fitting
    c. until the tubing will not move inside the fitting
    d. just finger-tight

20. A leak test is performed on completed tubing installations using a(n) _____.
    a. soap solution
    b. pressure gauge
    c. dye tracer
    d. sniffer

Trade Terms Quiz

Fill in the blank with the correct term that you learned from your study of this module.

1. The angle made at the enlarged end of a piece of tubing to match a fitting is called a ________.
2. Soft tubing that is pulled or stretched, by hand or by mechanically feeding it through a die, to reduce diameter and wall thickness is referred to as ________.
3. A device that is rotated around tubing to separate it by using a cutting wheel or wheels and rollers is a ________.
4. A tubing fitting used to close off a port of another fitting is called a ________.
5. A sleeve- or ring-type component of a tubing fitting, often referred to as a barrel, which forms an intricate part of the tubing connection and sealing process is called a ________.
6. Conditions within a pipeline system, identified by the operator, that may indicate (1) a malfunction of a component or (2) a deviation from normal operations that may indicate a condition exceeding design limits or result in a hazard to persons, property, or the environment are called ________.
7. A tubing fitting used to close off the end of tubing is called a ________.
8. A device used to bend tubing without kinking or crimping is a ________.
9. A device, such as an elbow, tee, or union, used with tubing to connect tubing to tubing, tubing to instruments, or tubing to piping, is a ________.
10. The process of mechanically connecting a length of tubing by compressing or squeezing a ferrule between a fitting and nut by tightening the nut is ________.
11. A device used to remove burrs and ridges from the inside of tubing is called a ________.

Trade Terms

Abnormal operating conditions (AOCs)
Cap
Compression
Drawn
Ferrule
Fitting
Flare
Plug
Reamer
Tubing bender
Tubing cutter

Trade Terms Introduced in This Module

Abnormal operating conditions (AOCs): Conditions within a pipeline system, identified by the operator, that may indicate (1) a malfunction of a component or (2) a deviation from normal operations that may indicate a condition exceeding design limits or result in a hazard to persons, property, or the environment.

Cap: A tubing fitting used to close off the end of tubing.

Compression: The process of mechanically connecting a length of tubing by compressing or squeezing a ferrule between a fitting and nut by tightening the nut.

Drawn: Soft tubing that is pulled or stretched, by hand or by mechanically feeding it through a die, to reduce diameter and wall thickness.

Ferrule: A sleeve- or ring-type component of a tubing fitting, often referred to as a barrel, which forms an intricate part of the tubing connection and sealing process.

Fitting: A device, such as an elbow, tee, or union, used with tubing to connect tubing to tubing, tubing to instruments, or tubing to piping. Fittings are also used to close off tubing and/or to branch tubing.

Flare: The angle made at the enlarged end of a piece of tubing to match a fitting.

Plug: A tubing fitting used to close off a port of another fitting.

Reamer: A device used to remove burrs and ridges from the inside of tubing.

Tubing bender: A device used to bend tubing without kinking or crimping. Tubing benders may be hand-operated, table-mounted, or hydraulically powered.

Tubing cutter: A device that is rotated around tubing to separate it by using a cutting wheel or wheels and rollers.

Additional Resources

The following recommended resources can provide additional helpful information related to the requirements for this covered task:

API Recommended Practice 1161, Third Edition, © 2014 American Petroleum Institute. API Publishing Services, 1220 L Street, NW, Washington, DC 20005. Distributed by Thompson Reuters (Scientific) LLC, **www.techstreet.com**.

ASTM A450, ASTM International, **www.astm.org**

ASME Standard B31.3, Process Piping, American Society of Mechanical Engineers (ASME), **www.asme.org**

Standards and Specifications for Tubing. Washington, DC: American National Standards Institute (ANSI), **www.ansi.org**

Standards and Specifications for Tubing. New York: American Society of Mechanical Engineers (ASME), **www.asme.org**

NCCER Module AOCCC-17, *Abnormal Operating Conditions Control Center*

NCCER Module AOCFG-17, *Abnormal Operating Conditions Field and Gas*

Section Review Answer Key

Answer	Section Reference	Objective
Section One		
1. c	1.1.0	1a
2. b	1.2.0	1b
3. b	1.3.1	1c
4. d	1.4.0	1d
5. b	1.6.0	1f
6. a	1.6.0	1f
Section Two		
1. d	2.1.2	2a
2. c	2.2.1	2b
3. b	2.3.3	2c
Section Three		
1. c	3.1.2	3a
2. c	3.2.1	3b
3. b	3.2.2	3b
4. b	3.2.5	3b

NCCER CURRICULA — USER UPDATE

NCCER makes every effort to keep its textbooks up-to-date and free of technical errors. We appreciate your help in this process. If you find an error, a typographical mistake, or an inaccuracy in NCCER's curricula, please fill out this form (or a photocopy), or complete the online form at **www.nccer.org/olf**. Be sure to include the exact module ID number, page number, a detailed description, and your recommended correction. Your input will be brought to the attention of the Authoring Team. Thank you for your assistance.

Instructors – If you have an idea for improving this textbook, or have found that additional materials were necessary to teach this module effectively, please let us know so that we may present your suggestions to the Authoring Team.

NCCER Product Development and Revision

13614 Progress Blvd., Alachua, FL 32615

Email: curriculum@nccer.org
Online: www.nccer.org/olf

❑ Trainee Guide ❑ Lesson Plans ❑ Exam ❑ PowerPoints Other ____________

Craft / Level: Copyright Date:

Module ID Number / Title:

Section Number(s):

Description:

Recommended Correction:

Your Name:

Address:

Email: Phone:

CTTP-17

Threaded Pipe Fabrication

Overview

One way that pipe and fittings can be joined is by threading. Threaded pipe systems (also referred to as *screwed fitting joint piping systems*), are relatively inexpensive to fabricate and are a common way to join pipe in many small-diameter mechanical and industrial applications. This module describes the materials used to fabricate a threaded pipe system, including types of pipe and their characteristics, pipe fittings, and tools. It also explains the methods used to measure pipe length and how to thread and assemble a pipe system.

Trainees with successful module completions may be eligible for credentialing through the NCCER Registry. To learn more, go to **www.nccer.org** or contact us at 1.888.622.3720. Our website has information on the latest product releases and training, as well as online versions of our Cornerstone magazine and Pearson's product catalog.

Your feedback is welcome. You may email your comments to **curriculum@nccer.org**, send general comments and inquiries to **info@nccer.org**, or fill in the User Update form at the back of this module.

This information is general in nature and intended for training purposes only. Actual performance of activities described in this manual requires compliance with all applicable operating, service, maintenance, and safety procedures under the direction of qualified personnel. References in this manual to patented or proprietary devices do not constitute a recommendation of their use.

CTTP-17

Threaded Pipe Fabrication

Objectives

When you have completed this module, you will be able to do the following:

1. Describe the materials used in threaded pipe systems.
   a. Describe the types of pipe used in threaded pipe systems.
   b. Explain the sizing of pipe used in threaded pipe systems.
   c. Identify schedules and wall thicknesses of pipe used in threaded pipe systems.
   d. Explain pipe threads used in threaded pipe systems.
2. Identify and explain the purpose of pipe fittings.
   a. Identify and explain the purpose of elbows, offsets, and return bends.
   b. Identify and explain the purpose of branch connections.
   c. Identify and explain the purpose of caps and plugs.
   d. Identify and explain the purpose of line connections.
   e. Identify and explain the purpose of nipples.
   f. Identify and explain the purpose of flanges.
3. Explain how to determine pipe lengths between fittings.
   a. Explain how to use the center-to-center method to determine pipe lengths between fittings.
   b. Explain how to use the center-to-face method to determine pipe lengths between fittings.
   c. Explain how to use the face-to-face method to determine pipe lengths between fittings.
   d. Explain how to calculate offsets.
4. Explain the processes and tools used to thread and assemble threaded pipe.
   a. Explain how to cut and ream pipe.
   b. Explain how to thread pipe manually and by machine.
   c. Identify types of pipe joint compounds and explain how to use them.
   d. Explain how to fit screwed pipe and fittings.
   e. Explain how to install threaded valves.
5. Describe the procedures used to thread and fit pipe (CTTP-17).
   a. Identify the prerequisites, competencies, and abnormal operating conditions (AOCs) associated with threaded pipe fabrication.
   b. Describe the procedures used to thread and fit pipe.

Performance Task

Under the supervision of your instructor, you should be able to do the following:

1. Thread and fit pipe (CTTP-17).
   - Identify potential abnormal operating conditions that may occur during performance of this CT, and know the appropriate actions to take in response to them.
   - Utilize the appropriate personal protective equipment according to relevant company procedures.
   - Verify type, size, length, and schedule (grade) of pipe.
   - Cut the pipe to the desired length and ream the ends.
   - Thread the pipe.
   - Select the proper size, shape, and type of fittings needed.
   - Clean the pipe and fittings thoroughly.
   - Test fit pipe to assure proper fit and thread compatibility.
   - Apply joint compound or tape to pipe threads.
   - Join piping and fittings as required.
   - Complete appropriate documentation as required by operator's procedures.

Trade Terms

Abnormal operating conditions (AOCs)
Branch
Bushing
Closenipple
Coupling
Elbow
Flange
Galling
National Pipe Thread (NPT)
Nipple
Offset
PTFE tape
Reducer
Run
Swage
Takeout
Tee
Thread engagement
Union
Wye

Industry-Recognized Credentials

If you're training through an NCCER-accredited sponsor, you may be eligible for credentials from NCCER's Registry. The ID number for this module is CTTP-17. Note that this module may have been used in other NCCER curricula and may apply to other level completions. Contact NCCER's Registry at 888.622.3720 or go to **www.nccer.org** for more information.

Contents

Figures and Tables

SECTION ONE

1.0.0 MATERIALS USED IN THREADED PIPING SYSTEMS

Objective

Describe the materials used in threaded pipe systems.

a. Describe the types of pipe used in threaded pipe systems.
b. Explain the sizing of pipe used in threaded pipe systems.
c. Identify schedules and wall thicknesses of pipe used in threaded pipe systems.
d. Explain pipe threads used in threaded pipe systems.

Trade Term

National Pipe Thread (NPT): The US standard for pipe thread; has a 1⁄16-inch taper per inch from back to front.

There are several advantages to a threaded piping system. These systems can be easily fabricated on the job site using pipe and various fittings. Portable threading equipment can be used almost anywhere. If a piping system must be installed near flammable liquids or gases, threaded pipe is much safer because it avoids the fire hazards of welding. Finally, threaded pipe can be cleaned on the inside before it is installed, reducing the possibility of metal particles becoming entrapped and damaging valves or strainers. Threaded pipe does have disadvantages. Threaded pipe joints are more prone to leaks than other types of joints. The strength of the pipe is reduced by the threading process because the wall thickness of the pipe is reduced at the threads. In addition, threaded pipe cannot be used if erosion, crevice corrosion, extreme shock, or vibration is anticipated.

Various types of materials can be used in threaded piping systems. These materials are classified according to the type of material, pipe size, and schedule or wall thickness of the pipe. In addition, by definition, threaded pipe also has threads used for connecting it. These threads follow thread sizing standards.

1.1.0 Types of Pipe

In the pipeline industry, personnel use carbon steel, stainless steel, and aluminum pipe. Carbon steel pipe is widely used in piping systems because it is ductile, durable, machinable, and less expensive than most other types of pipe. Stainless steel pipe is used less frequently because of the difficulty in correctly cutting the threads and making up leak-proof joints. Although it is more expensive than some other materials, stainless steel pipe is good for special applications such as high heat environments and areas that require the maximum resistance to corrosion, such as processing systems, instrument lines, and heat-transfer equipment. Stainless steel can also be kept sterile to prevent contamination.

Aluminum pipe is used in some specialized applications. Aluminum has the advantage of being relatively soft and easy to machine, but it can be damaged more easily than steel pipe. Aluminum, as used in pipe systems, is almost always an alloy, with widely varying levels of ductility and strength. Other alloys, such as titanium steel, are occasionally used for special-purpose piping.

1.2.0 Pipe Size

Pipe is listed in inches by its nominal size. For sizes of pipe up to and including 12 inches, nominal size is an approximation of the pipe's inside diameter. From 14 inches on, nominal size reflects the outside diameter of the pipe. *Figure 1* shows inside and outside diameters of pipe.

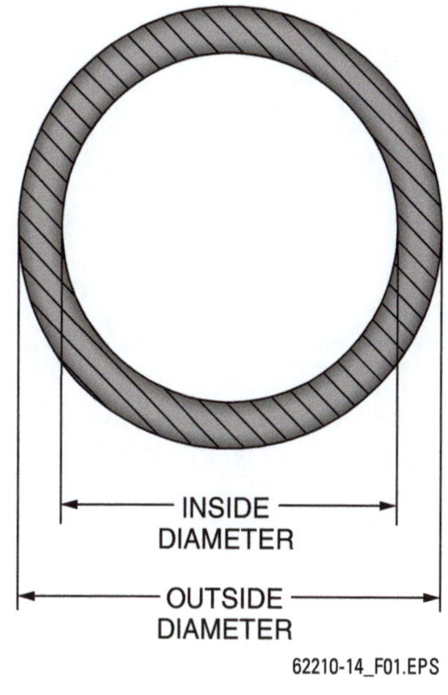

Figure 1 Inside and outside diameters of pipe.

There are times when the nominal size of a pipe and its actual inside or outside diameter differ greatly, but the nominal size is used to describe the pipe.

1.3.0 Schedules and Wall Thicknesses

Wall thickness can be described in two ways: by manufacturer's weight and by schedule number. As the schedule numbers increase, the wall thickness gets larger; therefore, the pipe is stronger and can withstand more pressure. Today, the pipeline industry commonly uses the following three classifications of manufacturer's weight, listed from smallest to largest:

- STD – Standard wall
- XS – Extra-strong wall
- XXS – Double extra-strong wall (also sometimes referred to as XXH for double extra-heavy)

Schedule numbers and wall thicknesses are somewhat interchangeable. Schedule numbers for pipe range from 5 to XXS, but no pipe with a wall thickness less than Schedule 80 should be threaded. It is important to remember that the schedule number refers only to the wall thickness of a pipe of a given nominal size. A ¾-inch Schedule 80 pipe will not have the same wall thickness as a 1-inch Schedule 80 pipe.

Common wall thicknesses of carbon steel pipe are Schedule 80, Schedule 160, and double extra-strong. Common wall thicknesses of stainless steel pipe range from Schedule 5 to Schedule 160. *Table 1* shows commercial pipe dimensions.

1.4.0 Pipe Threads

By definition, a pipe thread is an inclined plane wrapped around a cylinder. The thread angle is the angle formed by two inclined faces of the thread. The thread angle varies depending on the specific thread standard used. The pitch refers to the distance between adjacent screw threads when measured from center line to center line. Pitch also refers to threads per inch. The crest of the thread is the outermost part of the thread, or the top of the thread, when viewing it from the side. The root of the thread is the innermost part of the thread, or the bottom, if viewed from the side.

The **National Pipe Thread (NPT)** is the standard thread for pipe connections. The NPT has a thread angle of 60 degrees and is slightly flat at the crest and the root. While NPTs can be either straight or tapered, the tapered pipe thread, which has a tapered internal and external thread, is the most commonly used thread for pipe and fittings in industrial and commercial low-pressure applications. The taper used is $\frac{1}{16}$-inch per inch of threads.

All NPTs are designated by specifying in sequence the nominal size of the pipe, the number of threads per inch, and the symbols for the thread series. For example, 3/8-18-NPT, are National Pipe Threads for a ⅜-inch nominal-size pipe having 18 threads per inch.

Table 1 Commercial Pipe Dimensions

		Inside Diameter (inches)		
Nominal Size (inches)	**Outside Diameter (inches)**	**Extra-Strong, Schedule 80**	**Schedule 160**	**Double Extra-Strong**
1/8	0.405	0.215	—	—
1/4	0.540	0.302	—	—
3/8	0.675	0.423	—	—
1/2	0.840	0.546	0.466	0.252
3/4	1.050	0.742	0.614	0.434
1	1.315	0.957	0.815	0.599
1 1/4	1.660	1.278	1.160	0.896
1 1/2	1.900	1.500	1.338	1.100
2	2.375	1.939	1.689	1.503
2 1/2	2.875	2.323	1.885	1.771
3	3.500	2.900	2.625	2.300
3 1/2	4.000	3.364	—	–
4	4.500	3.826	3.438	3.152
5	5.562	4.813	4.313	4.063
6	6.625	5.761	5.187	4.897
8	8.625	7.625	6.813	6.875

Additional Resources

It is assumed that you will learn to perform these procedures under the direct supervision of your crew leader, foreman, or instructor. For independent study, refer to the recommended Additional Resources at the back of this module.

1.0.0 Section Review

1. One disadvantage of a threaded piping system is that it is more prone to _____.
   a. fire hazards
   b. leaks
   c. collecting dirt
   d. entrapping metal particles

2. The pipeline industry uses carbon steel, stainless steel, and _____.
   a. aluminum pipe
   b. copper pipe
   c. PVC pipe
   d. iron pipe

3. For sizes of pipe up to and including 12 inches, nominal size is an approximation of the pipe's _____.
   a. length
   b. width
   c. outside diameter
   d. inside diameter

4. Pipe has three classifications of manufacturer's weight: double extra-strong wall, extra-strong wall, and _____.
   a. standard wall
   b. normal wall
   c. strong wall
   d. heavy-duty wall

Section Two

2.0.0 Pipe Fittings

Objective

Identify and explain the purpose of pipe fittings.

a. Identify and explain the purpose of elbows, offsets, and return bends.
b. Identify and explain the purpose of branch connections.
c. Identify and explain the purpose of caps and plugs.
d. Identify and explain the purpose of line connections.
e. Identify and explain the purpose of nipples.
f. Identify and explain the purpose of flanges.

Trade Terms

Branch: The outlet or inlet of a fitting not in line with the run but which may make any angle.

Bushing: A pipe fitting that consists of a hollow plug with internal and external threads that suit the different diameters needed when connecting pipe with a fitting of a larger nominal size.

Close nipple: A nipple that is about twice the length of a standard thread and is threaded from end to end with no shoulder.

Coupling: A sleeve that is used to connect two straight pieces of pipe.

Elbow: A fitting that makes an angle between adjacent pipes; always 90 degrees unless another angle is stated. Also known as an ell.

Flange: A fitting on the end of a pipe that is shaped like a rim and allows the pipe to be bolted to another flanged pipe.

Nipple: A short piece of pipe that is threaded on both ends and is less than 12 inches long.

Offset: A lateral move that repositions the pipe out of its original line to another line that is parallel with the original line.

Reducer: A pipe fitting with female threads on each end, with one end being one or more sizes smaller than the other end.

Swage: A type of reducer with a male thread on each end used to join pipe of difference sizes in high-pressure applications.

Tee: A fitting that has a side outlet of 90 degrees to the run.

Union: A fitting consisting of two sleeves to be threaded on the ends of the pipes and a threaded coupling ring to draw the sleeves together; used to connect pipes where quick disconnection may be needed for maintenance.

Wye: A fitting that has a side outlet of 45 degrees to the run; also referred to as a lateral.

Piping systems consist of pipes, pumps, valves, and other parts, including fittings. Pipe fittings can be defined as fittings attached to pipe to change the direction of fluid flow, to connect a **branch** line to a main line, to close off the end of a line, or to join together two pipes of the same or different sizes. Pipe fittings come in various sizes, materials, strengths, and designs to match the various piping systems. Plain fittings are made of forged steel and used for pipeline applications. Pipe fittings can be grouped by their use into the following categories:

- Elbows, offsets, and return bends
- Branch connections
- Caps and plugs
- Line connections
- Nipples
- Flanges

2.1.0 Elbows, Offsets, and Return Bends

Elbows, offsets, and return bends are fittings used to change the direction of fluid flow in a piping system. An **elbow**, also known as an ell, L, 90, or 45, is a fitting that forms an angle between different connecting pipes. This angle is always 90 degrees unless another angle is stated. Elbows come in several different angles. The 90- and 45-degree elbows are the most common. The $5\frac{5}{8}$-, $11\frac{1}{4}$-, and 30-degree elbows are rarely used and hard to obtain. *Figure 2* shows elbows of various degrees.

An **offset** changes the direction of flow for a short distance. The offset moves the pipe run to one side and then continues it in the same direction. An offset can be made by using two fittings, one at each end of the offset. *Figure 3* shows an example of an offset.

A return bend is a U-shaped fitting that sends the fluid back in the same direction from which it came. Return bends are often used in systems in which the pipe must pass through the same area several times. Return bends are manufactured in various sizes and come in close, open, and wide patterns. *Figure 4* shows examples of return bends.

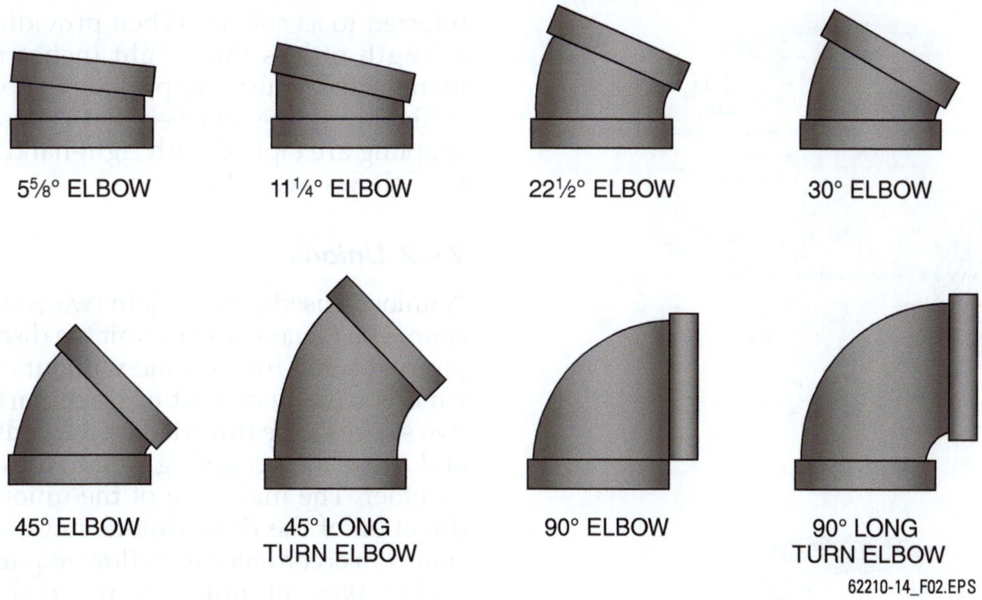

Figure 2 Elbows.

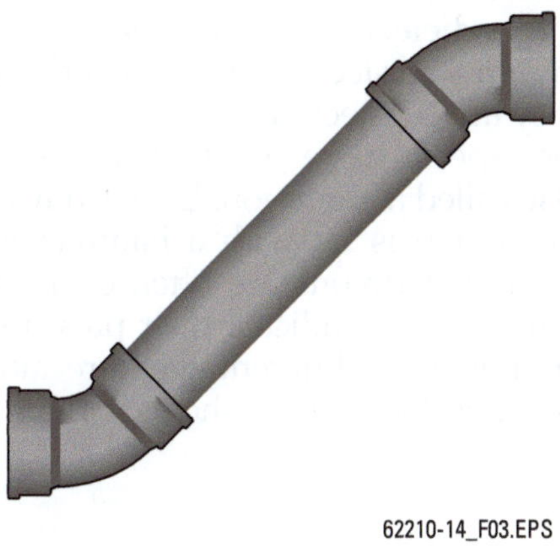

Figure 3 Offset.

2.2.0 Branch Connections

Branch connections are fittings that divide the flow and send the fluid in multiple directions. A **tee**, the most widely used type of branch connection, is used to make a 90-degree branch in the main pipe. A **wye** is another common branch connection that has a side opening set at a 45-degree angle. *Figure 5* shows some common branch connections.

2.3.0 Caps and Plugs

Caps and plugs are fittings used for closing off lines. Caps and plugs must have the same nominal wall thickness and ANSI pressure rating as the system in which they are installed. A cap fitting screws onto the outside of the pipe end, and a plug screws into the inside of the pipe or fitting end. Caps can either be plain caps with round heads or have a square projection on the body to allow for tightening with a wrench. Plugs can either be solid square-head plugs or countersink plugs, which have a square depression in which a square key can be inserted for tightening. A bull plug is a long shaft that is threaded on one end that may or may not have a hex head. Bull plugs are used to plug a pipeline so that the head will extend out of the insulation for easy removal. *Figure 6* shows caps and plugs.

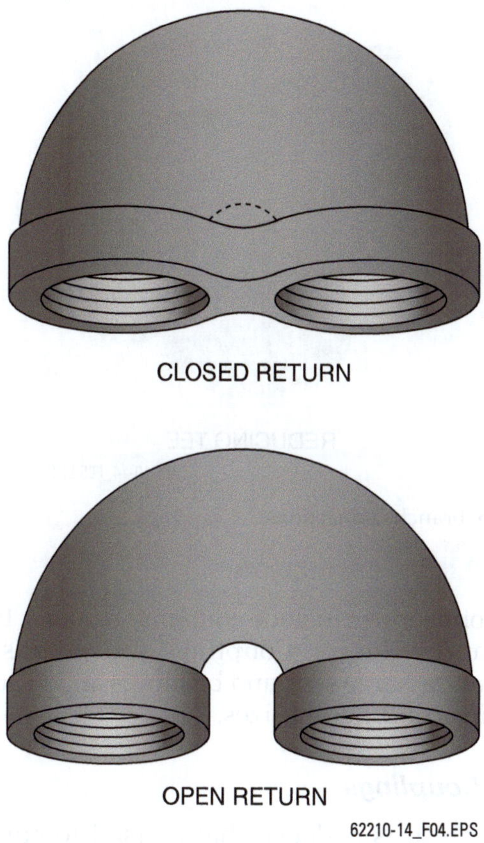

Figure 4 Return bends.

Figure 5 Branch connections.

2.4.0 Line Connections

Line connections include couplings, unions, bushings, and reducers. Couplings and unions join pipes of the same size, and bushings and reducers join pipes of different sizes.

2.4.1 Couplings

A **coupling** is a sleeve that is used to connect two straight pieces of pipe. Couplings are often referred to as collars. When providing pipe with a length of less than eight inches in diameter, manufacturers usually provide the pipe with one straight coupling attached. Both ends of a straight coupling are tapped with right-hand threads. *Figure 7* shows a coupling.

2.4.2 Unions

A union is used either to join two ends of pipe that cannot be turned or to permit the disconnecting of pipes at some future time without cutting them. Unions usually consist of three parts, including two sleeves to be threaded on the ends of the pipes and a threaded coupling ring to draw the sleeves together. The male side of the union goes in the direction of the flow. Unions must always be installed in accordance with flow requirements.

One type of union is the dielectric union, which is used to prevent the breakdown of metals caused by directly connecting two different metals. The dielectric union has an insulating ring between two sleeves that match the metal to which they are connected.

Another special type of union is the hammer union, also called a lug union. Unlike other unions, a hammer union is hit with a hammer to be installed. Hammer unions are often color-coded by the manufacturer to indicate their pressure rating. Hammer unions of different pressure ratings are not interchangeable. *Figure 8* shows various unions.

CAUTION

Never hit a hammer union with a hammer when it is under pressure.

2.4.3 Bushings

The function of a **bushing** is to connect the male end of a pipe to the female end of a fitting of a larger size. It consists of a hollow plug with male and female threads to suit the different diameters. Bushings are usually made with a hexagon head but can also be made without the head. Those without a head are known as face bushings or flush bushings and can be screwed into the fitting to form a neat, flush finish. Bushings are used in low-pressure systems because they restrict the flow in the system. *Figure 9* shows common bushings.

2.4.4 Reducers

A **reducer** is used to join the male ends of two different sizes (dimensions) of pipe. A reducer may be a separate fitting or may be a built-in part

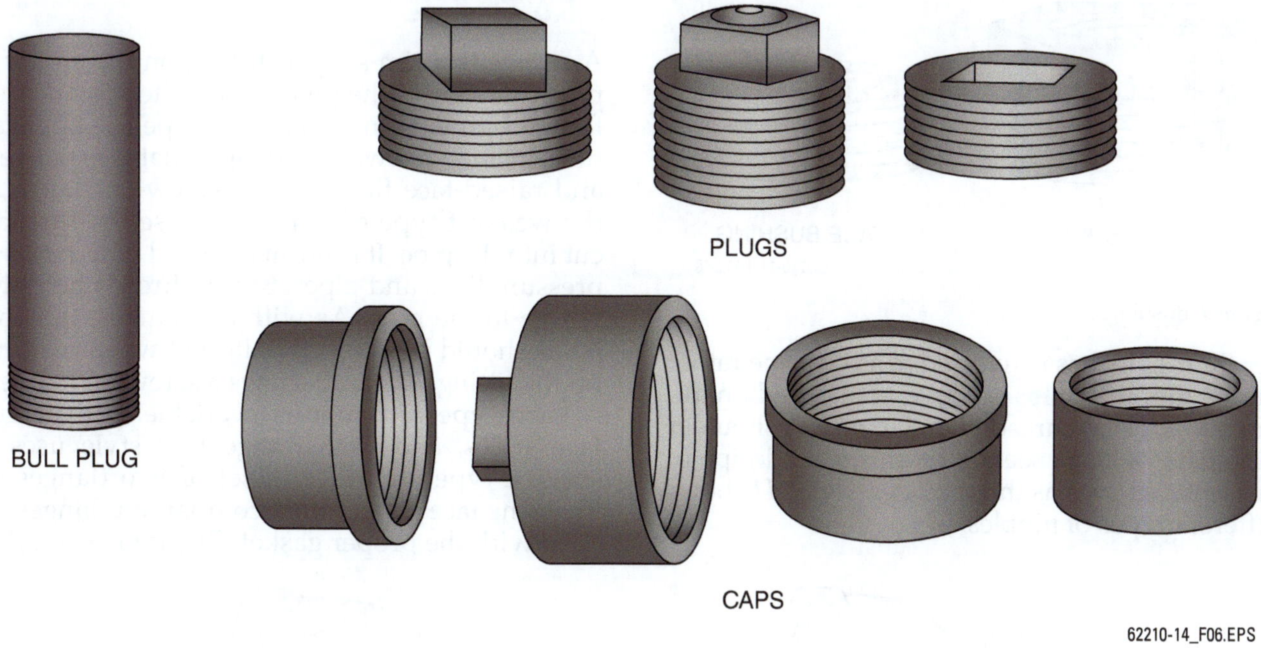

Figure 6 Caps and plugs.

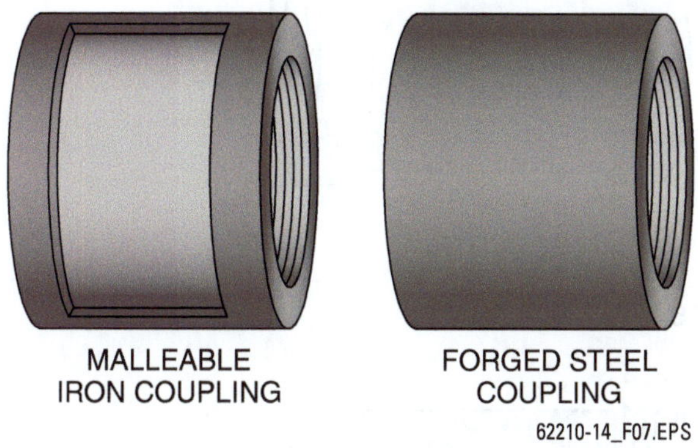

Figure 7 Coupling.

of another fitting. Most directional change and branch connection fittings are available with reduced outlet lines. A **swage** is a reducer with a male thread on each end and used to join pipe of difference sizes in high-pressure applications. Swages are made in concentric and eccentric types. Horizontal reducers are eccentric. Liquid reducers usually have a flat top and gas reducers have a flat bottom. A reducing coupling is a concentric high-pressure fitting that restricts flow. Reducing tees are available in forged steel. *Figure 10* shows reducers.

2.5.0 Nipples

A **nipple** is a short piece of pipe that is threaded on both ends and used to make close connections between fittings. Nipples are made in all sizes and types of pipe and are stocked in various lengths classified as close, shoulder, short, and long. The longest nipples are 12 inches long. Pipe that is lon-

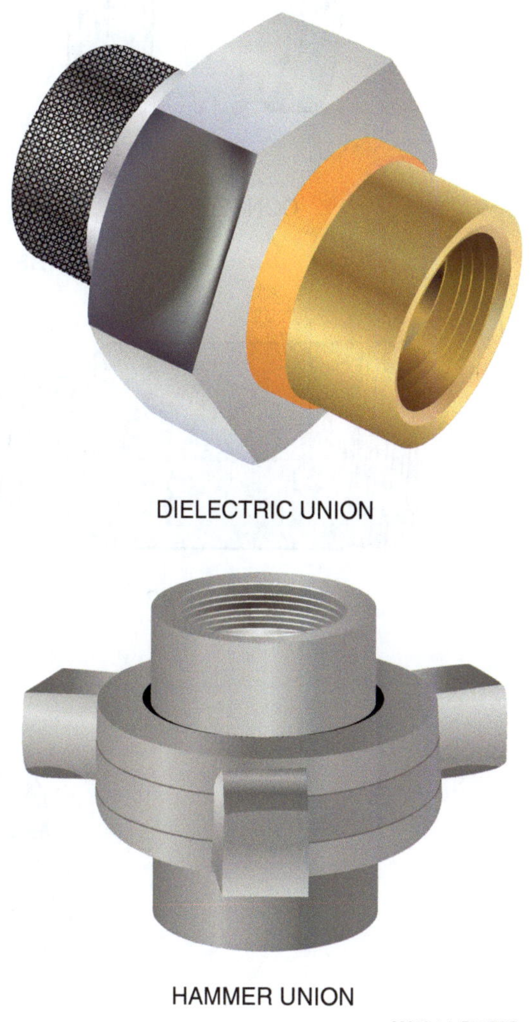

Figure 8 Unions.

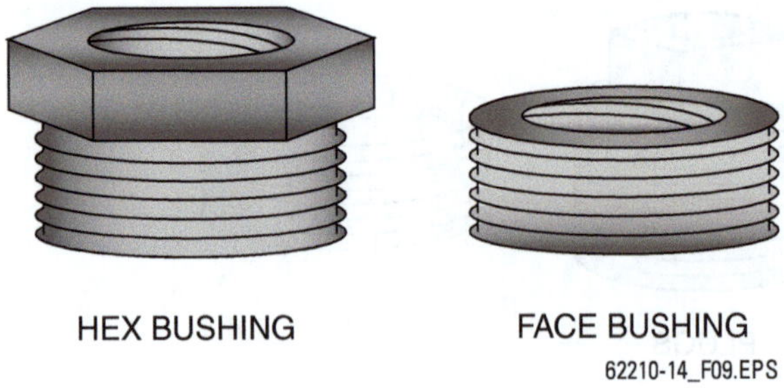

Figure 9 Bushings.

ger than 12 inches is referred to as cut pipe rather than a nipple. A **close nipple**, also called an all-thread nipple, is threaded end to end. A shoulder nipple has a short section of unthreaded pipe in the center known as the shoulder. *Figure 11* shows different types of nipples.

2.6.0 Flanges

A **flange** is a rim-shaped fitting on the end of a pipe that allows the pipe to be bolted to another flanged pipe. Flanges used in pipeline applications include screwed-on flanges, flat-face flanges, and raised-face flanges. The screw-on flange is the weakest type of flange because the threads cut into the pipe. It is normally used only for low-pressure lines and pipes. Screw threads hold the flange to the pipe. As with all flanges, the bolt holes should be carefully aligned when making up the fitting. *Figure 12* shows a screw-on flange.

Each type of flange is available in different face styles, and each flange face style uses a specific type of gasket. A set of two flanges of the same face style, called companion flanges, is used with the proper gasket. The two face styles

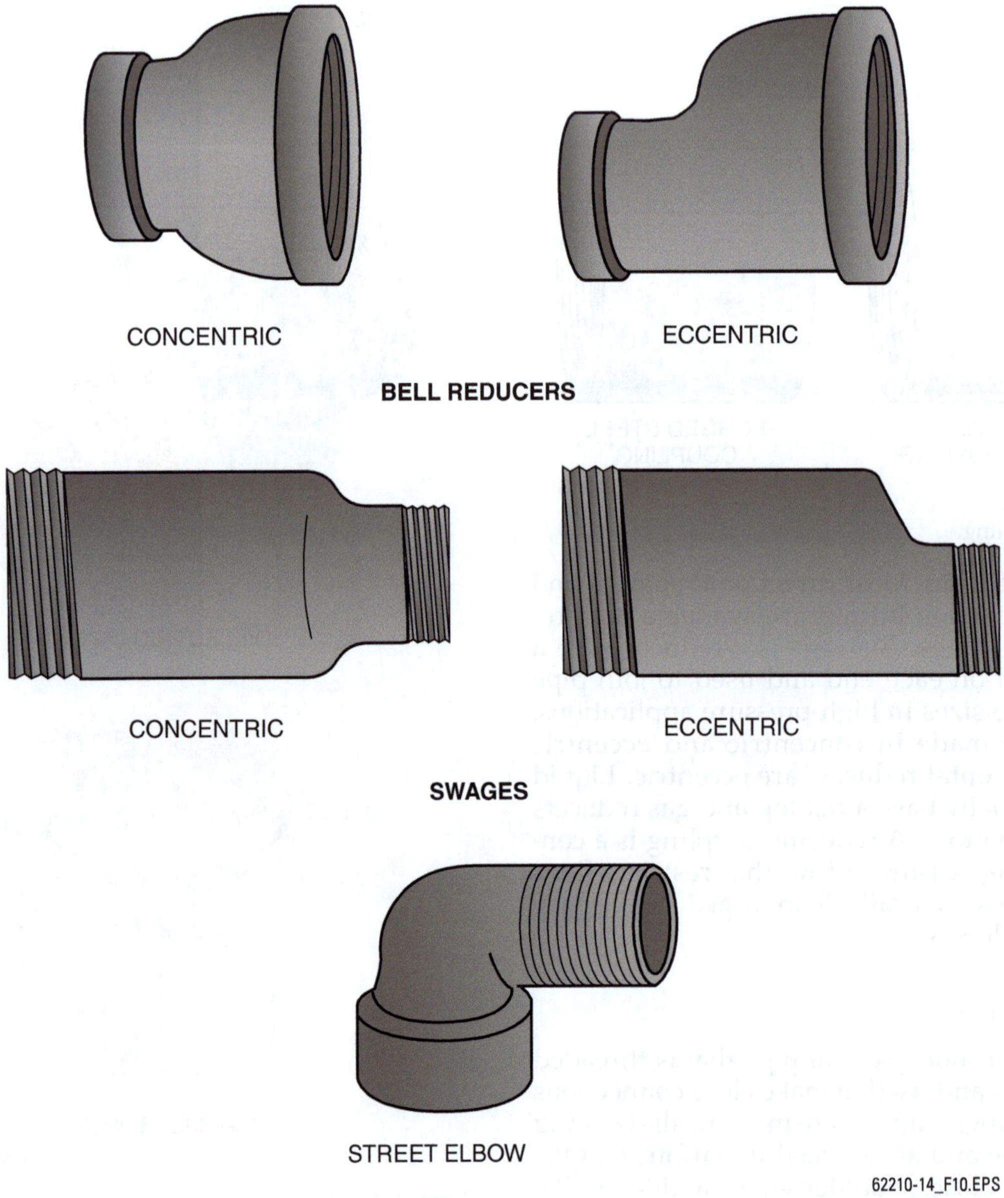

Figure 10 Reducers.

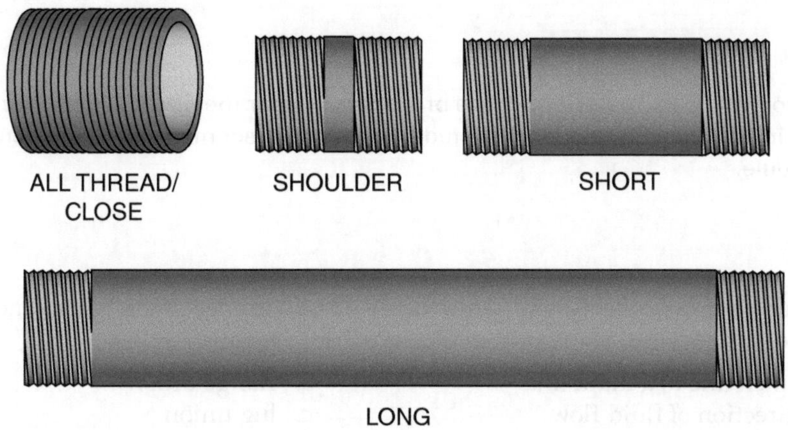

Figure 11 Nipples.

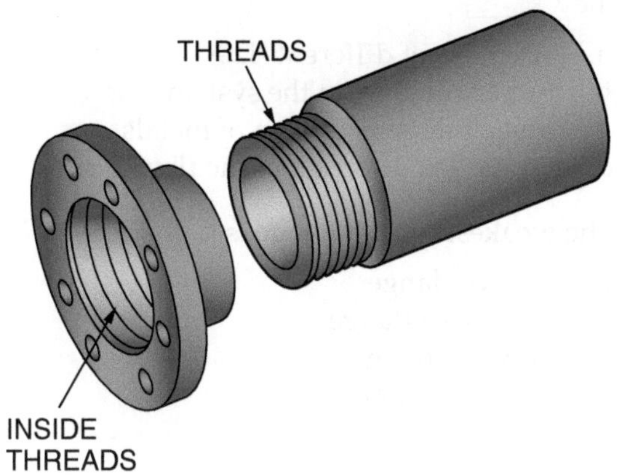

Figure 12 Screw-on flange.

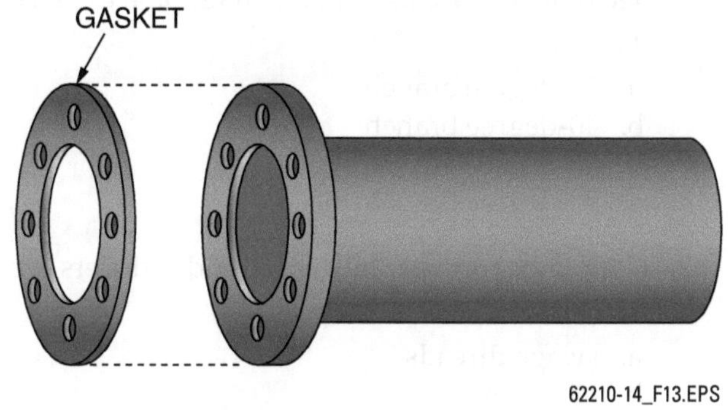

Figure 13 Flat-face flange and gasket.

of threaded flanges are the flat-face flange and the raised-face flange. Typically, personnel should never use two different face styles at one joint, but this may be necessary when connecting to manufactured equipment. If personnel must connect a flat-face flange to a raised-face flange, they must confirm that the flat-face flange is not cast iron. The pressure of the bolts can crack or break the cast iron flange. If the flat-face flange is cast iron, change the raised-face flange to a flat-face flange with engineering approval.

The flat-face flange is flat across the entire face. This flange uses a gasket that covers the entire flange face and has bolt holes cut in it. Flat-face flanges are mainly used when joining two flanges of dissimilar metals. *Figure 13* shows a flat-face flange and gasket.

The raised-face flange has a wide, raised rim around the center of the flange. This flange uses a wide, flat ring gasket that is squeezed tightly between the flanges. *Figure 14* shows a raised-face flange.

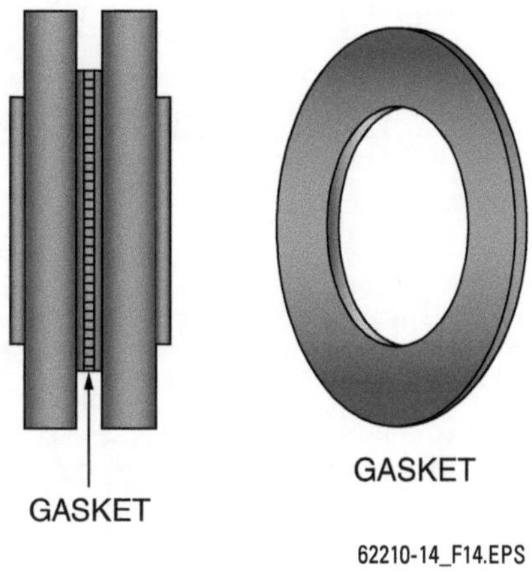

Figure 14 Raised-face flange.

Additional Resources

It is assumed that you will learn to perform these procedures under the direct supervision of your crew leader, foreman, or instructor. For independent study, refer to the recommended Additional Resources at the back of this module.

2.0.0 Section Review

1. In a piping system, elbows, offsets, and return bends are used to _____.
   a. connect a branch line to a main line
   b. change the direction of fluid flow
   c. close off the end of a line
   d. join two pipes of different sizes

2. Tee branches are used in the main pipe to make a _____.
   a. 30-degree branch
   b. 40-degree branch
   c. 45-degree branch
   d. 90-degree branch

3. Couplings, unions, bushings, and reducers are known as _____.
   a. swage threads
   b. elbow tees
   c. branch connections
   d. line connections

4. A hammer union is also known as a _____.
   a. bushing union
   b. flange union
   c. lug union
   d. wrench union

5. Low-pressure systems use bushings because they _____.
   a. join pipes of different sizes
   b. restrict the flow in the system
   c. prevent the breakdown of metals
   d. change the direction of the flow

6. The weakest type of flange is a _____.
   a. flat-face flange
   b. raised-face flange
   c. screw-on flange
   d. companion flange

Section Three

3.0.0 Determining Pipe Lengths Between Fittings

Objective

Explain how to determine pipe lengths between fittings.

a. Explain how to use the center-to-center method to determine pipe lengths between fittings.
b. Explain how to use the center-to-face method to determine pipe lengths between fittings.
c. Explain how to use the face-to-face method to determine pipe lengths between fittings.
d. Explain how to calculate offsets.

Trade Terms

Run: A length of pipe made up of more than one length of pipe.

Takeout: The dimension that the fittings take out of a center-to-center measurement; also known as takeup, takeoff, or makeup.

Thread engagement: The length of the thread overlap when the male threads of one pipe or fitting are screwed into the female threads of another pipe or fitting.

In order to replace a piece of pipe between two fittings, personnel must be able to determine the **takeout** of the fittings to find the length of pipe needed. Takeout, commonly referred to as *makeup*, *take-up*, or *takeoff*, refers to the distance from the end of the pipe to the center of the fitting (*Figure 15*). Knowing the takeout for the fittings is essential when figuring the lengths of straight pipe.

Piping drawings usually show the dimensions of the overall **run** of pipe, but it is up to the personnel who are fitting the pipe to determine the lengths of the individual straight pipe within the run. Personnel commonly use three methods for determining pipe lengths: center-to-center, center-to-face, and face-to-face. When it is necessary to change the position of a pipeline, personnel must also understand how to calculate offsets.

3.1.0 Center-to-Center Method

Most pipe drawings give the center-to-center dimensions, also known as center line, for a run of pipe. *Figure 16* displays a pipe drawing showing the center-to-center dimension.

When using the center-to-center method, measurements are taken from the center of one fitting to the center of the next fitting. The actual length of pipe between the fittings is determined by subtracting the center-to-face dimension of each fitting from the overall length and then adding the fitting **thread engagement** to obtain the actual length that the pipe needs to be. Thread engagement is the length of the thread overlap when the male threads of one pipe or fitting are screwed into the female threads of another pipe or fitting. *Figure 17* shows the dimensions of a common 90-degree elbow.

The thread engagement varies with the nominal size of the pipe. Thread engagement should always be measured for accuracy. Manufacturers of fittings give dimensions of screwed fittings in makeup charts. *Table 2* shows an example of a makeup chart.

Figure 18 shows the center-to-center method for determining the length of a pipe between two fittings.

Follow these steps to determine the length of a pipe between two fittings, using the center-to-center method:

Step 1 Find the center of each of the two fittings.

> **NOTE**
> The center of the fitting can be determined by using the manufacturer's makeup chart or by measuring the fitting using a rule.

Step 2 Measure the distance between the centers of the two fittings.

Step 3 Measure the fitting to determine the center-to-face length of each of the fittings.

Step 4 Subtract the center-to-face length of both fittings from the center-to-center measurement taken in Step 2.

Step 5 Determine the thread engagement of the fittings.

> **NOTE**
> The thread engagement can be determined by using the manufacturer's makeup chart or by measuring the threads using a rule. Larger fittings allow for inserting the rule inside the fitting to measure the thread engagement.

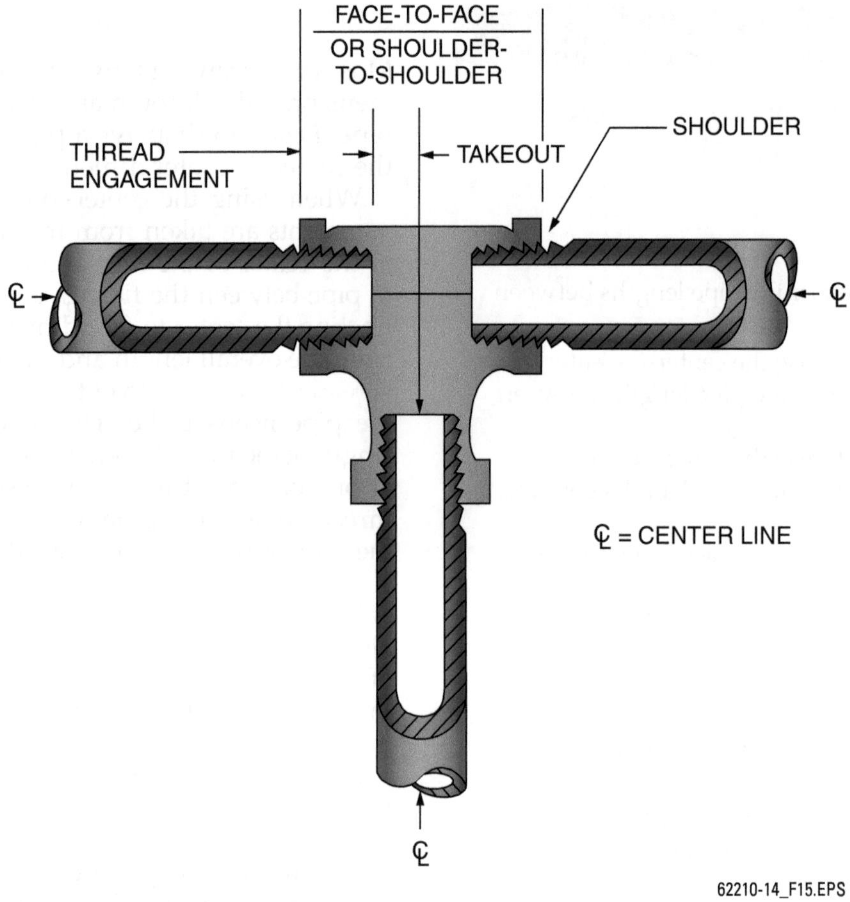

Figure 15 Takeout conventions.

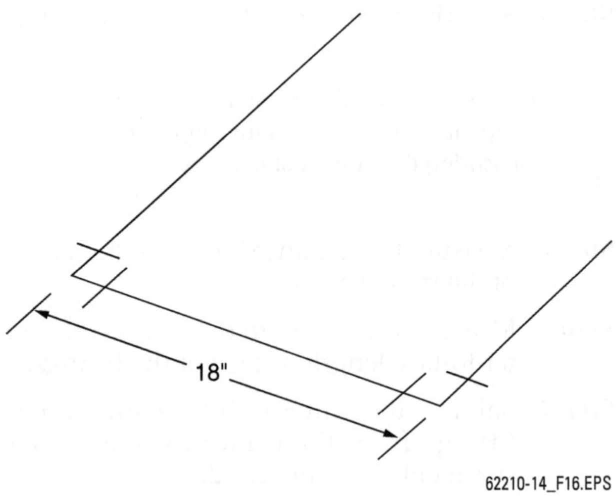

Figure 16 Center-to-center pipe drawing.

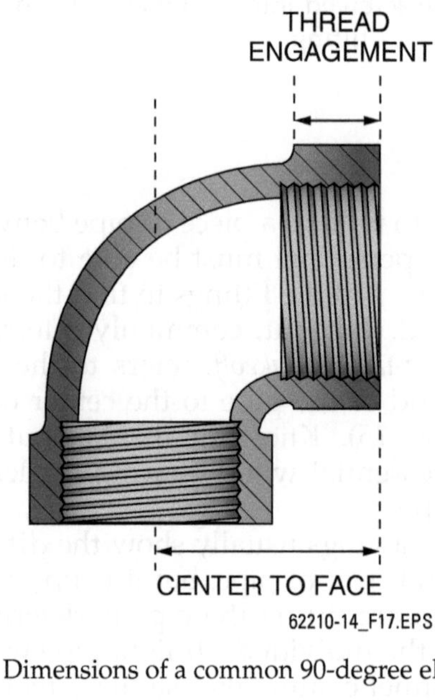

Figure 17 Dimensions of a common 90-degree elbow.

Table 2 Example of a Makeup Chart

Nominal Size	Actual OD	Threads Per Inch	Length Hand Tight	Hand Turns	Makeup Turns	Length of Effective Thread
1/8"	0.405	27	0.18	4.5	2.5	2.6
1/4"	0.54	18	0.2	4	3	0.401
3/8"	0.675	18	0.24	4.5	3	0.408
1/2"	0.84	14	0.32	4.5	3	0.534
3/4"	1.058	14	0.34	4.5	3	0.546
1"	1.315	11.5	0.4	4.5	3.25	0.682
1 1/4"	1.66	11.5	0.42	4.5	3.25	0.707
1 1/2"	1.9	11.5	0.42	4.5	3.25	0.724
2"	2.375	11.5	0.436	5	3	0.756
2 1/2"	2.875	8	0.682	5.5	3	1.136
3"	3.5	8	0.766	6	3	1.2

62210-14_T02.EPS

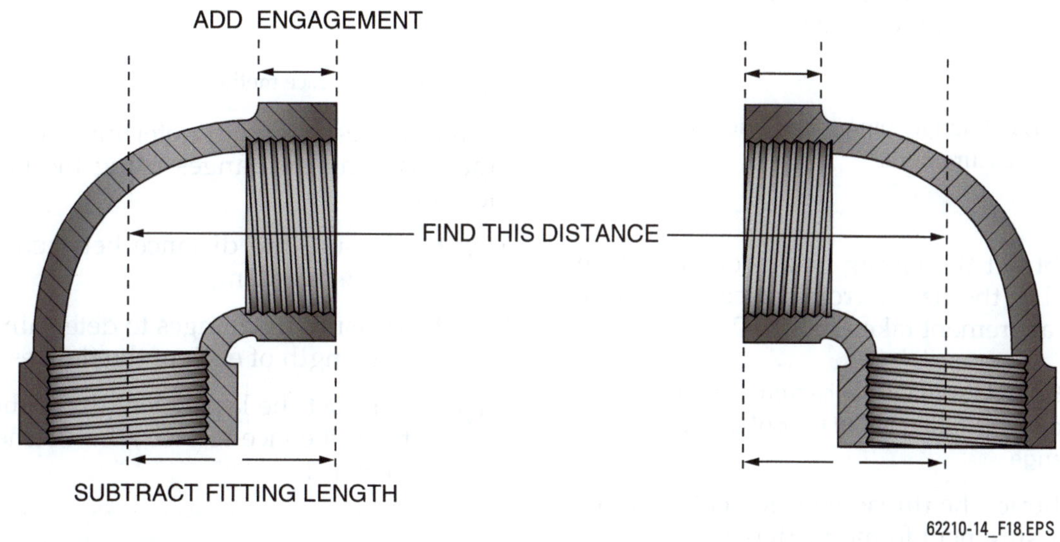

Figure 18 Center-to-center method.

Step 6 Add the thread engagement of both fittings to the length found in Step 4 to determine the length of pipe needed.

3.2.0 Center-to-Face Method

The center-to-face method is used when the drawing gives the length from the center of a fitting, such as a 90-degree ell or a tee, to the face of a flange. This method is used only in low-pressure applications. To find the length of pipe needed, subtract the laying lengths of the fitting, the flange, and the thread engagement. *Figure 19* shows the center-to-face method.

Follow these steps to determine the length of pipe between the center of a fitting to the face of a flange using the center-to-face method:

Step 1 Find the center of the fitting.

> **NOTE**
>
> The center of the fitting can be determined by using the manufacturer's dimensions chart or by measuring the fitting using a rule. Avoid using a standard dimensions chart *not* supplied by the fitting manufacturer; instead, physically measure the fitting.

Step 2 Measure the distance from the center of the fitting to the face of the flange.

Step 3 Measure the fitting to determine the laying length of the fitting.

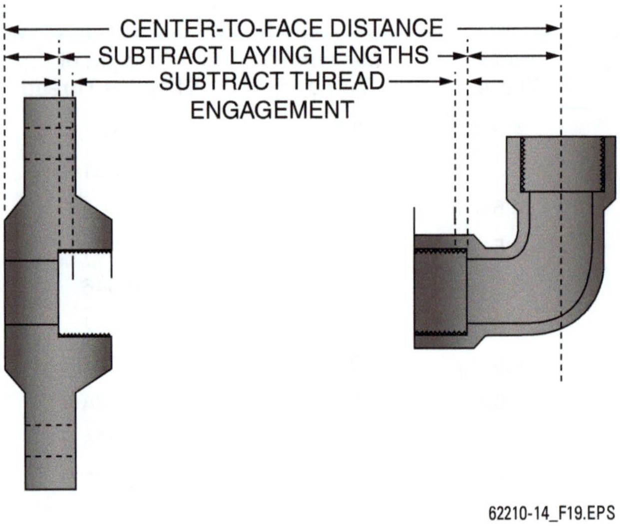

Figure 19 Center-to-face method.

Step 4 Measure the flange to determine the laying length of the flange.

> **NOTE**
> Thickness of flange minus socket depth equals takeoff measurement.

Step 5 Subtract the laying length of the fitting and of the flange from the center-to-face measurement taken in Step 2.

Note

Step 5 provides the distance from the back of the thread on the fitting to the back of the socket on the flange.

Step 6 Subtract the thread engagement from the measurement found in Step 5.

> **NOTE**
> This is for the thread engagement of each threaded fitting. The length found in Step 6 is the final length of the pipe between the fitting and the flange.

3.3.0 Face-to-Face Method

The face-to-face method is used to determine the length of a pipe that has a flange on each end. This method is used only in low-pressure applications. The measurement is taken from the face of one flange to the face of the other flange. To find the length of pipe needed, subtract the laying lengths of both flanges and the ⅛-inch gap required at each weld. *Figure 20* shows the face-to-face method.

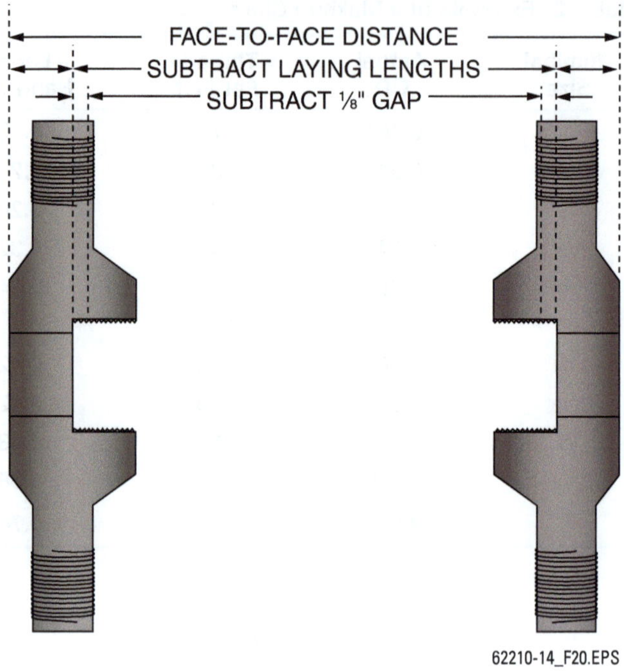

Figure 20 Face-to-face method.

Follow these steps to determine the length of pipe between two flanges using the face-to-face method:

Step 1 Measure the distance between the faces of the two flanges.

Step 2 Measure the flanges to determine the laying length of each of the flanges.

Step 3 Subtract the laying lengths of the flanges from the face-to-face measurement taken in Step 1.

> **NOTE**
> Step 3 provides the distance from the back of the socket on one flange to the back of the socket on the other flange.

Step 4 Subtract ¼ inch from the measurement found in Step 3.

> **NOTE**
> This is for the ⅛ inch that the ends of the pipe have to be recessed from the backs of the sockets of each flange. The length found in Step 4 is the final length of the pipe between the two flanges.

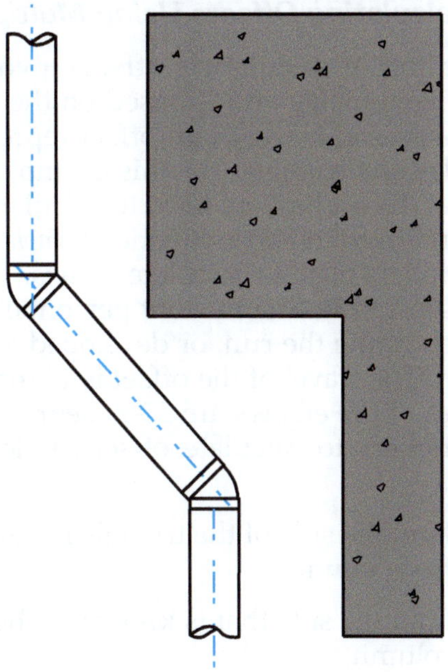

62210-14_F21.EPS

Figure 21 Simple offset around wall.

3.4.0 Calculating Offsets

An offset is a lateral move that moves the pipe out of its original line to another line that is parallel with the original line. Offsets are used when it is necessary to change the position of a pipeline in order to avoid an obstruction, such as a wall or a tank. *Figure 21* shows a simple offset around a wall.

When two lines are to be connected using elbows having an angle other than 90 degrees, personnel must use special calculations to determine the distance of the pipe between the two elbows. All piping offsets are based on the right triangle. The three basic sides of an offset are the set, run, and travel. The set and run are joined by a 90-degree angle, and the travel connects the span between their end points. *Figure 22* shows the components of an offset.

The set is the distance, measured center-to-center, that the pipeline is offset. The run is the total linear distance required for the offset. The travel is the center-to-center measurement of the offset piping. The angle of the fittings determines the number of degrees that the piping changes direction. The basic methods used for calculating offsets are calculating 45-degree offsets, calculating offsets when the set and elbow angle are known, calculating offsets using multipliers, and calculating travel of rolling offsets.

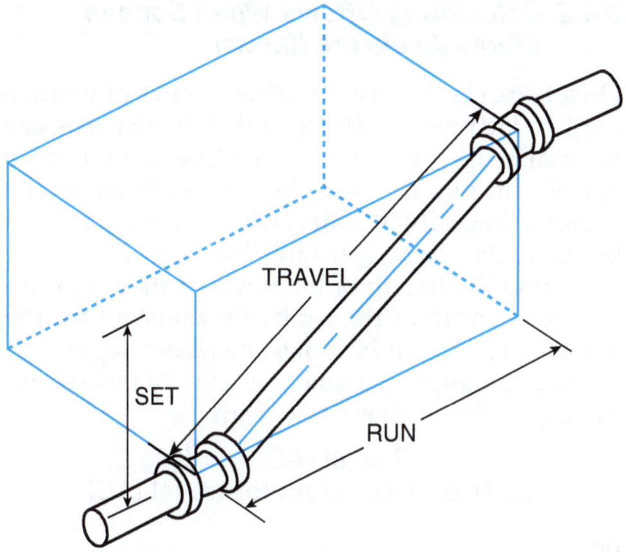

62210-14_F22.EPS

Figure 22 Offset components.

3.4.1 Calculating 45-Degree Offsets

When using 45-degree elbows to make an offset, use the following rule: multiply the length of the set by 1.414 to determine the travel, or the length between the centers of the elbows. This rule applies only for 45-degree elbows. *Figure 23* shows an offset using 45-degree elbows.

For this exercise, assume that the set is 20 inches. Follow these steps to find the length of the travel:

Step 1 Multiply the length of the set by 1.414: 20 × 1.414 = 28.28 inches. This is the length of the travel between the centers of the fittings.

Step 2 Follow the procedure for determining pipe lengths using the center-to-center method to determine the length of pipe needed for the travel.

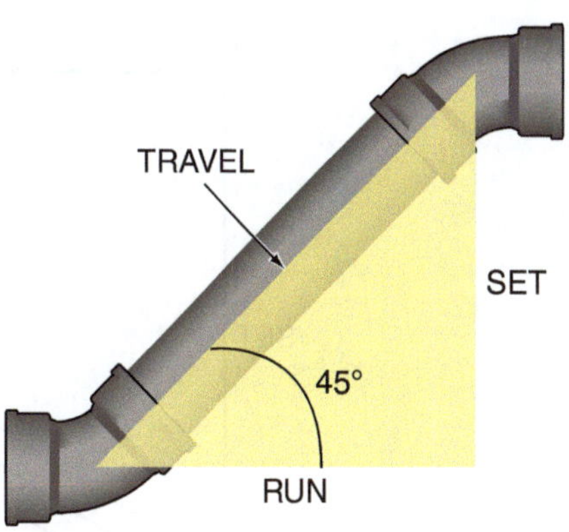

62210-14_F23.EPS

Figure 23 Offset using 45-degree elbows.

3.4.2 Calculating Offsets When Set and Elbow Angle are Known

Offsets can be fabricated using elbows of various angles. Fittings of 60, 30, and 22-$\frac{1}{2}$ degrees are normally used to offset a pipeline. When these various fittings are used, the distance between the centers can easily be found when the set is known by using the elbow constants listed in *Table 3*.

To find the length of the travel or the run, multiply the length of the set by the constant for the elbow used. *Figure 24* shows the elbow angle.

As an example, assume the set is 20 inches and the angle of the elbow is 30 degrees.

$$\text{Travel (AC)} = \text{set (AB)} \times \text{constant for travel (AC)}$$

or

$$\text{Travel} = 20 \times 2.00 = 40 \text{ inches}$$
$$\text{Run (BC)} = \text{set (AB)} \times \text{constant for run (BC)}$$

or

$$\text{Run} = 20 \times 1.73 = 34.6 \text{ inches}$$

Table 3 Elbow Constants

Elbow Angle	Travel (AC)	Run (BC)
60 degrees	1.15	0.58
45 degrees	1.41	1.00
30 degrees	2.0	1.73
22 $\frac{1}{2}$ degrees	2.61	2.41

62210-14_T03.EPS

3.4.3 Calculating Offsets Using Multipliers

This method of calculating offsets involves the use of trigonometry and is based on the fact that the run, set, and travel of an offset represent the three sides of a triangle. For this example, understand that the multipliers used to calculate offsets are natural trigonometric functions. *Table 4* shows the multipliers used to calculate offsets.

In this example, assume that personnel are trying to determine the run, or developed length, of the offset. The travel of the offset is 15 inches and the angles of the elbows are 22-$\frac{1}{2}$ degrees. To use the multipliers to calculate offsets, follow these steps:

Step 1 Find the side of the triangle needed in the first column.

Step 2 Find the side that is known in the second column.

Step 3 Continue across in that row to the column headed with the angle of fitting to be used, and read the multiplier.

Step 4 Multiply this number by the length of the known side. The answer is the length of the needed side.

3.4.4 Calculating Travel of a Rolling Offset

The rolling offset is similar to the simple offset except that there is one more factor to calculate: the roll of the offset. The roll is the displacement of the offset that is perpendicular to both the set and the run. While a triangle best represents a simple offset, a rectangular box best represents the rolling offset with the travel moving diagonally across the box from corner to corner. *Figure 25* shows the rolling offset.

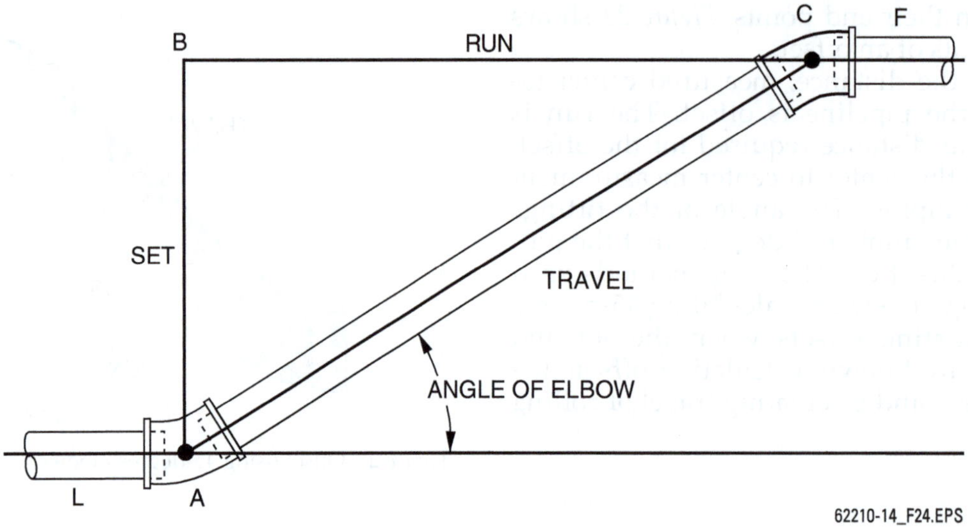

Figure 24 Elbow angle.

Table 4 Multipliers Used to Calculate Offsets

To Find Side	When Known Side Is	Multiply Side	Using 22½° Hand Tight	Using 30° Elbows	Using 45° Elbows	Using 60° Elbows
T	S	S	2.613	2.000	1.414	1.155
S	T	T	0.383	0.500	0.707	0.866
R	S	S	2.414	1.732	1.000	0.577
S	R	R	0.414	0.577	1.000	1.732
T	R	R	1.082	1.155	1.414	2.000
R	T	T	0.924	0.866	0.707	0.500

R = run S = set T = travel

62210-14_T04.EPS

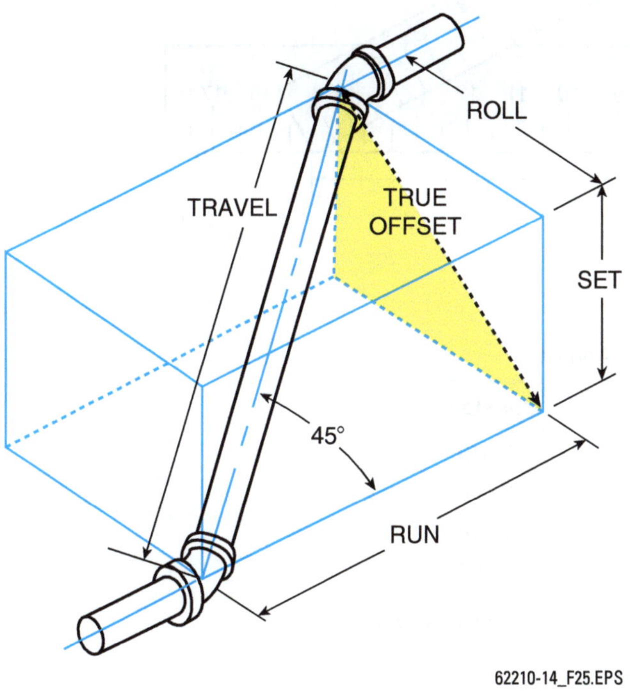

Figure 25 Rolling offset.

As an example, assume that the roll of a 60-degree offset is 15 inches and the height is 8 inches. To calculate the travel of a rolling offset, using a framing square, follow these steps:

Step 1 Lay out the roll on one side of the framing square and the set on the other side of the square. *Figure 26* shows laying out a rolling offset using a framing square.

Step 2 Measure the distance between these two points on the framing square. In this example, the measurement is 17 inches.

Step 3 Multiply the measurement by the constant for the angle of the fitting from *Table 5*.

NOTE

Since the measurement taken in Step 2 was 17 inches and the example uses 60-degree elbows, multiply 17 by 1.154 to get 19.6 or 19⅝ inches. This is the length of travel from center to center.

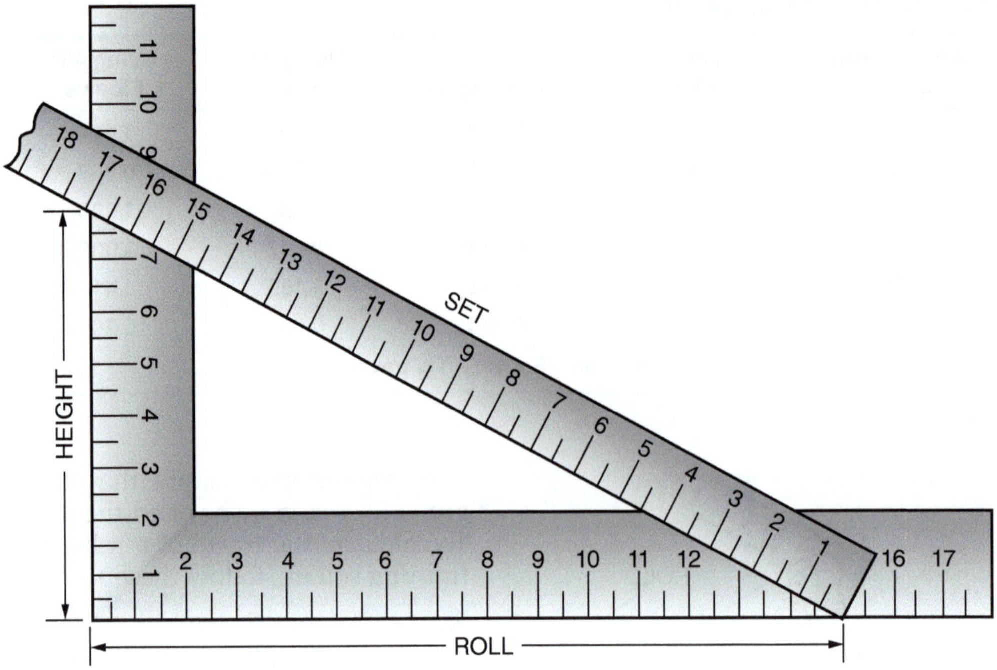

Figure 26 Laying out rolling offset using framing square.

Table 5 Constants for Common Angles

Angle	Constant
22 $^{1}/_{2}$ degrees	2.613
30 degrees	2.000
45 degrees	1.414
60 degrees	1.154

62210-14_T05.EPS

Additional Resources

It is assumed that you will learn to perform these procedures under the direct supervision of your crew leader, foreman, or instructor. For independent study, refer to the recommended Additional Resources at the back of this module.

3.0.0 Section Review

1. When the male threads of one pipe or fitting are screwed into the female threads of another pipe or fitting, the length of thread overlap is called _____.
   a. center line
   b. nominal line
   c. thread engagement
   d. pipe engagement

2. When using the center-to-face method to determine the length of a pipe between two fittings, the first step is to _____.
   a. measure the two fittings
   b. measure the flange
   c. find the center of the fitting
   d. find the laying length of the flange

3. The three basic sides of an offset are the _____.
   a. center, face, and length
   b. fitting, bushing, and reducer
   c. makeup, take-up, and takeoff
   d. set, run, and travel

4. A rolling offset is best represented by a(n) _____.
   a. equilateral triangle
   b. rectangular box
   c. hexagon
   d. pentagon

SECTION FOUR

4.0.0 THREADING AND ASSEMBLING PIPE

Objective

Explain the processes and tools used to thread and assemble threaded pipe.

a. Explain how to cut and ream pipe.
b. Explain how to thread pipe manually and by machine.
c. Identify types of pipe joint compounds and explain how to use them.
d. Explain how to fit screwed pipe and fittings.
e. Explain how to install threaded valves.

Trade Terms

Galling: A deformity of threads in which the threads are stripped.

PTFE tape: Special tape, made of polytetrafluoroethylene (PTFE), which is wrapped around the male threads of a pipe before it is screwed into a fitting to serve as a lubricant and a sealant; also referred to as thread seal tape.

Using pipe drawings as their guide, personnel must use the appropriate pipe and fittings to fabricate the threaded pipe system. As an overview of the process, it is first helpful to gain an understanding of the order of the main tasks that must be performed to fabricate a threaded pipe system.

This section describes the procedures for threading pipe, for using pipe joint compounds, and for selecting the proper tools as well as the process for fitting the screwed fittings onto the pipe.

4.1.0 Cutting and Reaming Pipe

After determining the type, size, and length of the pipe to be threaded, the pipe must next be cut and reamed. Follow these steps to cut and ream pipe:

Step 1 Place the cutter onto the pipe.

Step 2 Rotate the handwheel to move the cutter wheel to the cutting mark on the pipe.

Step 3 Rotate the cutter handle to tighten the cutter on the pipe.

Step 4 Direct the cutting fluid to the cutting area.

Step 5 Rotate the pipe cutter around the pipe while tightening the cutter head.

Step 6 Rotate the cutter handle to apply pressure to the cutter until the cut is complete.

CAUTION

Do not use a pipe cutter to cut pipe for a socket weld. Use a bandsaw. The pipe cutter will cause the pipe to flare. It may also interfere with the flow of material through the pipe.

Step 7 After the cut is complete, insert the reamer into the pipe and deburr the pipe.

CAUTION

Do not overream the pipe. Only ream enough to remove the burrs.

Figure 27 shows a pipe cutter and reamer.

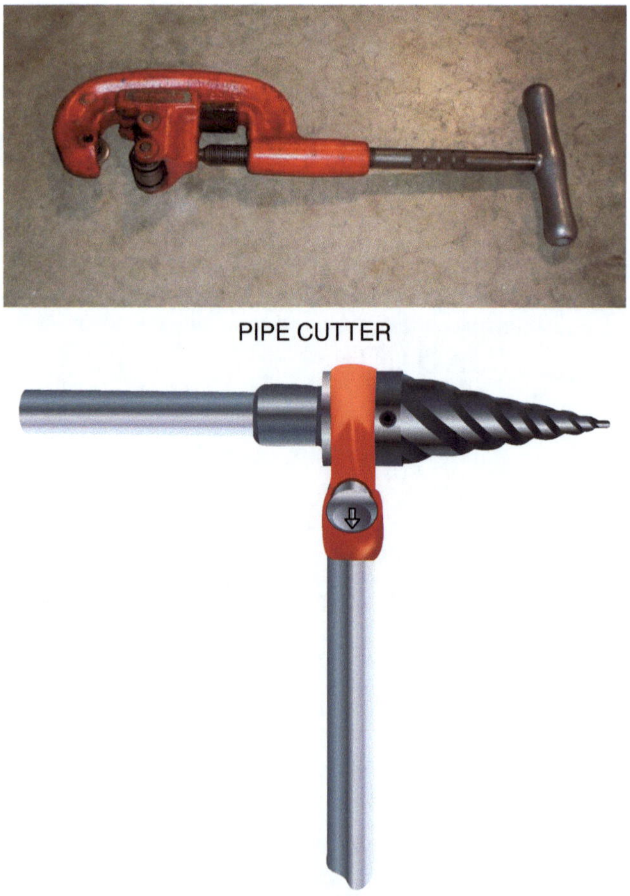

Figure 27 Pipe cutter and reamer.

4.2.0 Threading Pipe

Before pipe and fittings can be assembled, they must be threaded. Pipe can be threaded using a manual threader or a power threading machine.

4.2.1 Threading Pipe Using a Manual Pipe Threader

Figure 28 shows a manual pipe threader with threading dies. Follow these steps to thread pipe using a manual pipe threader:

Step 1 Identify the material being threaded, such as carbon steel or stainless steel.

Step 2 Identify the diameter of the material being threaded.

Step 3 Select the die head best suited for the threading process.

Step 4 Insert the die head into the pipe threader handle.

Step 5 Inspect the die and die handle to ensure that all parts are clean and in good condition.

Step 6 Secure the pipe in a vise.

Step 7 Make sure that the end of the pipe is square, clean, and deburred.

Step 8 Position and balance oneself, facing the work area, and within easy reach of the work area.

Step 9 Position the die flush on the end of the pipe (*Figure 29*).

Step 10 Press the die against the end of the pipe, and turn the die handle clockwise to start the threading process.

62210-14_F28.EPS

Figure 28 Manual pipe threader.

62210-14_F29.EP

Figure 29 Using a manual pipe threader.

Step 11 Turn the die handle two full rotations around the pipe. Add cutting fluid as necessary to keep a thin coat on the threads while cutting.

Step 12 Stop the work, and back the die off the pipe. This allows time to make sure the threads are properly started and allows the threads to be cleaned from the cutting die.

Step 13 Clean the cutting threads from the die if necessary.

Step 14 Rethread the die onto the pipe, and continue the process until the threading is complete. The threading is complete when two full threads extend past the back edge of the die. If the threading is correct, it should be possible to screw a fitting three-and-a-half revolutions by hand onto the new threads.

CAUTION

Do not thread the pipe too far. This can cause an improper fit between the fitting and the pipe, which will result in leaks.

Step 15 Back the die off the pipe.

Step 16 Inspect the new threads, and remove any debris from the threads using a rag.

WARNING!

Freshly cut threads are very sharp. Use care in handling to avoid cuts.

4.2.2 Threading Pipe Using a Threading Machine

Figure 30 shows a powered pipe-threading machine. When using a threading machine, be sure to check the cutting fluid level and prime the cutting fluid pump before threading pipe. Follow these steps to thread pipe:

Step 1 Load the pipe into the threading machine.

Step 2 Place pipe stands under the pipe as needed.

Step 3 Cut and ream the pipe to the required length.

Step 4 Make sure that the proper dies are in the die head.

Step 5 Loosen the clamp lever.

Step 6 Move the size bar to select the proper die setting for the size of pipe being threaded.

Step 7 Lock the clamp lever.

Step 8 Swing the die head down to the working position.

Step 9 Close the throw-out lever.

Step 10 Lower the lubrication arm, and direct the fluid supply onto the die.

Step 11 Turn the machine switch to the Forward position.

Step 12 Step on the foot switch to start the machine. When threading stainless steel or alloys, bump the foot pedal slowly and add fluid to prevent galling, a deformity of the threads in which the threads are stripped.

Step 13 Turn the carriage to bring the die against the end of the pipe.

Step 14 Apply light pressure on the handwheel to start the die.

CAUTION

Make sure that the die is flooded with fluid at all times while the die is cutting to prevent overheating the die and the pipe. Overheating can damage the die and the pipe threads.

NOTE

The dies feed onto the pipe automatically as they follow the newly cut threads.

Step 15 Release the handwheel once the dies have started to thread the pipe.

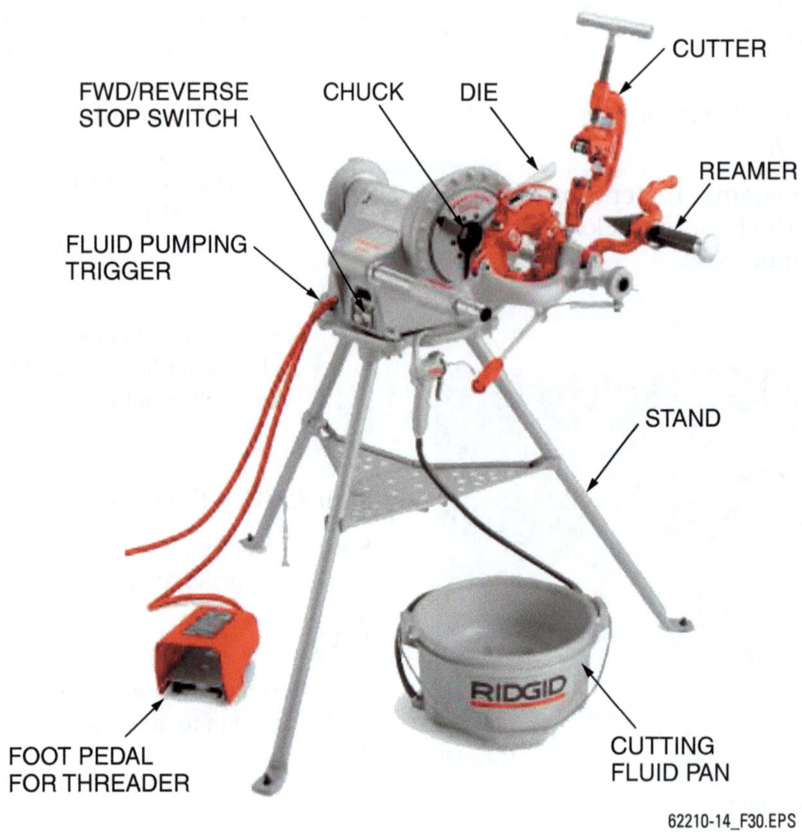

Figure 30 Powered pipe threader.

Step 16 Open the throw-out lever as soon as two full threads extend from the back of the dies.

Step 17 Release the foot switch to stop the machine.

Step 18 Turn the carriage handwheel to back the die off the pipe.

Step 19 Swing the die and the fluid spout up and out of the way.

Step 20 Screw a fitting onto the end of the pipe to check the threads. If the threads are correct, it should be possible to screw a fitting three-and-a-half revolutions by hand onto the new threads. If the threads are not deep enough, adjust the die with the clamp lever, and repeat the threading operation.

Step 21 Turn the machine switch to the Off position.

Step 22 Open the chuck, and remove the pipe.

4.3.0 Using Pipe Joint Compounds

A good grade of pipe joint compound must be applied to the male threads joint to lubricate the joint to prevent galling of the threads and to ease the makeup of the joint. To avoid product contamination or dangerous combinations of materials, always use caution when choosing a compound. Depending on the type of liquid being transferred through the pipes, compound may not be permitted. Always check with the engineering specifications and requirements before applying any type of pipe joint compound. Examples of pipe joint compounds include **PTFE tape**, liquid PTFE, and pipe dope (*Figure 31*).

4.3.1 PTFE Tape

PTFE tape is a special type of tape made of polytetrafluoroethylene (PTFE), which is wrapped around the male threads of a pipe before it is screwed into a fitting to prevent metal-to-metal contact and to create a leak-proof joint. It is also referred to as thread seal tape. Follow these steps to use PTFE tape:

> **NOTE**
> In the past, PTFE tape has been incorrectly referred to as Teflon© tape. Teflon© is a brand owned by DuPont that identifies products made with DuPont raw material(s), such as industrial coatings. DuPont does not make PTFE tape, and PTFE tape should never be described using the term *Teflon©*.

Figure 31 PTFE tape and pipe joining compound.

> **WARNING!**
> Never use PTFE tape on pipe that is to be welded or pipe that carries steam or other high-temperature service. When PTFE tape is heated, it emits a highly toxic gas that could be fatal.

Step 1 Remove all excess cutting fluid from the threads to improve the grip of the tape on the threads.

> **NOTE**
> Use ½-inch wide PTFE tape for pipe that is ¾-inch nominal size and smaller. Use ¾-inch wide tape for pipe that is 1-inch nominal size or larger.

Step 2 Start the tape from the end of the thread, leaving the first full thread bare to prevent the tape from overhanging at the end of the pipe.

Step 3 Wrap the tape around the pipe in the direction that the joint is to be assembled. Tape should be wrapped around right-hand threads in a clockwise direction and around left-hand threads in a counter-clockwise direction.

Step 4 Continue to wrap the tape around the joint, overlapping the edges of each wrap until all remaining threads have been covered.

Step 5 Press the tape against the threads to seal it to the threads and to prevent the tape from slipping off the threads when making up the joint.

4.3.2 Liquid PTFE

Liquid PTFE is a PTFE-based pipe joint compound that is applied to the male threads using a small brush. Always follow the manufacturer's recommendations regarding temperature constraints as well as the types of materials on which the compound is safe to use.

4.3.3 Pipe Dope

Pipe dope is another type of thread sealant that is applied with a brush. There are many brands of pipe dope available. Make sure that the pipe dope is suitable for the type of pipe on which it is to be used. Dopes are made specifically for stainless steel pipe and for carbon steel threads. Avoid dopes that claim to be suitable for both stainless steel and carbon steel because they tend to fail more often than others.

Pipe dope is relatively easy to apply. It is simply brushed onto the male threads to be fitted. Do not apply pipe dope to the female threads, and do not apply dope to the first thread of the pipe end. Following this precaution helps keep the pipe dope out of the pipeline and avoids contaminating the liquid inside the pipeline.

4.4.0 Fitting Screwed Pipe and Fittings

The proper fitting of pipe and fittings is crucial to fabricating a quality piping system and is a skill that must be learned through experience. Poor fitting practices can result in leaks and other defects in the system. In order to have a strong, leak-proof threaded joint, the threads must be clean and smoothly cut. They must have the correct pitch, lead, taper, and form, and the threads must be correctly sized. If these conditions are not met, the seal will leak.

The joining, or makeup, of a taper threaded pipe connection is performed first by screwing in the fitting by hand and then using a wrench to fully tighten the fitting and complete the makeup of the joint. Never use a power drive or threading machine to tighten fittings onto pipes. Personnel are usually able to make three and four complete revolutions by hand if the threads are cut properly. When rotating by hand, personnel have the opportunity to check the quality of threads after cutting. Personnel usually make about three turns with a wrench for a total of approximately seven revolutions for complete makeup. A greater number of turns is required for larger pipe. Threads should never bottom out of the fitting or adjacent pipe.

WARNING!

Never use a power drive or threading machine to tighten fittings onto pipes.

The standard for taper thread pipe makeup gives basic thread dimensions and the number of rotations needed by hand and with a wrench. Note that industry standards often refer to turns made by hand as *hand engagement* and turns made by a wrench as *wrench makeup*. The taper thread makeup can vary plus or minus one turn from the standard thread makeup allowances. *Table 6* shows the standard for taper pipe thread makeup. Note that these thread makeups are for perfect threads cut under perfect conditions, which seldom occurs.

When assembling threaded pipe joints, the pipe should either be secured in a vise or two pipe wrenches should be used: one to hold the pipe and the other to turn the fitting. Always choose the proper size pipe wrench for the pipe being used. Wrenches that are too short make the job much harder. Wrenches that are too long can strip the threads, crack the fitting, or crush the pipeline. Pipe wrenches should also be in good working condition and adjusted properly to avoid damaging the pipe. Never use a cheater bar on the end of a wrench or strike the wrench handle with a hammer.

WARNING!

Never use a cheater bar on the end of a wrench or strike the wrench handle with a hammer.

When tightening fittings onto a piece of pipe, personnel must be aware of the end position that the fitting must face. If the fitting is turned past the desired end position and then turned back, the fitting will leak. Always make sure the union is placed in the direction of flow.

4.5.0 Installing Threaded Valves

Special precautions must be taken when installing valves with threaded ends to avoid damaging the valve. Follow these guidelines when installing valves with threaded ends:

Table 6 Standard for Taper Pipe Thread Makeup

Nominal Pipe Size	Diameter of Pipe (D)	Threads (inches) (n)	Pitch of Thread	Pitch Diameter at Beginning of External Thread	Hand Tight Engagement Length (L1)			Effective Thread, External Length (L1)		
					Inch	Threads	Diameter (E1)	Inch	Threads	Diameter (E2)
1	2	3	4	5	6	7	8	9	10	11
1/2	0.840	14	0.07143	0.75843	0.320	4.48	0.77843	0.5337	7.47	0.79179
3/4	1.050	14	0.07143	0.96768	0.339	4.75	0.98887	0.5457	7.64	1.00179
1	1.315	11.5	0.08696	1.21363	0.400	4.60	1.23863	0.6828	7.85	1.25630
1 1/4	1.660	11.5	0.08696	1.55713	0.420	4.83	1.58338	0.7068	8.13	1.60130
1 1/2	1.900	11.5	0.08696	1.79609	0.420	4.83	1.82234	0.7235	8.32	1.84130
2	2.375	11.5	0.08696	2.26902	0.436	5.01	2.29627	0.7565	8.70	2.31630
2 1/2	2.875	8	0.12500	2.71953	0.682	5.46	2.76216	1.1375	9.10	2.79062
3	3.500	8	0.12500	3.34062	0.766	6.13	3.38850	1.2000	9.60	3.41562
3 1/2	4.000	8	0.12500	3.83750	0.821	6.57	3.88881	1.2500	10.00	3.91562
4	4.500	8	0.12500	4.33438	0.844	6.75	4.38712	1.3000	10.40	4.41562
5	5.563	8	0.12500	5.39073	0.937	7.50	5.44929	1.4063	11.25	5.47862
6	6.625	8	0.12500	6.44609	0.958	7.66	6.50597	1.5125	12.10	6.54062
8	8.625	8	0.12500	8.43359	1.063	8.50	8.50003	1.7125	13.70	8.54062
10	10.750	8	0.12500	10.54531	1.210	9.68	10.62094	1.9250	15.40	10.66562
12	12.750	8	0.12500	12.53281	1.360	10.88	12.61781	2.1250	17.00	12.66562

62210-14_T06.EPS

- Never place the valve in any type of vise. Always secure the pipe in a vise, and screw the valve onto the pipe.
- Always grip the valve with the pipe wrench on the side of the body of the valve closest to the pipe.
- Lubricate the male threads of the pipe. Do not lubricate the female threads inside the valve, and do not lubricate the thread nearest the end of the pipe.
- Do not run the male end of the pipe all the way into the valve. This will damage the seat and cause the valve to leak. Always leave three threads showing outside the valve. *Figure 32* shows proper thread makeup for a valve.

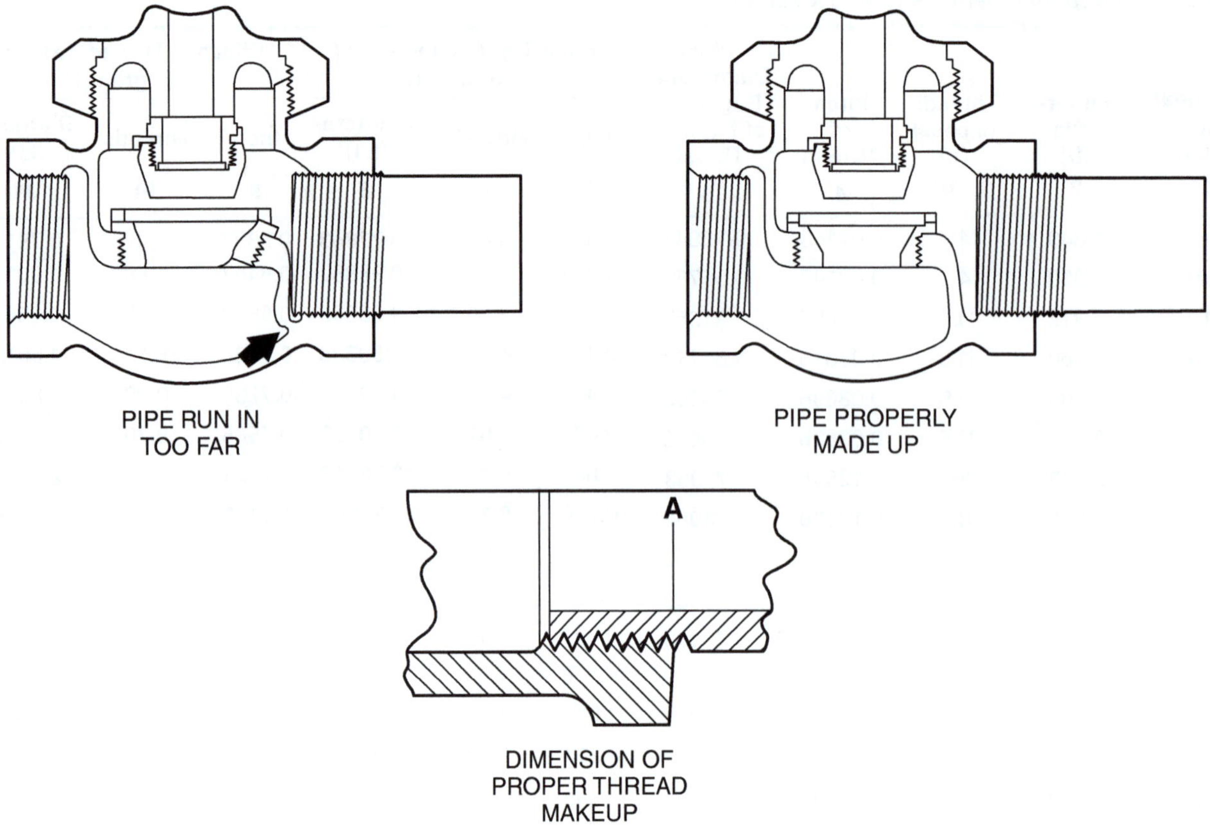

Dimensions, In Inches Dimensions Given Do Not Allow For Variations In Tapping or Threading									
SIZE	1/8	1/4	3/8	1/2	3/4	1	1 1/4	1 1/2	2	2 1/2
A	1/4	3/8	3/8	1/2	9/16	11/16	11/16	11/16	3/4	15/16
SIZE	3	3 1/2	4	5	6	8	10	12	14	
A	1	1 1/16	1 1/8	1 1/4	1 5/16	1 7/16	1 5/8	1 3/4		

62210-14_F32.EPS

Figure 32 Proper thread makeup for a valve.

Additional Resources

It is assumed that you will learn to perform these procedures under the direct supervision of your crew leader, foreman, or instructor. For independent study, refer to the recommended Additional Resources at the back of this module.

4.0.0 Section Review

1. When threading pipe using a manual pipe threader, the first step is to _____.
   a. make sure the pipe is deburred
   b. position the die flush
   c. identify the material being threaded
   d. secure the pipe in a vise

2. Galling is a deformity of threads in which the threads are _____.
   a. heated
   b. cut
   c. taped
   d. stripped

3. Examples of pipe joint compounds include PTFE tape, liquid PTFE, and _____.
   a. pipe dope
   b. pipe sealant
   c. pipe die
   d. brush tape

4. When tightening fittings onto pipes, never use a _____.
   a. wrench
   b. joint compound
   c. power drive
   d. carriage handwheel

5. Industry standards often refer to turns made by hand as _____.
   a. hand engagement
   b. hand turning
   c. hand adjustment
   d. hand tightening

6. When installing valves with threaded ends, there should be _____.
   a. no threads showing outside the valve
   b. one thread showing outside the valve
   c. two threads showing outside the valve
   d. three threads showing outside the valve

Section Five

5.0.0 Pipe Threading and Installation

Objective

Describe the procedures used to thread and fit pipe (CTTP-17).

a. Identify the prerequisites, competencies, and abnormal operating conditions (AOCs) associated with threaded pipe fabrication.
b. Describe the procedures used to thread and fit pipe.

Performance Task

1. Thread and fit pipe (CTTP-17).
   - Identify potential abnormal operating conditions that may occur during performance of this CT, and know the appropriate actions to take in response to them.
   - Utilize the appropriate personal protective equipment according to relevant company procedures.
   - Verify type, size, length, and schedule (grade) of pipe.
   - Cut the pipe to the desired length and ream the ends.
   - Thread the pipe.
   - Select the proper size, shape, and type of fittings needed.
   - Clean the pipe and fittings thoroughly.
   - Test fit pipe to assure proper fit and thread compatibility.
   - Apply joint compound or tape to pipe threads.
   - Join piping and fittings as required.
   - Complete appropriate documentation as required by operator's procedures.

Trade Terms

Abnormal operating conditions (AOCs): Conditions within a pipeline system, identified by the operator, that may indicate (1) a malfunction of a component or (2) a deviation from normal operations that may indicate a condition exceeding design limits or result in a hazard to persons, property, or the environment.

Personnel must verify the correct pipe, components, and compounds used before fabricating threaded pipe. They must carefully follow installation instructions for the fittings being used. They must also inspect the pipe and threads to ensure that is safe for operation.

5.1.0 Prerequisites, Competencies, and Abnormal Operating Conditions

Personnel must possess a variety of skills in order to successfully thread and fit pipe. The various operations require a series of steps, which technicians must be competent to perform. Personnel must also be aware of the particular **abnormal operating conditions (AOCs)** that may occur during the procedure and know the proper responses to them.

5.1.1 Prerequisites and Competencies

Personnel must possess the following basic knowledge and skills prior to threading and fitting pipe:

- Be competent with standard hand tools and any specialized tools recommended by the manufacturer.
- Be able to follow instructions and interpret pipe drawings and schematics.
- Be familiar with the types of piping and associated components.
- Be familiar with entry permits and procedures for a confined space.
- Know how to properly use personal protective equipment (PPE)

5.1.2 Abnormal Operating Conditions

Personnel threading and fitting pipe must be able to identify the following potential AOCs and know the proper response to each, should they occur:

- *A leak of hazardous gases or liquids occurs* – Take appropriate action using the proper PPE to protect people and the environment. Notify the proper authorities and company personnel, including 911 if necessary.
- *A fire and/or explosion or ignition of hazardous, flammable liquids or gases occurs* – Evacuate the area and take only defensive action using the proper PPE. Contact the control center to isolate or shut down the line as quickly as possible. Engage the proper emergency response services.

5.2.0 Threading and Fitting Pipe

The following procedure is a basic guideline for threading and fitting pipe. Always follow company-specific procedures. If you have any questions, contact your supervisor.

> **CAUTION**
>
> Perform any preliminary tasks required to ensure safe working conditions. Always follow company policy and procedures both before and during the fabrication and installation.

5.2.1 Procedure

Personnel should follow this general procedure for threading and fitting pipe:

Step 1 Determine the type, size, and schedule of pipe being used in the system.

Step 2 Determine the length of pipe needed.

Step 3 Cut the pipe to the desired length.

Step 4 Ream the pipe to remove all internal burrs.

Step 5 Thread the pipe, taking precautions not to cut the threads too deep.

Step 6 Select the proper size, shape, and type of fittings needed.

Step 7 Clean the pipe and fittings thoroughly, inside and out. All sand, dirt, and oil must be removed from the inside of the pipe and the fittings to avoid contaminating the system or clogging the line. The fittings and pipe ends may be cleaned with a clean rag soaked in nonflammable solvent.

Step 8 Check the threads on the pipe and the fitting to ensure that they are properly cut and not damaged.

Step 9 Apply joint compound to the pipe threads.

Step 10 Start the fitting on the pipe by hand.

Step 11 Tighten the fitting slowly using a pipe wrench.

> **CAUTION**
>
> Tightening the fitting quickly causes excessive heat due to friction, which could cause the threads or the fitting to expand. Do not allow the threads to bottom out into the fitting. This can damage the threads and the fitting.

Step 12 Complete appropriate documentation as required by operator's procedures.

Additional Resources

It is assumed that you will learn to perform these procedures under the direct supervision of your crew leader, foreman, or instructor. For independent study, refer to the recommended Additional Resources at the back of this module.

5.0.0 Section Review

1. When taking defensive action in response to a fire or explosion, personnel must always use _____.
   a. a fire extinguisher
   b. the appropriate emergency checklist
   c. the proper PPE
   d. a hazardous vapor detector

2. The proper method to clean fittings and pipe requires the use of _____.
   a. sandpaper
   b. a clean rag
   c. bare hands
   d. a scouring pad

SUMMARY

This module presented the basics of installing threaded pipe. Threaded pipe systems are a versatile, widely used joining method employed in the pipeline industry. You learned about the types of materials used in threaded pipe systems and the many fittings used to join pipe.

An important step in the installation process is determining the pipe length between fittings. This module introduced you to the three commonly used methods to do so: center-to-center, center-to-face, and face-to-face. You also learned about calculating offsets.

Pipe must be threaded according to national standards. The tapered pipe thread, or NPT, is the most common thread used on pipe and fittings for pipeline applications. To assemble the piping system, threaded pipe requires the use of a sealant and proper joining techniques to ensure a tight, leak-proof joint.

Review Questions

1. One advantage of a threaded piping system is that _____.
   a. the strength and integrity of the pipe is never compromised
   b. it does not require welding
   c. it can be used if erosion is anticipated
   d. it is leak-proof

2. Stainless steel pipe is used less frequently than other materials because of difficulty with _____.
   a. cutting the threads correctly
   b. obtaining a sufficient supply
   c. keeping the pipe sterile
   d. ensuring it doesn't corrode

3. Nominal size is used for sizes of pipe up to and including _____.
   a. 12 inches
   b. 14 inches
   c. 16 inches
   d. 18 inches

4. Wall thickness can be described by manufacturer's weight and by _____.
   a. pressure weight
   b. strength percentage
   c. table measure
   d. schedule number

5. When measured from center line to center line, the distance between adjacent screw threads is referred to as the _____.
   a. angle
   b. pitch
   c. root
   d. crest

6. In industrial and commercial low-pressure applications, the most commonly used thread for pipe and fittings is the _____.
   a. internal pipe thread
   b. external pipe thread
   c. straight pipe thread
   d. tapered pipe thread

7. Plain pipe fittings are made of _____.
   a. aluminum
   b. copper
   c. forged steel
   d. PVP piping

8. Unless otherwise stated, the angle of an elbow is always _____.
   a. 90 degrees
   b. 45 degrees
   c. 35 degrees
   d. 30 degrees

9. Manufactured in various sizes, return bends come in closed, open, and _____.
   a. broad patterns
   b. wide patterns
   c. narrow connections
   d. tapered connections

10. The fittings known as tees are the most widely used of the _____.
    a. caps and plugs
    b. return bends
    c. branch connections
    d. wye fittings

11. Fittings used for closing off pipelines are called _____.
    a. elbows and ells
    b. offsets and returns
    c. caps and plugs
    d. lids and stops

12. Pipes of the same size are joined by _____.
    a. couplings and unions
    b. couplings and collars
    c. unions and bushings
    d. reducers and bushings

13. Manufacturers often color-code hammer unions to indicate their _____.
    a. length
    b. diameter
    c. wall thickness
    d. pressure rating

14. Fitting manufacturers provide dimensions of screwed fittings in a _____.
    a. takeup chart
    b. makeup chart
    c. takeoff chart
    d. offset chart

15. The center-to-face method for determining pipe length is used only in _____.
    a. heated applications
    b. unheated applications
    c. high-pressure applications
    d. low-pressure applications

16. When using the face-to-face method, the first step is to _____
    a. determine the laying length of each flange
    b. subtract the laying lengths of the flanges
    c. measure the distance between the faces of the two flanges
    d. measure the distance between the two sockets

17. Offsets are used when a pipeline needs to _____.
    a. avoid an obstruction
    b. counter corrosion
    c. pass by a hazardous area
    d. stop abruptly

18. When calculating offsets, the rule about multiplying the length of the set by 1.414 only applies for _____.
    a. 90-degree elbows
    b. 60-degree elbows
    c. 45-degree elbows
    d. 30-degree elbows

19. A set 40 inches long with an elbow angle of 60 degrees has a travel that equals _____.
    a. 46 inches
    b. 56.4 inches
    c. 80 inches
    d. 104.4 inches

20. A set 20 inches long with an elbow angle of 22 $\frac{1}{2}$ degrees has a run that equals _____.
    a. 17.4 inches
    b. 30 inches
    c. 48.2 inches
    d. 51.9 inches

21. To calculate a rolling offset, the first step is to lay out the roll on one side of the _____.
    a. framing angle
    b. framing square
    c. framing triangle
    d. framing rectangle

22. When fabricating a threaded pipe system, the final step is to _____.
    a. ream the pipe to remove all internal burrs
    b. tighten the fitting slowly using a pipe wrench
    c. clean the pipe and fittings thoroughly
    d. strip the threads

23. Once you have determined the type, size, and length of the pipe to be threaded, the pipe must next be _____.
    a. cut and reamed
    b. rotated and tightened
    c. cleaned and deburred
    d. threaded and secured

24. When threading pipe using a manual pipe threader, the die handle should be turned _____.
    a. one full rotation around the pipe
    b. two full rotations around the pipe
    c. three full rotations around the pipe
    d. four full rotations around the pipe

25. When using a threading machine to thread pipe, be sure to prime the _____.
    a. cutting fluid pump
    b. joint compound pump
    c. die lubricant pump
    d. galling liquid pump

26. PTFE tape is wrapped around the male threads of a pipe to prevent _____.
    a. corrosion
    b. excess movement
    c. metal-to-metal contact
    d. metal fatigue

27. When using liquid PTFE, always follow the manufacturer's recommendations regarding _____.
    a. failure rate
    b. excess pressure
    c. contamination
    d. temperature constraints

28. Pipe dope should always be applied with a _____.

a. brush
b. clean rag
c. reamer
d. bushing

29. The pipeline industry refers to turns made by a wrench as _____.

a. wrench engagement
b. wrench makeup
c. wrench adjustment
d. wrench tightening

30. When installing valves with threaded ends, be sure to lubricate the _____.

a. male threads of the pipe
b. female threads inside the valve
c. thread nearest the end of the pipe
d. valve closest to the pipe

31. If a fire or explosion occurs during installation, the first step is to _____.

a. contact the control center
b. use the proper PPE
c. evacuate the area
d. shut down the pipeline

32. Tightening the fitting too quickly could cause the threads or fitting to _____.

a. crack
b. contract
c. break
d. expand

Trade Terms Quiz

Fill in the blank with the correct term that you learned from your study of this module.

1. A fitting consisting of two sleeves to be threaded on the ends of the pipes and a threaded coupling ring to draw the sleeves together is a(n) ________.
2. The dimension that the fittings take out of a center-to-center measurement is called the takeup, takeoff, or ________.
3. A length of pipe made up of more than one length of pipe is a(n) ________.
4. Special tape that is wrapped around the male threads of a pipe before it is screwed into a fitting to serve as a lubricant and a sealant is known as ________.
5. A fitting that has a side outlet of 45 degrees to the run is a(n) ________.
6. A lateral move that moves the pipe out of its original line to another line that is parallel with the original line is called a(n) ________.
7. A fitting that has a side outlet of 90 degrees to the run is a(n) ________.
8. A pipe fitting that consists of a hollow plug with internal and external threads that suit the different diameters needed when connecting pipe with a fitting of a larger nominal size is a(n) ________.
9. A fitting on the end of a pipe that is shaped like a rim and allows the pipe to be bolted to another flanged pipe is a(n) ________.
10. A short piece of pipe that is threaded on both ends and is less than 12 inches long is a called a(n) ________.
11. A fitting that makes an angle between adjacent pipes is a(n) ________.
12. A nipple that is about twice the length of a standard thread and is threaded from end to end with no shoulder is called a(n) ________.
13. The outlet or inlet of a fitting not in line with the run but which may make any angle is a(n) ________.
14. The US standard for pipe thread, which has a 1/16-inch taper per inch from back to front, is the ________.
15. A deformity of threads in which the threads are stripped is called ________.
16. A pipe fitting with female threads on each end, with one end being one or more sizes smaller than the other end, is a(n) ________.
17. Conditions within a pipeline system, identified by the operator, that may indicate (1) a malfunction of a component or (2) a deviation from normal operations that may indicate a condition exceeding design limits or result in a hazard to persons, property, or the environment are called ________.
18. A type of reducer with a male thread on each end used to join pipe of difference sizes in high-pressure applications is called a(n) ________.
19. The length of the thread overlap when the male threads of one pipe or fitting are screwed into the female threads of another pipe or fitting is the ________.
20. A sleeve that connects two straight pieces of pipe is a ________.

Trade Terms

Abnormal operating conditions (AOCs)
Branch
Bushing
Close nipple
Coupling
Elbow
Flange
Galling
National Pipe Thread (NPT)
Nipple
Offset
PTFE tape
Reducer
Run
Swage
Takeout
Tee
Thread engagement
Union
Wye

Trade Terms Introduced in This Module

Abnormal operating conditions (AOCs): Conditions within a pipeline system, identified by the operator, that may indicate (1) a malfunction of a component or (2) a deviation from normal operations that may indicate a condition exceeding design limits or result in a hazard to persons, property, or the environment.

Branch: The outlet or inlet of a fitting not in line with the run but which may make any angle.

Bushing: A pipe fitting that consists of a hollow plug with internal and external threads that suit the different diameters needed when connecting pipe with a fitting of a larger nominal size.

Close nipple: A nipple that is about twice the length of a standard thread and is threaded from end to end with no shoulder.

Coupling: A sleeve that is used to connect two straight pieces of pipe.

Elbow: A fitting that makes an angle between adjacent pipes; always 90 degrees unless another angle is stated. Also known as an ell.

Flange: A fitting on the end of a pipe that is shaped like a rim and allows the pipe to be bolted to another flanged pipe.

Galling: A deformity of threads in which the threads are stripped.

National Pipe Thread (NPT): The US standard for pipe thread; has a 1/16-inch taper per inch from back to front.

Nipple: A short piece of pipe that is threaded on both ends and is less than 12 inches long.

Offset: A lateral move that moves the pipe out of its original line to another line that is parallel with the original line.

PTFE tape: Special tape, made of polytetrafluoroethylene (PTFE), which is wrapped around the male threads of a pipe before it is screwed into a fitting to serve as a lubricant and a sealant; also referred to as thread seal tape.

Reducer: A pipe fitting with female threads on each end, with one end being one or more sizes smaller than the other end.

Run: A length of pipe made up of more than one length of pipe.

Swage: A type of reducer with a male thread on each end used to join pipe of difference sizes in high-pressure applications.

Takeout: The dimension that the fittings take out of a center-to-center measurement; also known as takeup, takeoff, or makeup.

Tee: A fitting that has a side outlet of 90 degrees to the run.

Thread engagement: The length of the thread overlap when the male threads of one pipe or fitting are screwed into the female threads of another pipe or fitting.

Union: A fitting consisting of two sleeves to be threaded on the ends of the pipes and a threaded coupling ring to draw the sleeves together; used to connect pipes where quick disconnection may be needed for maintenance.

Wye: A fitting that has a side outlet of 45 degrees to the run; also referred to as a lateral.

Additional Resources

The following recommended resources can provide additional helpful information related to the requirements for this covered task:

API Recommended Practice 1161, Third Edition, © 2014 American Petroleum Institute. API Publishing Services, 1220 L Street, NW, Washington, DC 20005. Distributed by Thompson Reuters (Scientific) LLC, **www.techstreet.com**.

NCCER Module AOCCC-17, *Abnormal Operating Conditions Control Center*

NCCER Module AOCFG-17, *Abnormal Operating Conditions Field and Gas*

Audel Mechanical Trades Pocket Manual. Thomas B. Davis, Carl A. Nelson. New York, NY: Macmillan & Company.

IPT's Pipe Trades Handbook. Robert A. Lee. Clinton, NC: Construction Trades Press.

Figure Credits

Bill Emery, Module Opener, Figure 27a, Figures 28-29, Figure 31

Courtesy of RIDGID© 2014, Figure 30

Section Review Answer Key

Answer	Section Reference	Objective
Section One		
1. b	1.0.0	1
2. a	1.1.0	1a
3. d	1.2.0	1b
4. a	1.3.0	1c
Section Two		
1. b	2.1.0	2a
2. d	2.2.0	2b
3. d	2.4.0	2d
4. c	2.4.2	2d
5. b	2.4.3	2d
6. c	2.6.0	2f
Section Three		
1. c	3.1.0	3a
2. c	3.2.0	3b
3. d	3.4.0	3d
4. b	3.4.4	3d
Section Four		
1. c	4.2.1	4b
2. d	4.2.2	4b
3. a	4.3.0	4c
4. c	4.4.0	4d
5. a	4.4.0	4d
6. d	4.5.0	4e
Section Five		
1. c	5.1.2	5a
2. b	5.2.1	5b

NCCER CURRICULA — USER UPDATE

NCCER makes every effort to keep its textbooks up-to-date and free of technical errors. We appreciate your help in this process. If you find an error, a typographical mistake, or an inaccuracy in NCCER's curricula, please fill out this form (or a photocopy), or complete the online form at **www.nccer.org/olf**. Be sure to include the exact module ID number, page number, a detailed description, and your recommended correction. Your input will be brought to the attention of the Authoring Team. Thank you for your assistance.

Instructors – If you have an idea for improving this textbook, or have found that additional materials were necessary to teach this module effectively, please let us know so that we may present your suggestions to the Authoring Team.

NCCER Product Development and Revision

13614 Progress Blvd., Alachua, FL 32615

Email: curriculum@nccer.org
Online: www.nccer.org/olf

❑ Trainee Guide ❑ Lesson Plans ❑ Exam ❑ PowerPoints Other ______________________

Craft / Level: Copyright Date:

Module ID Number / Title:

Section Number(s):

Description:

Recommended Correction:

Your Name:

Address:

Email: Phone: